# Natural-Based Biodegradable Polymeric Materials

# Natural-Based Biodegradable Polymeric Materials

Editors

**Vsevolod Aleksandrovich Zhuikov**
**Rosane Michele Duarte Soares**

MDPI

Basel • Beijing • Wuhan • Barcelona • Belgrade • Novi Sad • Cluj • Manchester

*Editors*
Vsevolod Aleksandrovich Zhuikov
Research Center of Biotechnology of the Russian Academy of Sciences
Moscow, Russia

Rosane Michele Duarte Soares
Institute of Chemistry, Universidade Federal do Rio grande do Sul (UFRGS)
Porto Alegre, Brazil

*Editorial Office*
MDPI
St. Alban-Anlage 66
4052 Basel, Switzerland

This is a reprint of articles from the Special Issue published online in the open access journal *Polymers* (ISSN 2073-4360) (available at: https://www.mdpi.com/journal/polymers/special_issues/E0J0CM2L2D).

For citation purposes, cite each article independently as indicated on the article page online and as indicated below:

Lastname, A.A.; Lastname, B.B. Article Title. *Journal Name* **Year**, *Volume Number*, Page Range.

**ISBN 978-3-0365-8898-8 (Hbk)**
**ISBN 978-3-0365-8899-5 (PDF)**
**doi.org/10.3390/books978-3-0365-8899-5**

# Contents

polymers

MDPI

Article

# Modification of Poly(lactic acid) with Orange Peel Powder as Biodegradable Composite

**Nonni Soraya Sambudi [1,*], Wai Yi Lin [2], Noorfidza Yub Harun [2,*] and Dhani Mutiari [3]**

[1] Department of Chemical Engineering, Universitas Pertamina, Simprug, Jakarta 12220, Indonesia
[2] Department of Chemical Engineering, Universiti Teknologi PETRONAS, Seri Iskandar 32610, Perak, Malaysia
[3] Department of Architecture, Universitas Muhammadiyah Surakarta, Jl. A Yani, Mendungan, Kartasura 57169, Indonesia
* Correspondence: nonni.ss@universitaspertamina.ac.id (N.S.S.); noorfidza.yub@utp.edu.my (N.Y.H.)

**Abstract:** Traditional fossil-based plastic usage and disposal has been one of the largest environmental concerns due to its non-biodegradable nature and high energy consumption during the manufacturing process. Poly(lactic acid) (PLA) as a renewable polymer derived from natural sources with properties comparable to classical plastics and low environmental cost has gained much attention as a safer alternative. Abundantly generated orange peel waste is rich in valuable components and there is still limited study on the potential uses of orange peel waste in reinforcing the PLA matrix. In this study, orange peel fine powder (OPP) synthesized from dried orange peel waste was added into PLA solution. PLA/OPP solutions at different OPP loadings, i.e., 0, 10, 20, 40, and 60 wt% were then casted out as thin films through solution casting method. Fourier-transform infrared spectroscopy (FTIR) analysis has shown that the OPP is incorporated into the PLA matrix, with OH groups and C=C stretching from OPP can be observed in the spectra. Tensile test results have reviewed that the addition of OPP has decreased the tensile strength and Young's modulus of PLA, but significantly improve the elongation at break by 49 to 737%. Water contact angle analysis shows that hydrophilic OPP has modified the surface hydrophobicity of PLA with a contact angle ranging from 70.12° to 88.18°, but higher loadings lead to decrease of surface energy. It is proven that addition of OPP improves the biodegradability of PLA, where PLA/60 wt% OPP composite shows the best biodegradation performance after 28 days with 60.43% weight loss. Lastly, all PLA/OPP composites have better absorption in alkaline solution.

**Keywords:** poly(lactic acid); orange peel powder; biodegradation; tensile strength; Young's modulus; biocomposites

**Citation:** Sambudi, N.S.; Lin, W.Y.; Harun, N.Y.; Mutiari, D. Modification of Poly(lactic acid) with Orange Peel Powder as Biodegradable Composite. *Polymers* **2022**, *14*, 4126. https://doi.org/10.3390/polym14194126

Academic Editors: Vsevolod Aleksandrovich Zhuikov and Rosane Michele Duarte Soares

Received: 12 September 2022
Accepted: 26 September 2022
Published: 2 October 2022



## 1. Introduction

Plastic is a durable, inexpensive, and lightweight polymer that has a broad range of applications ranging from food packaging to electronic goods. Over 381 million tonnes of plastic are produced annually, and it was reported that around 55% was for single-use purpose only, 25% was incinerated, and 20% was recycled [1]. The overwhelming usage of conventional petroleum-based plastic which is highly non-biodegradable leads to environmental challenges including pollution and depletion of non-renewable resources [2]. The manufacturing process of plastic products consumes intensive energy, while the common ways of disposing plastic waste such as landfill, incineration and dumping into the oceans have led to adverse impacts on the environment [3,4]. Shrinking land capacity is one of the consequences of plastic waste spread in the natural environment as plastic products usually take up to several hundreds of years to decompose completely in landfills [5,6]. It was also estimated that, up to 14.5 metric tonnes of plastic ends up in the ocean per year, causing irreversible impacts on marine life and contamination of water sources [7]. Disposing plastic wastes through incineration could significantly reduce its volume by 80 to 95% [3]. However, burning of plastics releases toxic heavy metal and harmful gases like dioxins and methane into the atmosphere [3,8].

Due to the increasing concerns on plastic pollution, biopolymer has attracted extensive interest in recent studies [9,10]. The environmental-friendly biopolymer derived from renewable sources or being able to decompose naturally are preferred as safer alternatives [11–14]. Agricultural crop-based feedstocks and biomass waste are the common sources in the production of bioplastics [3]. The former, is however, less favourable due to its competition for land, water sources, and food production [3]. Poly(lactic acid) (PLA), also known as Polylactide, is a renewable polymer synthesized through esterification of lactic acid derived from sugar fermentation [15,16]. It can be derived from agricultural sources such as corn starch, wheat, and sugar cane [3]. Being viewed as the mostly potentially biopolymer to replace the conventional petroleum-derived plastic, PLA received significant attention due to its excellent properties, including biocompatibility [3], good mechanical properties and processability into various products, as well as low carbon footprint [17]. However, drawbacks such as high brittleness and low barrier properties have limited some of the applications of PLA [17]. Under controlled and specific industrial composting conditions, PLA can be degraded within a few days or up to several months, whereas it can take up to hundreds of years to decompose in landfill [6].

Furthermore, citrus waste disposal has been challenging as it is generated abundantly but underutilized. According to Raimondo et al., the volume of citrus fruits processed is nearly 31.2 million tons per year, while 50 to 60% of the original mass become the residue [18,19]. The disposal of citrus waste in landfill remains unsatisfactory as its low pH may cause soil salinity as well as other harmful effects to ruminants [19]. Among the citrus waste, orange peel is rich in valuable components such as pectin and cellulose fibres. Pectin with good gelling ability and cellulose fibres that give strength could act as new building block in products [18], while cellulose is naturally degradable by microbial activities [19]. These benefits of orange peel waste have attracted attention for their potential uses in reinforcing the conventional petrochemical or biobased matrices [19]. The incorporation of orange peel powder into renewable PLA is believed can enhance the biodegradability of the biobased polymer while improving its mechanical properties such as flexibility [6]. Bassani et al. conducted research on the incorporation of antioxidant extract from orange peels to develop an innovative PLA-based active packaging [17]. Phenolic compounds in orange peels are one of the valuable active materials that exhibits good antioxidant property [20]. However, the study regarding orange peel modification in PLA matrix limits to its extract, and so far didn't address the mechanical strength, surface properties and biodegradability of composites. Hence, there is still need of study in combining orange peel powder with PLA to form biocomposites.

In this study, orange peel powder is incorporated into pure PLA to improve its biodegradability in natural environment while maintaining its good physical properties. Orange peel waste without prior chemical modification is converted into fine powder through mechanical grinding, then mixed at different concentrations with PLA in chloroform solution. The homogeneous solution is then casted onto flat surface to obtain smooth composite films through solution casting method. The biodegradability of the biocomposite film is studied by conducting soil burial degradation experiment over 28 days.

## 2. Materials and Methods

### 2.1. Materials

PLA (IngeoTM biopolymer 3251D) was purchased from NatureWorks LLC (Plymouth, MN, USA). Sweet orange (Citrus sinensis) was purchased from local supermarket in Seri Iskandar, Perak. Chloroform (CAS 67–66–3) was purchased from R&M Chemicals (Edmonton, AL, Canada). Hydrochloric acid (HCl) (CAS 7647–01–0) was purchased from Thermo Fisher Scientific (Waltham, MA, USA). Sodium hydroxide (NaOH) was purchased from Sigma Aldrich (St. Louis, MO, USA). Deionized (DI) water was utilized throughout the experiment. The chemicals are utilized without further purification or treatment.

### 2.2. Pre-Treatment and Synthesis of Orange Peel Powder

Sweet orange (Citrus sinensis) peel collected was washed with tap water to remove impurities on the surface. Orange peel was dried under the sun for 20 h, followed by drying in oven for 18 h at 60 °C according to previous method by Farahmandfar [21]. Dried orange peel was then reduced into smaller size using pestle and mortar, then milled into fine powder using electric blender. The powder was sieved using sieve shaker to get uniform powder size of 100 μm, then stored in plastic container at room temperature until use.

### 2.3. Synthesis of Composite Films

PLA pellets were dissolved in chloroform under constant stirring for 1 h at room temperature to make 10 wt% PLA solution. Around 0.5 g orange peel powder (OPP) was added into the PLA solution and stirred for another 1 h to prepare PLA/10 wt% OPP solution. The solution was casted onto flat glass plate and four petri dishes, respectively and dried for 24 h at room temperature, which is similar to experiment done by Musa et al., in synthesis of PLA/banana fibres film [22]. The procedures were repeated by varying the amount of orange peel powder to prepare PLA/OPP solutions containing 20 wt%, 40 wt%, and 60 wt% OPP. Previous studies have shown that high loading of cellulose increased the strength and stiffness of composite [23–25].

### 2.4. Characterization of Composite Films

The morphology of film was captured by using stereoscopic microscope (Leica S8 APO). The sample functional groups were characterized using Fourier Transform Infrared Spectroscopy (FTIR) Perkin Elmer Spectrum One. Blank PLA sample of size 2 × 2 $cm^2$ was placed on the FTIR. The intensity of transmitted light was computed in a wave range of 4000–500 $cm^{-1}$. For water contact angle analysis, same size PLA film was mounted on goniometer. A single drop of deionized water was released from the syringe on top onto the specimen surface. The angle between the specimen surface and the edge of the water drop was measured. The mechanical strength analysis by using universal testing machine (UTM, Instron, Norwood, MA, USA). Blank PLA film was prepared into sample of dumbbell shape following the ASTM D638 standard [26]. Geometry of the sample, including width, gauge length, and thickness were recorded into the software, while sample was loaded onto the grip of 5 kN Universal Testing Machine. The grip separation was started and run at a constant speed of 5 mm/min until sample break. Tensile strength (MPa), Young's Modulus (MPa), and elongation at break (%) of the sample were analysed from graph generated.

### 2.5. Swelling Test

Three samples with size of 4 × 4 $cm^2$ were prepared from each composite and weighed to record respective initial weight as $W_{dry}$. Samples were immersed into three petri dishes containing different pH solutions, i.e., 0.1 M HCl, distilled water, and 0.1 M NaOH, respectively [27]. After 1 h, samples were removed from the solutions and liquid droplets on the surface were wiped off. The weight after immersion was recorded as $W_{swollen}$. The swelling index was calculated Equation (1) as below

$$\text{Swelling index } (\%) = \frac{W_{swollen} - W_{dry}}{W_{dry}} \times 100\% \quad (1)$$

### 2.6. Biodegradability Test

Blank PLA and different concentrations of PLA/OPP composite samples casted in petri dish of diameter 8 cm were weighed to measure respective initial weight ($W_1$). Each sample was buried under 2 cm of moist garden soil in a container and kept for 28 days in an open environment. The soil burial method is adapted from previous studies by Park et al., and Lertphirun et al., with some modification [28,29]. The soil was kept moist by spraying sufficient amount of water. Degradation of the samples was observed at an interval of 7 days for continuous 4 weeks period. The sample residues were collected from soil, rinsed with water to remove the soil on its surface, followed by drying for 30 min at 40 °C in oven. Weight after ($W_2$) of samples were recorded. The biodegradability of the composites was measured using Equation (2) as follows:

$$\text{Biodegradability } (\%) = \frac{W_2 - W_1}{W_1} \times 100\% \quad (2)$$

## 3. Results

### *3.1. Morphology, Functional Groups and Contact Angle*

Blank PLA film (Figure 1a) retained its high transparency property and displayed a smooth, shiny surface. As OPP was introduced at low loadings, i.e., 10 and 20 wt%, the composites (Figure 1b,c) turned slightly yellowish but maintained characteristics of blank PLA where the film still appeared as a shiny and transparent surface. More distinct difference was observed when the OPP′s loadings increased to 40 and 60 wt%, the composite appeared as bright orange films and lost its surface shininess. Furthermore, increased roughness was detected on the composites′ surfaces due to the increased concentration of OPP.

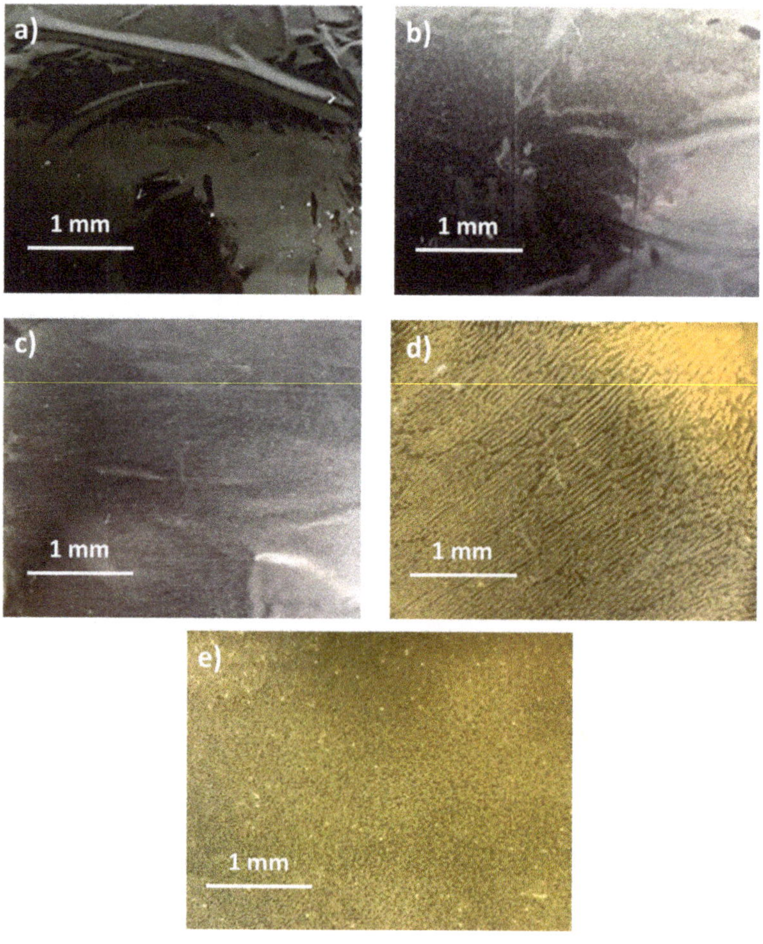


**Figure 1.** Surface appearance of (**a**) PLA film, (**b**) PLA/10 wt% OPP, (**c**) PLA/20 wt% OPP, (**d**) PLA/40 wt% OPP, and (**e**) PLA/60 wt% OPP composites.

FTIR characterization was conducted to determine the functional groups present in the blank PLA matrix and after incorporation of OPP. Figure 2 shows the FTIR spectra ranging from wavenumber 4000 $cm^{-1}$ to 550 $cm^{-1}$. The small peaks at 2946.09 to 2996.32 $cm^{-1}$ are attributed to the CH stretching vibration, and sharp absorption peaks at 1749.29 $cm^{-1}$ belongs to the C=O stretching which can be observed in all samples [30,31].

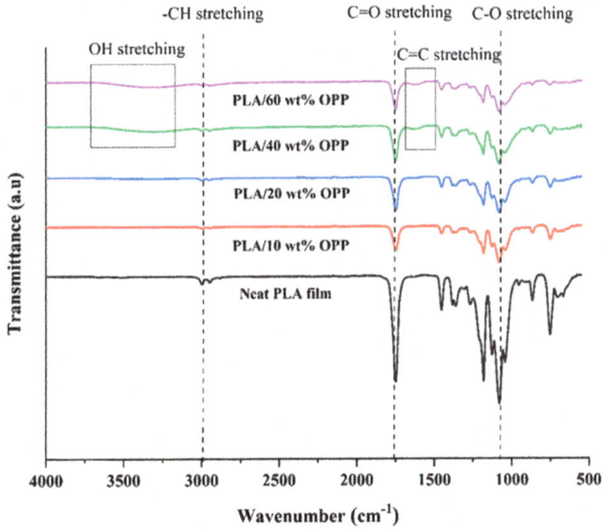


**Figure 2.** FTIR spectra of samples.

Water contact angle between water droplet and the composite film's surface was measured to study its hydrophobicity or known as surface wettability. Bar chart in Figure 3 illustrates the water contact angle (θ) obtained for each film. The blank PLA film shows an angle of 73.82°, corresponding to its hydrophobic nature which has a static contact angle within the range of 60 to 85° [32]. This value is also close to the measured angle of blank PLA film (67.27°) prepared using same solution casting method as reported by Alakrach et al. [33].

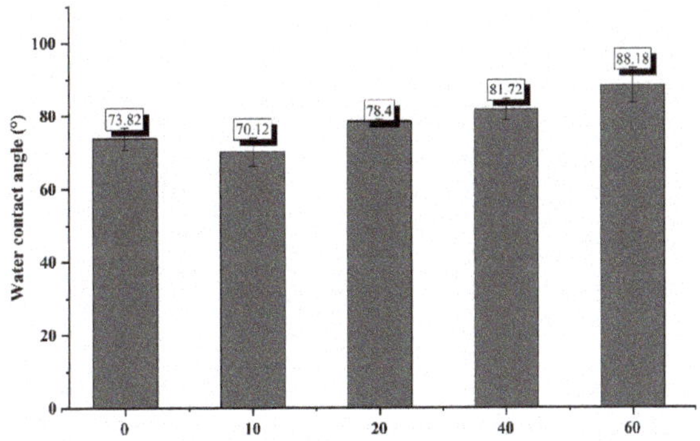


**Figure 3.** Water contact angle of PLA, PLA/10 wt%, PLA/20 wt%, PLA/40 wt%, and PLA/60 wt% OPP composites.

### 3.2. Mechanical Properties

It has been observed that the incorporation of OPP in the PLA matrix has weakened its mechanical properties, i.e., tensile strength and young's modulus decrease when compared to the blank PLA film (Figure 4). The initial tensile and modulus exhibited by blank PLA film are 13.21 MPa and 1646 MPa, respectively. The PLA/OPP composites of all loadings display lower tensile strength within the range of 2.59 to 6.9 MPa, and lower modulus within the range of 145 to 652 MPa. However, there is a significant improvement on the elongation at break. PLA/20 wt% OPP, PLA/40 wt% OPP, and PLA/60 wt% OPP composites show an improved elongation (15.32, 9.98, and 10.6 mm) compared to blank PLA film (2.73 mm).

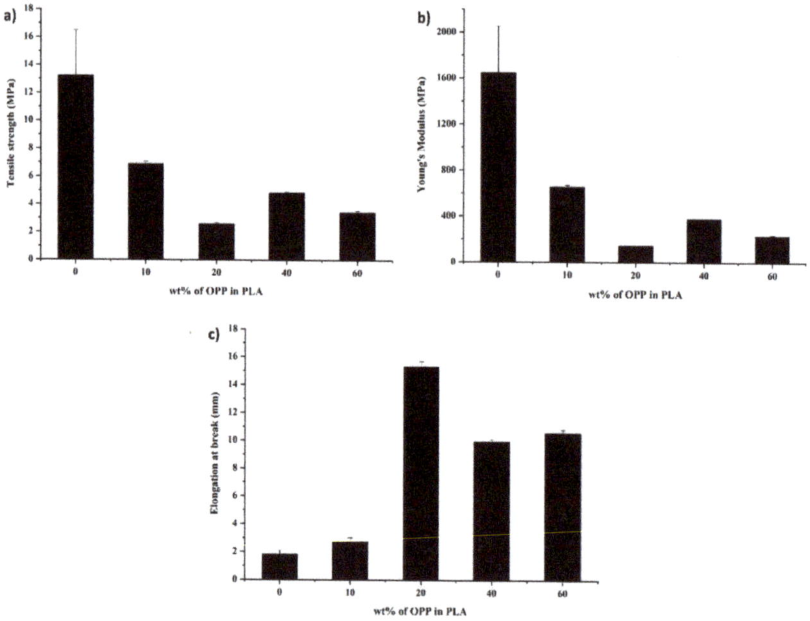


**Figure 4.** (**a**) Tensile strength, (**b**) Young's modulus, and (**c**) Elongation at break of PLA, PLA/10 wt%, PLA/20 wt%, PLA/40 wt%, and PLA/60 wt% OPP composites.

### 3.3. Swelling

Swelling rate (%) is calculated by comparing the weight before and after immersing each composite sample in acid, alkaline, and neutral distilled water at pH 1, 13, and 7, respectively. Based on Figure 5, it is observed that the samples' swelling percentage generally increases with increased OPP loadings, regardless of the pH of the solution. The blank PLA film shows the lowest swelling rate among all samples, which is corresponding to its hydrophobic nature [29]. A significant increase in the swelling percentage can be observed in PLA/10 wt%, PLA/20 wt%, PLA/40 wt%, and PLA/60 wt% composites.

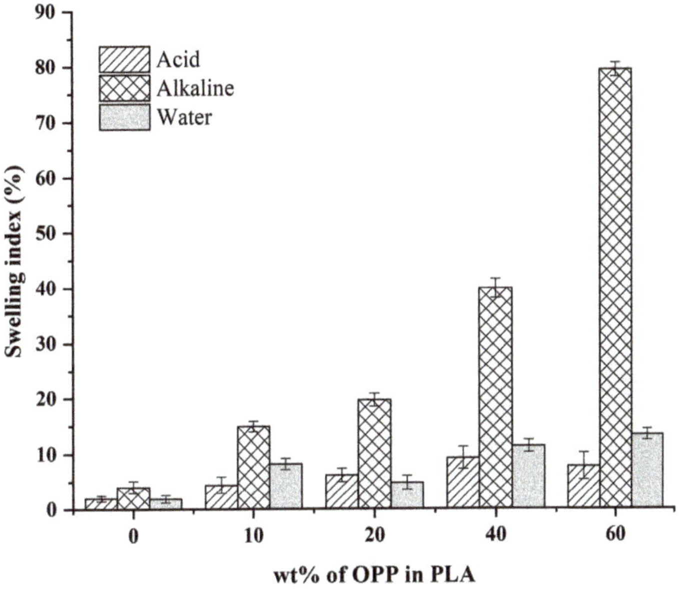


**Figure 5.** Swelling percentage of PLA, PLA/10 wt%, PLA/20 wt%, PLA/40 wt%, and PLA/60 wt% OPP composites in acid, alkaline and distilled water.

### 3.4. Degradation

Soil burial experiment was conducted for a total period of 28 days, where the samples were taken out, weighed, and observed at every 7-day interval. Figure 6 shows the graph of weight loss percentage of each film against days after soil burial, while Table 1 presents and compares the films' appearance before and after soil burial.

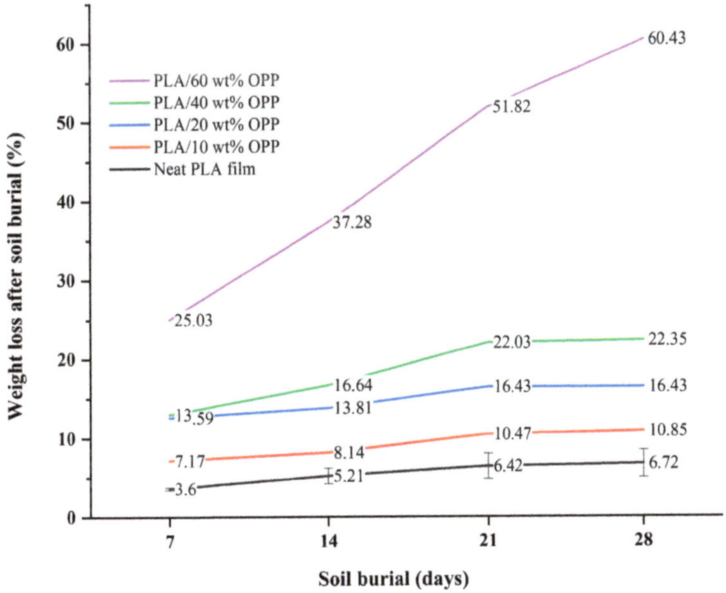


**Figure 6.** Weight loss percentage of composites over days after soil burial.

**Table 1.** Physical appearance of composites before and after soil burial test.

| Composite | Before Soil Burial | After 7 Days | After 14 Days | After 21 Days | After 28 Days |
|---|---|---|---|---|---|
| PLA | 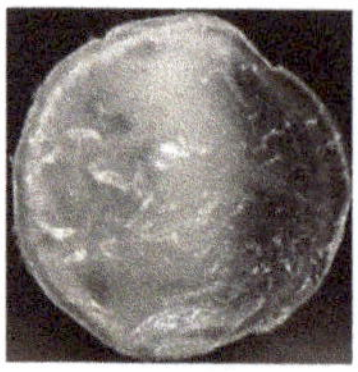 | 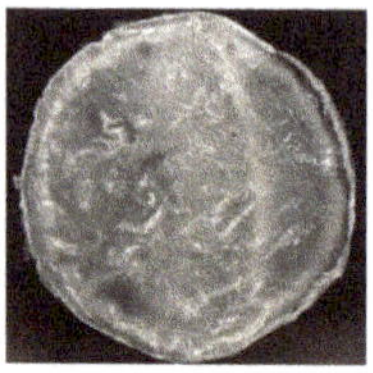 | 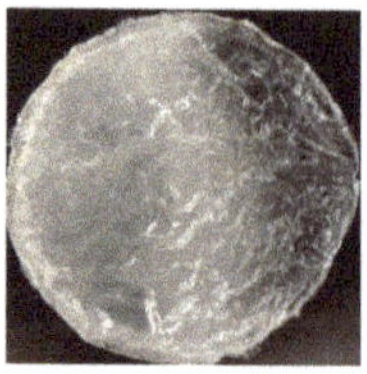 | 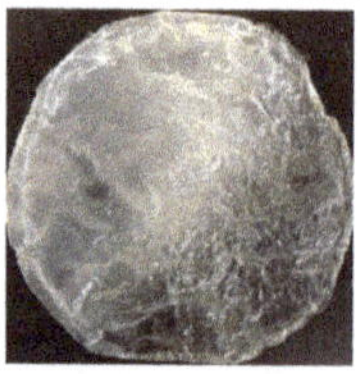 |  |
| PLA/ 10 wt% OPP | 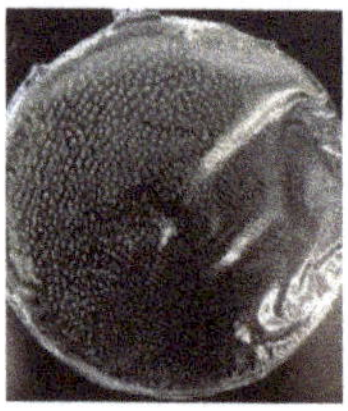 | 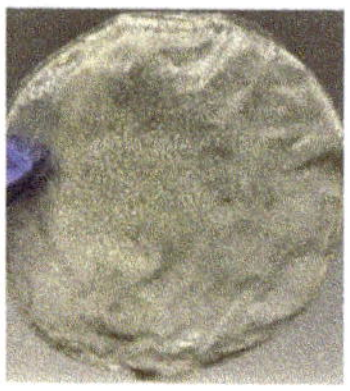 | 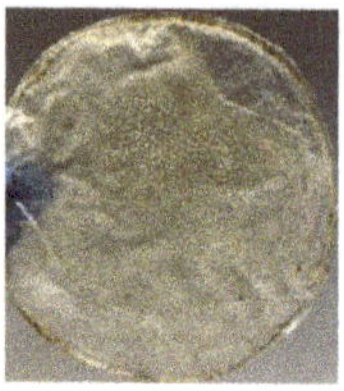 | 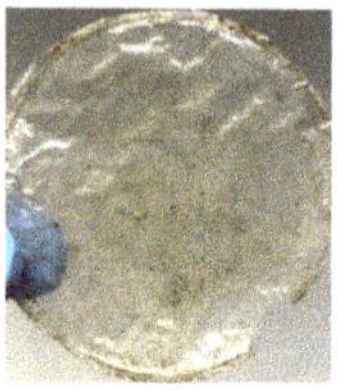 | 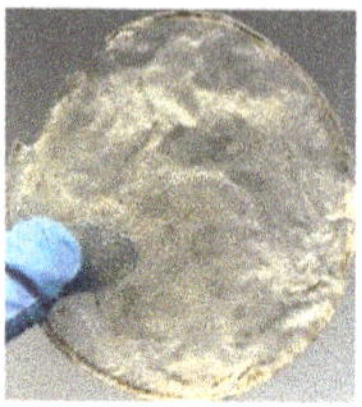 |
| PLA/ 20 wt% OPP | 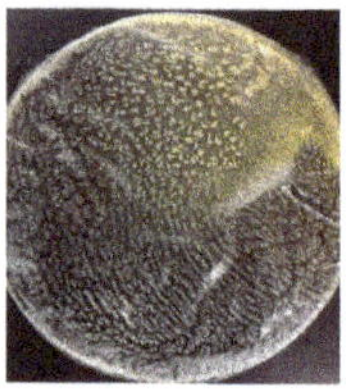 |  |  |  |  |
| PLA/ 40 wt% OPP | 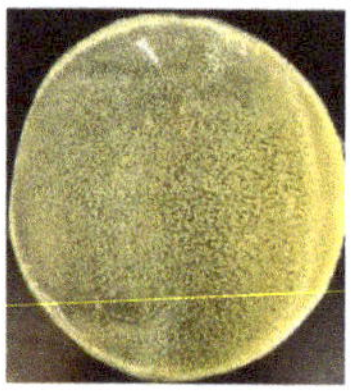 |  |  | 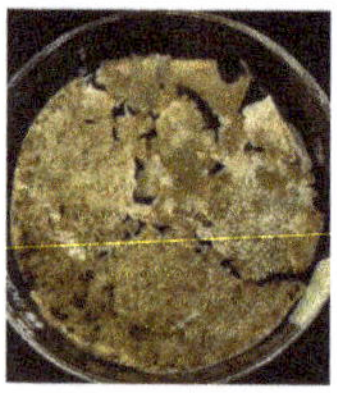 | 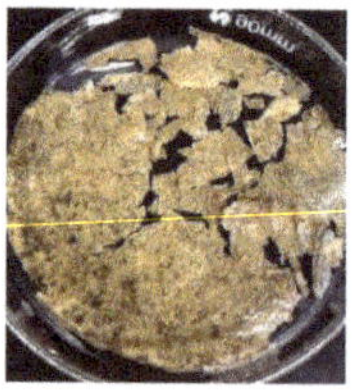 |
| PLA/ 60 wt% OPP | 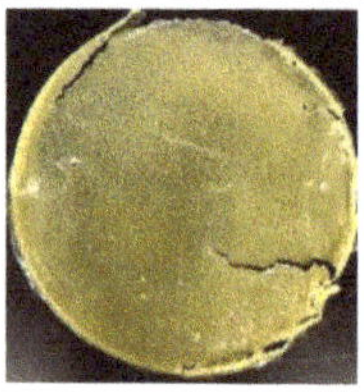 |  | 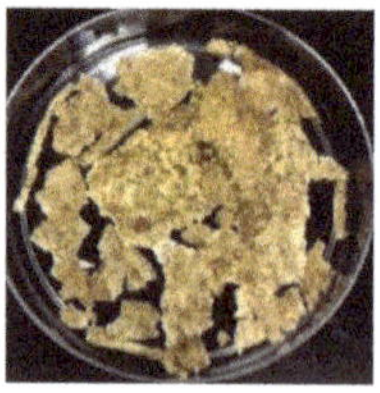 |  | 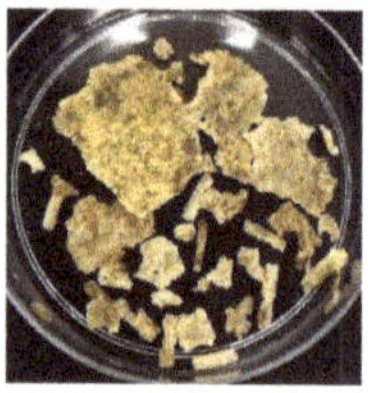 |

The blank PLA film has shown the least weight loss throughout the soil burial experiment, with low percentage ranging from 3.73 to 5.49%. From Table 1, it can also be observed that the blank PLA film experiences very little changes in appearance due to its hydrophobic nature and slow degradation rate. Similar result was obtained by Lertphirun and Srikulkit [29] for blank PLA degradation study in soil burial test. With increased loadings of OPP in PLA, the composites show an increase trend of weight loss percentage and degradation. The highest weight loss is observed in PLA/60 wt% OPP composite that shows a weight loss percentage in the range of 25.03 to 60.43%, which is 8 to 12 times higher than that of blank PLA film. Moreover, it can be observed from Table 1 that cracks, holes and fungi formation appear on the surface of PLA/10 wt% OPP, PLA/20 wt% OPP, PLA/40 wt% and PLA/60 wt% OPP composites and they have become more brittle after soil burial. These features increase by time for all composites while more obvious changes are clearly visible in higher loadings PLA/OPP composites.

## 4. Discussion

As observed from Figure 1b,c, good and even dispersion of OPP particles in PLA matrix was achieved at relatively low loadings at 10 and 20 wt%. When OPP loading increased to 40 wt%, it could be clearly seen that the OPP particles no longer distributed evenly on the surface, agglomeration and some aggregates were observed especially at the top right corner of the film (Figure 1d). At the highest concentration of 60 wt% OPP, the whole composite has become denser, with very little or no space in between could be observed (Figure 1e). For FTIR spectra in Figure 2, a broad peak in the high energy region (2944.72 to 3309.94 $cm^{-1}$) indicates the presence of hydroxyl group, which is only observed at PLA/OPP composites at higher loadings [31]. This is due to the large amount of OH groups present in carbohydrates and lignin in the orange peel [31,34,35], and is more visible when the OPP loadings are high as the fibre content is increased. Similar trend is observed for the C=C stretching at 1608 $cm^{-1}$ due to the presence of unsaturated aromatic compounds in orange peel [34,35]. Lastly, the peaks at 1081.31 to 1181.32 $cm^{-1}$ indicates the presence of C-O stretching in all spectra [36,37]. Addition of hydrophilic filler OPP which is rich in cellulose fibres has slightly modified the hydrophobic nature of PLA. PLA/10 wt% OPP composite reports a decreased contact angle (70.12°), which means that it is now more hydrophilic. The presence of OH groups in OPP increases the polarity and surface energy of the composite film [33]. High surface energy creates a strong attractive force which make water droplet to spread out better on the surface, thus showing a good wettability. However, at 20, 40, and 60 wt% OPP loadings, water contact angle increases to 78.4°, 81.72°, and 88.18°, respectively, showing that the composites have again shifted towards hydrophobic nature. This is because of the high loadings of OPP in PLA matrix increases the surface roughness, which subsequently reduces the surface energy [33], causing the composites to become slightly hydrophobic.

It is believed that the cellulose components in OPP have helped in modifying the toughness by improving the elongation. The poor interfacial adhesion between cellulose fibres and polymer can cause agglomerations due to hydrophilic nature of OPP and hydrophobic nature of PLA, hence resulting in low tensile and modulus [38,39]. While addition of OPP can seem to overcome the brittleness of PLA by the increasing elongation at break (Figure 4c). Similar reduction in tensile strength and increasing in elongation at break were observed in previous study by modifying PLA with cellulose as fillers [31,40,41], where the addition of cellulose can interlock with PLA matrix and ease the movement of composite upon stress, hence increasing the elongation.

Furthermore, the optimum loading is obtained at 20 wt% OPP that yields the highest elongation at break which gives the polymer a good ductility to not break easily. Nonetheless, at even higher OPP loadings (40 wt% and 60 wt%), all three mechanical properties, i.e., tensile strength, Young's modulus, and elongation at break are found decreased again. This is probably due to the high OPP loadings that cause poor distribution and agglomeration of the filler particles, resulting in embrittlement of the composites. The mechanical strength results show similar agreement to study by Singh et al., whereas high loading of cellulose in PLA reduced tensile strength and modulus, while elongation at break was increased [42]. The orange peel powder might act as plasticizer, to improve the ductility of material, as can be seen by the increasing of elongation at break at around 49 to 737%, usually by sacrificing tensile strength and stiffness [42]. The imperfect distribution of orange peel powder and weak interfacial adhesion between orange peel powder and PLA can be other factors that contribute to the lowering tensile strength and modulus [43].

The addition of OPP with hydrophilic cellulose fibre content has helped to bond the water molecules together through the affinity of OH groups, resulting in higher absorption or swelling rate [44,45]. Comparing all three solutions of different pH, the composites experience the highest swelling rate in alkaline NaOH, with the maximum rate 80.22%

achieved by PLA/60 wt% OPP composite. Moreover, cracking was observed after 1 h immersion in the alkaline solution. Similar to the study of Moliner et al. [44], the PLA-Sisal bio-composites also show the highest water uptake in alkaline medium (pH 8). This shows the potential of the composites to be treated with alkaline hydrolysis for degradation treatment as it is found to be easily degraded in strong basic solutions [44].

An increase in OPP loadings has led to better biodegradability of the composites due to the presence of higher amount of cellulose content. The high cellulose content (9.19 to 22%) in orange peel has favoured the biodegradation process through hydrolysis and microbial activities [46,47]. Hydrophilic nature of cellulose fibres facilitates the action of water as a carrier for microbes which eventually promotes the enzymatic hydrolysis of cellulose in the composites [29]. Thus, it is believed that chemical composition of the composite plays an important role in determining the microbes and moisture access, which further determines the biodegradability [28,48].

## 5. Conclusions

In this study, orange peel waste transformed into fine powder was incorporated into PLA to develop a biobased, biodegradable polymer. It was proven that OPP has been incorporated into the PLA matrix by the presence of hydroxyl group in the high energy region, and C=C functional group at wavenumber 1608 $cm^{-1}$. Moreover, it was found that addition of OPP into PLA has generally decreased the tensile and modulus, but significant increase of elongation was observed at low loadings of OPP. It was suggested that the OPP which contains high cellulose fibre components modified the toughness and gave a better ductility to the composites. Water contact angle analysis has shown that the hydrophilic OPP significantly modified the surface hydrophobicity of PLA, with higher OPP loadings resulted in rough surface of composites and low surface energy. From soil burial test over a consecutive study of 28 days, it was concluded that the presence of OPP has improved the biodegradability of PLA. Better biodegradation was observed with increasing concentration of OPP.

**Author Contributions:** Conceptualization, N.S.S. and N.Y.H.; methodology, N.S.S. and W.Y.L.; validation, N.S.S., W.Y.L. and N.Y.H.; formal analysis, W.Y.L. and N.S.S.; investigation, W.Y.L. and N.S.S.; resources, N.S.S., N.Y.H. and D.M.; data curation, W.Y.L.; writing—original draft preparation, N.S.S. and W.Y.L.; writing—review and editing, N.S.S. and N.Y.H.; visualization, W.Y.L.; supervision, N.S.S. and N.Y.H.; project administration, N.S.S.; funding acquisition, N.S.S., N.Y.H. and D.M. All authors have read and agreed to the published version of the manuscript.

**Funding:** The research work was financially supported by UTP—UMS International Collaborative Research Grant with cost centre of 015ME0-256 and 015ME0-257.

**Institutional Review Board Statement:** Not applicable.

**Informed Consent Statement:** Not applicable.

**Data Availability Statement:** Not applicable.

**Acknowledgments:** We would like to thank Universitas Muhammadiyah Surakarta and Universiti Teknologi PETRONAS for providing the necessary fund and research facilities to accommodate the experiments and completion of this research project.

**Conflicts of Interest:** Declare conflicts of interest or state. The authors declare no conflict of interest. The funders had no role in the design of the study; in the collection, analyses, or interpretation of data; in the writing of the manuscript; or in the decision to publish the results.

## References

1. Geyer, R.; Jambeck, J.R.; Law, K.L. Production, Use, and Fate of All Plastics Ever Made. *Sci. Adv.* **2017**, 3, e1700782. [CrossRef] [PubMed]
2. Moshood, T.D.; Nawanir, G.; Mahmud, F.; Mohamad, F.; Ahmad, M.H.; AbdulGhani, A. Sustainability of Biodegradable Plastics: New Problem or Solution to Solve the Global Plastic Pollution? *Curr. Res. Green Sustain. Chem.* **2022**, 5, 100273. [CrossRef]

3. Acquavia, M.; Pascale, R.; Martelli, G.; Bondoni, M.; Bianco, G. Natural Polymeric Materials: A Solution to Plastic Pollution from the Agro-Food Sector. *Polymers* **2021**, *13*, 158. [CrossRef] [PubMed]
4. Ncube, L.; Ude, A.; Ogunmuyiwa, E.; Zulkifli, R.; Beas, I. An Overview of Plastic Waste Generation and Management in Food Packaging Industries. *Recycling* **2021**, *6*, 12. [CrossRef]
5. Rhodes, C.J. Solving the Plastic Problem: From Cradle to Grave, to Reincarnation. *Sci. Prog.* **2019**, *102*, 218–248. [CrossRef]
6. Chamas, A.; Moon, H.; Zheng, J.; Qiu, Y.; Tabassum, T.; Jang, J.H.; Abu-Omar, M.; Scott, S.L.; Suh, S. Degradation Rates of Plastics in the Environment. *ACS Sustain. Chem. Eng.* **2020**, *8*, 3494–3511. [CrossRef]
7. Wayman, C.; Niemann, H. The Fate of Plastic in the Ocean Environment—A Minireview. *Environ. Sci. Process. Impacts* **2021**, *23*, 198–212. [CrossRef]
8. Royer, S.-J.; Ferrón, S.; Wilson, S.T.; Karl, D.M. Production of Methane and Ethylene from Plastic in the Environment. *PLoS ONE* **2018**, *13*, e0200574. [CrossRef]
9. Sanjay, M.R.; Yogesha, B. Studies on Natural/Glass Fiber Reinforced Polymer Hybrid Composites: An Evolution. *Mater. Today Proc.* **2017**, *4*, 2739–2747. [CrossRef]
10. Mohammed, A.S.; Meincken, M. Properties of Low-Cost WPCs Made from Alien Invasive Trees, and Ldpe for Interior Use in Social Housing. *Polymers* **2021**, *13*, 2436. [CrossRef]
11. Balart, R.; Garcia-Garcia, D.; Fombuena, V.; Quiles-Carrillo, L.; Arrieta, M.P. Biopolymers from Natural Resources. *Polymers* **2021**, *13*, 2532. [CrossRef] [PubMed]
12. Kovačević, Z.; Grgac, S.F.; Bischof, S. Progress in Biodegradable Flame Retardant Nano-Biocomposites. *Polymers* **2021**, *3*, 741. [CrossRef] [PubMed]
13. Mahmud, S.; Hasan, K.M.F.; Jahid, A.; Mohiuddin, K.; Zhang, R.; Zhu, J. Comprehensive Review on Plant Fiber-Reinforced Polymeric Biocomposites. *J. Mater. Sci.* **2021**, *56*, 7231–7264. [CrossRef]
14. Agustiany, E.A.; Ridho, M.R.; Rahmi D.N., M.; Madyaratri, E.W.; Falah, F.; Lubis, M.A.R.; Solihat, N.N.; Syamani, F.A.; Karungamye, P.; Sohail, A.; et al. Recent Developments in Lignin Modification and Its Application in Lignin-Based Green Composites: A Review. *Polym. Compos.* **2022**, *43*, 4848–4865. [CrossRef]
15. Balla, E.; Daniilidis, V.; Karlioti, G.; Kalamas, T.; Stefanidou, M.; Bikiaris, N.D.; Vlachopoulos, A.; Koumentakou, I.; Bikiaris, D.N. Poly(lactic acid): A Versatile Biobased Polymer for the Future with Multifunctional Properties—From Monomer Synthesis, Polymerization Techniques and Molecular Weight Increase to Pla Applications. *Polymers* **2021**, *13*, 1822. [CrossRef]
16. Singhvi, M.S.; Zinjarde, S.S.; Gokhale, D.V. Polylactic Acid: Synthesis and Biomedical Applications. *J. Appl. Microbiol.* **2019**, *127*, 1612–1626. [CrossRef] [PubMed]
17. Bassani, A.; Montes, S.; Jubete, E.; Palenzuela, J.; Sanjuán, A.; Spigno, G. Incorporation of Waste Orange Peels Extracts into Pla Films. *Chem. Eng. Trans.* **2019**, *74*, 1063–1068.
18. Raimondo, M.; Caracciolo, F.; Cembalo, L.; Chinnici, G.; Pecorino, B.; D'Amico, M. Making Virtue out of Necessity: Managing the Citrus Waste Supply Chain for Bioeconomy Applications. *Sustainability* **2018**, *10*, 4821. [CrossRef]
19. Bátori, V.; Jabbari, M.; Åkesson, D.; Lennartsson, P.R.; Taherzadeh, M.J.; Zamani, A. Production of Pectin-Cellulose Biofilms: A New Approach for Citrus Waste Recycling. *Int. J. Polym. Sci.* **2017**, *2017*, 9732329. [CrossRef]
20. Rafiq, S.; Kaul, R.; Sofi, S.; Bashir, N.; Nazir, F.; Nayik, G.A. Citrus Peel as a Source of Functional Ingredient: A Review. *J. Saudi Soc. Agric. Sci.* **2018**, *17*, 351–358. [CrossRef]
21. Farahmandfar, R.; Tirgarian, B.; Dehghan, B.; Nemati, A. Comparison of Different Drying Methods on Bitter Orange (*Citrus aurantium* L.) Peel Waste: Changes in Physical (Density and Color) and Essential Oil (Yield, Composition, Antioxidant and Antibacterial) Properties of Powders. *J. Food Meas. Charact.* **2020**, *14*, 862–875. [CrossRef]
22. Luqman, M.; Vesuanathan, T.A.; Salleh, M.N. Isolation and Characterization of Microcrystalline Cellulose Extracted from Banana Fiber in Poly(lactic acid) Biocomposite Produced from Solvent Casting Technique. *IOP Conf. Ser. Mater. Sci. Eng.* **2020**, *957*, 012005. [CrossRef]
23. Cruz, A.G.; Scullin, C.; Mu, C.; Cheng, G.; Stavila, V.; Varanasi, P.; Xu, D.; Mentel, J.; Chuang, Y.-D.; Simmons, B.A.; et al. Impact of High Biomass Loading on Ionic Liquid Pretreatment. *Biotechnol. Biofuels* **2013**, *6*, 52. [CrossRef] [PubMed]
24. Chang, H.; Luo, J.; Liu, H.C.; Davijani, A.A.B.; Wang, P.-H.; Lolov, G.S.; Dwyer, R.M.; Kumar, S. Ductile Polyacrylonitrile Fibers with High Cellulose Nanocrystals Loading. *Polymer* **2017**, *122*, 332–339. [CrossRef]
25. Lee, W.J.; Clancy, A.; Kontturi, E.; Bismarck, A.; Shaffer, M.S.P. Strong and Stiff: High-Performance Cellulose Nanocrystal/Poly(Vinyl Alcohol) Composite Fibers. *ACS Appl. Mater. Interfaces* **2016**, *8*, 31500–31504. [CrossRef]
26. *Astm D638-14, 17*; Standard Test Method for Tensile Properties of Plastics. ASTM International: West Conshohocken, PA, USA, 2022.
27. Sampath, U.; Chee, C.Y.; Chuah, C.H.; Rahman, N.; Nai-Shang, L. Ph-Responsive Poly(lactic acid)/Sodium Carboxymethyl Cellulose Film for Enhanced Delivery of Curcumin in Vitro. *J. Drug Deliv. Sci. Technol.* **2020**, *58*, 101787. [CrossRef]
28. Park, C.; Kim, E.Y.; Yoo, Y.T.; Im, S.S. Effect of Hydrophilicity on the Biodegradability of Polyesteramides. *J. Appl. Polym. Sci.* **2003**, *90*, 2708–2714. [CrossRef]
29. Lertphirun, K.; Srikulkit, K. Properties of Poly(lactic acid) Filled with Hydrophobic Cellulose/Sio2 Composites. *Int. J. Polym. Sci.* **2019**, *2019*, 7835172. [CrossRef]
30. Chi, H.Y.; Chan, V.; Li, C.; Hsieh, J.H.; Lin, P.H.; Tsai, Y.-H.; Chen, Y. Fabrication of Polylactic Acid/Paclitaxel Nano Fibers by Electrospinning for Cancer Therapeutics. *BMC Chem.* **2020**, *14*, 63. [CrossRef]

31. Paul, U.C.; Fragouli, D.; Bayer, I.S.; Zych, A.; Athanassiou, A. Effect of Green Plasticizer on the Performance of Microcrystalline Cellulose/Polylactic Acid Biocomposites. *ACS Appl. Polym. Mater.* **2021**, *3*, 3071–3081. [CrossRef]
32. Baran, E.H.; Erbil, H.Y. Surface Modification of 3d Printed Pla Objects by Fused Deposition Modeling: A Review. *Colloids Interfaces* **2019**, *3*, 43. [CrossRef]
33. Alakrach, A.M.; Noriman, N.Z.; Dahham, O.S.; Hamzah, R.; Alsaadi, M.A.; Shayfull, Z.; Idrus, S.Z.S. Chemical and Hydrophobic Properties of Pla/Hnts-Zro2 Bionanocomposites. *J. Phys. Conf. Ser.* **2018**, *1019*, 012065. [CrossRef]
34. Dey, S.; Basha, S.; Babu, G.; Nagendra, T. Characteristic and Biosorption Capacities of Orange Peels Biosorbents for Removal of Ammonia and Nitrate from Contaminated Water. *Clean. Mater.* **2021**, *1*, 100001. [CrossRef]
35. Zapata, B.; Balmaseda, J.; Fregoso-Israel, E.; Torres-García, E. Thermo-Kinetics Study of Orange Peel in Air. *J. Therm. Anal. Calorim.* **2009**, *98*, 309. [CrossRef]
36. Shen, W.; Zhang, G.; Ge, X.; Li, Y.; Fan, G. Effect on Electrospun Fibres by Synthesis of High Branching Polylactic Acid. *R. Soc. Open Sci.* **2018**, *5*, 180134. [CrossRef] [PubMed]
37. Nugraha, M.W.; Wirzal, M.D.H.; Ali, F.; Roza, L.; Sambudi, N.S. Electrospun Polylactic Acid/ Tungsten Oxide/ Amino-Functionalized Carbon Quantum Dots (Pla/Wo3/N-Cqds) Fibers for Oil/Water Separation and Photocatalytic Decolorization. *J. Environ. Chem. Eng.* **2021**, *9*, 106033. [CrossRef]
38. Izzati, A.N.A.; John, W.C.; Fazita, M.R.N.; Najieha, N.; Azniwati, A.A.; Khalil, H.P.S.A. Effect of Empty Fruit Bunches Microcrystalline Cellulose (Mcc) on the Thermal, Mechanical and Morphological Properties of Biodegradable Poly (Lactic Acid) (Pla) and Polybutylene Adipate Terephthalate (Pbat) Composites. *Mater. Res. Express* **2020**, *7*, 015336. [CrossRef]
39. Sousa, S.; Costa, A.; Silva, A.; Simões, R. Poly(lactic acid)/Cellulose Films Produced from Composite Spheres Prepared by Emulsion-Solvent Evaporation Method. *Polymers* **2019**, *11*, 66. [CrossRef]
40. El-Hadi, A.M. Increase the Elongation at Break of Poly (Lactic Acid) Composites for Use in Food Packaging Films. *Sci. Rep.* **2017**, *7*, 46767. [CrossRef]
41. Xu, L.; Zhao, J.; Qian, S.; Zhu, X.; Takahashi, J. Green-Plasticized Poly(lactic acid)/Nanofibrillated Cellulose Biocomposites with High Strength, Good Toughness and Excellent Heat Resistance. *Compos. Sci. Technol.* **2021**, *203*, 108613. [CrossRef]
42. Singh, A.A.; Genovese, M.E.; Mancini, G.; Marini, L.; Athanassiou, A. Green Processing Route for Polylactic Acid–Cellulose Fiber Biocomposites. *ACS Sustain. Chem. Eng.* **2020**, *8*, 4128–4136. [CrossRef]
43. Wang, Q.; Ji, C.; Sun, J.; Zhu, Q.; Liu, J. Structure and Properties of Polylactic Acid Biocomposite Films Reinforced with Cellulose Nanofibrils. *Molecules* **2020**, *25*, 3306. [CrossRef] [PubMed]
44. Moliner, C.; Finocchio, E.; Arato, E.; Ramis, G.; Lagazzo, A. Influence of the Degradation Medium on Water Uptake, Morphology, and Chemical Structure of Poly(lactic acid)-Sisal Bio-Composites. *Materials* **2020**, *13*, 3974. [CrossRef] [PubMed]
45. Aouat, T.; Kaci, M.; Lopez-Cuesta, J.-M.; Devaux, E. Investigation on the Durability of Pla Bionanocomposite Fibers under Hygrothermal Conditions. *Front. Mater.* **2019**, *6*, 323. [CrossRef]
46. Orozco, R.S.; Hernández, P.B.; Morales, G.R.; Nuñez, F.U.; Villafuerte, J.O.; Lugo, V.L.; Ramírez, N.F.; Díaz, C.E.B.; Vázquez, P.C. Characterization of Lignocellulosic Fruit Waste as an Alternative Feedstock for Bioethanol Production. *Bioresources* **2014**, *9*, 1873.
47. Ayala, J.; Montero, G.; Coronado, M.; García, C.; Curiel-Alvarez, M.; León, J.; Sagaste, C.; Montes, D. Characterization of Orange Peel Waste and Valorization to Obtain Reducing Sugars. *Molecules* **2021**, *26*, 1348. [CrossRef]
48. Teixeira, S.; Eblagon, K.M.; Miranda, F.; Pereira, M.F.R.; Figueiredo, J.L. Towards Controlled Degradation of Poly(Lactic) Acid in Technical Applications. *C* **2021**, *7*, 42. [CrossRef]

*polymers*

MDPI

*Article*

# Effect of Epoxidized and Maleinized Corn Oil on Properties of Polylactic Acid (PLA) and Polyhydroxybutyrate (PHB) Blend

**Jaume Sempere-Torregrosa, Jose Miguel Ferri, Harrison de la Rosa-Ramírez, Cristina Pavon and Maria Dolores Samper ***

Instituto de Tecnología de Materiales (ITM), Universitat Politècnica de València (UPV), Plaza Ferrándiz y Carbonell 1, 03801 Alcoy, Alicante, Spain
* Correspondence: masammad@upv.es; Tel.: +34-96-652-8478

**Abstract:** The present work analyzes the influence of modified, epoxidized and maleinized corn oil as a plasticizing and/or compatibilizing agent in the PLA–PHB blend (75% PLA and 25% PHB wt.%). The chemical modification processes of corn oil were successfully carried out and different quantities were used, between 0 and 10% wt.%. The different blends obtained were characterized by thermal, mechanical, morphological, and disintegration tests under composting conditions. It was observed that to achieve the same plasticizing effect, less maleinized corn oil (MCO) is needed than epoxidized corn oil (ECO). Both oils improve the ductile properties of the PLA–PHB blend, such as elongation at break and impact absorb energy, however, the strength properties decrease. The ones that show the highest ductility values are those that contain 10% ECO and 5% MCO, improving the elongation of the break of the PLA–PHB blend by more than 400% and by more than 800% for the sample PLA.

**Keywords:** polylactic acid; polyhydroxybutyrate; blend; modified corn oil

**Citation:** Sempere-Torregrosa, J.; Ferri, J.M.; de la Rosa-Ramírez, H.; Pavon, C.; Samper, M.D. Effect of Epoxidized and Maleinized Corn Oil on Properties of Polylactic Acid (PLA) and Polyhydroxybutyrate (PHB) Blend. *Polymers* **2022**, *14*, 4205. https://doi.org/10.3390/polym14194205

Academic Editors: Rosane Michele Duarte Soares and Vsevolod Aleksandrovich Zhuikov

Received: 13 September 2022
Accepted: 3 October 2022
Published: 7 October 2022



## 1. Introduction

Nowadays, polymeric materials have replaced a significant part of the different materials employed in the engineering sector [1], as well as other sectors such as construction [2], key parts in electronics appliances [3–5], and food packaging [6,7]. Consequently, the demand for these resources has considerably increased in recent years [8].

These resources, mostly of petrochemical origin, are obtained from non-renewable and limited sources which increase their depletion and the price of energy resources [9]. Additionally, the production of these resources has a devastating impact on our planet, causing environmental problems and large amounts of $CO_2$ emissions which directly increase global warming and destroy ecosystems [10]. Another problem derived from the materials of petrochemical origin is the accumulation of plastic residues that mostly end up in landfills or in the oceans, causing microplastic migrations that contaminate and spoil wildlife [11,12]. The strategies and directives proposed by the European Commission give priority to sustainable and non-toxic single-use products, aiming first and foremost to reduce the quantity of plastic waste [13].

Therefore, one conceivable solution for this massive problem is the recycling and reuse of these materials with the aim of using less virgin material [8]. This is a good way to fight against this problem, however, today, this is not enough and another solution is required. Another possible way to reduce the number of materials of petrochemical origin would be to increase the use of biodegradable materials. Nevertheless, polymers still need research to improve their properties and facilitate their manufacture for them be comparable to other polymers of petrochemical origin [14,15].

One of the most widely used biodegradable materials today is polylactic acid (PLA) [16,17]. PLA is a linear aliphatic polyester which is produced from a fermentation process of corn starch, wheat starch, among others, meaning that it comes from renewable resources [18,19]. Considering the outstanding properties of PLA such as biocompatibility and high strength, it also has some slight disadvantages such as its low ductility which reduce its possible applications [20–22].

Due to the brittleness of PLA, it is often modified with other polymers: thanks to the compatibility of PLA, it can be blended with other polymers to improve its properties. In addition, such a combination can also have other benefits for both polymers, so it is interesting to make such modifications. Some of the petroleum-based materials with which PLA has good miscibility are polyethylene terephthalate (PET) [23,24], polyvinyl chloride (PVC) [25], thermoplastic elastomers (TPEs) [26,27], polypropylene (PP) [28,29] and polyethylene (PE) [30–32]—all of which are commodity plastics.

Because of the aforementioned problems, biopolymers for blending with PLA such as polyhydroxyalkanoates (PHAs) [33–35], poly (E-caprolactone) (PCL) [36–38], or polybutylene adipate-co-terephthalate (PBAT) [39–41] have been tested and can achieve improvements in mechanical properties or degradation stability.

Plasticizers, low-molecular-weight molecules that generally improve the ductility and processing of polymers, can be used to improve the miscibility between materials. There are two types of plasticizers, depending on their origin, namely petrochemical, such as tar pitch polymerization resins, etc. Some of these are used because they are environmentally friendly and non-toxic plasticizers, such as poly (ethylene glycol) (PEG) [42–46], poly (propylene glycol) (PPG) [47], lactic acid oligomer (OLA) [48], triethyl citrate (TEC) [49], and tributyl citrate (TBC) [50–52]. In addition, a second option of natural origin is vegetable and animal oils, acids, etc. Some of the most commonly used are epoxidized palm oil (EPO) [53], linseed oil (ELO) [54], octyl epoxy stearate (OES) [55], karanja oil (EKO) [56], epoxidized soybean oil (ESBO) [53], or epoxidized cottonseed oil (ECSO) [57].

Overall due to PLA's compatibility, it can be blended with other polymers such as poly-B-hydroxybutyrate (PHB) [58–60], among other starch acetylated thermoplastics discussed above, to improved its properties and achieve significant improvements in the mechanical properties or the degradation stability.

Thus, the aim of the work was to improve the properties of PLA by blending it with PHB at 75–25%, respectively, to improve the miscibility and reduce the melting temperatures to promote its use in certain industries. For this purpose, different percentages of epoxidized corn oil and maleinized corn oil, 1, 2.5, 5, 7.5, and 10 wt%, were used as plasticizers. The processes of the maleinization and epoxidation of vegetable oils provide greater reactivity to the oil and greater thermal stability with respect to unmodified oils. Therefore, the compatibility of the oils with PLA is improved and the plasticizing effect is increased [56,61]. These oils were modified from corn oil due to certain characteristics that make it interesting compared to other oils from cereals or seeds. Corn oil is interesting since it has an interesting lipid profile for functionalization and has the advantage, compared to other modifiable oils, of being able to be produced in large quantities and in many countries of the world, as it is economic and very stable to oxidation. Its good oxidative stability is related to its low $\alpha$-linolenic acid content, which is the case of corn oil between 0–2% and depends on the genotype [62]. Among the fatty acids contained in the oil, 85% are unsaturated, which means a significant amount of double bonds that allow its functionalization by chemical processes such as epoxidation and maleinization. Therefore, this work evaluates the potential of epoxidized corn oil (ECO) and maleinized corn oil (MCO) as bio-based plasticizers to improve the ductile properties of PLA–PHB blends.

## 2. Materials and Methods

### *2.1. Materials*

Poly (lactic acid), PLA commercial grade Biopolymer 2003D, was supplied by Nature Works LLC (Minnetonka, MN, USA) in pellet form with a density of 1.24 g cm$^{-3}$. The PLA–PHB blend containing 25% PHB was supplied by Biomer (Krailling, Germany) in pellet form with a density of 1.25 g cm$^{-3}$ with commercial grade P226E. The vegetable oil used to carry out the epoxidation and maleinization process was food-grade corn vegetable oil (CO—corn oil).

### 2.2. Corn Oil Modification

The corn oil modification processes were carried out in a three-necked round bottom flask with a capacity of 500 mL in order to incorporate all the necessary elements. One of them was used to introduce the reagents, another one for temperature control and the central one to connect the reflux condenser.

#### 2.2.1. Epoxidation Process

The epoxidation process started by adding 182.0 g of corn oil into the flask until 50–55 °C was reached with magnetic stirring. Then, 18.64 mL of acetic acid was added until the temperature was stable at 60 °C, then a mixture of 182.70 mL of hydrogen peroxide (3:1 M ratio (peroxide:oil)) with 1.46 mL of sulfuric acid was added dropwise, and the slow addition of this mixture was performed to avoid an exothermic reaction and with the heating mantle off. Once the addition of this mixture was finished and the reaction was controlled, the working temperature was increased between 80 and 85 °C for 8 h. Samples were collected at each hour of the epoxidation process according to ASTMD1652-97 [63] and ISO 3961:2009 [64] oxirane oxygen methods. Prior to the test, the samples were washed with distilled water and purified for 30 min in a centrifuge at 4000 rpm. Figure 1 shows a schematic representation of the epoxidation process.

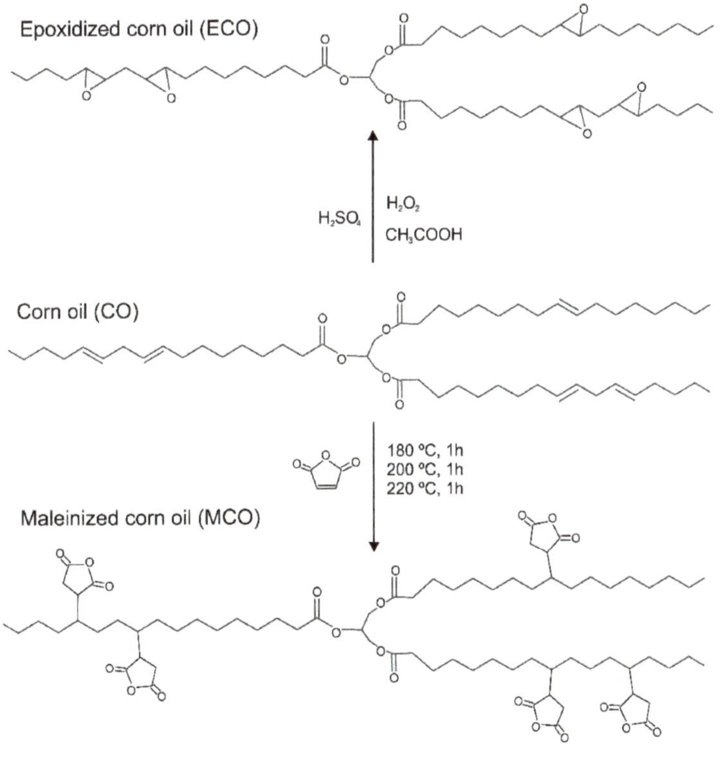


**Figure 1.** Schematic representation of the corn oil epoxidation and maleinization (MCO) processes.

#### 2.2.2. Maleinization Process

The maleinization ratio used was 2.4:1, following the recommendations of previous works [65,66]. A total of 27 g of maleic anhydrive (MA) and 300 g of corn oil were used, the maleinization process was carried out in three stages at different temperatures—namely 180, 200, and 220 °C—and the samples were taken for acid value (AV) calculations every 60 min.

Initially, the oil was introduced in the round flask and when it reached the temperature of 180 °C, 9 g of MA (1/3 of the total) was added. The temperature of 180 °C was maintained for one hour and increased to 200 °C, before another 9 g of MA was added and maintained again for 1 h and the same process was carried out at 220 °C. Finally, the mixture was cooled to room temperature. The degree of maleinization was determined following the guidelines of ISO 660:2009 [67] and using the following expression:

$$Acid\ Value = \frac{56.1 \cdot c \cdot V}{m} \quad (1)$$

where $c$ is the exact concentration of the KOH standard solution used (mol-$L^{-1}$), $V$ is the volume of the KOH standard solution used (mL), and $m$ is the analyzed mass (g). A schematic representation of the maleinization process can be seen in Figure 1.

### 2.3. Sample Preparation

The samples used in the different experimental techniques and tests were obtained by mixing processes in the laboratory. Before handling the PLA and PHB, both materials were dried at 60 °C for 24 h to eliminate the percentage of moisture indicated by the supplier.

First, different mixtures were made with 75% PLA, 25% PHB and different percentages of epoxidized and maleinized corn oil, as shown in Table 1. The mixtures were left to dry for a minimum of 24 h at 60 °C.

**Table 1.** Processed blends.

| Code | PLA (wt%) | PHB (wt%) | ECO (phr *) | MCO (phr *) |
|---|---|---|---|---|
| PLA | 100 | - | - | - |
| B (blend) | 75 | 25 | - | - |
| B-1ECO | 75 | 25 | 1 | - |
| B-2.5ECO | 75 | 25 | 2.5 | - |
| B-5ECO | 75 | 25 | 5 | - |
| B-7.5ECO | 75 | 25 | 7.5 | - |
| B-10ECO | 75 | 25 | 10 | - |
| B-1MCO | 75 | 25 | - | 1 |
| B-2.5MCO | 75 | 25 | - | 2.5 |
| B-5MCO | 75 | 25 | - | 5 |
| B-7.5MCO | 75 | 25 | - | 7.5 |
| B-10MCO | 75 | 25 | - | 10 |

* phr (per hundred resin) represents de weighted parts of the oil particles added to one hundred weight parts of the base PLA–PHB blend.

To achieve homogeneity in the mixtures, an extrusion process was carried out. The temperature range program used for the different zones of the extruder was the following: 175–180–180–185 °C, from the hopper to the nozzle, respectively, at constant speed of 40 rpm. At the end of the extrusion process, the material was cut until it was reduced to pellet size.

Subsequently, the material in the pellets form was injected to obtain standard specimens for subsequent characterization. The chosen temperature program was 175–180–185 °C, from the hopper to the nozzle, respectively. The temperatures used in the manipulation process were those recommended by the material supplier.

### 2.4. Characterization Techniques

#### 2.4.1. Mechanical Properties

The mechanical characterization was carried out with tensile, flexural, impact, and hardness tests. The tensile and flexural tests were performed on an ELIB 30 universal testing machine (S.A.E. Ibertest, Madrid, Spain). A minimum of five different specimens were tested using a 5 KN load cell. The speed used in the tests was 5 mm $min^{-1}$, at room temperature. For the tensile test, the parameters defined in the ISO 527-1:2019 standard [68] were followed, and for the flexural test, the ISO 178:2019 standard [69] was used. In addition, from the integral of the stress–strain curves obtained in the tensile tests, the toughness modulus was calculated.

To determine the amount of energy absorbed by the material in the case of impact, the test was carried out with a 6 J Charpy pendulum (Metrotec, San Sebastian, Spain). Five measurements have were taken, with bending type specimens and without notch. The parameters followed in the test are those defined in the ISO 179-1:2010 standard [70].

Finally, the materials surface hardness was measured. Five measurements were taken at different points of the surface of several specimens. A Shore D scale hardness tester, JBA S.A. model 673-D (Instruments JBot, S.A. Cabrils, Barcelona, Spain), was used. The parameters followed in the test are those defined in the ISO 868:2003 standard [71].

#### 2.4.2. Thermo-Mechanical Properties

To evaluate the effects of temperature on the mechanical properties of the materials, a study of softening temperature (VICAT) and heat deflection temperature (HDT) was carried out. Both tests were carried out with a Metrotec model (San Sebastian, Spain). The parameters for the VICAT test were those of the B50 method, as defined in the ISO 306 standard [72]. The standard establishes a load of 50 N and raises the temperature to 50 C $h^{-1}$ until the penetrator penetrates 1 mm into the sample. For the HDT test, procedure B of the ISO 75-1:2013 standard [73] was used. The standard determines the use of 0.45 MPa load, using a weight of 74 g, and carrying out the test at 120 °C $min^{-1}$ until reaching 0.31 mm of deformation.

#### 2.4.3. Thermal Properties

The materials' miscibility was evaluated through the analysis of their thermal properties, using the differential scanning calorimetry technique (DSC). The thermoanalytical tests were carried out in a Mettler Toledo 882e (Mettler Toledo S.A.E., Barcelona, Spain). The weight range of the samples was 5–10 mg, with an initial heating program from −50 to 180 °C to remove the thermal history, followed by a cooling program from 180 to −50 °C, and finally a second heating program from −50 to 220 °C, at a heating rate of 10 °C $min^{-1}$ in a nitrogen atmosphere with a nitrogen flux of 66 mL $min^{-1}$. With this technique, the glass transition temperature ($T_g$), cold crystallization peak ($T_{cc}$), and the melt peak temperature ($T_m$) of the PLA blends formulations were identified. The degree of crystallization (X%) was calculated with the formula by Equation (2).

$$X_C(\%) = 100 \times \frac{|\Delta H_{CC} + \Delta H_m|}{\Delta H_m(100\%)} \times \frac{1}{W_{PLA}} \quad (2)$$

where $\Delta H_{CC}$ is the cold crystallization enthalpy, $\Delta H_m$ is the melt enthalpy, $\Delta H_m$ (100%) is the theorical melt enthalpy for a fully crystalline PLA structure (93 J $g^{-1}$) [74], and WPLA is the PLA weight fraction.

Samples were tested by thermogravimetric analysis. The thermogravimetric analysis (TGA) was carried out in a TGA PT1000 equipment from Linseis Inc. (Selb, Germany). The analyzed samples, with an average weight between 15 and 20 mg, were subjected to a temperature program from 30 °C to 700 °C at a speed of 10 °C $min^{-1}$, with a constant nitrogen flow of 30 mL $min^{-1}$.

#### 2.4.4. Microscopic Characterization

The morphology of the blends was analyzed with a ZEISS ULTRA microscope from Oxford Instruments (Oxfordshire, UK) using a 1.5 KV accelerating voltage. The analyzed samples were the broken impact test specimens, which were sputtered with a thin layer of platinum, using a Leica Microsystems (Buffalo Grove, IL, USA) EM MED0200 high vacuum sputter coater.

2.4.5. Disintegration under Composting Conditions

The study of the materials' disintegration behavior, under composting conditions, was carried out according to the ISO 20200:2015 standard [75]. The tested materials were cut into fragments of 25 × 25 mm, and eight fragments of each type of material were made. The samples were buried in a compost reactor. The degree of disintegration (*D*) was calculated in percentage with the following equation:

$$D = \frac{m_i - m_r}{m_i} \quad (3)$$

where $m_i$ is the initial dry mass of the test material and $m_r$ is the dry mass of the residual test material recovered from the sieving.

## 3. Results

### *3.1. Optimization of the Epoxidation Process Conditions of Epoxidized Corn Oil (ECO)*

The epoxidation of corn oil was performed with a peroxide-to-corn oil molar ratio of 3:1, according to previous experience in the epoxidation process of other vegetable oils [76]. The epoxidation process was confirmed by analyzing the oxirane oxygen number and iodine number. Figure 2a shows how the conversion of epoxidizable double bonds into oxirane rings takes place in the epoxidation process, as it was observed that the value of oxirane oxygen before the process is 0 and rapidly increases during the first hours due to the wide availability of double bonds, and from the fourth hour, it slows down as most of the double bonds have already reacted.

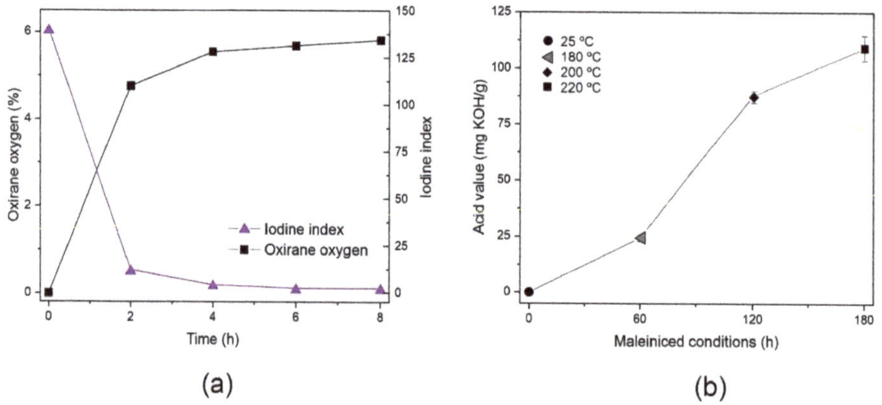


**Figure 2.** (**a**) Evolution of the oxirane oxygen index and iodine index in the corn oil epoxidation process and (**b**) evolution of the acid value in the different stages of the corn oil epoxidation process.

At the end of the process, at 8 h, an oxirane oxygen index of 5.82 was obtained. Comparing this result with the theoretical rate, obtained from the amount of fatty acids in the corn oil, a conversion of 85.9% was achieved. In addition, the conversion of double bonds represented by the iodine index rapidly decreased during the first 2 h from values close to 140 until the end at 1.5 after 8 h of treatment, which reflects a greater decrease in double bonds than that reflected by the oxygen oxirane. This is due to reactions parallel to the conversion of double bonds into epoxy groups [56]. Analyzing the data obtained in the epoxidation process, it is possible to reduce the epoxidation process to 6 h, since the results of the conversion of double bonds into oxirane rings are obtained at 8 h.

### *3.2. Synthesis of Maleinized Corn Oil (MCO)*

Figure 2b shows the evolution of the acid number values at the end of each one of the stages at different temperatures, namely 180, 200, and 220 °C. Initially, the AV is

0.15 mg KOH $g^{-1}$ and at the end of the first hour at 180 °C, it is observed that it increases up to 24.4 mg KOH $g^{-1}$, indicating that the maleinization reaction is occurring. After the second hour, at 200 °C, another large increase is observed, reaching an AV of 87.8 mg KOH $g^{-1}$, and finally, after the last stage, at 220 °C, the AV reaches 109.2 mg KOH $g^{-1}$. These results are in total agreement with values obtained by A. Perez-Nakai et al. with an AV of 105 for the maleinized hemp seed and 130 for the maleinized Brazil nut at the end of the epoxidation process [67]. In addition, Quiles Carrillo et al. indicated that commercial-grade maleinized linseed oil has an AV of between 105 and 130 mg KOH $g^{-1}$ [77].

### 3.3. Mechanical Properties

PLA is a brittle material, and with the intention of improving this characteristic, mixtures were made with PHB. Moreover, different percentages of MCO and ECO were used as plasticizers and compatibilizers. The tensile strength, elongation at break and impact absorbed energy values of the studied materials are represented in Figure 3 and the results obtained in the flexural characterization and Young's modulus and toughness modulus are represented in Table 2. It is observed that, when mixing 25% of PHB with PLA, the strength properties such as tensile and flexural strength, Young's modulus, and flexural modulus are considerably reduced and ductile properties such as elongation at break and impact absorbed energy are improved [78]. By incorporating ECO in the PLA–PHB blend, the tensile and flexural strength decrease slightly with respect to the PLA–PHB blend and the ductile properties increase as the amount of ECO in the system increases, achieving an elongation at break of 127% when incorporating 10 phr of ECO, which is 450% more elongation compared to the PLA–PHB blend and 810% more compared to PLA. The impact absorbed energy and toughness modulus also increases when incorporating ECO, achieving the best result with 10 phr of ECO. The behavior of the materials when incorporating MCO is similar, however, the tensile strength value decreases to a lesser extent when incorporating ECO, and it remains between 30 and 33 MPa with amounts lower than 7.5 phr of MCO and the flexural strength values are also slightly higher when incorporating ECO. In addition, it is also observed that the ductile properties increase only up to 5 phr of MCO, and at higher amounts, both the elongation at break and the impact absorbed energy decrease. The B-5MCO material has similar mechanical properties to B-10ECO with an increase in elongation at break of approximately 460% more than PLA–PHB blends and 830% more than PLA.

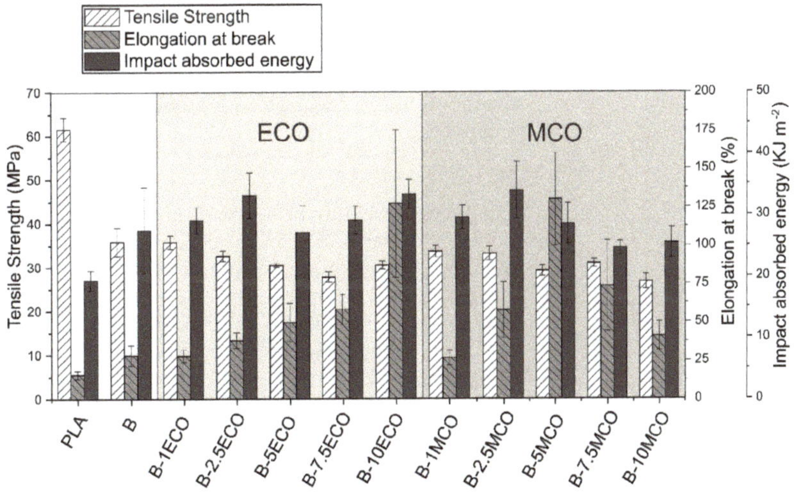


**Figure 3.** Effect of epoxidized and maleinized corn oil content on the tensile strength, elongation at break and impact energy of the PLA–PHB blend.

**Table 2.** Summary of tensile and flexural properties and HDT/VST results of neat PLA, PLA–PHB blend and plasticized blends with a different ECO and MCO content.

| Sample | Tensile Strength (MPa) | Young's Modulus (MPa) | Elongation at Break (%) | Toughness Modulus ($MJ/m^3$) | Impact Absorbed Energy ($kJ/m^2$) | Flexural Strength (MPa) | Flexural Modulus (MPa) | HDT (°C) | VST (°C) |
|---|---|---|---|---|---|---|---|---|---|
| PLA | 61.6 ± 2.7 | 3470 ± 17 | 15.7 ± 2.7 | 6.0 ± 0.2 | 19.3 ± 1.6 | 99.8 ± 6.5 | 3290 ± 84 | 55.8 | 58.5 |
| B (blend) | 35.9 ±3.2 | 533 ± 102 | 28.3 ± 6.7 | 5.5 ± 1.0 | 27.5 ± 7.0 | 68.7 ± 4.0 | 2041 ± 92 | 52.5 | 53.8 |
| B-1ECO | 35.8 ± 1.6 | 1038 ± 168 | 28.0 ± 4.0 | 5.4 ± 0.4 | 29.1 ± 2.1 | 64.9 ± 2.9 | 2323 ± 154 | 54.2 | 53.5 |
| B-2.5ECO | 32.6 ± 1.2 | 1521 ± 22 | 38.0 ± 5.1 | 8.5 ± 1.0 | 33.1 ± 3.7 | 60.2 ± 0.7 | 2198 ± 195 | 54.5 | 53.7 |
| B-5ECO | 30.5 ± 0.4 | 1131 ± 182 | 49.6 ± 12.7 | 11.3 ± 3.8 | 27.1 ± 4.3 | 54.0 ± 3.6 | 2257 ± 84 | 52.4 | 54.3 |
| B-7.5ECO | 27.8 ± 1.2 | 1668 ± 207 | 58.1 ± 9.9 | 11.1 ± 1.0 | 29.1 ± 2.2 | 51.0 ± 0.6 | 2233 ± 115 | 54.2 | 53.3 |
| B-10ECO | 20.5 ± 0.9 | 1707 ± 48 | 127.2 ± 48.1 | 29.9 ±4.7 | 33.3 ± 2.4 | 54.7 ± 0.6 | 2392 ± 64 | 53.8 | 54.0 |
| B-1MCO | 33.6 ± 1.3 | 1189 ± 115 | 26.4 ± 4.8 | 6.4 ± 0.8 | 29.6 ± 1.9 | 64.8 ± 3.1 | 2291 ± 50 | 55.5 | 53.6 |
| B-2.5MCO | 33.1 ± 1.6 | 1403 ± 237 | 57.6 ± 18.3 | 11.3 ± 3.6 | 34.0 ± 4.6 | 61.6 ± 1.5 | 2235 ± 91 | 54.4 | 53.3 |
| B-5MCO | 29.2 ± 1.2 | 1372 ± 213 | 130.0 ±30.0 | 30.3 ± 4.4 | 28.6 ± 3.3 | 59.7 ± 1.0 | 2274 ± 261 | 54.4 | 53.0 |
| B-7.5MCO | 30.9 ± 0.9 | 1645 ± 54 | 73.5 ± 29.6 | 6.8 ± 2.7 | 24.6 ± 1.1 | 54.2 ± 2.1 | 2216 ± 100 | 54.1 | 53.7 |
| B-10MCO | 26.7 ± 1.7 | 1267 ± 102 | 40.8 ± 9.3 | 6.0 ± 1.1 | 25.5 ± 2.5 | 49.5 ± 1.5 | 2011 ± 93 | 55.0 | 52.6 |

Both plasticizers achieved an increase in the ductility of the initial formulation, as can be seen in the values of the elongation at break, toughness modulus, and impact absorbed energy. In the case of the tensile strength, a variation was observed, with the increasing percentage of plasticizer in the formulation, decreasing tensile strength values were obtained. The values of Young's modulus obtained a variation, although in both cases, with values above the initial formulation. Additionally, both plasticizers managed to reduce the maximum flexural strength of the initial formulation, in such a way that the higher the plasticizer percentage, the lower the flexural strength values. In the case of flexural modulus, both plasticizers increased these values, except for the formulation with 10 wt% of MCO.

### 3.4. *Thermo-Mechanical Properties*

Regarding the thermo-mechanical properties, as shown in Table 2, the incorporation of PHB causes a decrease in both the Vicat softening temperature (VST) and heat deflection temperature (HDT), up to 53.8 and 52.5 °C, respectively, with respect to PLA (58.5 °C VST and 55.8 °C HDT). When ECO and MCO are added to blend PLA–PHB, it is observed that the HDT and VST values increase slightly. This is ascribed to the fact that the incorporation of the biomaterial and the plasticizers reduces the glass transition temperature, as shown below in the DSC section, and thus the deformation under high temperatures. Finally, in the VST results, it is observed that the incorporation of biomaterials and plasticizers reduces the VST values obtained by PLA. The incorporation of these materials means that the hardness is affected when PLA is raised to temperatures close to the glass transition temperature.

### 3.5. *Thermal Properties*

The thermal stability of the PLA, PLA–PHB blend, and plasticized formulations with different contents of ECO and MCO was assessed by thermogravimetry analysis (TGA). Table 3 shows some thermal parameters such as the degradation onset temperature (T5%), which indicates the temperature at which a 5% weight loss occurs, the maximum degradation rate ($T_{max}$), which corresponds to the peak of the first derivative curve, and the endset (T90%), which indicates the temperature at which a 90% weight loss occurs. Neat PLA possesses a thermal stability value with a T5% of 311.3 °C, a $T_{max}$ of 330.5 °C, and a T90% of 342.6 °C. However, by adding PHB to PLA, the $T_{max}$ value increases considerably (by approximately 20 °C), whilst the addition of certain amounts of ECO or MCO to blend (B) formulations causes a slight decrease in the T5%, T90%, and $T_{max}$ values. In both modified oils, these values are lower when the quantity is larger. These values are lower for the formulations with MCO. For example, the addition of 10 phr of ECO (B-10ECO) results in a T5% reduction of 15.5 °C, while with 10 phr of MCO (B-10MCO), the reduction is 21.3 °C with respect to the B formulation value (286.1 °C). Regarding the $T_{max}$

temperature, it can be observed that both plasticizers lead to a decrease in temperature, obtaining the lowest values at approximately 20 °C and 26 °C lower for B-10ECO and B-10MCO, respectively. A similar tendency was observed by Garcia-Campo et al. [79] for PLA/PHB/PCL blends compatibilized with epoxidized soybean oil (ELO). The addition of ELO into the formulations resulted in a reduction of T5% by 19.1 °C and $T_{max}$ by 20 °C. For the specific case of maleinized oils, the work of Perez-Nakai et al. [67] shows how the addition of 10 phr of maleinized hemp seed oil (MHO) decreases the $T_{max}$ by 10 °C with respect to the value of neat PLA.

**Table 3.** Summary of the main thermal parameters of the neat PLA, PLA–PHB blend, and plasticized blend with different ECO and MCO contents.

| Code | TGA Parameters | | | DSC Parameters | | | | | | |
|---|---|---|---|---|---|---|---|---|---|---|
| | $T_{5\%}$ (°C) | $T_{max}$ (°C) | $T_{90\%}$ (°C) | $T_g$ (°C) | $T_{cc}$ (°C) | $\Delta H_{cc}$ (J/g) | $T_{mPLA}$ (°C) | $\Delta H_m$ (J/g) | Xc (%) | $T_{mPHB}$ (°C) |
| PLA | 311.3 | 330.5 | 342.6 | 61.3 | 121.9 | 9.1 | 151.2 | 15.7 | 7.1 | - |
| B (blend) | 286.4 | 349.1 | 359.4 | 56.1 | 124.3 | 2.2 | 151.5 | 2.7 | 0.7 | 173.3 |
| B-1ECO | 276.4 | 342.8 | 350.4 | 54.4 | 125.6 | 2.3 | 151.0 | 2.7 | 0.6 | 172.5 |
| B-2.5ECO | 278.1 | 339.1 | 349.6 | 54.4 | 125.1 | 2.6 | 150.6 | 3.8 | 1.7 | 171.7 |
| B-5ECO | 275.5 | 338.2 | 349.4 | 54.1 | 125.2 | 1.5 | 151.2 | 2.6 | 1.6 | 173.9 |
| B-7.5ECO | 275.5 | 338.2 | 347.6 | 54.2 | 127.1 | 1.2 | 150.3 | 2.5 | 1.9 | 172.7 |
| B-10ECO | 270.9 | 328.7 | 338.7 | 54.6 | 126.7 | 2.6 | 151.8 | 4.1 | 2.7 | 175.7 |
| B-1MCO | 271.0 | 327.2 | 347.9 | 52.1 | 122.8 | 1.7 | 149.6 | 1.4 | 0.3 | 169.3 |
| B-2.5MCO | 270.1 | 329.2 | 348.0 | 47.3 | 122.4 | 0.9 | 149.8 | 1.2 | 0.4 | 166.1 |
| B-5MCO | 269.2 | 325.6 | 346.8 | 49.7 | 121.8 | 2.3 | 148.2 | 2.7 | 0.5 | 165.5 |
| B-7.5MCO | 269.1 | 323.2 | 345.9 | 49.6 | 122.1 | 2.6 | 148.2 | 4.1 | 2.4 | 165.1 |
| B-10MCO | 265.1 | 322.8 | 345.3 | 48.2 | 121.6 | 1.6 | 148.8 | 2.4 | 1.6 | 159.6 |

Regarding to the DSC results, Table 3 also shows the main thermal properties and Figure 4 shows the calorimetric graphs of samples B and B-5MCO. The effect of the addition of PHB to the PLA matrix reduces the glass transition temperature of PLA to a value of 56.1 °C (B formulation). As can be seen, both modified oils have a direct effect on some thermal properties of PLA. Unplasticized PLA has a glass transition temperature ($T_g$) of 61.3 °C, a cold crystallization temperature ($T_{cc}$) of 121.9 °C, and a melting temperature ($T_m$) of 151.2 °C. Both plasticizers—ECO and MCO—decrease the Tg of PLA, which indicates an increase in the mobility of the polymer chains at lower temperatures, evidencing the plasticizing effect of both modified oils [80]. However, the greatest plasticizing effect is shown by the formulations with MCO, obtaining a decrease in $T_g$ of up to almost 10 °C for the specific case of B-2.5MCO and in reference to the formulation B. The same plasticizing effect was observed by Dominguez-Candela et al. [81], who employed epoxidized chia oil in PLA. Regarding to the Tm, no significant variations were observed in Tm for the formulation with ECO. However, the formulations with MCO shows a slight decrease (of approximately 2–3 °C).

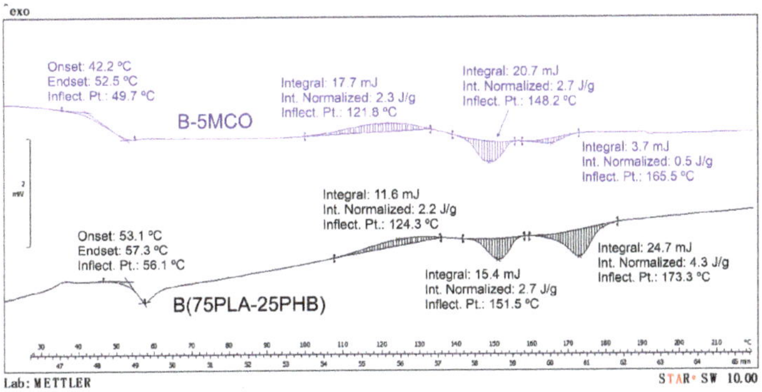


**Figure 4.** Calorimetric graphs of samples B (PLA–PHB) and B-5MCO.

With respect to crystallinity, it is observed that PLA without plasticizing has a crystallinity of 7.1%, whereas as PHB is added, the PLA is totally amorphous. The PHB molecular chains intercalated between the PLA chains do not allow the PLA chains to be ordered. The incorporation of ECO or MCO shows a slight increase, however, the values are negligible. The $T_{cc}$ peak is observed at higher temperatures with the incorporation of ECO or MCO. In fact, the $T_{cc}$ peak temperature increases from 121.9 °C (neat PLA) to 127.1 °C for the B-7.5ECO formulation. This is due to the steric impediment exerted by the ECO on the PLA chains.

### 3.6. Field Emission Scanning Electron Microscopy (FESEM)

In this characterization technique, the breaking surface of the specimens was observed, in this case, as those obtained in the impact test. The PLA morphology, as shown in Figures 5a and 6a, showed a typical surface of brittle materials with a smooth fracture surface and angular fracture flanks. When making the PLA blend with 25% PHB, as shown in Figure 5b, the morphology of the fracture surface was completely modified since it is no longer smooth, and the high roughness and rounded fracture flanks characteristic of ductile materials are observed. In addition, in Figure 6b, made at higher magnifications, a blend without gaps is observed due to the low miscibility and with the PLA acting as a matrix and the dispersed domains of PHB [82]. The samples of blends that contain ECO can be seen in Figure 5c–g, where the incorporation of ECO is progressive, and the fracture morphology of all of them is similar to the PLA–PHB blend with high roughness and rounded fracture flanks. In addition, the appearance of holes and cavities is also observed due to some type of plasticizer saturation for ECO contents and above on the blend (Figure 5e) and these holes increase in number as the amount of ECO in the blend increases up to 10 phr (Figure 5g) [83]. On the other hand, in Figure 6c–e, this effect is observed at higher magnifications, and, in addition, it allows us to appreciate that the gaps due to the accumulation of oil are mainly generated in the PHB domains.

The morphology of the fracture surface of the blends containing MCO can be seen in Figures 5h–l and 6f–h. As occurs in the samples with ECO, the fracture surface remains smooth with rounded ridges, typical of ductile materials. Additionally, the voids also increase due to oil saturation when the amount of oil in the matrix increases, however, the first voids are observed at smaller amounts of MCO, starting at 2.5 phr (Figure 5i) and considerably increase these gaps up to 10 phr of MCO. These voids or cavities negatively affect the ductile properties of the material, as seen previously. When observing the images at 10,000× magnification (Figure 6f–h), it can be seen that in addition to the plasticizing effect, it also has a compatibilizing effect between PLA and PHB, since the dispersed domains or dispersed phase increase in number and represent more than 25% of the images, which possibly, due to these domains, can be a PLA–PHB blend thanks to the influence of the MCO.

As was verified by the mechanical properties of the different blends, the saturation of the oil negatively affects the ductile properties of the blends. In the PLA–PHB–ECO system, the sample with the best ductile behavior is the sample with 10 phr of ECO. However, in the PLA–PHB–MCO system, saturation occurs at lower amounts of oil, at approximately 5 phr of MCO, which has a similar morphology to sample B-10ECO. Therefore, to obtain the same plasticizing effect, it is achieved with less amount of MCO than ECO.

### 3.7. Disintegration under Composting Conditions

Finally, to evaluate the disintegration behavior of the developed materials, they were buried in a composting reactor carried out at a laboratory scale. The variation of the mass loss of the different materials under composting conditions can be seen in Figure 7. The disintegration of the PLA–PHB blends with ECO and MLO is characterized by presenting an initial induction period, where mass loss barely occurs, at approximately 7 days for PLA–PHB–ECO materials and 10 days for PLA–PHB–MCO materials. After this initial state, the disintegration of the different materials accelerates. In the case of the disintegration of the PLA–PHB blend, it does not occur until reaching the first 15 days, at which time the disintegration accelerates and disintegrates until reaching 80% at the end of the test. The incorporation of the modified corn oils into the blend disrupts the disintegration of the PLA–PHB blend. In the case of the incorporation of the MCO, it is observed that at higher percentages, the disintegration of the blend slows down, reaching between 50 and 60% loss of mass with low percentages of MCO (less than 5%) and decreasing below 30% of mass loss with percentages higher than 5% of MCO at the end of the test. However, the incorporation of ECO to the blend generates a different behavior regarding disintegration under composting conditions, since the degree of disintegration at the end of the test decreases very clearly with any of the percentages of ECO used, reaching a loss of mass between 20 and 35%.

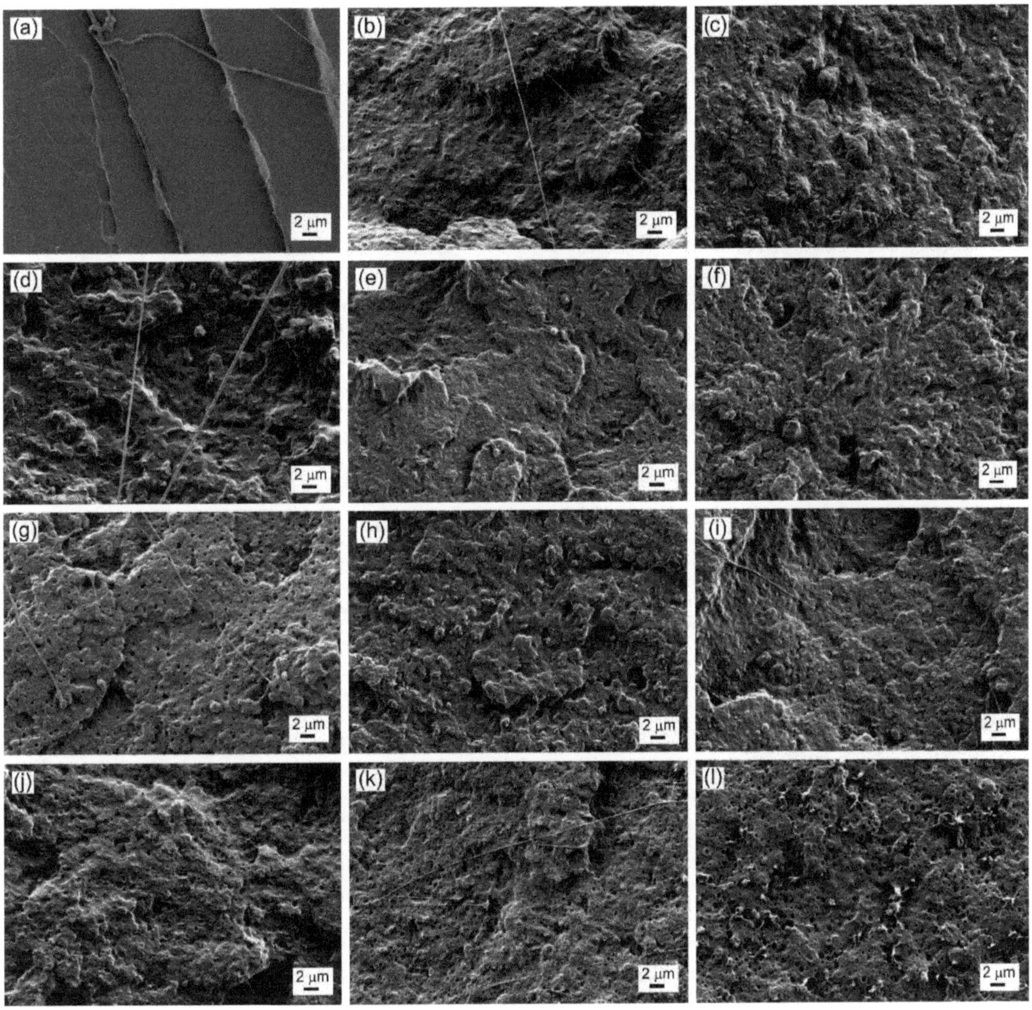


**Figure 5.** FESEM images at 2500× of fracture surface of (**a**) PLA; (**b**) PLA–PHB blend (B); (**c**) B-1ECO; (**d**) B-2.5ECO; (**e**) B-5ECO; (**f**) B-7.5ECO; (**g**) B-10ECO; (**h**) B-1MCO; (**i**) B-2.5MCO (**j**) B-5MCO; (**k**) B-7.5MCO; and (**l**) B-10MCO.

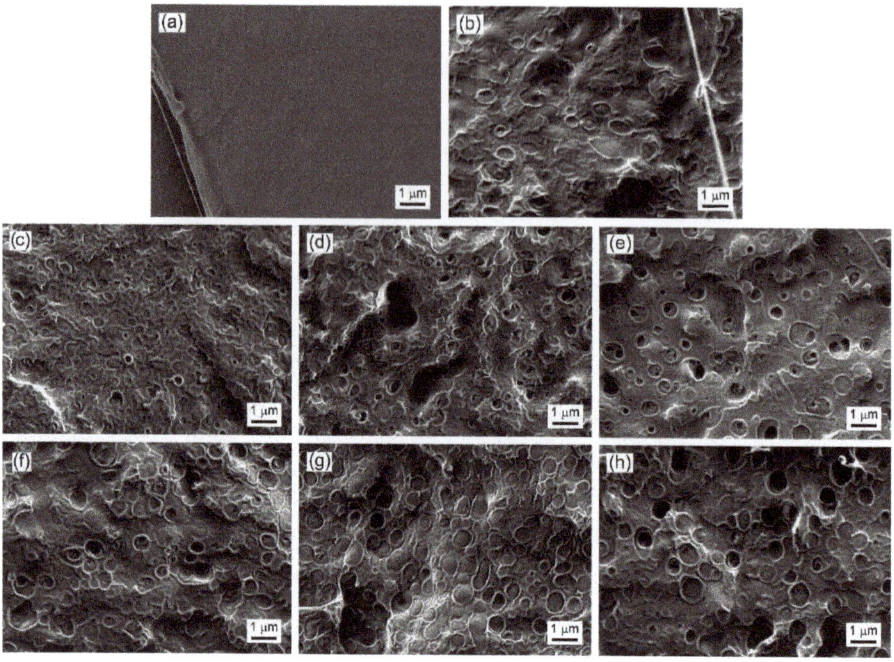


**Figure 6.** FESEM images at 10,000× of fracture surface of (**a**) PLA; (**b**) PLA–PHB blend (B); (**c**) B-5ECO; (**d**) B-7.5ECO; (**e**) B-10ECO; (**f**) B-5MCO; (**g**) B-7.5MCO; and (**h**) B-10MCO.

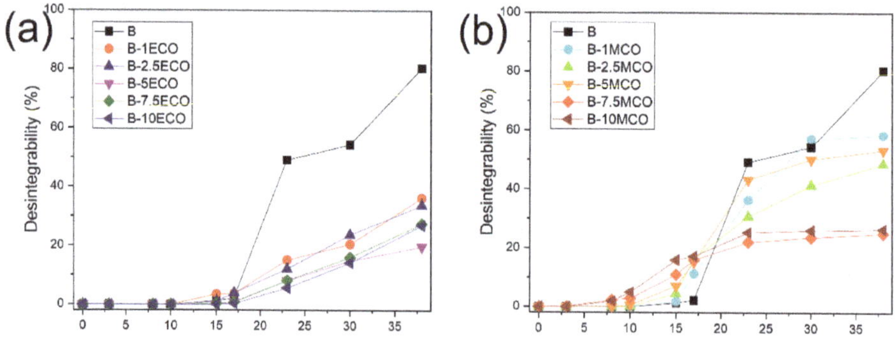


**Figure 7.** Degree of disintegration under composting conditions expressed as the weight loss as a function of time of (**a**) PLA–PHB blend and plasticized blend with different ECO content and (**b**) PLA–PHB blend and plasticized blend with different MCO content.

## 4. Conclusions

This work analyzes the usefulness of the use of modified—epoxidized (ECO) and maleinized (MCO)— corn oil as plasticizing and/or compatibilizing agents of the blend 75 wt% PLA–25 wt% PHB. Both chemical modifications of corn oil were carried out at the laboratory level and both processes were carried out correctly according to the results obtained in the analysis of the number of oxirane oxygen and iodine number for the ECO and according to the evolution of the acid number value for the MCO.

The incorporation of both modified oils, ECO and MCO, allows to obtain PLA–PHB formulations with better ductile properties, requiring less MCO to achieve the same plasticizing effect as with ECO. The plasticizing effect was observed by differential scanning calorimetry due to the decrease in the glass transition temperature (Tg), decreasing by approximately 2 °C when using ECO and between 4 and 8 °C when using MCO in the PLA–PHB blend. Additionally, the plasticizing effect is observed to a greater extent in the mechanical properties, since the elongation at break and the impact absorbed energy improve considerably when using the oils in the PLA–PHB formulations, increasing the elongation at break by more than 400% when using 10% ECO and 5% MCO. The FESEM analysis reveals a particular morphology of the samples plasticized with oil, presenting spherical cavities dispersed in the matrix and that increase with the increase in oil in the formulations. Finally, the oils delay the disintegration in composting conditions with respect to the PLA–PHB formulation, delaying said degradation to a greater extent than the ECO and it is also affected by the amount of oil used.

Therefore, although for this specific study, the MCO shows better plasticization and compatibilization results, two new oils with greater commercial potential were obtained by chemical modification. Compared to other oils obtained from cereals or seeds, CO presents a higher abundance and greater stability against oxidation, which makes it a very interesting raw material from the industrial point of view for applications in both bio-based and synthetic polymers.

**Author Contributions:** Conceptualization, M.D.S. and J.M.F.; methodology, C.P.; validation, J.S.-T. and H.d.l.R.-R.; formal analysis, M.D.S.; investigation, J.S.-T., J.M.F., and M.D.S.; resources, M.D.S.; writing—original draft preparation, J.S.-T. and M.D.S.; writing—review and editing, J.M.F. and H.d.l.R.-R.; visualization, C.P.; supervision, M.D.S.; project administration, M.D.S.; funding acquisition, M.D.S. All authors have read and agreed to the published version of the manuscript.

**Funding:** This research received no external funding.

**Institutional Review Board Statement:** Not applicable.

**Data Availability Statement:** Not applicable.

**Acknowledgments:** This work was supported by the Spanish Ministry of Science and Innovation, NANOCIRCOIL (PID2021-123753NA-C33).

**Conflicts of Interest:** The authors declare no conflict of interest.

## References


1. Gupta, G.; Kumar, A.; Tyagi, R.; Kumar, S. Application and future of composite materials: A review. *Int. J. Innov. Res. Sci. Eng. Technol.* **2016**, *5*, 6907–6911.
2. Nik Md Noordin Kahar, N.N.F.; Osman, A.F.; Alosime, E.; Arsat, N.; Mohammad Azman, N.A.; Syamsir, A.; Itam, Z.; Abdul Hamid, Z.A. The versatility of polymeric materials as self-healing agents for various types of applications: A review. *Polymers* **2021**, *13*, 1194. [CrossRef]
3. Ji, D.; Li, T.; Hu, W.; Fuchs, H. Recent Progress in Aromatic Polyimide Dielectrics for Organic Electronic Devices and Circuits. *Adv. Mater.* **2019**, *31*, 1806070. [CrossRef]
4. Tiwari, N.; Ho, F.; Ankit, A.; Mathews, N. A rapid low temperature self-healable polymeric composite for flexible electronic devices. *Camb. R. Soc Chem. J. Mater. Chem. A Mater. Energy Sustain.* **2018**, *6*, 21428–21434. [CrossRef]
5. Dingler, C.; Dirnberger, K.; Ludwigs, S. Semiconducting Polymer Spherulites—From Fundamentals to Polymer Electronics. *Weinh. Wiley-V C H Verl. Gmbh. Macromol. Rapid Commun.* **2019**, *40*, 1800601. [CrossRef]
6. Arrieta, M.P.; López Martínez, J.; Hernández, A.; Rayón Encinas, E. *Ternary PLA-PHB-Limonene Blends Intended for Biodegradable Food Packaging Applications*; Elsevier: Amsterdam, The Netherlands, 2014.
7. Arrieta, M.P.; López Martínez, J.; Ferrándiz Bou, S.; Peltzer, M.A. *Characterization of PLA-Limonene Blends for Food Packaging Applications*; Elsevier: Amsterdam, The Netherlands, 2013.
8. Plastics the Facts 2020. An Analysis of European Plastics Production, Demand and Waste Data—Plastics Europe—Association of Plastics Manufacturers. Available online: https://plasticseurope.org/knowledge-hub/plastics-the-facts-2020/ (accessed on 13 November 2019).
9. Gonçalves, A.L.; Pires, J.C.M.; Simões, M. Green fuel production: Processes applied to microalgae. *Environ. Chem. Lett.* **2013**, *11*, 315–324. [CrossRef]
10. Baena-Moreno, F.M.; le Saché, E.; Pastor-Pérez, L.; Reina, T.R. Membrane-based technologies for biogas upgrading: A review. *Int. Publishing. Environ. Chem. Lett.* **2020**, *18*, 1649–1658. [CrossRef]

11. Hale, R.C.; Seeley, M.E.; la Guardia, M.J.; Mai, L.; Zeng, E.Y. A Global Perspective on Microplastics. *Wash. Amer Geophys. Union J. Geophys. Res. Ocean.* **2020**, *125*, e2018JC014719. [CrossRef]
12. Bollaín Pastor, C.; Vicente Agulló, D. Presencia de microplásticos en aguas y su potencial impacto en la salud pública. *Revista Española de Salud Pública* **2020**, *93*, e201908064.
13. DIRECTIVE (EU) 2019/904 OF THE EUROPEAN PARLIAMENT AND OF THE COUNCIL of 5 June 2019 on the reduction of the impact of certain plastic products on the environment. Official Journal of the European Union]. Available online: https://eur-lex.europa.eu/eli/dir/2019/904/oj (accessed on 10 September 2022).
14. Sikora, J.; Majewski, Ł.; Puszka, A. Modern biodegradable plastics-processing and properties: Part I. *Materials* **2020**, *13*, 1986. [CrossRef] [PubMed]
15. Sikora, J.W.; Majewski, Ł.; Puszka, A. Modern biodegradable plastics—processing and properties part II. *Materials* **2021**, *14*, 2523. [CrossRef] [PubMed]
16. Drumright, R.E.; Gruber, P.R.; Henton, D.E. Polylactic acid technogy. *Adv. Mater.* **2000**, *12*, 1841–1846. [CrossRef]
17. Södergard, A.; Stolt, M. Properties of lactic acid based polymers and their correlation with composition. *Prog. Polym. Sci.* **2002**, *27*, 1123–1163. [CrossRef]
18. Jamshidian, M.; Tehrany, E.A.; Imran, M.; Jacquot, M.; Desobry, S. Poly-Lactic Acid: Productior Applications, Nanocomposites, and Release Studies. *Compr. Rev. Food Sci. Food Saf.* **2010**, *9*, 552–571. [CrossRef] [PubMed]
19. Vink, E.T.H.; Rábago, K.R.; Glassner, D.A.; Gruber, P.R. Applications of life cycle assessment to NatureWorks polylactide (PLA) production. *Polym. Degrad. Stab.* **2003**, *80*, 403–419. [CrossRef]
20. Nofar, M.; Sacligil, D.; Carreau, P.J.; Kamal, M.R.; Heuzey, M.-C. Poly (lactic acid) blends: Processing, properties and applications. *Int. J. Biol. Macromol.* **2019**, *125*, 307–360. [CrossRef] [PubMed]
21. Farto-Vaamonde, X.; Auriemma, G.; Aquino, R.P.; Concheiro, A.; Alvarez-Lorenzo, C. Post-manufacture loading of filaments and 3D printed PLA sca_olds with prednisolone and dexamethasone for tissue regeneration applications. *Eur. J. Pharm. Biopharm.* **2019**, *141*, 100–110. [CrossRef] [PubMed]
22. Fajardo, J.; Valarezo, L.; López, L.; Sarmiento, A. Experiencies in obtaining polymeric composites reinforced with natural fiber from Ecuador. *Ingenius* **2013**, *9*, 28–35. [CrossRef]
23. Gere, D.; Czigany, T. Future trends of plastic bottle recycling: Compatibilization of PET and PLA. *Polym. Test.* **2020**, *81*, 106160. [CrossRef]
24. Palma-Ramírez, D.; Torres-Huerta, A.; Domínguez-Crespo, M.; Del Angel-López, D.; Flores-Vela, A.; de la Fuente, D. Data supporting the morphological/topographical properties and the degradability on PET/PLA and PET/chitosan blends. *Data Brief* **2019**, *25*, 104012. [CrossRef]
25. Hachemi, R.; Belhaneche-Bensemra, N.; Massardier, V. Elaboration and characterization of bioblends based on PVC/PLA. *J. Appl. Polym. Sci.* **2014**, *131*, 40045. [CrossRef]
26. Nehra, R.; Maiti, S.N.; Jacob, J. Analytical interpretations of static and dynamic mechanical properties of thermoplastic elastomer toughened PLA blends. *J. Appl. Polym. Sci.* **2018**, *135*, 45644. [CrossRef]
27. Jašo, V.; Cvetinov, M.; Raki´c, S.; Petrovi´c, Z.S. Bio-plastics and elastomers from polylactic acid/thermoplastic polyurethane blends. *J. Appl. Polym. Sci.* **2014**, *131*, 41104. [CrossRef]
28. Mandal, D.K.; Bhunia, H.; Bajpai, P.K. Thermal degradation kinetics of PP/PLA nanocomposite blends. *J. Thermoplast. Compos. Mater.* **2019**, *32*, 1714–1730. [CrossRef]
29. Azizi, S.; Azizi, M.; Sabetzadeh, M. The role of multiwalled carbon nanotubes in the mechanical, thermal, rheological, and electrical properties of PP/PLA/MWCNTs nanocomposites. *J. Compos. Sci.* **2019**, *3*, 64. [CrossRef]
30. Quiles-Carrillo, L.; Montanes, N.; Jorda-Vilaplana, A.; Balart, R.; Torres-Giner, S. A comparative study on the e_ect of di_erent reactive compatibilizers on injection-molded pieces of bio-based high-density polyethylene/polylactide blends. *J. Appl. Polym. Sci.* **2019**, *136*, 47396. [CrossRef]
31. Torres-Huerta, A.; Domínguez-Crespo, M.; Palma-Ramírez, D.; Flores-Vela, A.; Castellanos-Alvarez, E.; Angel-López, D.D. Preparation and degradation study of HDPE/PLA polymer blends for packaging applications. *Rev. Mex. Ing. Química* **2019**, *18*, 251–271. [CrossRef]
32. Ferri, J.M.; Garcia-Garcia, D.; Rayón, E.; Samper, M.D.; Balart, R. Compatibilization and Characterization of Polylactide and Biopolyethylene Binary Blends by Non-Reactive and Reactive Compatibilization Approaches. *Polymers* **2020**, *12*, 1344. [CrossRef]
33. Arrieta, M.P.; López, J.; Ferrándiz, S.; Peltzer, M.A. Characterization of PLA-limonene blends for food packaging applications. *Polym. Test.* **2013**, *32*, 760–768. [CrossRef]
34. Arrieta, M.P.; López, J.; Hernández, A.; Rayón, E. Ternary PLA–PHB–Limonene blends intended for biodegradable food packaging applications. *Eur. Polym. J.* **2014**, *50*, 255–270. [CrossRef]
35. Montes, M.I.; Cyras, V.; Manfredi, L.; Pettarín, V.; Fasce, L. Fracture evaluation of plasticized polylactic acid/poly (3-HYDROXYBUTYRATE) blends for commodities replacement in packaging applications. *Polym. Test.* **2020**, *84*, 106375. [CrossRef]
36. Ferri, J.M.; Fenollar, O.; Jorda-Vilaplana, A.; García-Sanoguera, D.; Balart, R. E_ect of miscibility on mechanical and thermal properties of poly (lactic acid)/polycaprolactone blends. *Polym. Int.* **2016**, *65*, 453–463. [CrossRef]
37. Mittal, V.; Akhtar, T.; Matsko, N. Mechanical, thermal, rheological and morphological properties of binary and ternary blends of PLA, TPS and PCL. *Macromol. Mater. Eng.* **2015**, *300*, 423–435. [CrossRef]

38. Rao, R.U.; Venkatanarayana, B.; Suman, K. Enhancement of mechanical properties of PLA/PCL (80/20) blend by reinforcing with MMT nanoclay. *Mater. Today Proc.* **2019**, *18*, 85–97.
39. Carbonell-Verdu, A.; Ferri, J.; Dominici, F.; Boronat, T.; Sanchez-Nacher, L.; Balart, R.; Torre, L. Manufacturing and compatibilization of PLA/PBAT binary blends by cottonseed oil-based derivatives. *Express Polym. Lett.* **2018**, *12*, 808–823. [CrossRef]
40. Wang, X.; Peng, S.; Chen, H.; Yu, X.; Zhao, X. Mechanical properties, rheological behaviors, and phase morphologies of high-toughness PLA/PBAT blends by in-situ reactive compatibilization. *Compos. Part. B Eng.* **2019**, *173*, 107028. [CrossRef]
41. Kilic, N.T.; Can, B.N.; Kodal, M.; Ozkoc, G. Compatibilization of PLA/PBAT blends by using Epoxy-POSS. *J. Appl. Polym. Sci.* **2019**, *136*, 47217. [CrossRef]
42. Chieng, B.W.; Ibrahim, N.A.; Yunus, W.M.Z.W.; Hussein, M.Z. Plasticized poly(lactic acid) with low molecular weight poly(ethylene glycol): Mechanical, thermal, and morphology properties. *J. Appl Polym. Sci.* **2013**, *130*, 4576–4580. [CrossRef]
43. Yu, Y.; Cheng, Y.; Ren, J.; Cao, E.; Fu, X.; Guo, W. Plasticizing effect of poly(ethylene glycol)s with different molecular weights in poly(lactic acid)/starch blends. *J. Appl. Polym. Sci.* **2015**, *132*, 41808. [CrossRef]
44. Nazari, T.; Garmabi, H. Polylactic acid/polyethylene glycol blend fibres prepared via melt electrospinning: Effect of polyethylene glycol content. *Micro Nano Lett.* **2014**, *9*, 686–690. [CrossRef]
45. Ke, W.; Li, X.; Miao, M.; Liu, B.; Zhang, X.; Liu, T. Fabrication and Properties of Electrospun and Electrosprayed Polyethylene Glycol/Polylactic Acid (PEG/PLA) Films. *Coatings* **2021**, *11*, 790. [CrossRef]
46. Ruan, G.; Feng, S.S. Preparation and characterization of poly (lactic acid)–poly (ethylene glycol)–poly (lactic acid)(PLA–PEG–PLA) microspheres for controlled release of paclitaxel. *Biomaterials* **2003**, *24*, 5037–5044. [CrossRef]
47. Piorkowska, E.; Kulinski, Z.; Galeski, A.; Masirek, R. Plasticization of semicrystalline poly(L-lactide) with poly(propylene glycol). *Polymer* **2006**, *47*, 7178–7188. [CrossRef]
48. Burgos, N.; Martino, V.P.; Jimenez, A. Characterization and ageing study of poly(lactic acid) films plasticized with oligomeric lactic acid. *Polym. Degrad. Stab.* **2013**, *98*, 651–658. [CrossRef]
49. Arrieta, M.; de Dicastillo, C.L.; Garrido, L.; Roa, K.; Galotto, M.J. Electrospun PVA fibers loaded with antioxidant fillers extracted from Durvillaea Antarctica algae and their effect on plasticized PLA bionanocomposites. *Eur. Polym. J.* **2018**, *103*, 145–157. [CrossRef]
50. Hassouna, F.; Raquez, J.-M.; Addiego, F.; Toniazzo, V.; Dubois, P.; Ruch, D. New development on plasticized poly(lactide): Chemical grafting of citrate on PLA by reactive extrusion. *Eur. Polym. J.* **2012**, *48*, 404–415. [CrossRef]
51. Jing, J.; Qiao, Q.A.; Jin, Y.; Ma, C.; Cai, H.; Meng, Y.; Cai, Z.; Feng, D. Molecular and mesoscopic dynamics simulations on the compatibility of PLA/plasticizer blends. *Chin. J. Chem* **2012**, *30*, 133–138. [CrossRef]
52. Notta-Cuvier, D.; Murariu, M.; Odent, J.; Delille, R.; Bouzouita, A.; Raquez, J.-M.; Lauro, F.; Dubois, P. Tailoring polylactide properties for automotive applications: Effects of coaddition of halloysite Nanotubes and selected plasticizer. *Macromol. Mater. Eng.* **2015**, *300*, 684–698. [CrossRef]
53. Chieng, B.W.; Ibrahim, N.A.; Then, Y.Y.; Loo, Y.Y. Epoxidized vegetable oils plasticized poly(lactic acid)biocomposites: Mechanical, thermal and morphology properties. *Molecules* **2014**, *19*, 16024–16038. [CrossRef]
54. Alam, J.; Alam, M.; Raja, M.; Abduljaleel, Z.; Dass, L.A. MWCNTs-reinforced epoxidized linseed oil plasticized polylactic acid nanocomposite and its electroactive shape memory behaviour. *Int. J. Mol. Sci.* **2014**, *15*, 19924–19937. [CrossRef]
55. Ferri, J.M.; Samper, M.D.; García-Sanoguera, D.; Reig, M.J.; Fenollar, O.; Balart, R. Plasticizing effect of biobased epoxidized fatty acid esters on mechanical and thermal properties of poly(lactic acid). *J. Mater. Sci.* **2016**, *51*, 5356–5366. [CrossRef]
56. Garcia-Garcia, D.; Carbonell-Verdu, A.; Arrieta, M.P.; Lopez-Martínez, J.; Samper, M.D. Improvement of PLA film ductility by plasticization with epoxidized karanja oil. *Polym. Degrad. Stab.* **2020**, *179*, 109259. [CrossRef]
57. Carbonell-Verdu, A.; Samper, M.D.; Garcia-Garcia, D.; Sanchez-Nacher, L.; Balart, R. Plasticization effect of epoxidized cottonseed oil (ECSO) on poly(lactic acid). *Ind. Crops Prod.* **2017**, *104*, 278–286. [CrossRef]
58. Erceg, M.; Kovacic, T.; Klaric, I. Thermal degradation of poly( 3-hydroxybutyrate) plasticized with acetyl tributyl citrate. *Polym. Degrad. Stab.* **2005**, *90*, 31–318. [CrossRef]
59. Weng, Y.X.; Wang, L.; Zhang, M.; Wang, X.L.; Wang, Y.Z. Biodegradation behavior of P(3HB,4HB)/PLA blends in real soil environments. *Polym. Test.* **2013**, *32*, 60–70. [CrossRef]
60. Sanchez-Garcia, M.D.; Gimenez, E.; Lagaron, J.M. Morphology and Barrier Properties of Nanobiocomposites of Poly (3-hydroxybutyrate) and Layered Silicates. *J. Appl. Polym. Sci.* **2008**, *108*, 2787–2801. [CrossRef]
61. Lerma-Canto, A.; Gomez-Caturla, J.; Herrero-Herrero, M.; Garcia-Garcia, D.; Fombuena, V. Development of polylactic acid thermoplastics starch formulations using malenized hemp oil as biobased plasticizer. *Polymers* **2021**, *13*, 1392. [CrossRef]
62. Barrera-Arellano, D.; Badan-Ribeiro, A.P.; Serna-Saldivar, S.O. *Chapter 21-Corn Oil: Composition, Processing and Utilization in Corn, Chemistry and Technology*, 3rd ed.; Serna-Saldivar, S.O., Ed.; AACC International Elsevier: Eagan, MN, USA, 2019; pp. 519–613.
63. *ASTM, D. 1652-97*; Standard test methods for epoxy content of epoxy resins. ASTM International: West Conshohocken, PA, USA, 2004.
64. *ISO 3961:2009*; Animal and vegetable fats and oils—Determination of iodine value. International Organization for Standardization: Geneva, Switzerland, 2009.
65. Carbonell-Verdu, A.; Garcia-Garcia, D.; Dominici, F.; Torre, L.; Sanchez-Nacher, L.; Balart, R. PLA films with improved flexibility properties by using maleinized cottonseed oil. *Eur. Polym. J.* **2017**, *91*, 248–259. [CrossRef]

66. Perez-Nakai, A.; Lerma-Canto, A.; Domingez-Candela, I.; Garcia-Garcia, D.; Ferri, J.M.; Fombuena, V. Comparative Study of the Properties of Plasticized Polylactic Acid with Maleinized Hemp Seed Oil and a Novel Maleinized Brazil Nut Seed Oil. *Polymers* **2021**, *13*, 2376. [CrossRef] [PubMed]
67. *ISO 3961:2009*; Animal and vegetable fats and oils—Determination of acid value and acidity. International Organization for Standardization: Geneva, Switzerland, 2009.
68. *ISO 527-1:2019*; Plastics—Determination of tensile properties—Part 1: General principles. International Organization for Standardization: Geneva, Switzerland, 2019.
69. *ISO 178:2019*; Plastics—Determination of flexural properties. International Organization for Standardization: Geneva, Switzerland, 2019.
70. *ISO 179-1:2010*; Plastics—Determination of Charpy impact properties—Part 1: Non—instrumented impact test. International Organization for Standardization: Geneva, Switzerland, 2010.
71. *ISO 868:2003*; Plastics and ebonite—Determination of indentation hardness by means of a durometer (Shore hardness). International Organization for Standardization: Geneva, Switzerland, 2003.
72. *ISO 75-1:2013*; Plastics—Determination of temperature of deflection under load—Part 1: General test method. International Organization for Standardization: Geneva, Switzerland, 2013.
73. *ISO 306:2015*; Plastics—Thermoplastics materials - Determination of Vicat softening temperature (VST). International Organization for Standardization: Geneva, Switzerland, 2015.
74. Yeh, J.T.; Wu, C.J.; Tsou, C.H.; Chai, W.L.; Chow, J.D.; Huang, C.Y.; Chen, K.N.; Wu, C.S. Study on the crystallization, miscibility, morphology, properties of poly(lactic acid)/poly(ε-caprolactone) blends. *Polym.-Plast. Technol. Eng.* **2009**, *48*, 571–578. [CrossRef]
75. *ISO 20200:2015*; Plastics—Determination of the degree of disintegration of plastics materials under simulated conditions in a laboratory-scale test.
76. Carbonell-Verdu, A.; Bernardi, L.; Garcia-Garcia, D.; Sanchez-Nacher, L.; Balart, R. *Development of Environmentally Friendly Composite Matrices from Epoxidized Cottonseed Oil*; Elsevier: Amsterdam, The Netherlands, 2015.
77. Quiles-Carrillo, L.; Montanes, N.; Sammon, C.; Balart, R.; Torres-Giner, S. Compatibilization of highly sustainable polylactide/almond shellflourcomposites by reactive extrusion with maleinized linseed oil. *Ind. Crops Prod.* **2018**, *111*, 878–888. [CrossRef]
78. Arrieta, M.P.; Samper, M.D.; Aldas, M.; López, J. On the use of PLA-PHB blends for sustainable food packaging applications. *Materials* **2017**, *10*, 1008. [CrossRef] [PubMed]
79. García Campo, M.J.; Quiles-Carrillo, L.; Masiá Vañó, J.; Reig-Pérez, M.J.; Montanes, N.; Balart, R. Environmentally Friendly Compatibilizers from Soybean Oil for Ternary Blends of Poly(lactic acid)-PLA, Poly(e-caprolactone)-PCL and Poly(3-hydroxybutyrate)-PHB. *Materials* **2017**, *10*, 1339. [CrossRef] [PubMed]
80. Dobircau, L.; Delpouve, N.; Herbinet, R.; Domenek, S.; Le Pluart, L.; Delbreilh, L.; Ducruet, V.; Dargent, E.J.P.E. Molecular mobility and physical ageing of plasticized poly (lactide). *Science* **2015**, *55*, 858–865. [CrossRef]
81. Dominguez-Candela, I.; Ferri, J.M.; Cardona, S.C.; Lora, J.; Fombuena, V.J.P. Dual Plasticizer/Thermal Stabilizer Effect of Epoxidized Chia Seed Oil (Salvia hispanica L.) to Improve Ductility and Thermal Properties of Poly (Lactic Acid). *Polymers* **2021**, *13*, 1283. [CrossRef] [PubMed]
82. Zhang, M.; Thomas, N.L. Blending polylactic acid with polyhydroxybutyrate: The effect on thermal, mechanical, and biodegradation properties. *Adv. Polym. Technol.* **2011**, *30*, 67–79. [CrossRef]
83. Ferri, J.M.; Garcia-Garcia, D.; Montanes, N.; Fenollar, O.; Balart, R. The effect of maleinized linseed oil as biobased plasticizer in poly (lactic acid)-based formulations. *Polym. Int.* **2017**, *66*, 882–891. [CrossRef]

polymers

MDPI

Review

# A Critical Review on Hygrothermal and Sound Absorption Behavior of Natural-Fiber-Reinforced Polymer Composites

V. Bhuvaneswari [1], Balaji Devarajan [1], B. Arulmurugan [1], R. Mahendran [2], S. Rajkumar [3], Shubham Sharma [4,5,*], Kuwar Mausam [6], Changhe Li [5] and Elsayed Tag Eldin [7,*]

1 Department of Mechanical Engineering, KPR Institute of Engineering and Technology, Arasur 641407, India
2 Department of Mechanical Engineering, Jai Shriram Engineering College, Avinashipalayam 638660, India
3 Department of Mechanical Engineering, Faculty of Manufacturing, Institute of Technology, Hawassa University, Awasa 3870006, Ethiopia
4 University Centre of Research and Development, Mechanical Engineering Department, Chandigarh University, Mohali 140413, India
5 School of Mechanical and Automotive Engineering, Qingdao University of Technology, Qingdao 266520, China
6 Department of Mechanical Engineering, GLA University, Mathura 281406, India
7 Faculty of Engineering and Technology, Future University in Egypt, New Cairo 11835, Egypt
* Correspondence: shubham543sharma@gmail.com or shubhamsharmacsirclri@gmail.com (S.S.); elsayed.tageldin@fue.edu.eg (E.T.E.)

**Abstract:** Increasing global environmental problems and awareness towards the utilization of eco-friendly resources enhanced the progress of research towards the development of next-generation biodegradable and environmentally friendly material. The development of natural-based composite material has led to various advantages such as a reduction in greenhouse gases and carbon footprints. In spite of the various advantages obtained from green materials, there are also a few disadvantages, such as poor interfacial compatibility between the polymer matrix and natural reinforcements and the high hydrophilicity of composites due to the reinforcement of hydrophilic natural fibers. This review focuses on various moisture-absorbing and sound-absorbing natural fiber polymer composites along with the synopsis of preparation methods of natural fiber polymer composites. It was stated in various studies that natural fibers are durable with a long life but their moisture absorption behavior depends on various factors. Such natural fibers possess different moisture absorption behavior rates and different moisture absorption behavior. The conversion of hydrophilic fibers into hydrophobic is deemed very important in improving the mechanical, thermal, and physical properties of the natural-fiber-reinforced polymer composites. One more physical property that requires the involvement of natural fibers in place of synthetic fibers is the sound absorption behavior. Various researchers have made experiments using natural-fiber-reinforced polymer composites as sound-absorbing materials. It was found from various studies that composites with higher thickness, porosity, and density behaved as better sound-absorbing materials.

**Keywords:** natural fiber; polymer matrix; bio-composites; moisture absorption; sound absorption

**Citation:** Bhuvaneswari, V.; Devarajan, B.; Arulmurugan, B.; Mahendran, R.; Rajkumar, S.; Sharma, S.; Mausam, K.; Li, C.; Eldin, E.T. A Critical Review on Hygrothermal and Sound Absorption Behavior of Natural-Fiber-Reinforced Polymer Composites. *Polymers* **2022**, *14*, 4727. https://doi.org/10.3390/polym14214727

Academic Editors: Vsevolod Aleksandrovich Zhuikov and Rosane Michele Duarte Soares

Received: 28 August 2022
Accepted: 24 October 2022
Published: 4 November 2022



## 1. Introduction

Reducing the use of fossil fuels, particularly in the construction industry, is a major focus of scientific investigation. Buildings accounted for 21% of global final energy consumption in 2017, placing them third behind industry and transportation. Global final consumption energy, as well as unused, dissipated at a rate of 9.1% [1]. Energy consumption in buildings has enhanced over the last few decades, which necessitates more power generation and, consequently, $CO_2$ emissions. Climate policy has undergone radical transformations in the last three decades, particularly since the United Nations' publication [2] in 1987, also recognized as the Brundtland Report. One of the most important aspects of sustainable development is protecting the environment and climate, and also bringing the global environment and development issues into the official concerns of all nations. With this as a backdrop, the UNFCCC [3] came into being as a means of addressing both global climate change and economic growth, particularly in energy-intensive industries such as construction. Ecosystem stability and efficiency are therefore the focus of numerous efforts at the global and national levels, rather than those made in Europe.

Categorizations of international conventions and protocols, including the 1987 Montréal and Kyoto Protocols, as well as the 1997 editions [4,5] of both, and the 2016 Paris agreement [6], were developed to classify these endeavors. One of their objectives is to maintain greenhouse gas levels at a level that protects biodiversity as well as enabling the environment to participate in climate change in a traditional manner. Buildings seem to be Europe's second-largest energy consumer, consuming up to 26.1% of the EU's energy consumption [7]. A few new trends are currently being selected to minimize elevated greenhouse gas emissions as well as minimize energy use. Building heating energy accounts for approximately 63.6% [8] of the residential building sector's overall energy consumption. As a result, two major trends have emerged in recent years to combat rising energy consumption as well as greenhouse gas emissions: improving building structure envelope insulation efficiency to minimize energy consumption as well as developing new low carbon materials, and reducing the carbon dioxide footprint. It is possible to achieve these trends through the use of porous materials such as concrete [9,10], fibers concrete [11–15], bio-based material [16,17] as natural fiber components, straw and rammed earth. In comparison to denser concrete, the lower thermal conductivity that these porous materials achieve is the primary benefit of their use.

Although mechanical strength decreases with porousness, an acceptable balance must be accomplished between mechanical and thermal properties. Buildings have the ability to trap moisture as well as potentially relieve it depending on climate and outside conditions when this mechanical thermal stability is met, due to their greater porosity and tortuosity. They must be considered when measuring and simulating the entire building's power requirements [18]. Regulations and labeling, as well as decision-making, are made more difficult by effective power simulation of this type of building construction. We currently lack a reference model for this envelope type's transient hygrothermal behavior. Furthermore, no model has been developed yet to account for the influence of coupled heat and mass transfer concepts on the world's energy efficiency of this construction type. Coupling heat and moisture transfers of construction elements, as well as envelopes, have become increasingly important for simulation methods in the past few decades [19]. Figure 1 shows the schematic of the degradation of the composites' interface due to moisture absorption. It clearly portrays the absorption of water molecules into the voids present in between the fiber–matrix interface and the development of matrix cracks due to excessive water absorption.

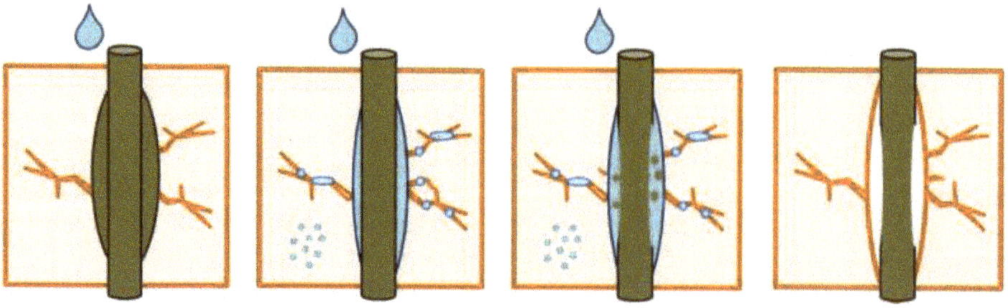

**Figure 1.** Degradation of composite interface due to moisture absorption (Adapted from the reference [4]).

Further research into potential low-carbon buildings' hygrothermal effectiveness or the effect of different weather conditions on their hygroscopic properties would be extremely beneficial. Both Building Energy Simulation (BES) and Computational Fluid Dynamics (CFD) are being used in the construction sector to determine energy balance requirements. It is possible to conduct long-term unsteady analysis using the BES methodology because the profile of physicochemical characteristics is consistent all across a given region, enabling this. When compared to the BES methodology, the CFD methodology involves breaking down the overall building volume into smaller elements or volumes through the usages of numerical meshing. These findings lead to a significant increase in computing power during unsteady state simulation studies. As a result, it is critical that the construction industry takes substantial steps to developing models as well as methodologies that can be used to forecast potential improvements. Building simulation tools must integrate efficient computing techniques while taking into account unique aspects, such as the coupled heat, as well as the moisture transfer through all of the envelope. The wide range of computational methods used by scientists and engineers depends on the application [20–24].

Improving and evaluating energy performance necessitates the use of a precise hygrothermal model. One can distinguish between white-box models, which are based on a physical understanding of the system, and energy balance equations. Black-box models, including the COMSOL Multiphysics and Transient System Simulation (TRNSYS) programs, which use only measured input/output data, as well as statistical estimation methods, including Artificial Neural Networks (ANN), without physics-dependent models, have been commonly used to obtain these results. Each method has pros and cons of its own. M. H. Benzaama and co-researchers [25] conducted a comparative study of these models. Hearing loss, sleep disorders, fatigue, heart disease, and other physiological and psychological issues are just a few of the health effects that are now widely recognized as being associated with excessive noise in the workplace and in daily life [26,27]. As a result, it is critical to keep household noise levels under control. Noise pollution can be minimized by using fibers' sound absorption substances. Acoustic comfort (for example, speech intelligibility) could also be improved by using sound absorption materials to control reverberation time in concert halls, exhibition halls, workplaces, opera houses, and some others [28,29].

The construction industry relies heavily on fibrous material with a dual insulating material (sound along with thermal) [30,31]. As a result of their high specific surface area, elevated acoustical performance, and affordable price, some traditional fibrous insulators such as glass fiber and mineral wool are commonly used in sound absorption applications. A total of 60% of Europe's market for insulating materials in 2005 was made up of glass fiber and mineral wool. In addition, organic foamy substances (such as polystyrene and polyurethane) account for 27% of the market and other materials account for 13%. Although the acoustic and thermal insulation properties of glass fiber and mineral wool are significantly superior, it could be neglected that breathing fibers and particles could indeed cause irritation as well as lay-down within lung alveoli, which can cause a few potential health problems [32–35]. Natural fibers, however, have a lower influence on the atmosphere than synthetic substances. When it comes to synthetic fibers, high-temperature manufacturing processes, as well as petrochemical sources, are often used to produce synthetic fibers, resulting in a substantial carbon footprint [36]. According to a Life Cycle Assessment (LCA), synthetic materials use more energy and have higher global warming possibilities from cradle to platform installation. Since conventional sound absorbers are harmful to the ecosystem, it is important to look into ecologically friendly alternatives. The low toxicity of natural fibers and the fact that they are safe for humans make them ideal substitutes for traditional sound absorbers. Because of their biodegradability, outstanding sustainability, abundance, and environmentally friendly nature, natural fibers have been referred to as green materials. Not to mention they have a substantially lower carbon footprint than synthetic equivalents [37–40], which is another advantage of organic fibrous acoustic absorbers. According to research, mineral wool is considered to be a natural fiber [41–43]. With all the above facts and figures in mind, the current review focuses on the hygrothermal and acoustic behavior of natural-fiber-reinforced polymer composites. Animal and vegetable fibers make up the bulk of the natural fibers considered in this study. Hybridization of fibers with synthetic fibers was also carried out in many studies. Such consolidation has also been made along with the synthesis and preparation of hybrid fiber-reinforced polymer composites.

## 2. Hygrothermal Behavior of Polymer Composites

### 2.1. Carbon Nanofibers Hybridized with Natural Fibers

Flax fibers are considered to be the most hydrophobic fibers among various natural fibers. The consequences of carbon nanofibers' (CNFs) composition on flax-fiber-reinforced epoxy (FFRE) thermoplastic composite hygrothermal aging behaviors and mechanisms were examined in many of the previous studies. It took 180 days to test CNF/FFRE laminates comprising 0.25–2.0 wt.% CNFs. The properties of water absorption and tensile strength, as well as thermodynamics, have all been examined. According to the findings of the study, CNFs had a significant impact on the FFRE laminates' hygrothermal properties. Due to the water obstacle characteristics of CNFs with FFRE laminates, water uptake was considerably lowered. Due to increased matrix stability, as well as improved interface bonding, CNFs/FFRE laminates had better tensile and thermodynamic characteristics. FFRE laminates' hygrothermal durability can be enhanced by having the appropriate balance of CNFs, as shown in this analysis [44].

FFRE laminates were reinforced with bi-directional flax fiber fabrics. 240 $g/m^2$ area density, as well as a standardized thickness of 0.16 mm, were measured for flax fabrics in this study. For the binder of the FFRE laminates, an epoxy resin containing the main agent as well as the curing agent was used. In the FFRE laminates, CNFs were used as nanofillers. CNFs had an average size of 100 nm in diameter and 50–200 m in length. A graphic depiction of the manufacturing process of CNF-altered FFRE composites is demonstrated and also discussed in detail. For 30 min, acetone was mixed with four different concentrations of CNFs (that is, 0.25%, 0.50%, 1.0%, and 2.0% by mass of said epoxy resin) that were dehydrated inside an oven to eliminate whatever moisture they had before being weighed. Second, the CNFs with the acetone mixture were sonicated for 6 h at 60 °C before being placed inside an ultrasonic cleaning solution for scattering. For uniform CNFs with epoxy dispersion, the acetone/CNFS mixture was added to the epoxy primary agent and agitated mechanically before ultrasonic dispersion. Mechanical stirring entirely volatilized the mixture's acetone [45,46].

The CNFs with the epoxy mixture were kept inside a drying oven for around 20 min to eliminate the bubbles. The curing agent was incorporated into the CNF/epoxy combination and stirred for 5 min at a low speed (a rotational speed of 500 rpm). For the next 20 min, the mixture was held in a vacuum oven to remove any residual gas. Wet lay-up was used to make the CNFs with FFRE laminates of two plies of bi-directional flax fiber fabrics with 350 mm × 350 mm measurements. Fabrics made from flax fibers were woven in the same direction of laying. CNFs with FFRE laminates had first been dried for 24 h at room temperature, then for another 24 h at 60 °C in the oven. In order to age as well as test the CNFs/FFRE laminates, they were kept in distilled water at various temperatures for 168 h (7 days). In the curing process, CNFs with FFRE laminates measured 0.2 to 1.3 mm in thickness. About 0.25 seemed to be the fiber volume ratio of said laminates. The epoxy binder to flax fabric laminate mass ratio was approximately 9:4. F1.0, F0.25, F0, F0.5, and F2.0 were used to indicate the FFRE as well as CNFs with FFRE laminates comprising 0.25%, 0.50%, 1.0%, and 2.0% of CNFs, respectively [42,47,48]. The laminates were cut into samples for every category of experiment in this analysis. Four types of samples for the water absorption, dynamic mechanical analysis (DMA), tensile, and FTIR experiments are displayed in Figure 2.

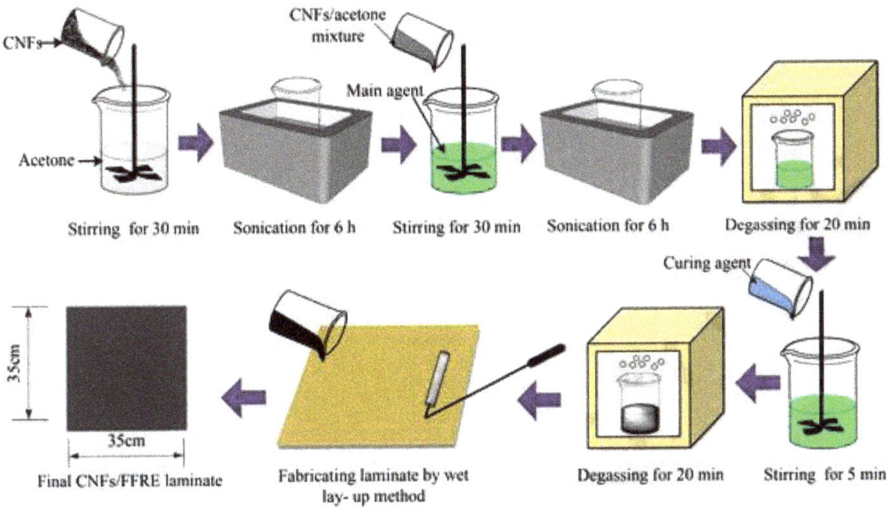


**Figure 2.** Fabrication of flax carbon nanofibers polymer composite (Adapted from the reference [42]).

The hygrothermal aging behavior of CNF/FFRE composite laminates since adding CNFs throughout distilled water were explored, as well as the aging mechanism itself. The following inferences can be derived from this research: The hygrothermal aging behavior of FFRE laminates was improved by the addition of CNFs. Of the CNF/FFRE laminates tested in this analysis, those containing 1.0 wt.% CNFs had the best hygrothermal aging behavior. However, even with a 2.0 wt.% rise in the quantity of CNFs used, the absorption of water of the CNF/FFRE composites remained less than with the genuine FFRE laminates (1.0 wt.% CNFs). As a result of the hydrophobic CNFs filling in the matrix and creating a tortuous diffusion way for water molecules, the FFPR laminates were more resistant to water penetration. Tensile properties, as well as the elastic modulus of CNF with FFRE laminated under hygrothermal conditioning, improved as the CNF content rose to 1.0 wt.%.

As a result of this, SEM images of fractured samples showed that the addition of CNFs to the epoxy matrix not only increased epoxy matrix stability but also strengthened interface bonding between flax fibers and the epoxy matrix. The mechanical properties of CNFs with FFRE laminates before as well as after hygrothermal aging were higher than that of genuine FFRE laminates. Since CNFs made an obstruction within the epoxy matrix of FFE laminates, the cross-linking density of the epoxy matrix increased. Due to the hydrogen bonds formed by type II bound water, the glass transition temperature (Tg) of CNFs with FFRE, as well as genuine FFRE, significantly increased after immersion in water. There were slower aging rates for CNFs with FFRE laminates subjected to a hygrothermal environment, as evidenced by changes in relative bands as well as functional groups in the FTIR spectrum. Due to CNFs ability to strengthen the epoxy matrix as well as inhibit water diffusion in laminates, one such result was achieved. In this analysis, CNFs at a concentration of 1.0 wt.% had the greatest impact on trying to slow the aging of FFRE laminates. Consequently, FFRE laminates can be improved in their long-term hygrothermal durability by incorporating an adequate quantity of CNFs into laminates [49,50].

### 2.2. Palm and Hemp Fibers

Though fully biodegradable natural fibers are hydrophilic in nature, their hydrophobicity was improved by the addition of nanofillers in natural-fiber-reinforced polymer composites. Natural-fiber-reinforced mortars were tested experimentally for hygrothermal and mechanical properties in many studies. The percentages of fibers derived from palm stems (PS) in nano scale, as well as hemp, (HF) were compared. Using a scanning electron microscope (SEM), we discovered that the PS fibers had rough surfaces and complex microstructures. The fibers were treated to reduce their hydrophilicity before being incorporated into the mortar. Both PS and HF fibers had their water absorption reduced significantly thanks to the treatments used. Low thermal conductivity, as well as excellent moisture buffering, were also found in the mortar mixtures containing these fibers. The investigated mixtures had moisture buffer values (MBVs) ranging from 2.7 g/% HR·m$^2$ to 3.1 g/% HR·m$^2$, indicating their good moisture regulator properties. When the fiber mortar mixtures were aged for 28 days, the expected outcomes emerged: extremely high porosity as well as low compressive strength (between 0.6 and 0.9 MPa). Materials developed for thermal insulation and construction filling in this analysis had low environmental footprints [51].

The results of the experiments described in this paper led to the development of a new natural fiber. After successfully extracting and characterizing the fiber from the date palm, it was incorporated into mortar mixtures at varying dosages in various ways. The effect of these fibers on the hydrothermal characteristics of mortar was studied. Lignin, hemicellulose, and pectin deposit, were visible on the surface of the PS fibers. When examined under a microscope, it was discovered that the PS fibers have a morphology comparable to coir fibers, as well as a microstructure composed of an arrangement of elementary fibers that had decided to open on their surfaces. In addition, the new PS fibers are hygroscopic and have a high capillary condensation. Up to 50% of their water absorption was diminished as a result of the addition of the fibers', and implantation in hydrophobic resin also decreased their hydrophilic properties. In comparison, 185 mW per (m·K) was the thermal conductivity of a control mix in the dry mortar mixes containing 5% PS and HF fibers. To put this in perspective, a 245% and 200% increase over the control mortar was seen in the wet mixture. In addition to the percentage and orientation of fibers in the matrix, water content has an impact on the heat conductivity of such a mixture. Regardless of the fiber content, the fiber mortar blends demonstrated good porosity as well as water absorption compared to the control mixture. Porosity, as well as water absorption, increased as a result of fiber content. Improved moisture buffer capacity, as well as lower thermal conductivity and compressive strength, were achieved as a result of these changes. Regardless of fiber content, all of the investigated fiber mortar mixtures had an MBV exceeding 2 g/% HR·m$^2$. The better moisture absorption behavior was observed with 5% of natural fibers [52,53].

The creep/recovery behavior of three high-grade composites manufactured of green epoxy, flax, and hemp fibers were investigated in the literature. The stress level and hygrothermal conditions were examined. In this study, it was found that as load and environmental severity increased, so did the levels of extremely rapid, time-delayed and residual strains. Material time-delayed tensile strains seem to be higher during creeping than during recovery. Moreover, a stiffness effect is observed as the body begins to heal, and the stiffening effect is observed. Using the recovery function, an anisotropic viscoelastic existing legislation can identify the material's viscoelastic properties after creep. Stress levels and environmental conditions do not affect the relaxation time function. For example, the viscosity parameter rises dramatically with increasing stress levels and harsh environmental conditions. Stiffening and irreversible mechanisms, rather than viscoelastic behavior, are to blame for the recovery behavior's stress-level dependence [54].

### 2.3. Wood Fiber

There are a slew of new insulation materials currently being designed, tested, and refined in order to enhance the energy efficiency of both newly constructed and existing structures. To save energy as well as reduce environmental impact, bio-sourced materials have emerged as one of the new insulation options available today. In recent years, wood fiber has become one of the most popular natural insulation materials. There are numerous benefits to using this type of insulation, including its ability to store moisture, which improves indoor air quality, as well as its natural tendency to accumulate and mitigate noise.

Few studies investigated hygrothermal modeling and the performance of wood nanofiber insulation in constructing implementations in order to assess its efficiency. A mathematical model was used to describe heat as well as mass transfer within wood fibers as porous media using numerical methods. Experimental data, thermal conductivity, heat capacity, and sorption/desorption isotherms of the wood fiber material will be first determined for this purpose. In order to verify the accuracy of the numerical model, an iterative procedure is used in both controlled as well as uncontrolled environments. For this experiment, a wood fiber sample measuring 50 cm by 50 cm and measuring 8 cm thick was used. Samples at x = 2 cm and x = 4 cm show very good agreement between the measured and modeled temperatures as well as relative humidity evolutions, with mean differences between the two of 0, 21 °C at 4 cm and 1 °C at 2 cm. The measured maximum distinctions again for relative humidity seem to be: 5.5% at x = 2 cm but also 4.5% at 4 cm, respectively. This is consistent with the results of the model. Predicting wood fiber insulation's heat transfer as well as hygroscopic behavior will help researchers learn more about the effectiveness of natural insulation substances [55,56]. Figure 3a,b denotes the moisture absorption rate curve for different polymer composites based on the traditional method of measurement and Fick's model.

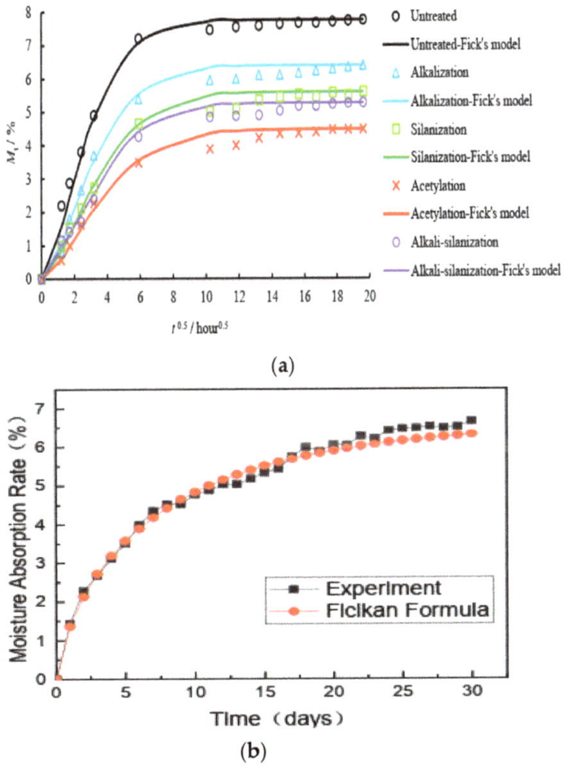


**Figure 3.** (**a**,**b**) Hygrothermal aging curves of polymer composites (Adapted from the references [49,55]).

The hygrothermal properties of wood fiber insulation was studied both experimentally and numerically in this study. The heat and mass transfer in the porous media model developed by a few researchers [57] were used for the above bio-based material with a complex structure. This material's response to microclimatic conditions as well as its hygroscopic map were first determined using a water vapor sorption curve. Experimentation was used to determine the material's thermophysical and hygroscopic characteristics in simulating the material's behavior under real-world conditions. Thermal conductivity increased between 0.045 and 0.06 W per m1.K1 for an increase in water content from 2 to 18%. Correspondingly, for almost the same moisture content range, volumetric heat capacity seemed to have an increasing shape ranging between 250,000 and 350,000 J/$m^3$ K. Using thermocouples and humidity sensors, we experimented with various environmental conditions on a representative wood fiber sample.

A numerical model was developed and tested against experimental data from controlled and uncontrolled ambiances using boundary conditions of air temperature as well as relative humidity there at the interaction. First, the temperature was kept constant, while the relative humidity was kept constant. Then, a transient temperature was applied, while the relative humidity was kept constant. The third scenario studied a coupling of temperature and humidity variations in the air. Validation of wood fiber behavior under controlled conditions was confirmed by comparing calculated and observed temperatures and relative humidity there in three configurations. Due to an inability to forecast the hygrothermal behavior of the wood fibers in real ambient conditions, the sample was tested under uncontrolled conditions. The temperature and relative humidity changes in the sample were in excellent agreement, with an upper limit difference of less than 5% for the relative humidity as well as a maximum difference of 0.21 °C again for temperature distributions. With these findings, we can verify the mathematical model for the stochastic heat and mass transfer within wood fiber insulation in a real-world setting, highlighting the effect of coupling within bio-based materials. More accurate predictions of the hygrothermal behavior of wood fibers can be made using numerical and experimental approaches. To understand better the wood's nature, additional characterizations, such as water vapor permeability persistence or porosity analysis, can be performed. Previous studies predicted the use of hysteresis models to describe the isothermal behavior of sorption as well as desorption [58,59].

### 2.4. Flax-Fiber-Based Composites

Due to their own renewable green origin as well as high tensile strength, flax fibers are becoming a popular research topic. Because of their water absorption as well as poor adhesion to the polymer matrix, composites made from natural fibers have low interfacial strength. For better flax/polypropylene composite performance, this study uses a hybrid chemical therapeutic technique that combines alkali (sodium hydroxide) and silane treatments. SEM, FTIR, AFM, XRD, and a microfiber tester were used to examine the surface morphology, microstructure, chemical composition, wettability, crystallinity, and tensile properties of a single flax fiber before and after chemical treatments. Due to alkalization, hemicellulose, and lignin being removed from the fiber surface, there was a reduction in moisture absorption in the composites. The polypropylene matrix compatibility was improved after alkali-treated flax fibers were exposed to silane treatment, and the composites' moisture absorption was further reduced following alkali–silane hybrid chemical treatment. At about the same time, the strength of the interfacial bond between flax and polypropylene was significantly improved. It is clear from these findings that the hybrid chemical therapeutic approach for flax with polypropylene composite materials has a lot to offer the plant fiber composite company, particularly in terms of expanding the use of chemical treatment methodologies in the field [60,61].

Due to the other environmental and economic advantages, organic FRPs (fiber-reinforced polymers) are being targeted in a variety of industries. One of the most common natural FRPs is FFRPs (flax FRPs). A key factor in the FFRP's practical engineering application is its long-term durability, as well as its performance. FFRP's creep, as well as its dynamic mechanical properties, were investigated experimentally. Cranking up its creep as well as dynamic mechanical performance saw less progress. An investigation into the effects of surface treatment on the creep as well as the dynamic mechanical behavior of FFRP was conducted in this article under hygrothermal aging conditions. These findings demonstrated enhanced creep, dynamic effectiveness, and reduced moisture content in the FFRP after surface treatment. For the substances studied, a fractional-order creep framework was developed. The analysis indicates that fractional calculus seems to be an effective tool for characterizing FFRP's creep behavior accurately and precisely with fewer features than the conventional creep model does [62]. Figure 4 shows the surface morphology of flax-fiber-reinforced polypropylene composites, before being subjected to hygrothermal aging. Figure 4($a_1$,$a_2$) represents the untreated composites, Figure 4($b_1$,$b_2$) represents the alkali-treated composites, and Figure 4($c_1$,$c_2$) represents the alkali silane-treated composites.

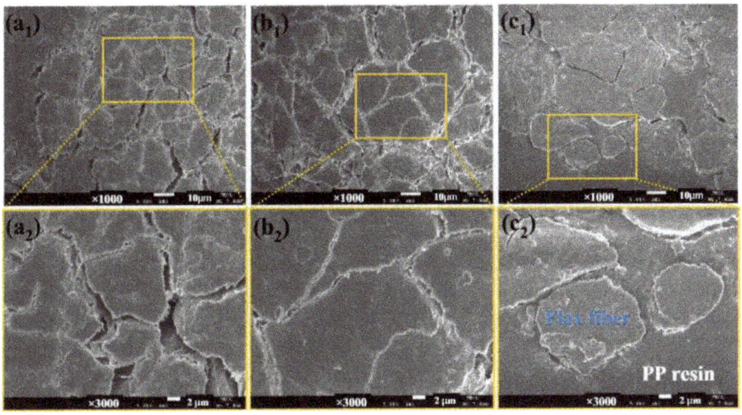


**Figure 4.** Morphology of flax/PP composites: (**$a_1$,$a_2$**). untreated flax/PP composites, (**$b_1$,$b_2$**). alkali-treated flax/PP composites, and (**$c_1$,$c_2$**). alkali silane-treated flax/PP composites (Adapted from the reference [48]).

### 2.5. Mycelium-Based Composites

In order to better understand mycelium-based composites (MBCs), this study will evaluate their effectiveness as a foam-like wall insulation. The composite's performance has been maximized by using a variety of substrates. A longer growing period resulted in a denser outer layer of mycelium in MBC, which improved water resistance owing to mycelium's hydrophobicity. More than any other parameter of MBC, substrate choice has the greatest impact on thermal conductivity, as well as mechanical characteristics. The effects of aging and MBC's moisture buffer capacity were also examined in this study. The accelerated aging test (that is, drying and wetting cycles) showed that MBC not only preserved its functional performance but also constituted excellent moisture buffering capacity. To inactively monitor and control indoor relative humidity as well as thermal comfort, MBC can be used as an internal wall insulation layer, which is a layer in vapor-permeable built environment systems [63].

### 2.6. Coconut Fiber

As an "exotic" insulator, coconut fiber insulators clash with the skepticism of their thermohygrometric behavior, especially in the context of covering technology including green roofs, which are a workable alternative often implemented in the case of green solutions or nearly zero energy structures. Green roofs are a viable option for both new and existing buildings because of their high thermal performance. As an alternative to synthetic insulators, coconut fiberboards (CF) were used to study the thermohygrometric behavior of concrete (CLS) and cross-laminated wood (CLT), respectively, on ten different green roof scenarios. In the end, the results indicate that coconut fiber insulations seem to be significant compared to organic and inorganic materials, and that the doubts about their applications, including green roofs, are connected to engineering solutions for application in the market and their own diffusion between both the building materials, rather than their own hygrothermal characteristics [64].

### 2.7. Jute-Fiber-Based Composites

Reinforced composites made using VARI technology were evaluated for changes in hygroscopic properties and the impact of hygrothermal aging on their mechanical properties. The findings demonstrate that the composites' first-stage moisture absorption follows Fickian diffusion closely. The Fickian equation's diffusion coefficient-temperature correlation was corrected using experimental data. This suggests that the correction was successful, given the small discrepancy between the experimentally based moisture uptake curve and the one anticipated by the altered equation. Temperature also has an effect on the composite strength because the resin-dissolving process becomes more intense. At stage one of this analysis, the hygroscopic conduct of jute fabric composite materials corresponded well to Fickian diffusion. At 40 °C, there was some minor deviation between the modified equation's hygroscopic curve and the experimental-data-based curve, which proves that the correction is effective, as well as its possible utilization to predict other temperature-dependent hygroscopic curves of jute fabric composites. As the temperatures rise, the resin appears to dissolve, resulting in degradation of the composites' tensile properties, as shown by the crucial tensile loads measured at 25 °C, 40 °C, 55 °C, and 70 °C, which were 56.42 MPa, 52.17 MPa, 33.51 MPa, and 16.60 MPa, respectively [65].

Polylactide (PLA) composite materials were tested for their aging characteristics in the hygrothermal environment. To create the material, the film-layering hot-pressed technique was used as a fabrication process. Saturated vapor at 70 °C was used to age both uncoated and adhesive-tape-coated samples. While the samples were aged, the rate at which they absorbed moisture was charted. There were three stages of moisture absorption in uncoated samples: a stage of short and rapid moisture uptaking; a stage of slow, steady uptaking; and a stage of an abrupt, extremely quick uptaking. Different stages of the aging of the samples were observed in their microstructures. Pores, microcracks, delamination, and complete structural relaxation were among the most common aging-related defects. The coating appears to be an effective way to slow down the aging and moisture absorption processes. The gel permission chromatography (GPC) results indicated that the PLA matrix had been severely degraded in a hygrothermal atmosphere. After aging, tensile strength reduced significantly [66].

### 2.8. Rice-Husk-Based Composites

For in-house development, NFPCs (natural fiber plastic composites) are most commonly used as deck boards. The moisture adsorption or desorption behavior and hygrothermal dimensional stability of these composite materials with high fiber content as well as hollow profiles are put into question when exposed to a wide range of climate conditions. Commercial decking panels made of rice hull and high-density polyethylene (HDPE) were tested under simulated adverse climate environmental conditions in this analysis. When exposed to 93% RH and 40 °C for 2000 h, the specimens gained 4.5% of their original weight in water content. A significant (7.1%) swelling of the samples' walls occurred at the same time, which resulted in a longitudinal bowing of about 5 mm (based on 61 cm longboards). After another 2000 h of exposure to 20% RH and 40 °C, both expansions, as well as bowing, partially managed to recover. When exposed to a variety of weathering conditions, the samples took longer to reach a new humidity balance than a new dimensional equilibrium. Deformation, including swelling and bowing, was largely due to the presence of moisture. Additionally, temperature affected the rate and amount of moisture adsorption, as well as causing straightforward thermal expansion or contraction [67,68].

### 2.9. Agave-Fiber-Based Composites

The longevity of agave organic fiber-reinforced polymer composite materials was tested under hygrothermal aging conditions using three distinct pre-treatment methods: alkali hornification, liquid hornification, and alkali treatment. Moisture diffusion analysis was performed on the composite specimens using standard test procedures. Distinct hygrothermal aging conditions were applied to the standard test samples taken to examine the effects of different treatment conditions. Alkali hornification contributed to a 27% decrease in water preservation in the agave fibers after four consecutive hornification cycles. Water hornification reduced water retention by only 6% in similar circumstances. In direct contact with liquid, the alkali-hornified composite materials gained 4.1%, 4.3%, and 4.7% mass gain, respectively, between 25 °C and 75 °C. At the same temperatures, alkali-treated composites gained mass at rates of 4.3%, 4.5%, and 5.1%, while water hornification gained mass at rates of 5.5%, 5.9%, and 6.1%. There were mass gains ranging from 0.9% to 1.3% for the alkali-hornified composites when humidity levels were kept constant between 25 °C and 75 °C. Comparable temperatures led to mass gains of 0.6%, 1.05%, and 1.2% for alkali-treated composites, as well as 0.66%, 1.07%, and 1.3% for water-hornified composites. Fiber pre-treatment methods have an impact on the moisture resistance of agave natural fiber composites, as shown in the current study. Microstructural investigations bolster the findings [69].

### *2.10. Silk- and Ramie-Fiber-Based Composites*

Bio-composites reinforced with silk or henequen natural fiber were evaluated for their hygrothermal properties in the literature. Compression molding was used to create the bio-composites. For 1000 h, the bio-composites were kept at 60 °C and 85% RH. Deterioration of the bio-composites' achievement was primarily due to PBS matrix degradation, and the bio-composites deteriorated more slowly than the PBS matrix. For the bio-composites exposed to 60 °C and 85% RH for 1000 h, the storage modulus of the fibers decreased by 20% and 50%, correspondingly, when compared to the initial specimens [70]. Short ramie fibers and poly (lactic acid) (PLA) composites were prepared by a combination of extrusion and injection molding. For the ramie with PLA composites at 60 °C, water absorption and aging were studied. In this study, the water absorption and mechanical properties of something like the ramie with PLA composite materials with immersion time were revealed and evaluated by comparing them to those of PLA alone. Differential scanning calorimetry along with gel permeation chromatography was used to measure the crystallinity and molecular weight of neat PLA as well as of ramie with PLA composite materials. The water absorption and mechanical characteristics of the short ramie fiber were reported to be influenced by the results. As a result of the hydrophilic nature of short ramie natural fibers, composites of ramie with PLA showed greater saturation weight gains and diffusion coefficients than did neat PLA alone. Age-related deterioration in tensile and flexural strength was extreme. Furthermore, PLA degradation was found to be amplified by ramie in a hygrothermal environment. Scanning electronic microscopy was used to examine the fracture surface morphologies of clean PLA as well as of ramie with PLA composites, with varying immersion times. PLA degradation and ramie with PLA bond de-bonding led to the microcracks and voids [71,72].

### *2.11. Luffa Cylindrica Fiber-Based Composites*

The mechanical and hygrothermal properties of polyester with luffa composites were examined in relation to luffa fiber surface chemical modification. The matrix was made up of unsaturated polyester resin. Luffa fibers that were untreated, alkali-treated, combined-processed, and acetylated were used. The fibers of the luffa plant were studied using scanning electron microscopy as well as infrared spectroscopy. Tests for the composites' mechanical characteristics were conducted using a three-point bend test. Saturation in filtered water at 25 °C was used to test the fibers and composite materials for water absorption. Improved mechanical characteristics were achieved through acetylation treatment. The infrared assessment showed that the procedure reduced the hydrophilic behavior of the luffa fibers, which improved their bonding to the polyester matrix. The diffusion coefficient and maximum water absorption of luffa fibers were both reduced as a result of the chemical modifications made to their surfaces. For the fibers tested in this analysis, the diffusion methodology was "Fickian" at the beginning of immersion but became more complicated at the end. Composite materials showed similar results when immersed in water at previous phases. Composite materials exposed to external loads were found to affect diffusion. Water absorbed by the body increases in volume at an even faster rate as the load is an excessive amount [73]. Table 1 consolidates the various studies carried out on different natural-fiber-reinforced polymer composites along with the test conditions and moisture absorption values.

**Table 1.** Hygrothermal properties of natural fiber polymer composites.

| S. No. | Reinforcements | Weight Fraction (%) | Matrix | Time of Immersion (h/days) and Immersion Temperature (°C) | Moisture Absorption (%) | References |
|---|---|---|---|---|---|---|
| 1 | Bamboo/jute/ glass fibers | 2.5/2.5/5 | Polyester | 144 h and 25 °C | 23.23 | [74] |
| 2 | Waste hemp fibers | 10 | Polybenzoxazine | 90 h and 30 °C | 7.05 | [75] |
| 3 | Jute/basalt fibers | 5/15 | Epoxy | 120 h and 40 °C | 6.95 | [76] |
| 4 | Roselle/sugar palm | 5/5 | Polyurethane | 24 h and 25 °C | 8.48 | [77] |
| 5 | Jute fiber/rosewood and padauk wood dust | 10/2.5 | Epoxy | 15 days and 32 °C | 4.42 | [78] |
| 6 | Flax/nano $TiO_2$ | 10/1.5 | Epoxy | 30 h and 25 °C | 1.2 | [79] |
| 7 | Continuous bamboo fibers | 20 | Epoxy | 4 h and 100 °C | 19 | [80] |
| 8 | Sugar palm/ glass fiber | 10/20 | Thermoplastic Polyurethane | 168 h and 25 °C | 9.78 | [81] |
| 9 | Waste corn husk flour | 25 | Polyurethane | 30 days and 25 °C | 9.5 | [82] |
| 10 | Hemp/sisal fibers | 15/15 | Epoxy | 42 days and 25 °C | 11.6 | [83] |
| 11 | Abaca fiber | 25 | Polypropylene | 80 days and 50 °C | 15.09 | [84] |
| 12 | Wood flour | 20 | Polypropylene | 96 h and 25 °C | 1.09 | [85] |
| 13 | Kenaf fiber | 40 | Polypropylene | 24 h and 25 °C | 1.05 | [86] |
| 14 | Wood flour | 35 | Polypropylene | 48 h and 25 °C | 11.57 | [87] |
| 15 | Olive stone flour | 30 | Polypropylene | 48 h and 25 °C | 9.55 | [87] |
| 16 | Rice husk ash filler | 40 | Polypropylene | 24 h and 25 °C | 15.31 | [88] |
| 17 | Luffa cylindria fiber | 30 | Polypropylene | 960 h and 25 °C | 28.4 | [89] |
| 18 | Jute fiber | 40 | Polypropylene | 18 h and 23 °C | 21.5 | [90] |
| 19 | Jute fiber | 30 | Epoxy | 336 h and 25 °C | 8 | [91] |
| 20 | Alfa pulps filler | 35 | Low-density Polyethylene | 480 h and 25 °C | 25.71 | [92] |
| 21 | Bamboo mat | 25 | Polyester | 1440 h | 50.31 | [93] |
| 22 | Ijuk fiber | 30 | Polypropylene | 480 h and 23 °C | 5.22 | [94] |
| 23 | Hemp/glass hybrid | 15/20 | Epoxy | 3600 h | 21.31 | [95] |
| 24 | Jute/glass hybrid | 10/30 | Unsaturated Polyester | 504 | 58.36 | [96] |
| 25 | Sisal/banana hybrid | 20/15 | Epoxy | 50 | 11.48 | [97] |
| 26 | Coir/glass hybrid | 15/15 | Epoxy | 1440 | 39.16 | [98] |
| 27 | Short snake grass fiber | 25 | Isopthallic polyester | 120 days and 60 °C | 29.79 | [99] |
| 28 | *Phoenix* sp. fiber | 25 | Epoxy | 100 days and 30 °C | 17.52 | [100] |
| 29 | Calotropis gigantea fiber | 15 | Polyester | 15 days and 70 °C | 18.12 | [101] |
| 30 | Calotropis gigantea fiber | 20 | Epoxy | 72 h and 25 °C | 17.44 | [102] |
| 31 | Tindora tendril fiber filled with haritaki nanopowder | 20/7.5 | Epoxy | 8 h and 30 °C | 5.87 | [103] |

## 3. Sound Absorption Behavior of Polymer Composites

Nowadays, humans are at risk of developing serious diseases as a result of increased noise pollution caused by external factors. A filler or panel made from bio-based materials can help reduce undesirable noise in workplaces and also homes. An acoustic absorber made from eco waste fibers (remains upon harvesting) has been developed and tested. The gleaning methodology, which is the method of acquiring field leftovers, is used to gather the eco waste fibers [104–106]. Figure 5 shows the methodology of measuring sound absorption along with the equipment and different mathematical models. The following sections deal with the determination of the sound absorption coefficient (SAC) of various natural-fiber-reinforced polymer composites and the possibility of using such composites in acoustic insulation applications.

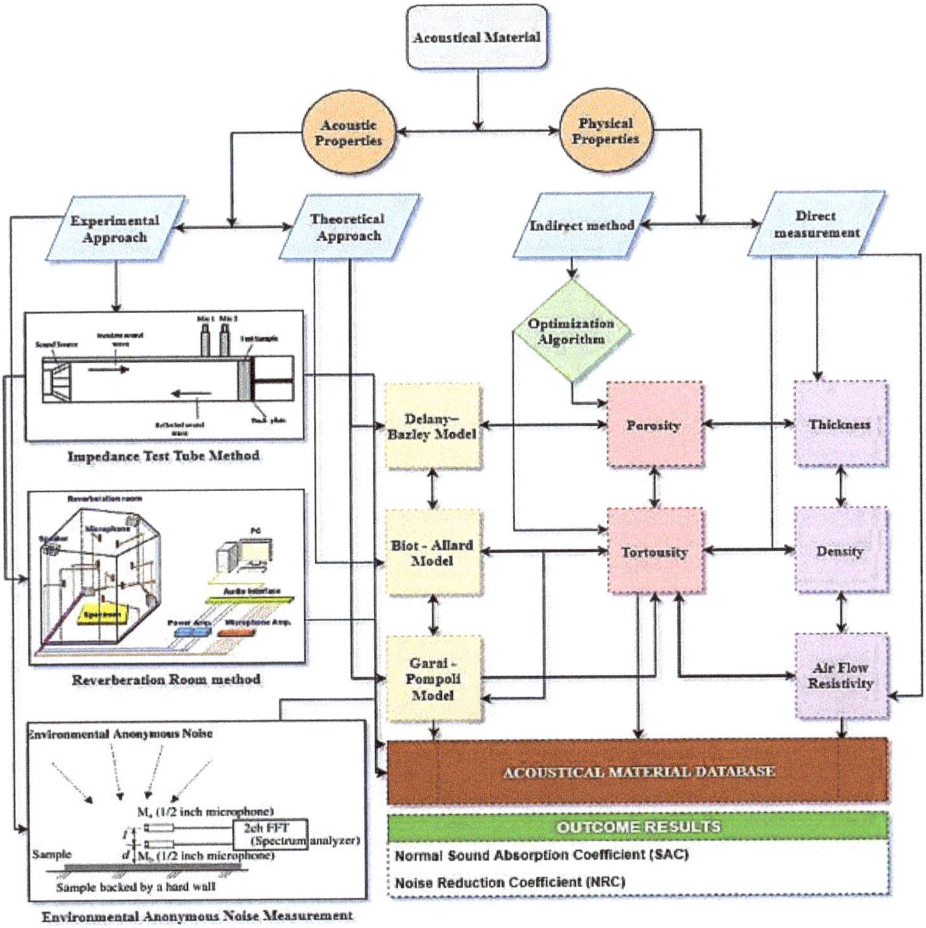


**Figure 5.** Methods of measuring sound absorption properties (Adapted from the reference [104]).

### 3.1. Yucca-Gloriosa-Based Composites

Natural fabrics, with their numerous environmental, physical, mechanical, and sound-absorbing benefits, have revolutionized the manufacturing of organic fibers. Investigators are paying expanding consideration to the acoustic behavior analysis of natural fiber composites, known as "The Green Fibers," because of the additional funding they can generate by absorbing sound. Using a mathematical imitation as well as an optimization approach, this analysis aims to optimize and replicate the sound absorption behavior of Yucca Gloriosa (YG) composites. The alkaline treatment of the fibers was used in this investigational cross-sectional study to manufacture the natural acoustic composites. In order to enhance sound absorption, Response Surface Methodology encouraged the design of experiments as well as the determination of the optimal quantity of alkaline treatment specifications (NaOH concentration as well as immersion time) (RSM). Additionally, an impedance tube structure was used to determine the YG fiber's sound absorption coefficient (SAC) (ISO10534-2 standard). Delany–Bazley (DB) and Miki analytical models were tested in MATLAB software to see if they could be applied to predicting the SAC of natural composites. At all frequencies, a comparison of the procured SAC values showed that the optimized composites had higher values than the untreated ones. There was an 18.92% increase in the Sound Absorption Average (SAA) Index, especially when compared to the raw composites. Moreover, the empirical models, as well as the experimental data in the low and mid-range of the one-third octave band, were found to be in good agreement. Optimum alkaline treatment and empirical-model-based SAC forecasting are deemed appropriate strategies for acoustic implementations because of the prominent advantages of natural materials and their widespread use [107].

### 3.2. Oil Palm Trunk

Because they are more sustainable and can be replenished, natural fibers are increasingly being used to replace synthetics in sound-absorbing materials. Research on oil palm trunk (OPT) fiber's SAC as an acoustic material is presented here. All OPT specimens were examined for SAC using the impedance tube method (ITM). Including an average density of around 100 kg per m$^3$, OPT's natural fiber was used to create three different thicknesses of panels: 8 mm, 12 mm, and 16 mm. OPT's acoustic efficiency was evaluated to be very good, with all samples almost reaching unity (0.9) at a high rate above 3000 Hz. The results also show that SAC varies between 0.5–0.85 at low frequencies below 500 Hz. 0.99 at a frequency range of 3000 to 6000, and 6400 Hz was found in the thickest natural fiber panel of 12 mm, making oil palm trunk an extremely promising natural fiber for use as a sound-absorbing content. Fiberglass, despite its excellent acoustical absorption properties, is not a good choice. Fiberglass has been linked to a number of serious health issues, including skin irritation and redness; eye, nose, and throat irritation; and even cancer. Fiberglass dust can cause bronchitis, difficulty in breathing, coughing, and perhaps even lung disease if it is inhaled too much. Only the acoustic properties, including the use of OPT natural fibers as a replacement for fiberglass, were examined in this study [108].

Additionally, natural fibers play an essential part in the design of ergonomic products. They reduce both the noise level as well as the health risks while also preserving a pleasant working environment. Using a frequency range of 0 to 6400 Hz, OPT fiber exhibits excellent sound absorption properties. SAC (alpha) = 0.99 was reached by some OPT fiber samples in this manner. OPT fiber was able to absorb 99% of the incident sound, while only 1% of the sound was reflected. The influence of OPT fiber thickness was discovered through sample characterization. The results clearly show that increasing the thickness of a material significantly reduces its absorption rate. Porosity decreases with increasing sample thickness because more fiber is present, reducing the sample's porosity. In order for the OPT fiber to perform at a high SAC (alpha) when the frequency raises, the porosity of the fiber must lessen. Thus, the ideal OPT fiber thickness was found to be 12 mm. OPT fiber's ideal density and thickness were the primary focus of this investigation. Because of this, a fiber with a density of around 100 kg per m$^3$ was found to be the best OPT fiber. The sample's maximum SAC (alpha) value was 0.99 under these circumstances. When compared to synthetic-based commercial products, the OPT experimental trial findings reveal that it has excellent acoustic characteristics [109].

Oil palm timber is one of several solid wastes that can be used for non-structural purposes. Numerous researchers concentrated on the strength of their materials and indeed the binder they use. The use of oil palm timber as an insulation material has only been examined in a few research findings. In order to better understand oil palm timber binderless panels' thermal and acoustic properties, this study was conducted. Various particle sizes and pressing times were used to make panels from oil palm timber. Binderless panel properties were affected by particle size, but pressing periods were not, according to the findings. The greater the particle size, the greater the resistance to heat and sound, but the lower the density, the greater the water resistance, as well as the lower the bending strength. Large granules also resulted in the lowest heat conductivity (0.050 W/mK) and, indeed, the highest SAC (0.33). Flexible strength and liquid absorption usually range from 4.21 to 8.18 MPa and 84.51%–119.06%, respectively. This study's research results suggest that oil palm timber binderless boards can be used in acoustic insulation applications [110].

### 3.3. *Sugarcane-Bagasse-Based Composites*

Sugarcane bagasse fibers are considered to be the most porous of all fibers and can render better acoustic behavior when reinforced in composites. Few studies examined the thermal and acoustic properties of sugarcane bagasse and bamboo charcoal insulation specimens for use in the automotive industry. For thermal and sound insulation applications, sugarcane bagasse and bamboo charcoal fiber could be viable raw material sources. The primary use for natural fibers is sound absorption, and this is one of the most common applications. At this time, the natural fiber hybrid composite is more sought after by industry because of its advantages, such as low-cost, biodegradability, appropriate physical characteristics, etc. Bamboo charcoal and sugarcane bagasse fibers have been used to develop environmentally friendly sound-absorbing composite materials. Compression bonding was used to create five different types of natural fiber green composite from these fibers. An important factor in the widespread use of sound absorbers made from natural composites is their noise-control performance. The impedance tube technique was used to measure the SAC in accordance with ASTM E 1050. Using the ASTM standard, the physical characteristics of natural fiber composites were evaluated for all specimens, including the thermal conductivity, thickness, air permeability, porosity, and density. Natural fiber green composites absorb more than 70% of the sound resistance and provide the greatest acoustic absorption characteristics; these composite materials possess appropriate moisture resistance in wet environments without influencing the insulation or acoustic characteristics among these composites [111].

### 3.4. *Kenaf with Waste Tea Leaf Fiber Composites*

Waste tea leaf fibers (WTLF) are most commonly used as a nanofiller in composites to increase the porosity of the composites. The purpose of this study is to examine the structural and sound absorption properties of manufacturing waste tea leaf fiber, kenaf, and E-glass fiber-reinforced combination epoxy composites. WTLF and kenaf fibers were first treated with sodium hydroxide at a concentration of 5%. Compression molding was used to create hybrid composites with a 40-to-60-to-1 fiber-to-matrix ratio. Mechanical and sound absorption tests were conducted on the manufactured hybrid composites in accordance with ASTM standards. The composites containing 25% kenaf and 5% WTLF had better mechanical properties, while the composites containing 25% WTLF and 5% kenaf had better sound absorption properties. Scanning electron microscopy was used to study the surface morphology of something like the shattered samples, such as fiber pullout and matrix crack. Research on alkali-treated hybrid composites showed that polymer and fiber interfacial bonds were much stronger than those in untreated composites [112]. Figure 6 denotes the sound absorption coefficient variation of various polymer composites. In Figure 6, L denotes oil palm timber particles of different particle sizes, G denotes glass, K denotes kenaf, and T denotes WTLF.

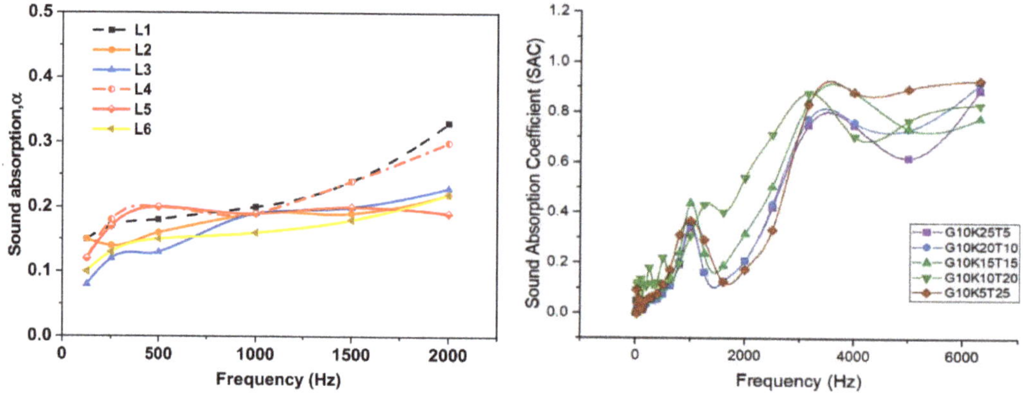


**Figure 6.** Sound absorption coefficients of polymer composites (Adapted from the references [110,112]).

The acoustic properties of a natural fiber panel are determined by the fiber's physical properties. In this study, a chemical treatment was used to enhance the sound absorption properties of kenaf fibers. Sodium hydroxide (NaOH) treatment from 1% to 8% concentration levels seems to be the goal. SEM was used to observe the effects of NaOH treatment on the kenaf fiber strands. Measurements were made using an impedance tube to determine the SAC and noise reduction coefficient (NRC) values, respectively. The diameter of kenaf fiber decreased as the NaOH concentration increased, according to the results. FTIR analysis proved that the strand diameter decreased as a result of the removal of the hemicellulose and the lignin layer of the strands. Treatment with 6%, 7%, and 8% NaOH resulted in elevated SAC at high frequencies (>2500 Hz) than untreated fiber. Specimens with NRC values of 0.67 were found. Thinner strands of kenaf absorber were found to improve sound absorption, while a 6% NaOH concentration was found to be optimal for treating kenaf fiber for sound absorption [113].

### 3.5. *Sisal and Palm Fibers-Based Composites*

Hybrid composite materials made of palm fiber and sisal fiber were examined in this study for tensile, flexural, impact, and sound absorption properties. A compression molding technique was used to create three distinct hybrid composites, each with a different weight ratio with 5%, 10%, and 15% of palm fibers. The tensile, flexural, and impact tests for the three hybrid composites were carried out in accordance with the appropriate ASTM standards. According to these results: The combination of epoxy resin (65%), cellulose (20%), sisal (15%), and palm fiber (15%) yielded the highest tensile strength, flexural strength (22%), and impact strength (36%). Images of fractography showed fiber to fracture, surface, and fiber pull out, as well as the deformations of tightly packed fiber and matrix stickiness. An impedance test on hybrid composites was used in accordance with ASTM E1050 standards to determine their sound absorption behavior. Due to the greater weight proportion of palm fiber (65% epoxy resin, 20% sisal fiber, and 15% palm) in the composite, it rendered a high SAC at frequencies of 1600 Hz, 2000 Hz, 2500 Hz, 3150 Hz, and 4000 Hz. Above that, the above hybrid composite exhibited a maximum SAC difference of 76.61% in comparison to the other three hybrid composites at a frequency of 4000 Hz [114].

### 3.6. *Jute and Luffa Fiber-Based Composites*

Before utilizing natural fiber-reinforced composite materials in different industrial applications, such as sound and vibration isolation, it is necessary to determine their acoustic characteristics. The acoustic effectiveness of jute and luffa fiber-reinforced biomaterials is examined in this study by varying the sample thickness as well as the fiber and resin ratio. The impedance tube method was used to measure the SACs and transmission losses (TLs) of jute and luffa composite samples of various thicknesses, fibers, and epoxy ratios.

The thickness-dependent characteristics of the SAC/TL of jute and luffa composites were identified in the low-, medium-, and high-frequency ranges. The acoustic characteristics of jute and luffa composite materials as a function of frequency were ascertained by determining the fiber/epoxy ratio. In addition, mathematical techniques were used to estimate the SACs and TLs of various natural-fiber-based specimens with varying thicknesses, and with the conceptual and experimental findings were compared and evaluated [115].

### 3.7. Coconut Coir and Oil Palm Fruit Bunches, with Pineapple Leaf

Three different types of natural fiber composites were compared for their sound absorption properties. To make the fibers, oil palm waste and pineapple leaves were combined with coconut coir and pineapple leaf fiber. Sample thicknesses ranged from 10 to 20 mm, and two fiber density options were used to determine the SAC. For the evaluation, a non-sophisticated impedance tube with 2 microphones was utilized. The transfer function of the two microphones was well within the range of 200 Hz to 3000 Hz, and was used to determine the SAC. Three types of fibers were found to reduce reflected sound at a higher frequency than previously thought. The high point of an absorption coefficient is shifted lower in frequency as the density and thickness of such fibrous material raises. Moreover, the SAC of pineapple leaf fiber is the highest among several other fibers. Pineapple fiber absorbs sound energy due to its small and uniform fiber diameter. Because of this, the SAC of coconut coir, as well as palm hollow fruit bunches fiber, is likely to be lower, particularly within the frequency range of measurement [116].

The composite material microperforated panel (MPP) created from coconut fiber and polylactic acid (PLA) bio-composite material was for the bio-composite microperforated panel (BMPP). PLA served as a matrix for the BMPP specimens made from coconut fiber. The SA performance of the BMPP specimen was determined using an impedance tube technique, whereas the porosity of the specimen was determined using a porosity tester. The SA performance of BMPP and steel MPP was compared. The BMPP with different percentages of coconut fiber and PLA had different sound absorption performances because of the presence of pores and tortuous structures in the specimen. BMPP specimens were subjected to a scanning electron microscope (SEM) evaluation to better understand their structure [96]. Using both experimental and theoretical methods, this study examines the sound absorption properties of coconut coir fiber. Sound absorption coefficients were measured using samples of coconut coir fiber with different thicknesses and densities. ISO 10534-2 and ASTM E1050-98 guidelines were used to set up an impedance tube to measure the SAC. Thickness and density were studied using the Delany–Bazley model. This paper presents and compares the results of theoretical and experimental studies on SAC. When compared with experimental results, the Delany–Bazley model appeared promising, but further deviations were found. It was found that a 35 mm thick sample with a density of 220 kg/m$^3$ had a SAC of 0.84 (2900 Hz). The transfer function technique of the impedance tube was used to estimate sound transmission loss (STL) for samples with 220 kg/m$^3$ density (21 mm, 28 mm, 35 mm). Coconut coir fiber samples were found to effectively dissipate sound energy, resulting in acceptable sound absorption properties [117].

### 3.8. Finger Millet Straw and Darbha, with Ripe Bulrush Fibers

Few experimental trials focused on testing the sound absorption capabilities of natural fibers extracted from Eleusine coracana, Desmostachya bipinnata (Darbha), and Typha domingensis, or hybrid configurations There are a number of variables that influence the material's ability to absorb sound, including its density, porosity, thickness, flow resistance, and tortuosity. For individual fibers or hybrid combinations, the length and type of the fibers play important roles. The hybridized fiber configuration was tested to see if stacking had an effect on sound absorption. All pairings had a lower SAC (alpha) in the 1000 Hz to 2500 Hz frequency range. For a 50 mm thickness, the darbha fiber had the best NRC of 0.86, whereas the ripe bulrush and darbha combination had the best NRC of 0.90, which is much more responsible for absorbing sound there in the crucial frequency spectrum of 500 to

2000 Hz. Natural fiber fillers of this type are excellent sound absorbers and are found in a variety of settings, including classrooms, recording studios, and theatres [118–120]. Table 2 shows the sound absorption coefficient of various fiber-reinforced polymer composites along with the density or porosity of the composites, and the measuring frequency.

**Table 2.** Acoustic properties of natural fiber polymer composites.

| S.No | Composite Material | Bulk density (kg/m$^3$)/Porosity (%) | Measuring Frequency Range (Hz) | Sound Absorption Coefficient | References |
|---|---|---|---|---|---|
| 1 | Pineapple leaf fiber in epoxy matrix | - | >1000 | 0.9 | [121] |
| 2 | Rubber crumbs waste fiber in polyester matrix | - | 1000–6000 | 0.93 | [122] |
| 3 | Kapok fiber in epoxy matrix | 8.3 kg/m$^3$ | 100–6300 | 0.98 | [123] |
| 4 | Flax, ramie, and jute fibers in epoxy matrix | - | 1000–10,000 | 0.88 | [124] |
| 5 | Rice waste fiber in polyurethane foam matrix | 8.35 kg/m$^3$ | 400–6400 | 0.9 | [125] |
| 6 | Sugarcane bagasse in epoxy matrix | 10.26 kg/m$^3$ | 0–1600 | 0.17 | [126] |
| 7 | Date palm fiber powder in cement matrix | - | 200–2000 | 0.78 | [127] |
| 8 | Banana fiber in polyester matrix | 88.4% | 500–6000 | 0.97 | [128] |
| 9 | Mineralized wood flour in epoxy matrix | - | 2000 | 0.4 | [129] |
| 10 | Ijuk fiber in polyurethane matrix | 83.5% | 3000–4500 | 0.9 | [130] |
| 11 | Grass fiber in epoxy matrix | - | 2000 | 0.98 | [131] |
| 12 | Flax fibers in epoxy matrix | 8.5 kg/m$^3$ | 63–6300 | 0.9 | [132] |
| 13 | Kapok fiber in epoxy matrix | 20 kg/m$^3$ | 125–4000 | 0.405 | [133] |
| 14 | Cotton fabric in epoxy matrix | 92% | 3000–3500 | 0.92 | [134] |
| 15 | Nonwoven cotton fiber in polyester matrix | - | 125–3000 | 0.638 | [135] |
| 16 | Betelnut fiber in polypropylene matrix | 85% | 6000 | 0.42 | [136] |
| 17 | Hemp fiber in polyester matrix | 141 kg/m$^3$ | 1000–4500 | 0.58 | [137] |
| 18 | Sugarcane bagasse in polyvinyl alcohol matrix | 200.8 kg/m$^3$ | 172–2000 | 0.75 | [137] |
| 19 | Sisal fiber in polyvinyl alcohol matrix | 214.7 kg/m$^3$ | 172–2000 | 0.69 | [138] |
| 20 | Rice husk ash in glue matrix | 170 kg/m$^3$ | 200–6400 | 0.81 | [139] |
| 21 | Yucca gloriosa fiber in epoxy matrix | 200 kg/m$^3$ | 63–6300 | 0.95 | [140] |
| 22 | Coarse wool with binding fibers | 249.55 kg/m$^3$ | 60–6300 | 0.84 | [141] |
| 23 | Pineapple leaf fiber in epoxy matrix | 117 kg/m$^3$ | 500–4500 | 0.91 | [142] |
| 24 | Corn husk fiber in epoxy matrix | 92% | 1600–3250 | 0.88 | [143–145] |
| 25 | Kenaf fiber in polylactic acid matrix | 82% | 1450–1522 | 0.8 | [144–146] |
| 26 | Polyurethane foam in Coffee grounds polyol | 88% | 150–4000 | 0.96 | [145–147] |
| 27 | Sugarcane bagasse in urea formaldehyde resin matrix | 78% | 500–4000 | 0.75 | [146–149] |
| 28 | Date palm fiber in lime matrix | 81% | 750–6300 | 0.55 | [147–149] |
| 29 | Date palm branch powder filled urea formaldehyde | - | 800–1250 | 0.4 | [148–150] |
| 30 | Broom fiber in epoxy matrix | 73% | >1650 | 0.9 | [149–151] |

Hence, the natural fiber fillers of this type are excellent sound absorbers and are found in a variety of settings, including classrooms, recording studios, and theatres [151–153].

## 4. Summary and Conclusions

Moisture absorption and sound absorption properties of various polymer composites were reviewed. The necessity of natural-fiber-reinforced polymer composites in various applications has a wider scope these days, but their poor compatibility and water absorption behavior retards the application spectrum. It was stated in many of the studies that the moisture absorption of the natural fibers and their composites is due to the hydrophilic nature of the natural fibers when they come into contact with moisture content. The moisture absorption of natural fiber composites increases with the content of natural fiber and this deteriorates the strength of natural fiber composites by paving the way for the initiation of microcracks at the interfacial region of matrix and reinforcements. Moisture absorption is linearly proportional with time but saturates after reaching the saturation point beyond which the moisture absorption capability of the composites remains constant. As natural fibers absorb more moisture relatively, the use of synthetic matrix materials is encouraged for many instances. Various physical and chemical treatment methods help in retarding the moisture absorption capability of natural fiber polymer composites.

However, utilization of natural materials as acoustic insulators is booming these days in spite of various prevalent disadvantages with such natural materials, which includes poor flammability and thickness reduction. Solving these issues through the use of nanomaterials as hybrid reinforcements mitigates the aforesaid disadvantages and expands the scope of natural-fiber-reinforced composites in soundproofing applications. Natural fiber and particulate-based porous composite materials are widely used in many applications, including building layouts, automotive applications, marine and aviation applications, sound-insulating industrial cabinets, classroom environments, and home theatres. In order to continuously obtain good acoustical performance from the natural fiber polymer composites, it is necessary to address various physical parameters associated with the manufacturing of composites and the measuring of sound absorption, which is completely aligned with the environmental point of view. Such water- and sound-resistant polymer composites find their applications in various parts of marine, automobile, aeronautical, and sports applications.

**Author Contributions:** Conceptualization, V.B., B.D., B.A., R.M., S.R. and S.S.; formal analysis, V.B., B.D., B.A., R.M., S.R. and S.S.; investigation, V.B., B.D., B.A., R.M., S.R. and S.S.; writing—original draft preparation, V.B., B.D., B.A., R.M., S.R. and S.S.; writing—review and editing, S.S., K.M., C.L. and E.T.E.; supervision, S.S., K.M., C.L. and E.T.E.; project administration, S.S. and E.T.E.; funding acquisition, S.S. and E.T.E. All authors have read and agreed to the published version of the manuscript.

**Funding:** This research received no external funding.

**Conflicts of Interest:** The authors declare no conflict of interest.

## References

1. United Nations and Department of Economic and Social Affairs. *Energy Statistics Pocketbook 2020*; United Nations and Department of Economic and Social Affairs: New York, NY, USA, 2020.
2. Bachchan, A.A.; Das, P.P.; Chaudhary, V. Effect of moisture absorption on the properties of natural fiber reinforced polymer composites: A review. *Mater. Today Proc.* **2021**, *49*, 3403–3408. [CrossRef]
3. Sands, P. *United Nations Framework Convention on Climate Change*; General Assembly: New York, NY, USA, 1992.
4. Al-Maharma, A.Y.; Al-Huniti, N. Critical Review of the Parameters Affecting the Effectiveness of Moisture Absorption Treatments Used for Natural Composites. *J. Compos. Sci.* **2019**, *3*, 27. [CrossRef]
5. Kyoto Protocol to the United Nations Framework Convention on Climate Change. 1997. Available online: http://unfccc.int/resource/docs/cop3/07a01.pdf (accessed on 4 August 2022).
6. Gholampour, A.; Ozbakkaloglu, T. A review of natural fiber composites: Properties, modification and processing techniques, characterization, applications. *J. Mater. Sci.* **2020**, *55*, 829–892. [CrossRef]

7. Energy, E.; Eurostat. Energy Statistics—Quantities. *Eur. Comm. Database* **2020**. Available online: https://ec.europa.eu/eurostat/en/web/main/data/database (accessed on 6 August 2022).
8. Eurostat. Energy Consumption and Use by Households. *Eur. Comm. Database* **2020**. Available online: https://ec.europa.eu/eurostat/statistics-explained/index.php?title=Energy_consumption_in_households (accessed on 6 August 2022).
9. Berger, J.; Mazuroski, W.; Mendes, N.; Guernouti, S.; Woloszyn, M. 2D whole-building hygrothermal simulation analysis based on a PGD reduced order model. *Energy Build.* **2016**, *112*, 49–61. [CrossRef]
10. Steeman, H.J.; Janssens, A.; Carmeliet, J.; Paepe, M.D. Modelling indoor air and hygrothermal wall interaction in building simulation: Comparison between CFD and a well-mixed zonal model. *Build. Environ.* **2009**, *12*, 572–583. [CrossRef]
11. Rahim, M.; Le, A.T.; Douzane, O.; Promis, G.; Langlet, T. Numerical investigation of the effect of non-isotherme sorption characteristics on hygrothermal behavior of two bio-based building walls. *J. Build. Eng.* **2016**, *7*, 263–272. [CrossRef]
12. Seng, B.; Lorente, S.; Magniont, C. Scale analysis of heat and moisture transfer through bio-based materials—Application to hemp concrete. *Energy Build.* **2017**, *155*, 546–558. [CrossRef]
13. Deepa, C.; Rajeshkumar, L.; Ramesh, M. Preparation, synthesis, properties and characterization of graphene-based 2D nano-materials for biosensors and bioelectronics. *J. Mater. Res. Technol.* **2022**, *19*, 2657–2694. [CrossRef]
14. Ramesh, M.; Rajeshkumar, L.; Bhoopathi, R. Carbon substrates: A review on fabrication, properties and applications. *Carbon Lett.* **2021**, *31*, 557–580. [CrossRef]
15. Promis, G.; Douzane, O.; Le, A.T.; Langlet, T. Moisture hysteresis influence on mass transfer through bio-based building materials in dynamic state. *Energy Build.* **2018**, *166*, 450–459. [CrossRef]
16. Alioua, T.; Agoudjil, B.; Chennouf, N.; Boudenne, A.; Benzarti, K. Investigation on heat and moisture transfer in bio-based building wall with consideration of the hysteresis effect. *Build. Environ.* **2019**, *163*, 106333. [CrossRef]
17. Ramesh, M.; Deepa, C.; Kumar, L.R.; Sanjay, M.R.; Siengchin, S. Life-cycle and environmental impact assessments on processing of plant fibres and its bio-composites: A critical review. *J. Ind. Text.* **2020**, *51*, 5518S–5542S. [CrossRef]
18. Khoukhi, M. The combined effect of heat and moisture transfer dependent thermal conductivity of polystyrene insulation material: Impact on building energy performance. *Energy Build.* **2018**, *169*, 228–235. [CrossRef]
19. Delgado, J.M.P.Q.; Barreira, E.; Ramos, N.M.M.; de Freitas, V.P. *Hygrothermal Numerical Simulation Tools Applied to Building Physics*; Springer: Berlin/Heidelberg, Germany, 2013. [CrossRef]
20. Bai, B.; Xu, T.; Nie, Q.; Li, P. Temperature-driven migration of heavy metal Pb2+ along with moisture movement in unsaturated soils. *Int. J. Heat Mass Transf.* **2020**, *153*, 119573. [CrossRef]
21. Ramesh, M.; Rajeshkumar, L.; Deepa, C.; Selvan, M.T.; Kushvaha, V.; Asrofi, M. Impact of Silane Treatment on Characterization of *Ipomoea Staphylina* Plant Fiber Reinforced Epoxy Composites. *J. Nat. Fibers* **2021**, 1–12. [CrossRef]
22. Ramesh, M.; Rajeshkumar, L.; Balaji, D. Influence of Process Parameters on the Properties of Additively Manufactured Fiber-Reinforced Polymer Composite Materials: A Review. *J. Mater. Eng. Perform.* **2021**, *30*, 4792–4807. [CrossRef]
23. Ramesh, M.; Deepa, C.; Rajeshkumar, L.; Selvan, M.T.; Balaji, D. Influence of fiber surface treatment on the tribological properties of *Calotropis gigantea* plant fiber reinforced polymer composites. *Polym. Compos.* **2021**, *42*, 4308–4317. [CrossRef]
24. Vidhyashankar, R.; Vinze, R.; Nagarathinam, S.; Natrajan, V.K. Modelling spatial variations in thermal comfort in indoor open-plan spaces using a whole-building simulation tool. *J. Build. Eng.* **2021**, *46*, 103727. [CrossRef]
25. Benzaama, M.; Rajaoarisoa, L.; Ajib, B.; Lecoeuche, S. A data-driven methodology to predict thermal behavior of residential buildings using piecewise linear models. *J. Build. Eng.* **2020**, *32*, 101523. [CrossRef]
26. Marquis-Favre, C.; Premat, E.; Aubrée, D.; Vallet, M. Noise and its effects—A review on qualitative aspects of sound. Part I: Notions and acoustic ratings. *Acta Acust. United Acust.* **2005**, *91*, 613–625.
27. Ramesh, M.; Kumar, L.R. Bioadhesives. In *Green Adhesives*; Inamuddin, R., Boddula, M.I., Ahamed and Asiri, A.M., Eds.; Wiley: Hoboken, NJ, USA, 2020; pp. 145–164. [CrossRef]
28. Marquis-Favre, C.; Premat, E.; Aubrée, D. Noise and its effects—A review on qualitative aspects of sound. Part II: Noise and annoyance. *Acta Acust. United Acust.* **2005**, *91*, 626–642.
29. Cao, L.; Fu, Q.; Si, Y.; Ding, B.; Yu, J. Porous materials for sound absorption. *Compos. Commun.* **2018**, *10*, 25–35. [CrossRef]
30. Na, Y.; Lancaster, J.; Casali, J.; Cho, G. Sound Absorption Coefficients of Micro-fiber Fabrics by Reverberation Room Method. *Text. Res. J.* **2007**, *77*, 330–335. [CrossRef]
31. Yang, T.; Xiong, X.; Mishra, R.; Novák, J.; Militký, J. Acoustic evaluation of Struto nonwovens and their relationship with thermal properties. *Text. Res. J.* **2016**, *88*, 426–437. [CrossRef]
32. Papadopoulos, A. State of the art in thermal insulation materials and aims for future developments. *Energy Build.* **2005**, *37*, 77–86. [CrossRef]
33. Sahayaraj, A.F.; Muthukrishnan, M.; Ramesh, M.; Rajeshkumar, L. Effect of hybridization on properties of tamarind (*Tamarindus indica* L.) seed nano-powder incorporated jute-hemp fibers reinforced epoxy composites. *Polym. Compos.* **2021**, *42*, 6611–6620. [CrossRef]
34. Deepa, C.; Rajeshkumar, L.; Ramesh, M. Thermal Properties of Kenaf Fiber-Based Hybrid Composites. In *Natural Fiber-Reinforced Composites: Thermal Properties and Applications*; John Wiley & Sons, Inc.: Hoboken, NJ, USA, 2021; pp. 167–182. [CrossRef]
35. Zhu, X.; Kim, B.-J.; Wang, Q.; Wu, Q. Recent Advances in the Sound Insulation Properties of Bio-based Materials. *BioResources* **2013**, *9*, 1764–1786. [CrossRef]

36. Devarajan, B.; LakshmiNarasimhan, R.; Venkateswaran, B.; Rangappa, S.M.; Siengchin, S. Additive manufacturing of jute fiber reinforced polymer composites: A concise review of material forms and methods. *Polym. Compos.* **2022**, *43*, 6735. [CrossRef]
37. Arenas, J.P.; Crocker, M.J. Recent trends in porous sound-absorbing materials. *Sound Vib.* **2010**, *44*, 12–18.
38. Ramesh, M. Aerogels for Insulation Applications. *Mater. Res. Found* **2021**, *98*, 57–76. [CrossRef]
39. Kumar, T.S.; Kumar, S.S.; Kumar, L.R. Jute fibers, their composites and applications. In *Plant Fibers, Their Composites, and Applications*; Woodhead Publishing: Sawston, UK, 2022; pp. 253–282. [CrossRef]
40. Priyadharshini, M.; Balaji, D.; Bhuvaneswari, V.; Rajeshkumar, L.; Sanjay, M.R.; Siengchin, S. Fiber Reinforced Composite Manufacturing With the Aid of Artificial Intelligence—A State-of-the-Art Review. *Arch. Comput. Methods Eng.* **2022**, 1–14. [CrossRef]
41. Wang, Y.; Zhu, W.; Wan, B.; Meng, Z.; Han, B. Hygrothermal ageing behavior and mechanism of carbon nanofibers modified flax fiber-reinforced epoxy laminates. *Compos. Part A Appl. Sci. Manuf.* **2020**, *140*, 106142. [CrossRef]
42. Zouaoui, Y.; Benmahiddine, F.; Yahia, A.; Belarbi, R. Hygrothermal and Mechanical Behaviors of Fiber Mortar: Comparative Study between Palm and Hemp Fibers. *Energies* **2021**, *14*, 7110. [CrossRef]
43. Sala, B.; Gabrion, X.; Trivaudey, F.; Guicheret-Retel, V.; Placet, V. Influence of the stress level and hygrothermal conditions on the creep/recovery behaviour of high-grade flax and hemp fibre reinforced GreenPoxy matrix composites. *Compos. Part A Appl. Sci. Manuf.* **2020**, *141*, 106204. [CrossRef]
44. Asli, M.; Sassine, E.; Brachelet, F.; Antczak, E. Hygrothermal behavior of wood fiber insulation, numerical and experimental approach. *Heat Mass Transf.* **2021**, *57*, 1069–1085. [CrossRef]
45. Balaji, D.; Ranga, J.; Bhuvaneswari, V.; Arulmurugan, B.; Rajeshkumar, L.; Manimohan, M.P.; Devi, G.R.; Ramya, G.; Masi, C. Additive Manufacturing for Aerospace from Inception to Certification. *J. Nanomater.* **2022**, *2022*, 7226852. [CrossRef]
46. Ramesh, M.; Balaji, D.; Rajeshkumar, L.; Bhuvaneswari, V. Manufacturing methods of elastomer blends and composites. In *Elastomer Blends and Composites*; Elsevier: Amsterdam, The Netherlands, 2022; pp. 11–32. [CrossRef]
47. Ramesh, M.; Rajeshkumar, L.; Balaji, D.; Bhuvaneswari, V. Influence of Moisture Absorption on Mechanical properties of Biocomposites reinforced Surface Modified Natural Fibers. In *Aging Effects on Natural Fiber-Reinforced Polymer Composites*; Springer: Berlin/Heidelberg, Germany, 2022; pp. 17–34. [CrossRef]
48. Xiao, X.; Zhong, Y.; Cheng, M.; Sheng, L.; Wang, D.; Li, S. Improved hygrothermal durability of flax/polypropylene composites after chemical treatments through a hybrid approach. *Cellulose* **2021**, *28*, 11209–11229. [CrossRef]
49. Ramesh, M.; Rajeshkumar, L.; Balaji, D.; Bhuvaneswari, V. Keratin-based biofibers and their composites. In *Advances in Bio-Based Fiber*; Woodhead Publishing: Sawston, UK, 2022; pp. 315–334. [CrossRef]
50. Gauvin, F.; Tsao, V.; Vette, J.; Brouwers, H.J.H. Physical properties and hygrothermal behavior of mycelium-based composites as foam-like wall insulation material. In *Construction Technologies and Architecture*; Trans Tech Publications Ltd.: Wollerau, Switzerland, 2022; Volume 1, pp. 643–651.
51. Ramesh, M.; Rajeshkumar, L.; Bhuvaneswari, V. Leaf fibres as reinforcements in green composites: A review on processing, properties and applications. *Emergent Mater.* **2021**, *5*, 833–857. [CrossRef]
52. Dayo, A.Q.; Babar, A.A.; Qin, Q.-R.; Kiran, S.; Wang, J.; Shah, A.H.; Zegaoui, A.; Ghouti, H.A.; Liu, W.-B. Effects of accelerated weathering on the mechanical properties of hemp fibre/polybenzoxazine based green composites. *Compos. Part A Appl. Sci. Manuf.* **2019**, *128*, 105653. [CrossRef]
53. Luo, C. Research on Hygrothermal Property of Jute Woven Fabric Reinforced Composites. In *IOP Conference Series: Earth and Environmental Science*; IOP Publishing: Bristol, UK, 2021; Volume 719, p. 022087.
54. Ramesh, M.; Rajeshkumar, L.; Balaji, D.; Priyadharshini, M. Properties and Characterization Techniques for Waterborne Polyurethanes. In *Sustainable Production and Applications of Waterborne Polyurethanes*; Springer: Cham, Switzerland, 2021; pp. 109–123. [CrossRef]
55. Hu, R.H.; Sun, M.Y.; Lim, J.K. Moisture absorption, tensile strength and microstructure evolution of short jute fiber/polylactide composite in hygrothermal environment. *Mater. Des.* **2010**, *31*, 3167–3173. [CrossRef]
56. Ramesh, M.; Rajeshkumar, L.; Sasikala, G.; Balaji, D.; Saravanakumar, A.; Bhuvaneswari, V.; Bhoopathi, R. A Critical Review on Wood-Based Polymer Composites: Processing, Properties, and Prospects. *Polymers* **2022**, *14*, 589. [CrossRef] [PubMed]
57. Wang, W.; Sain, M.; Cooper, P. Hygrothermal weathering of rice hull/HDPE composites under extreme climatic conditions. *Polym. Degrad. Stab.* **2005**, *90*, 540–545. [CrossRef]
58. Ramesh, M.; Balaji, D.; Rajeshkumar, L.; Bhuvaneswari, V.; Saravanakumar, R.; Khan, A.; Asiri, A.M. Tribological behavior of glass/sisal fiber reinforced polyester composites. In *Vegetable Fiber Composites and Their Technological Applications*; Springer: Singapore, 2021; pp. 445–459. [CrossRef]
59. Moudood, A.; Rahman, A.; Öchsner, A.; Islam, M. Francucci GFlax fiber its composites: An overview of water moisture absorption impact on their performance. *J. Reinf. Plast. Compos.* **2018**, *38*, 323–339. [CrossRef]
60. Ramesh, M.; Rajeshkumar, L. Case-Studies on Green Corrosion Inhibitors. In *Sustainable Corrosion Inhibitors*; Inamuddin, Ed.; Materials Research Fourm LLC: Millersville, PA, USA, 2021; Volume 107, pp. 204–221.
61. Habibi, M.; Laperrière, L.; Hassanabadi, H.M. Effect of moisture absorption and temperature on quasi-static and fatigue behavior of nonwoven flax epoxy composite. *Compos. Part B Eng.* **2019**, *166*, 31–40. [CrossRef]

62. Wang, X.; Petrů, M. The effects of surface treatment on creep and dynamic mechanical behavior of flax fiber reinforced composites under hygrothermal aging conditions. In *Surface Treatment Methods of Natural Fibres and Their Effects on Biocomposites*; Woodhead Publishing: Sawston, UK, 2022; pp. 203–242.
63. Elsacker, E.; Vandelook, S.; Van Wylick, A.; Ruytinx, J.; De Laet, L.; Peeters, E. A comprehensive framework for the production of mycelium-based lignocellulosic composites. *Sci. Total Environ.* **2020**, *725*, 138431. [CrossRef]
64. Mittal, M.; Chaudhary, R. Experimental investigation on the mechanical properties and water absorption behavior of randomly oriented short pineapple/coir fiber-reinforced hybrid epoxy composites. *Mater. Res. Express* **2018**, *6*, 015313. [CrossRef]
65. Deng, K.; Cheng, H.; Suo, H.; Liang, B.; Li, Y.; Zhang, K. Aging damage mechanism and mechanical properties degradation of 3D orthogonal woven thermoset composites subjected to cyclic hygrothermal environment. *Eng. Fail. Anal.* **2022**, *140*, 106629. [CrossRef]
66. Aouat, T.; Kaci, M.; Lopez-Cuesta, J.-M.; Devaux, E. Investigation on the Durability of PLA Bionanocomposite Fibers under Hygrothermal Conditions. *Front. Mater.* **2019**, *6*, 323. [CrossRef]
67. Jindal, B.B.; Jangra, P.; Garg, A. Effects of ultra fine slag as mineral admixture on the compressive strength, water absorption and permeability of rice husk ash based geopolymer concrete. *Mater. Today Proc.* **2020**, *32*, 871–877. [CrossRef]
68. Ramesh, M.; Rajeshkumar, L.; Balaji, D.; Bhuvaneswari, V.; Sivalingam, S. Self-Healable Conductive Materials. *Self-Heal. Smart Mater. Allied Appl.* **2021**, 297–319. [CrossRef]
69. Kamboj, I.; Jain, R.; Jain, D.; Bera, T.K. Effect of Fiber Pre-treatment Methods on Hygrothermal Aging Behavior of Agave Fiber Reinforced Polymer Composites. *J. Nat. Fibers* **2020**, *19*, 2929–2942. [CrossRef]
70. Han, S.O.; Ahn, H.J.; Cho, D. Hygrothermal effect on henequen or silk fiber reinforced poly(butylene succinate) biocomposites. *Compos. Part B Eng.* **2010**, *41*, 491–497. [CrossRef]
71. Ramesh, M.; Rajeshkumar, L.; Balaji, D. Mechanical and dynamic properties of ramie fiber-reinforced composites. *Mech. Dyn. Prop. Biocompos.* **2021**, 275–291. [CrossRef]
72. Yu, T.; Sun, F.; Lu, M.; Li, Y. Water absorption and hygrothermal aging behavior of short ramie fiber-reinforced poly(lactic acid) composites. *Polym. Compos.* **2016**, *39*, 1098–1104. [CrossRef]
73. Ghali, L.H.; Aloui, M.; Zidi, M.; Daly, H.B.; Msahli, S.; Sakli, F. Effect of chemical modification of luffa cylindrica fibers on the mechanical and hygrothermal behaviours of polyester/luffa composites. *BioResources* **2011**, *6*, 3836–3849.
74. Chandramohan, D.; Murali, B.; Vasantha-Srinivasan, P.; Dinesh Kumar, S. Mechanical, moisture absorption, and abrasion resistance properties of bamboo–jute–glass fiber composites. *J. Bio-Tribo-Corros.* **2019**, *5*, 1–8. [CrossRef]
75. Dayo, A.Q.; Zegaoui, A.; Nizamani, A.A.; Kiran, S.; Wang, J.; Derradji, M.; Cai, W.-A.; Liu, W.-B. The influence of different chemical treatments on the hemp fiber/polybenzoxazine based green composites: Mechanical, thermal and water absorption properties. *Mater. Chem. Phys.* **2018**, *217*, 270–277. [CrossRef]
76. Ma, G.; Yan, L.; Shen, W.; Zhu, D.; Huang, L.; Kasal, B. Effects of water, alkali solution and temperature ageing on water absorption, morphology and mechanical properties of natural FRP composites: Plant-based jute vs. mineral-based basalt. *Compos. Part B Eng.* **2018**, *153*, 398–412. [CrossRef]
77. Radzi, A.; Sapuan, S.; Jawaid, M.; Mansor, M. Water absorption, thickness swelling and thermal properties of roselle/sugar palm fibre reinforced thermoplastic polyurethane hybrid composites. *J. Mater. Res. Technol.* **2019**, *8*, 3988–3994. [CrossRef]
78. Dinesh, S.; Kumaran, P.; Mohanamurugan, S.; Vijay, R.; Singaravelu, D.L.; Vinod, A.; Sanjay, M.R.; Siengchin, S.; Bhat, K.S. Influence of wood dust fillers on the mechanical, thermal, water absorption and biodegradation characteristics of jute fiber epoxy composites. *J. Polym. Res.* **2020**, *27*, 9. [CrossRef]
79. Prasad, V.; Joseph, M.; Sekar, K. Investigation of mechanical, thermal and water absorption properties of flax fibre reinforced epoxy composite with nano $TiO_2$ addition. *Compos. Part A Appl. Sci. Manuf.* **2018**, *115*, 360–370. [CrossRef]
80. Huang, J.-K.; Young, W.-B. The mechanical, hygral, and interfacial strength of continuous bamboo fiber reinforced epoxy composites. *Compos. Part B Eng.* **2018**, *166*, 272–283. [CrossRef]
81. Atiqah, A.; Jawaid, M.; Sapuan, S.; Ishak, M.; Ansari, M.; Ilyas, R. Physical and thermal properties of treated sugar palm/glass fibre reinforced thermoplastic polyurethane hybrid composites. *J. Mater. Res. Technol.* **2019**, *8*, 3726–3732. [CrossRef]
82. Maslinda, A.; Majid, M.A.; Ridzuan, M.; Afendi, M.; Gibson, A. Effect of water absorption on the mechanical properties of hybrid interwoven cellulosic-cellulosic fibre reinforced epoxy composites. *Compos. Struct.* **2017**, *167*, 227–237. [CrossRef]
83. Thiagamani, S.M.K.; Krishnasamy, S.; Muthukumar, C.; Tengsuthiwat, J.; Nagarajan, R.; Siengchin, S.; Ismail, S.O. Investigation into mechanical, absorption and swelling behaviour of hemp/sisal fibre reinforced bioepoxy hybrid composites: Effects of stacking sequences. *Int. J. Biol. Macromol.* **2019**, *140*, 637–646. [CrossRef]
84. Bledzki, A.K.; Mamun, A.A.; Jaszkiewicz, A.; Erdmann, K. Polypropylene composites with enzyme modified abaca fibre. *Compos. Sci. Technol.* **2010**, *70*, 854–860. [CrossRef]
85. Lin, Q.; Zhou, X.; Dai, G. Effect of hydrothermal environment on moisture absorption and mechanical properties of wood flour-filled polypropylene composites. *J. Appl. Polym. Sci.* **2002**, *85*, 2824–2832. [CrossRef]
86. Ramesh, M.; Selvan, M.T.; Rajeshkumar, L.; Deepa, C.; Ahmad, A. Influence of *Vachellia nilotica* Subsp. *indica* Tree Trunk Bark Nano-powder on Properties of Milkweed Plant Fiber Reinforced Epoxy Composites. *J. Nat. Fibers* **2022**, 1–14. [CrossRef]
87. Naghmouchi, I.; Espinach, F.X.; Mutjé, P.; Boufi, S. Polypropylene composites based on lignocellulosic fillers: How the filler morphology affects the composite properties. *Mater. Des.* **2015**, *65*, 454–461. [CrossRef]

88. Yeh, S.-K.; Hsieh, C.-C.; Chang, H.-C.; Yen, C.C.; Chang, Y.-C. Synergistic effect of coupling agents and fiber treatments on mechanical properties and moisture absorption of polypropylene–rice husk composites and their foam. *Compos. Part A Appl. Sci. Manuf.* **2015**, *68*, 313–322. [CrossRef]
89. Demir, H.; Atikler, U.; Balköse, D.; Tıhmınlıoğlu, F. The effect of fiber surface treatments on the tensile and water sorption properties of polypropylene–luffa fiber composites. *Compos. Part A Appl. Sci. Manuf.* **2006**, *37*, 447–456. [CrossRef]
90. Gao, S.L.; Mäder, E. Jute/polypropylene composites I. Effect of matrix modification. *Compos. Sci. Technol.* **2006**, *66*, 952–963.
91. Gassan, J.; Bledzki, A.K. Effect of cyclic moisture absorption desorption on the mechanical properties of silanized jute-epoxy composites. *Polym. Compos.* **1999**, *20*, 604–611. [CrossRef]
92. Abdelmouleh, M.; Boufi, S.; Belgacem, M.N.; Dufresne, A. Short natural-fibre reinforced polyethylene and natural rubber composites: Effect of silane coupling agents and fibres loading. *Compos. Sci. Technol.* **2007**, *67*, 1627–1639. [CrossRef]
93. Kushwaha, P.K.; Kumar, R. Studies on Water Absorption of Bamboo-Polyester Composites: Effect of Silane Treatment of Mercerized Bamboo. *Polym. Technol. Eng.* **2009**, *49*, 45–52. [CrossRef]
94. Zahari, W.; Badri, R.; Ardyananta, H.; Kurniawan, D.; Nor, F. Mechanical Properties and Water Absorption Behavior of Polypropylene/Ijuk Fiber Composite by Using Silane Treatment. *Procedia Manuf.* **2015**, *2*, 573–578. [CrossRef]
95. Panthapulakkal, S.; Sain, M. Injection-molded short hemp fiber/glass fiber-reinforced polypropylene hybrid composites—Mechanical, water absorption and thermal properties. *J. Appl. Polym. Sci.* **2007**, *103*, 2432–2441. [CrossRef]
96. Zamri, M.H.; Akil, H.; Abu Bakar, A.; Ishak, Z.A.M.; Cheng, L.W. Effect of water absorption on pultruded jute/glass fiber-reinforced unsaturated polyester hybrid composites. *J. Compos. Mater.* **2011**, *46*, 51–61. [CrossRef]
97. Venkateshwaran, N.; ElayaPerumal, A.; Alavudeen, A.; Thiruchitrambalam, M. Mechanical and water absorption behaviour of banana/sisal reinforced hybrid composites. *Mater. Des.* **2011**, *32*, 4017–4021. [CrossRef]
98. Rout, J.; Misra, M.; Tripathy, S.; Nayak, S.; Mohanty, A. The influence of fibre treatment on the performance of coir-polyester composites. *Compos. Sci. Technol.* **2001**, *61*, 1303–1310. [CrossRef]
99. Sathishkumar, T.P.; Navaneethakrishnan, P.; Shankar, S.; Rajasekar, R. Mechanical properties and water absorption of short snake grass fiber reinforced isophthallic polyester composites. *Fibers Polym.* **2014**, *15*, 1927–1934. [CrossRef]
100. Rajeshkumar, G.; Hariharan, V.; Sathishkumar, T.P.; Fiore, V.; Scalici, T. Synergistic effect of fiber content and length on mechanical and water absorption behaviors of *Phoenix* sp. fiber-reinforced epoxy composites. *J. Ind. Text.* **2017**, *47*, 211–232. [CrossRef]
101. Velusamy, K.; Navaneethakrishnan, P.; RajeshKumar, G.; Sathishkumar, T. The influence of fiber content and length on mechanical and water absorption properties of *Calotropis gigantea* fiber reinforced epoxy composites. *J. Ind. Text.* **2018**, *48*, 1274–1290. [CrossRef]
102. Ramesh, M.; Deepa, C.; Selvan, M.T.; Rajeshkumar, L.; Balaji, D.; Bhuvaneswari, V. Mechanical and water absorption properties of *Calotropis gigantea* plant fibers reinforced polymer composites. *Mater. Today Proc.* **2021**, *46*, 3367–3372. [CrossRef]
103. Ramesh, M.; Deepa, C.; Niranjana, K.; Rajeshkumar, L.; Bhoopathi, R.; Balaji, D. Influence of Haritaki (*Terminalia chebula*) nano-powder on thermo-mechanical, water absorption and morphological properties of Tindora (*Coccinia grandis*) tendrils fiber reinforced epoxy composites. *J. Nat. Fibers* **2021**, 1–17. [CrossRef]
104. Gokulkumar, S.; Thyla, P.; Prabhu, L.; Sathish, S. Measuring Methods of Acoustic Properties and Influence of Physical Parameters on Natural Fibers: A Review. *J. Nat. Fibers* **2019**, *17*, 1719–1738. [CrossRef]
105. Yang, T.; Hu, L.; Xiong, X.; Petrů, M.; Noman, M.T.; Mishra, R.; Militký, J. Sound Absorption Properties of Natural Fibers: A Review. *Sustainability* **2020**, *12*, 8477. [CrossRef]
106. Bhingare, N.H.; Prakash, S.; Jatti, V.S. A review on natural and waste material composite as acoustic material. *Polym. Test.* **2019**, *80*, 106142. [CrossRef]
107. Aziz, M.A.A.; Sari, K.A.M. Comparison of sound absorption coefficient for natural fiber. *Prog. Eng. Appl. Technol.* **2021**, *2*, 157–163.
108. Samaei, S.E.; Mahabadi, H.A.; Mousavi, S.M.; Khavanin, A.; Faridan, M. Optimization and sound absorption modeling in Yucca Gloriosa natural fiber composite. *Iran Occup. Health J.* **2021**, *18*, 1–17. [CrossRef]
109. Kalaivani, R.; Ewe, L.S.; Zaroog, O.S.; Woon, H.S.; Ibrahim, Z. Acoustic properties of natural fiber of oil palm trunk. *Int. J. Adv. Appl. Sci.* **2018**, *5*, 88–92. [CrossRef]
110. Mawardi, I.; Aprilia, S.; Faisal, M.; Ikramullah; Rizal, S. An investigation of thermal conductivity and sound absorption from binderless panels made of oil palm wood as bio-insulation materials. *Results Eng.* **2021**, *13*, 100319. [CrossRef]
111. Sakthivel, S.; Kumar, S.S.; Solomon, E.; Getahun, G.; Admassu, Y.; Bogale, M.; Gedilu, M.; Aduna, A.; Abedom, F. Sound absorbing and insulating properties of natural fiber hybrid composites using sugarcane bagasse and bamboo charcoal. *J. Eng. Fibers Fabr.* **2021**, *16*, 15589250211044818. [CrossRef]
112. Prabhu, L.; Krishnaraj, V.; Gokulkumar, S.; Sathish, S.; Sanjay, M.R.; Siengchin, S. Mechanical, chemical and sound absorption properties of glass/kenaf/waste tea leaf fiber-reinforced hybrid epoxy composites. *J. Ind. Text.* **2020**, *51*, 1674–1700. [CrossRef]
113. Nasidi, I.N.; Ismail, L.H.; Samsudin, E.; Jaffar, M.I. Effects of Kenaf Fiber Strand Treatment by Sodium Hydroxide on Sound Absorption. *J. Nat. Fibers* **2021**, 1–10. [CrossRef]
114. Dhandapani, N.; Megalingam, A. Mechanical and Sound Absorption Behavior of Sisal and Palm Fiber Reinforced Hybrid Composites. *J. Nat. Fibers* **2021**, *19*, 4530–4543. [CrossRef]
115. Koruk, H.; Ozcan, A.C.; Genc, G.; Sanliturk, K.Y. Jute and Luffa Fiber-Reinforced Biocomposites: Effects of Sample Thickness and Fiber/Resin Ratio on Sound Absorption and Transmission Loss Performance. *J. Nat. Fibers* **2021**, 1–16. [CrossRef]

116. Rusli, M.; Irsyad, M.; Dahlan, H.; Gusriwandi; Bur, M. Sound absorption characteristics of the natural fibrous material from coconut coir, oil palm fruit bunches, and pineapple leaf. *IOP Conf. Ser. Mater. Sci. Eng.* **2019**, *602*, 012067. [CrossRef]
117. Sheng, D.D.C.V.; Bin Yahya, M.N.; Din, N.B.C. Sound Absorption of Microperforated Panel Made from Coconut Fiber and Polylactic Acid Composite. *J. Nat. Fibers* **2020**, *19*, 2719–2729. [CrossRef]
118. Bhingare, N.H.; Prakash, S. An experimental and theoretical investigation of coconut coir material for sound absorption characteristics. *Mater. Today Proc.* **2021**, *43*, 1545–1551. [CrossRef]
119. Rakesh, K.M.; Srinidhi, R.; Gokulkumar, S.; Nithin, K.S.; Madhavarao, S.; Sathish, S.; Karthick, A.; Muhibbullah, M.; Osman, S.M. Experimental Study on the Sound Absorption Properties of Finger Millet Straw, Darbha, and Ripe Bulrush Fibers. *Adv. Mater. Sci. Eng.* **2021**, *2021*, 1–12. [CrossRef]
120. Mohankumar, D.; Rajeshkumar, L.; Muthukumaran, N.; Ramesh, M.; Aravinth, P.; Anith, R.; Balaji, S. Effect of fiber orientation on tribological behaviour of *Typha angustifolia* natural fiber reinforced composites. *Mater. Today Proc.* **2022**, *62*, 1958–1964. [CrossRef]
121. El Hajj, N.; Mboumba-Mamboundou, B.; Dheilly, R.-M.; Aboura, Z.; Benzeggagh, M.; Queneudec, M. Development of thermal insulating and sound absorbing agro-sourced materials from auto linked flax-tows. *Ind. Crop. Prod.* **2011**, *34*, 921–928. [CrossRef]
122. Pfretzschner, J.; Rodriguez, R.M. Acoustic properties of rubber crumbs. *Polym. Test.* **1999**, *18*, 81–92. [CrossRef]
123. Xiang, H.-F.; Wang, D.; Liua, H.-C.; Zhao, N.; Xu, J. Investigation on sound absorption properties of kapok fibers. *Chin. J. Polym. Sci.* **2013**, *31*, 521–529. [CrossRef]
124. Zhang, S.; Li, Y.; Zheng, Z. Effect of physiochemical structure on energy absorption properties of plant fibers reinforced composites: Dielectric, thermal insulation, and sound absorption properties. *Compos. Commun.* **2018**, *10*, 163–167. [CrossRef]
125. Olcay, H.; Kocak, E.D. Rice plant waste reinforced polyurethane composites for use as the acoustic absorption material. *Appl. Acoust.* **2021**, *173*, 107733. [CrossRef]
126. Hassan, T.; Jamshaid, H.; Mishra, R.; Khan, M.Q.; Petru, M.; Novak, J.; Choteborsky, R.; Hromasova, M. Acoustic, Mechanical and Thermal Properties of Green Composites Reinforced with Natural Fibers Waste. *Polymers* **2020**, *12*, 654. [CrossRef]
127. Lahouioui, M.; Ben Arfi, R.; Fois, M.; Ibos, L.; Ghorbal, A. Investigation of Fiber Surface Treatment Effect on Thermal, Mechanical and Acoustical Properties of Date Palm Fiber-Reinforced Cementitious Composites. *Waste Biomass-Valorization* **2019**, *11*, 4441–4455. [CrossRef]
128. Bin Bakri, M.K.; Jayamani, E.; Soon, K.H.; Hamdan, S.; Kakar, A. An experimental and simulation studies on sound absorption coefficients of banana fibers and their reinforced composites. In *Nano Hybrids and Composites*; Trans Tech Publications Ltd.: Wollerau, Switzerland, 2017; Volume 12, pp. 9–20.
129. Berardi, U.; Iannace, G. Acoustic characterization of natural fibers for sound absorption applications. *Build. Environ.* **2015**, *94*, 840–852. [CrossRef]
130. Yahya, M.N.; Sambu, M.; Latif, H.A.; Junaid, T.M. August. A study of acoustics performance on natural fibre composite. In *IOP Conference Series: Materials Science and Engineering*; IOP Publishing: Bristol, UK, 2017; Volume 226, p. 012013.
131. Fouladi, M.H.; Nassir, M.H.; Ghassem, M.; Shamel, M.; Peng, S.Y.; Wen, S.Y.; Xin, P.Z.; Nor, M.J.M. Utilizing Malaysian natural fibers as sound absorber. In *Modeling And Measurement Methods for Acoustic Waves and for Acoustic Microdevices*; Intechopen: London, UK, 2013; pp. 161–170.
132. Lee, H.P.; Ng, B.M.P.; Rammohan, A.V.; Tran, L.Q.N. An Investigation of the Sound Absorption Properties of Flax/Epoxy Composites Compared with Glass/Epoxy Composites. *J. Nat. Fibers* **2016**, *14*, 71–77. [CrossRef]
133. Mohapatra, T.K.; Satapathy, S.; Panigrahi, I.; Mishra, D. Biodegradable Acoustic Proficiency in Sound Absorption Capacity After Water Treatment and Testing by Impendence Tube Method: Hot Water and Saline Treatment to Test Acoustic Property. In *Handbook of Research on Advancements in Manufacturing, Materials, and Mechanical Engineering*; IGI Global: Hershey, PA, USA, 2021; pp. 75–90.
134. Hassan, N.N.M.; Rus, A.Z.M. Influences of thickness and fabric for sound absorption of biopolymer composite. In *Applied Mechanics and Materials*; Trans Tech Publications Ltd.: Wollerau, Switzerland, 2013; Volume 393, pp. 102–107.
135. Santhanam, S.; Temesgen, S.; Atalie, D.; Ashagre, G. Recycling of cotton and polyester fibers to produce nonwoven fabric for functional sound absorption material. *J. Nat. Fibers* **2017**, *16*, 300–306. [CrossRef]
136. Jayamani, E.; Hamdan, S.; Rahman, M.R.; Bakri, M.K.B. Investigation of fiber surface treatment on mechanical, acoustical and thermal properties of betelnut fiber polyester composites. *Procedia Eng.* **2014**, *97*, 545–554. [CrossRef]
137. Santoni, A.; Bonfiglio, P.; Fausti, P.; Marescotti, C.; Mazzanti, V.; Mollica, F.; Pompoli, F. Improving the sound absorption performance of sustainable thermal insulation materials: Natural hemp fibres. *Appl. Acoust.* **2019**, *150*, 279–289. [CrossRef]
138. da Silva, C.C.B.; Terashima, F.J.H.; Barbieri, N.; de Lima, K.F. Sound absorption coefficient assessment of sisal, coconut husk and sugar cane fibers for low frequencies based on three different methods. *Appl. Acoust.* **2019**, *156*, 92–100. [CrossRef]
139. Buratti, C.; Belloni, E.; Lascaro, E.; Merli, F.; Ricciardi, P. Rice husk panels for building applications: Thermal, acoustic and environmental characterization and comparison with other innovative recycled waste materials. *Constr. Build. Mater.* **2018**, *171*, 338–349. [CrossRef]
140. Soltani, P.; Taban, E.; Faridan, M.; Samaei, S.E.; Amininasab, S. Experimental and computational investigation of sound absorption performance of sustainable porous material: Yucca Gloriosa fiber. *Appl. Acoust.* **2019**, *157*, 106999. [CrossRef]
141. Qui, H.; Enhui, Y. Effect of Thickness, Density and Cavity Depth on the Sound Absorption Properties of Wool Boards. *Autex Res. J.* **2018**, *18*, 203–208. [CrossRef]

142. Putra, A.; Or, K.H.; Selamat, M.Z.; Nor, M.J.M.; Hassan, M.H.; Prasetiyo, I. Sound absorption of extracted pineapple-leaf fibres. *Appl. Acoust.* **2018**, *136*, 9–15. [CrossRef]
143. Sari, N.H.; Wardana, I.N.G.; Irawan, Y.S.; Siswanto, E. Physical and Acoustical Properties of Corn Husk Fiber Panels. *Adv. Acoust. Vib.* **2016**, *2016*, 5971814. [CrossRef]
144. Chin, D.D.V.S.; Yahya, M.N.B.; Din, N.B.C.; Ong, P. Acoustic properties of biodegradable composite micro-perforated panel (BC-MPP) made from kenaf fibre and polylactic acid (PLA). *Appl. Acoust.* **2018**, *138*, 179–187. [CrossRef]
145. Gama, N.; Silva, R.; Carvalho, A.P.; Ferreira, A.; Barros-Timmons, A. Sound absorption properties of polyurethane foams derived from crude glycerol and liquefied coffee grounds polyol. *Polym. Test.* **2017**, *62*, 13–22. [CrossRef]
146. Othmani, C.; Taktak, M.; Zein, A.; Hentati, T.; Elnady, T.; Fakhfakh, T.; Haddar, M. Experimental and theoretical investigation of the acoustic performance of sugarcane wastes based material. *Appl. Acoust.* **2016**, *109*, 90–96. [CrossRef]
147. Belakroum, R.; Gherfi, A.; Kadja, M.; Maalouf, C.; Lachi, M.; El Wakil, N.; Mai, T. Design and properties of a new sustainable construction material based on date palm fibers and lime. *Constr. Build. Mater.* **2018**, *184*, 330–343. [CrossRef]
148. Ghofrani, M.; Ashori, A.; Mehrabi, R. Mechanical and acoustical properties of particleboards made with date palm branches and vermiculite. *Polym. Test.* **2017**, *60*, 153–159. [CrossRef]
149. Berardi, U.; Iannace, G.; Di Gabriele, M. The Acoustic Characterization of Broom Fibers. *J. Nat. Fibers* **2017**, *14*, 858–863. [CrossRef]
150. Sheng, C.; He, G.; Hu, Z.; Chou, C.; Shi, J.; Li, J.; Meng, Q.; Ning, X.; Wang, L.; Ning, F. Yarn on yarn abrasion failure mechanism of ultrahigh molecular weight polyethylene fiber. *J. Eng. Fibers Fabr.* **2021**, *16*, 1925832385. [CrossRef]
151. Ning, F.; He, G.; Sheng, C.; He, H.; Wang, J.; Zhou, R.; Ning, X. Yarn on yarn abrasion performance of high modulus polyethylene fiber improved by graphene/polyurethane composites coating. *J. Eng. Fibers Fabr.* **2021**, *16*, 1–10. [CrossRef]
152. Li, C.; Jiang, T.; Liu, S.; Han, Q. Dispersion and band gaps of elastic guided waves in the multi-scale periodic composite plates. *Aerosp. Sci. Technol.* **2022**, *124*, 107513. [CrossRef]
153. Chen, B.; Lu, Y.; Li, W.; Dai, X.; Hua, X.; Xu, J.; Wang, Z.; Zhang, C.; Gao, D.; Li, Y.; et al. DPM-LES investigation on flow field dynamic and acoustic characteristics of a twin-fluid nozzle by multi-field coupling method. *Int. J. Heat Mass Transf.* **2022**, *192*, 122927. [CrossRef]

*polymers*

*Review*

# Recent Progress on Tailoring the Biomass-Derived Cellulose Hybrid Composite Photocatalysts

**Yi Ding Chai [1], Yean Ling Pang [1,2,*], Steven Lim [1,2], Woon Chan Chong [1,2], Chin Wei Lai [3] and Ahmad Zuhairi Abdullah [4]**

1 Department of Chemical Engineering, Lee Kong Chian Faculty of Engineering and Science, Universiti Tunku Abdul Rahman, Kajang 43000, Malaysia
2 Centre for Photonics and Advanced Materials Research, Universiti Tunku Abdul Rahman, Kajang 43000, Malaysia
3 Nanotechnology & Catalysis Research Centre (NANOCAT), Institute for Advanced Studies, University of Malaya, Kuala Lumpur 50603, Malaysia
4 School of Chemical Engineering, Universiti Sains Malaysia, Nibong Tebal 14300, Malaysia
* Correspondence: pangyl@utar.edu.my or pangyeanling@hotmail.com; Tel.: +603-9086-0288; Fax: +603-9019-8868

**Abstract:** Biomass-derived cellulose hybrid composite materials are promising for application in the field of photocatalysis due to their excellent properties. The excellent properties between biomass-derived cellulose and photocatalyst materials was induced by biocompatibility and high hydrophilicity of the cellulose components. Biomass-derived cellulose exhibited huge amount of electron-rich hydroxyl group which could promote superior interaction with the photocatalyst. Hence, the original sources and types of cellulose, synthesizing methods, and fabrication cellulose composites together with applications are reviewed in this paper. Different types of biomasses such as biochar, activated carbon (AC), cellulose, chitosan, and chitin were discussed. Cellulose is categorized as plant cellulose, bacterial cellulose, algae cellulose, and tunicate cellulose. The extraction and purification steps of cellulose were explained in detail. Next, the common photocatalyst nanomaterials including titanium dioxide ($TiO_2$), zinc oxide (ZnO), graphitic carbon nitride (g-$C_3N_4$), and graphene, were introduced based on their distinct structures, advantages, and limitations in water treatment applications. The synthesizing method of $TiO_2$-based photocatalyst includes hydrothermal synthesis, sol-gel synthesis, and chemical vapor deposition synthesis. Different synthesizing methods contribute toward different $TiO_2$ forms in terms of structural phases and surface morphology. The fabrication and performance of cellulose composite catalysts give readers a better understanding of the incorporation of cellulose in the development of sustainable and robust photocatalysts. The modifications including metal doping, non-metal doping, and metal–organic frameworks (MOFs) showed improvements on the degradation performance of cellulose composite catalysts. The information and evidence on the fabrication techniques of biomass-derived cellulose hybrid photocatalyst and its recent application in the field of water treatment were reviewed thoroughly in this review paper.

**Keywords:** biomass; cellulose; photocatalysts; hybrid materials; degradation performance

**Citation:** Chai, Y.D.; Pang, Y.L.; Lim, S.; Chong, W.C.; Lai, C.W.; Abdullah, A.Z. Recent Progress on Tailoring the Biomass-Derived Cellulose Hybrid Composite Photocatalysts. *Polymers* **2022**, *14*, 5244. https://doi.org/10.3390/polym14235244

Academic Editors: Vsevolod Aleksandrovich Zhuikov and Rosane Michele Duarte Soares

Received: 29 October 2022
Accepted: 21 November 2022
Published: 1 December 2022



## 1. Introduction

Nowadays, the rising population on Earth has significantly amplified the production of organic solid waste [1]. The disposal of organic solid waste, such as biomass, creates a prolonged problem, especially in the agricultural industry [2]. Dumping solid waste into natural territory, such as landfills, has heightened the global waste level and rather induced certain risks regarding the handling method of solid waste [3]. At landfills, the degradation of biomass produces methane gases, and they are being released into the surroundings and further become a factor in the Greenhouse Gas (GHG) effect [4]. Such environmental issues have directed researchers towards discovering environmentally friendly and low cost methods to produce and commercialize potential materials from organic solid waste such as biomass [5]. In fact, organic solid waste such as biomass is also intrinsic to a high carbon value, so these types of waste should be redefined as "resources" [4]. Biomass has been regarded as a sustainable resource that could potentially minimize GHG emissions [5]. To acquire zero GHG emissions, there must be a balance between both the production of plant biomass and the management of its residual wastes. This can be acquired by establishing a circular economy which involves the activities of reuse, recycle, repurpose, and up-cycle [4,5].

Cellulose is a biopolymer that is widely available, renewable, and makes up the majority of biomass. Biomass materials including corn cobs, banana stems, sugarcane bagasse, and wheat straw are sources of cellulose. Depending on the source, lignocellulosic biomass contains 40–60% *w*/*w* cellulose, 15–30% *w*/*w* hemicellulose, and 10–25% *w*/*w* lignin [6]. The cellulose isolated from biomass is presented in the form of cellulose fibers. It has been claimed that the promising composition of agricultural residues, which contains more hemicellulose and less lignin than wood, promotes more effective nanofibrillation [7].

Among advanced oxidation processes, photocatalysis is commonly applied for the degradation of organic dye. It comprises a photo-oxidation reaction in the presence of photocatalysts under light irradiation. In photocatalysis, the photocatalyst is activated through the absorption of photon energy to accelerate the chemical reaction without being consumed [8]. When the energy received from light irradiation is equal to or exceeds the band gap energy of photocatalysts, the electrons in the valence band of photocatalysts will migrate to the conduction band of photocatalysts, leaving holes in the valence band of photocatalysts [9]. Simultaneously, the generated electrons and holes will carry out the reduction and oxidation reactions to produce superoxide anion radical and hydroxyl radicals, respectively. These reactive oxygen radicals may contribute to the oxidative pathways to degrade the organic pollutant molecules.

Despite the efficient removal of organic pollutants, conventional photocatalysis possesses limitations such as difficult recovery of the catalyst, generation of secondary pollution, and a high consumption of both catalysts and energy. To overcome these limitations, it is highly desirable to develop semiconductor photocatalysts with promising charge migration, high quantum efficiency, broad light spectral response, and good stability. Through the emergence of cellulose and its potential derivatives, the performance and sustainability of cellulose-based photocatalysts have advanced progressively in the past decade. Therefore, this review is able to provide insights and guidance for scientists who are looking for the development and functionality of cellulose-based nanostructured photocatalyst hybrids to address emerging environmental concerns.

## 2. Biomass

In general, biomass is known as a biological material obtained from plant-based or animal-based resources and their respective derived residues and wastes [5]. Biomass can be categorized into agricultural wastes, forestry wastes, industrial wastes, and municipal wastes [1,5,10,11]. Researchers tend to make use of biomass in researches via a facile route due to its recyclability and sustainability [3]. Both material costs and solid waste management risks can be lowered due to the utilization of renewable biomass as starting materials [12].

The global primary production of agricultural biomass is approximately 220 billion tonnes annually based on the dry weight basis [13]. Agricultural biomass mainly comprises cellulose, hemicellulose, and lignin as shown in Figure 1 [14]. Cellulose is a straight chain polymer comprising glucose monomers [15]. Hemicellulose has different short-chain polymers such as d-xylose, d-galactose, and d-glucose [16]. Lignin is a complex non-crystalline phenolic macromolecule with an amorphous nature and aromatic structures such as sinapyl, coniferyl, and coumaryl alcohols [15,16]. Furthermore, inorganic constituents such as calcium, potassium, silicon, magnesium, sodium, phosphorous, and chlorine can also be found in biomass [10,15,17]. Biomass contains about 51 wt% of carbon, 42 wt% of oxygen, and other remaining elements [10]. Biomass collected from agricultural waste is mostly lignocellulosic biomass and it can be utilized to synthesize carbon-based catalysts such as biochar and AC owing to high carbon content [2].

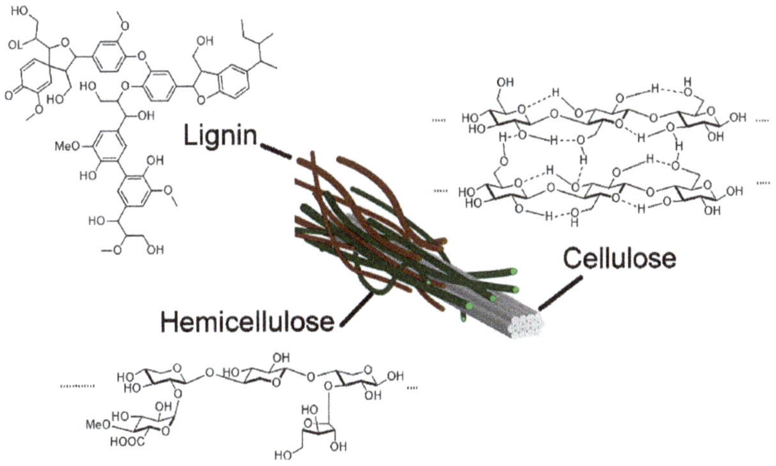


**Figure 1.** Cellulose, hemicellulose, and lignin contained inside biomass. Reproduced with permission from [14].

Looking into the marine biomass, it was reported that 17.4 million tonnes of mollusks and 8.4 million tonnes of crustaceans were produced globally in 2017 [18]. The examples of mollusks are mussel, clam, and oyster whereas the examples of crustaceans are lobster, shrimp, and crab. However, seafood wastes are often thrown back into the sea, burned, landfilled, or left aside to decompose [19]. The utilization of marine biomass has become an environmental priority and has amplified research regarding waste conversion into valuable products such as cellulose, chitosan, and chitin [19–21].

*2.1. Biochar*

Biochar is known as biomass-derived char and it is a carbon-rich material because it comprises about 60–90% carbon [22,23]. Biomass, such as agricultural residues (rice hulls and bagasse), garden residues, and municipal wastes, can be used to produce biochar [24]. The promising properties of biochar are the large specific surface area, high pore volume, high fertility, long-term stability structure, strong adsorption capacity, and enriched surface functional groups and mineral components [22,24,25].

Biochar can be produced via pyrolysis, gasification, and carbonization [26,27]. Pyrolysis is the conventional method to produce biochar and it is conducted within an oxygen-limited or oxygen-free environment at a temperature range of 300–800 °C [22,24,27]. During pyrolysis, the biomass components will experience both thermal reactions and molecular arrangements in order to construct a polymerized aromatic structure and this structure enables active compounds to functionalize biochar [23]. Meanwhile, gasification transforms biomass into a major gaseous product of syngas and a minor solid product of biochar at a temperature higher than 700 °C in the presence of an oxidizing agent [22,27]. Gases such as oxygen, air or steam can be applied as oxidizing agents [27].

Carbonization can be carried out prior to activation where biomass performs a thermal treatment (slow pyrolysis) to enrich the carbon content in the biomass [28,29]. The resultant solid residue (biochar) possesses low porosity and the formed pores are entrapped with tar-like materials [22,29]. The biochar can be further developed in activation processes even though the initial porosity of char is relatively low [28]. Carbonization parameters including carbonization temperature, heating rate, and residence time, contribute to the formation of the initial pore structure through the emissions of volatile matter from the carbon matrix [28,29].

Biochar is introduced into soil to improve soil fertility, enhance carbon sequestration, and mitigate greenhouse gas emission [22,24]. Other than that, biochar is also used as a precursor of synthesizing catalysts and contaminant adsorbents in both soil amendment and wastewater treatment [27]. The raw materials of biochar synthesis are cheap and abundant, while the preparation of biochar is cost-effective with lower energy requirements [24,25]. Thus, biochar is reported to be a potential low-cost and effective adsorbent in applications such as catalysis and soil remediation [25].

### 2.2. Activated Carbon (AC)

AC is defined as a form of amorphous carbonaceous materials with high surface area, high porosity nature, high adsorption capacity, and oxygenated functional groups [23,29–31]. The preparation of AC is close to the preparation of biochar except for the additional activation process and higher treatment temperatures are required [25,32]. The activation process can be carried out with three methods which are physical activation, chemical activation, and physiochemical activation [28–30]. The purpose of the activation process is to enlarge pore size, improve the pore volume, expand the pore diameter and increase the porosity [23,28]. Physical activation involves thermally treating the carbonaceous precursor in the presence of oxidizing agents such as inert gas, carbon dioxide, and steam, at elevated temperatures ranging from 400 to 1000 °C [29,30]. The introduction of an oxidizing agent promotes the internal porosity of AC [29].

For chemical activation, chemical agents will be added to the material prior to the same thermal treatment as physical activation [30]. The type and concentration of chemical agents, or named as activating agents, such as zinc chloride ($ZnCl_2$), potassium hydroxide (KOH), phosphoric acid ($H_3PO_4$), and sodium carbonate ($Na_2CO_3$), are taken into account for the surface properties of AC. Nayak et al. [33] reported that the micropore volume and surface area of $ZnCl_2$-prepared AC (0.61 cc/g and 2430.8 $m^2/g$) was higher than KOH-prepared AC (0.32 cc/g and 1506.2 $m^2/g$). Under similar activation conditions (precursor to activating agent mass ratio =1:0.5, 600 °C, and 1 h), $ZnCl_2$ promoted the microporosity on AC, whereas KOH induced the development of larger pores on AC. Shrestha and Rajbhandari [34] found out that the AC impregnated with $H_3PO_4$ exhibited the highest surface area (1269.5 $m^2/g$) as compared to AC impregnated with KOH (280.6 $m^2/g$) and $Na_2CO_3$ (58.9 $m^2/g$) at similar activation conditions (precursor to activating agent mass ratio =1:1, 400 °C, and 3 h) due to numerous mesopores and micropores. The activating agents such as KOH and $Na_2CO_3$ at the activation temperature of 400 °C were insufficient to induce a high porous structure on AC. The favorable activation temperature for acidic and alkaline activating agents were reported at 400–500 °C and 750–850 °C, respectively [35]. In addition, Zhang et al. [36] reported that the increasing mass ratio of sodium hydroxide (NaOH) to carbonized wheat bran from 1 to 5 at the activation temperature of 800 °C increased the mesoporosity of AC. Excessive usage of NaOH could damage the pore walls, widen the pore through violent etching, and destruct the micropore structure of AC.

Next, AC is usually utilized in applications of pollution control such as gaseous filter systems and wastewater treatment [23,30]. The high adsorption capacity of AC allows it to act as catalyst support, and to remove pollutants such as heavy metal and organic dyes from contaminated wastewater or dyeing unit effluent effectively [23,32,37]. Nevertheless, the drawbacks of commercial AC are high production cost, separation difficulty, low regeneration rate, and difficulty in reactivation process [31,32]. Similar to biochar, any low cost carbonaceous material can be used as promising precursor sources of AC and it can be obtained from woody biomass, agricultural waste, and forestry residues [29,30]. For example, coconut shells, bamboo, wood, silk cotton hull, coal, fruit peel, and especially biochar are used in the fabrication of AC [23,29,30,32,37].

Besides, Wickramaarachchi et al. [38,39] reported the fabrication of AC with hierarchical porous structure by using biomass such as mango seed husk and grape marc to develop sustainable supercapacitor materials. Both mango-seed-husk- and grape-marc-derived AC achieved high specific capacitance of 135 F/g and 139 F/g, respectively, at optimum conditions. Arun et al. [40] discovered that adding 0.1 wt% orange-peel-derived AC to the negative electrode of a lead acid battery cell increased capacity and raised charge acceptance along with lowering gassing voltages. When jute-fiber-derived AC was used as an anode material for lithium-ion batteries, it demonstrated a high specific capacity of 742.7 mA h/g after 100 cycles at 0.2 C [41]. This showed that the jute-fiber-derived AC was a stable and efficient anode material. Hence, biomass-derived AC can also be a potential material for energy storage applications.

### 2.3. Cellulose

Cellulose is defined as a linear chain of ringed glucose molecules forming a flat ribbon-like structure [42,43]. The molecular mass of cellulose ranges from $1.44 \times 10^6$ to $1.8 \times 10^6$ g and it shows thermal softening at a temperature of 231–253 °C [44]. Cellulose is a biopolymer made up of glucose monomers with the degree of polymerization ranging between $1 \times 10^4$ and $2 \times 10^4$ based on the source of cellulosic material [45,46]. Each glucose monomer contains three hydroxyl groups which control the crystalline packing and physical properties of cellulose [47]. The glucose linkages can be stabilized to form a linear cellulose chain because of the existence of hydrogen bonding between hydroxyl groups and oxygen atoms of the adjoining ring molecules [42]. These hydroxyl groups also define the chemical reactivity of cellulose, whereby their reactive sites enable the functionalization of cellulose materials [48]. Thus, chemical modifications of cellulose can enhance the adsorption property of cellulose towards pollutant removal by altering its physical and chemical properties [32].

Cellulose has degree of polymerization >2000, is insoluble in common solvents such as water, and is also weakly accessible to acid and enzymatic hydrolysis [32,49,50]. The material properties of cellulose are contributed by the phenomenon of aggregation via Van der Waals forces, and both inter- and intra-molecular hydrogen bonding inside the cellulose chains [46,51]. With the formation of inter-molecular hydrogen bonds, the cellulose chains are organized in a parallel arrangement to form stiff ribbon- or sheet-like structures through the hindering of free rotation of the rings on the glucose linkages [32,49,51]. The Van der Waals forces and weak CH–O bonds or intra-molecular hydrogen bonds hold these cellulosic sheets together in layers [44,52].

The linear cellulose chains, or named as cellulose fibrils, aggregate to form a microfibril, and subsequently the microfibrils are further assembled into the common cellulose fibers during the cellulose biosynthesis [46,51]. The range of microfibrils cross dimensions is between 2 and 20 nm, and this depends on the origin of cellulose [46]. Looking into the cellulose fibrils, cellulose chains are divided into crystalline and amorphous regions [43]. The cellulose chains in the crystalline region are arranged highly ordered, while the cellulose chains in the amorphous region are arranged disorderly. The crystalline region consists of a complex network of hydrogen bonds, which gives strength and toughness to cellulose fibrils [51]. The amorphous region, comprising cellulose chains with lower density compared to the crystalline region, has higher availability to form more hydrogen bonds with other molecules [53]. Eventually, the amorphous region can be easily hydrolyzed under harsh pre-treatment conditions to deliver nanosized cellulose [51,54].

Within the crystalline region, the variations of the inter- and intra-molecular hydrogen bonding and molecular orientations generate cellulose polymorphs or allomorphs [55]. The generated cellulose polymorphs are influenced by the source, extraction method, and treatment of cellulose [46]. Native cellulose is known as the primary cellulose produced from natural biosynthesis, which does not undergo any changes of form and it has the polymorph of cellulose I [47,56]. Cellulose I can be transformed into cellulose II by either mercerization in strong alkali medium or acid regeneration [56]. Cellulose I and cellulose II can also be converted into the respective cellulose $III_I$ and cellulose $III_{II}$ through ammoniacal treatments [57].

Cellulose I is commonly studied due to its high abundance in nature [56]. Cellulose I comprises two crystalline forms or suballomorphs which are Iα and Iβ lattices [46,49]. The Iα lattice has a one-chain triclinic structure while the Iβ lattice has a two-chain monoclinic structure [45,52]. Next, different cellulose sources vary in the fractions of Iα and Iβ crystal structures [45]. The Iα crystal structures predominate in bacterial cellulose and algae cellulose, whereas the Iβ crystal structures predominate in plant cellulose and tunicate cellulose [58].

It is interesting that both Iα and Iβ crystal structures comprise stacks of sheet-like cellulose chain layers to form three-dimensional (3D) crystals [52,56]. The Iα and Iβ crystal structures have comparable unit chain length, interchain distance, and intersheet distance, which are 10.4 Å, 8.2 Å, and 3.9 Å, respectively [52]. Nevertheless, the Iα and Iβ crystal structures differ from the framework of relative displacement between adjacent cellulosic layers [56]. The Iα crystal structure can be converted into Iβ crystal structure through annealing and it is an irreversible process [45,49]. Thus, this shows that the Iβ crystal structure is more stable compared to the meta-stable Iα crystal structure [49,56].

### *2.4. Chitosan and Chitin*

Chitosan is known as poly-β-(1,4)-2-amino-2-deoxy-glucopyranose whereby it is a polysaccharide consisting of glucosamine and N-acetylglucosamine as copolymers [59,60]. Chitosan is produced from the partial deacetylation of chitin [19,59]. Typically, chitosan is determined as the second abundant biomass-derived polysaccharide which is low cost, environmentally friendly, biocompatible, and biodegradable [60]. Chitosan has two monosaccharide units where their proportions will be affected by the subsequent alkaline treatment [20,21]. Chitosan consists of primary amine and free hydroxyl groups [21]. Besides the acetylamine or free amino groups which replace the hydroxyl group at the C-2 position of the cellulose structure, chitosan has a similar structure compared to cellulose as shown in Figure 2 [61–63]. The presence of highly reactive amino groups promotes the generation of intra- and inter-molecular hydrogen bonds with the abundant hydroxyl groups which form linear aggregates and rigid crystalline domains [20,21]. This resulted in the high viscosity and exhibition of polymorphism of chitosan [21,61].

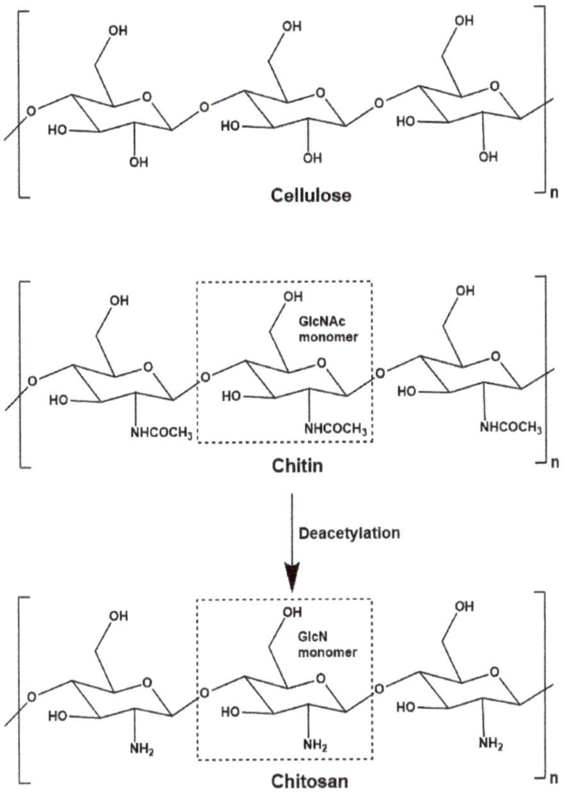


**Figure 2.** Structure comparison between cellulose, chitin, and chitosan. No special permission is required to reuse all or part of an article published by MDPI [64].

Crustacean shells are the primary chitin source for the industrial production of chitosan [19]. Generally, chitin is a structural element that can be found in crustaceans, exoskeletons of insects, and cell walls of fungi as shown in Figure 3 [60]. Although marine animals are still the main source of chitin, the fungal chitin obtained from mushrooms is gradually increasing nowadays. It is the second most abundant naturally occurring biopolymer after cellulose [65]. The chitin within these seafood wastes has a slow rate of biodegradation and this causes the yielding of large piles of processing discards from seafood processing plants [21].

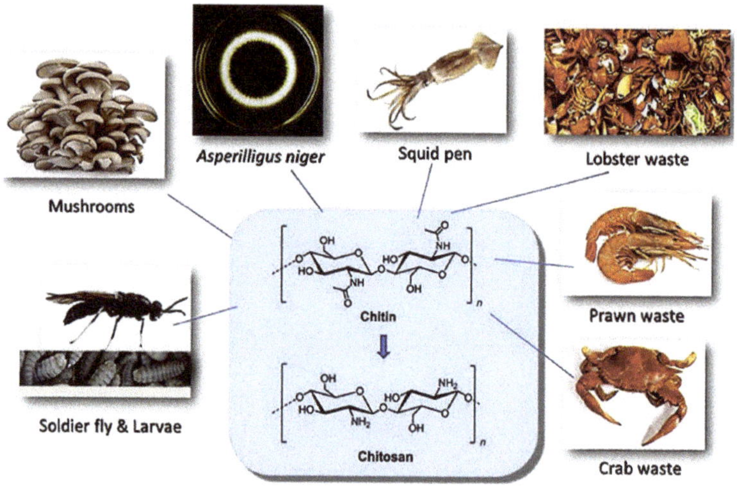


**Figure 3.** The chitin sources. Reproduced with permission from [66].

Chitin can be deacetylated into chitosan by chemical and enzymatic methods [67]. The chemical method includes demineralization, deproteinization, deacetylation, and decolorization [19]. The demineralization step removes calcium carbonate ($CaCO_3$) from the crustacean shells with dilute hydrochloric acid (HCl), whereas deproteinization solubilizes the protein with dilute aqueous NaOH [21]. Then, the produced chitin can be treated with hot concentrated NaOH for a long period to synthesize chitosan through heterogeneous and homogeneous processes with 85–99% and 48–55% degree of deacetylation, respectively [62,67,68]. The alkaline treatment hydrolyzes the acetyl groups and converts N-acetylglucosamine into glucosamine with free amino groups. The degree of deacetylation indicates the glucosamine to N-acetylglucosamine ratio whereby chitin transforms into chitosan. When the proportion of glucosamine is higher than N-acetylglucosamine, the produced compound is called chitosan, and vice versa for chitin [19]. The additional decolorization step can be carried out to remove color and improve physical appearance [21].

Furthermore, the enzymatic method for producing chitosan uses enzymes under mild conditions [62]. Enzymatic methods utilize enzymes extracted from *Mucor roxii* and *Absidia coerulea* for the deacetylation of chitin [67]. Chitinases or chitin deacetylases can be obtained from *Mucor roxii* and *Absidia coerulea*, and these enzymes have a good thermostability which could act optimally at 50 °C [69]. The utilization of chitin deacetylases is mainly to produce novel and well-defined chitosan oligomers [68]. In comparison with the enzymatic method, chemical methods have shorter processing time and higher suitability for mass production [67,68,70]. Therefore, the chemical method is usually preferred to produce chitosan.

Besides viscosity, both the degree of deacetylation and molecular weight of chitosan affect the solubility, reactivity of proteinaceous material coagulation, heavy metal ion chelation, and physical properties of chitosan films [62]. Chitosan is considered a multifunctional polymer that can be used in food preservatives, tissue engineering, biocatalysis, and anticancer applications [21,68,71]. The main drawback of chitosan is poor solubility at physiological pH value 7.4 owing to the partial protonation of the amino groups [21,62]. Hence, chitosan can be further modified with processes such as grafting, cross-linking, composites, and substituent incorporation [21]. In energy storage applications, Ramkumar and Minakshi [72] reported the fabrication of cobalt molybdate modified by using chitosan cross-linked with glutaraldehyde as a cathode material in a hybrid capacitor. Cross-linking has the main benefit of enhancing the surface functionality of the modified electrode [73]. The degree of amorphosity in the composite was impacted by the cross-linking of chitosan and glutaraldehyde, which reduced particle size and enhanced the development of cluster-like particles that produced a capacitance that was about four times greater than as-prepared cobalt molybdate. The modified electrode also exhibited outstanding cycling stability with 97% coulombic efficiency (over 2000 cycles), indicating that chitosan gel adheres firmly to the molybdate moiety of cobalt molybdate.

## 3. Types of Cellulose

Cellulose is classified as the most abundant and renewable biopolymer found in nature [49]. It is present in biomass such as plants, bacteria, and animals (tunicates) [42,44]. Cellulose produced from biomass in nature is around 1 trillion tonnes annually, and this proves that it is an inexhaustible source of raw material [74]. Regardless of the sources, cellulose is a biopolymer composed of β-d-glucopyranose (glucose) monomers held by linear β (1–4) linkages and the repeating unit is named as cellobiose in which a dimer of glucose [46,75]. The term β is determined from the position of the ether oxygen which is located on the same side of the glucose rings with the hydroxyl groups [76]. In the glucose rings, the hydroxyl groups are arranged in equatorial positions, and they possess important features such as controlling the crystalline packing, stabilizing the glucose linkages, and determining the chemical reactivity of cellulose [44,76].

Other than abundant and renewable properties, cellulose is extensively used in its natural purified state or derivatives because it is cheap, environmentally friendly, biocompatible, and readily available [42,49,50]. Cellulose fibers, microcrystalline cellulose (MCC), cellulose nanofibers (CNF), nanocrystalline cellulose (NCC), cellulose hydrogels and aerogels, and cellulosic composites are the examples of cellulosic derivatives [43,51]. High surface area, low density, high aspect ratio, good mechanical properties, low cost, and adaptable surface properties contribute towards the utilization of cellulose in composites, polymers, synthetic fibers, and antibodies [42,77]. Cellulose and its derivatives are used as contaminant adsorbents and stabilizers for active particles in water treatments to remove organic and inorganic pollutants [78]. Their global market value is predicted to hit $1.08 billion by 2020, which benefits the pharmaceutical and food divisions [48]. Hence, the cellulose in biomass is a suitable carbon source to replace commercial synthetic applications.

### 3.1. Plant Cellulose

Plants are the primary reserves of cellulose [55,74]. Plants are categorized into wood (e.g., hardwood, softwood, cotton linter) and non-wood types (e.g., agricultural biomass) [48]. Agricultural biomass is in high abundance and readily available. Based on the structure of agricultural biomass, lignin is located in the outer plant cell wall and cellulose is located within the lignin shell along with hemicellulose [79]. Hemicellulose and lignin are mostly bonded with cellulose via hydrogen bonds or covalent bonds [79,80]. Lignin is has a resistance towards biological attack and stiffens the plant stem to protect it from external forces (e.g., wind), whereas hemicellulose supports the compatibility between lignin and cellulose [81].

In general, the plant cell wall is divided into primary and secondary walls. The secondary wall is responsible for the overall characteristics of plant fibers as cellulose is mainly located in the secondary wall [75]. The secondary wall consists of three layers where the middle layer mainly contributes to the mechanical properties of cellulose fibers [81]. Inside the middle layer of secondary plant cell wall, the cellulose fibrils impart rigidity and maximum tensile and flexural strengths upon their alignments along the length of plant fibers [47]. There are usually 30–100 cellulose molecules aligned helically in the extended cellulose chain configuration [81]. In short, cellulose helps to maintain the plant cell wall structure with its appealing mechanical properties [46].

Cellulose fiber exists in terms of cellulose chain groups inside the lignin matrix as it does not occur naturally as an isolated molecule [47,75]. The synthesis or isolation of cellulose particles require purification and mechanical treatments in order to remove matrix materials (e.g., hemicellulose and lignin) partially or completely, isolate the cellulose fibers, and promote uniform reactions in the subsequent treatments [43]. Agricultural biomass, or in other words, lignocellulosic biomass, undergoes chemical pretreatments (e.g., acid hydrolysis, alkali treatment, acid-chlorite treatment, organosolv treatment), biological pretreatments, and mechanical pretreatments (e.g., homogenization, grinding processes) [82,83]. The final cellulose product is greatly influenced by the concentration of chemicals, reaction time, and temperature.

Upon chemical pretreatment, acid hydrolysis is widely applied in the production of MCC and NCC [83,84]. Table 1 shows the preparation of MCC and NCC from various sources using different acid hydrolysis methods. Acid hydrolysis uses mineral acids such as sulfuric acid ($H_2SO_4$), $H_3PO_4$, and HCl [46]. For example, weak $H_2SO_4$ (below 4 wt%) hydrolyzes the polysaccharide hemicellulose completely into monosaccharide xylose by breaking the xylosidic bonds [54]. Strong $H_2SO_4$ hydrolyzes the amorphous regions of cellulose fibrils through the esterification of hydroxyl groups by sulphate ions, and yields highly crystalline nanosized cellulose particles in the remaining treated solution [82,83].

**Table 1.** Preparation of MCC and NCC from various sources using different acid hydrolysis methods.

| Cellulose Form | Cellulose Source | Acid Hydrolysis Method | MCC/NCC Yield (%) | α-Cellulose Content (%) | Crystallinity (%) | Thermal Stability (°C) | Ref |
|---|---|---|---|---|---|---|---|
| MCC | *Ensete glaucum* (Roxb.) Cheesman | 2.5 M HCl at 105 °C for 15 min | 33 | 99 | 53.41 | - | [85] |
| | Sweet sorghum | 7 wt% HCl at 40 °C for 90 min | 81.8 | 93.2 | 75.19 | - | [86] |
| | Kans grass | 5% (w/w) $H_2SO_4$ at 50 °C for 120 min | 83 | 83.33 | 74.06 | 338 | [87] |
| | Date seeds | 2.5 M HCl at 105 °C for 45 min | 12.51 | - | 70 | 352.52 | [88] |
| | Conocarpus fiber | 2.5 M HCl at 80 °C for 30 min | 27 | - | 75.7 | 408.5 | [89] |
| NCC | Rice husk | 4 M $H_2SO_4$ at 60 °C for 60 min | 95 | 95 | 65 | - | [90] |
| | OPEFB | 3 M HCl at 80 °C 120 min | 21 | 94.6 | 65 | 358.5 | [91] |
| | Jackfruit peel | 65% (w/w) $H_2SO_4$ at 37 °C for 60 min | 7 | 20.08 | 83.42 | - | [92] |
| | Olive fiber | 35 wt% $H_2SO_4$ at 40–50 °C for 60 min | 16.4 | 86.2 | 83.1 | 363.8 | [93] |
| | Rice husk | 64 wt% $H_2SO_4$ at 45 °C for 30 min | 35–37 | - | 82.8 | 286 | [94] |

For acid hydrolysis, prolonged time is necessary to achieve a complete reaction [95]. However, a prolonged reaction time can hydrolyze hemicellulose and some extent of cellulosic materials as cellulose degrades into water-soluble glucose molecules [54,95]. It was reported that microwave-assisted acid pretreatment minimized the reaction time remarkably [54]. The subsequent washing by water or NaOH was usually used to neutralize the pH for the treated cellulose [83]. The main drawback of acid hydrolysis is the generation of acid-containing effluent, which requires additional treatment before disposal in the environment [95]. This is because acid is corrosive and toxic, and it is extremely harmful to the environment. Hence, the overall effect of acid hydrolysis towards lignocellulosic biomass is influenced by acid–biomass ratio, acid concentration, and reaction temperature and time [54].

Alkali treatment and acid-chlorite treatment are mainly employed to remove lignin from lignocellulosic biomass [83]. Alkali treatment uses mediums such as NaOH, potassium hydroxide (KOH), calcium hydroxide ($Ca(OH)_2$), sodium carbonate ($Na_2CO_3$), and ammonia [54]. Lignin and silica can be dissolved in alkali treatment by breaking down uronic and acetic esters linkages, which causes cellulose swelling [96]. In addition, alkali treatment also hydrolyzes hemicellulose partially to reduce the crystallinity of cellulose, and increases the internal area and porosity of cellulose [54]. Better delignification effect is also obtained from alkali treatment as compared to acid hydrolysis [54]. The efficiency of the alkaline treatment can be improved by using the combination of both alkali and acidic treatments with low amounts, and it is more economical and environmentally friendly [84].

Besides, acid-chlorite treatment involves the combination of sodium chlorite ($NaClO_2$) and glacial acetic acid ($CH_3COOH$) [83]. The acidified $NaClO_2$ delignifies and bleaches the plant fibers until the product becomes white and free of lignin. Normally, the bleaching process is employed when incomplete delignification of the lignocellulosic materials occurs, which yields brown-colored cellulosic products [54]. The purpose of bleaching is to eliminate the remaining lignin and hemicellulose contents embedded in the obtained cellulosic products. The commonly used bleaching agents are chlorine and hypochlorite compounds (e.g., sodium hypochlorite (NaClO), $NaClO_2$) due to their economical production of high bright chemical pulps [97]. Modifications to the bleaching stage are made to deal with the generation of effluents containing carcinogenic and mutagenic chlorinated compounds after the bleaching process [54]. Table 2 shows different combined treatment methods to extract cellulose from various lignocellulosic biomass sources.

**Table 2.** Extraction of cellulose from lignocellulosic biomass using different combined treatment methods.

| Cellulose Source | Cellulose Extraction Method | Cellulose Content (%) | Crystallinity (%) | Thermal Stability (°C) | Ref |
|---|---|---|---|---|---|
| Rice straw | • 1.25% acidified $NaClO_2$ at 75 °C for 1 h<br>• 5 wt% KOH at room temperature for 16 h followed by 90 °C for 2 h | 88.5 | 58.12 | 358 | [98] |
| Oil palm frond | • 15 wt% NaOH at 150 °C and 7 bar for 1 h<br>• 10% $H_2O_2$ at 90–100 °C for 1 h | 91.33 | 77.78 | 366.8 | [99] |
| Sugarcane bagasse | • 10% (*v*/*v*) $H_2SO_4$ at 100 °C for 1 h<br>• 5% (*m*/*v*) NaOH at 100 °C for 1 h<br>• 5% (*v*/*v*) $H_2O_2$ and 0.1% $MgSO_4$ (in polypropylene bags) at 70 °C for 1 h | 89.12 | 56.19 | 360 | [100] |
| Wheat straw | • 0.5 mL $CH_3COOH$ and 1 g $NaClO_2$ in 80 mL water at reflux (oil bath at 80 °C) for 4 h<br>• 17.5% (*w*/*v*) NaOH at room temperature for 30 min | 81.4 | 66.6 | 385 | [101] |
| | • 1% (*w*/*v*) NaOH and 20% (*w*/*v*) $H_2O_2$ at 121 °C for 35 min | 79 | 66.87 | 360 | |
| *Agave gigantea* | • 5% (*w*/*v*) NaOH at 80 °C for 2 h<br>• NaOH/$CH_3COOH$/water (27 g/75 mL/1 L) and 1.7 wt% $NaClO_2$ at 80 °C for 1 h | 89.39 | 70.94 | 362.59 | [102] |

Another alternative delignification method for lignocellulosic biomass is organosolv treatment or organosolv pulping process. Organosolv treatment uses organic solvents such as ethanol, $CH_3COOH$, and acetone [96]. For instance, OPEFB was treated with a mixture of aqueous ethanol and diluted $H_2SO_4$ at 120 °C for 1 h, and followed by the treatment of diluted hydrogen peroxide ($H_2O_2$) at 50 °C for 4 h to remove lignin and hemicellulose [103]. The distillation of organic solvent could isolate the lignin content from the lignocellulosic materials by using a mixture of organic solvent and water [96].

Biological pretreatment is employed to produce cellulose with high purity and crystallinity [54]. Typically, the removal of protective layers (lignin and hemicellulose) is required to allow easier access to cellulose. Microorganisms (e.g., white rot fungi, soft rot fungi) attack both lignin and hemicellulose, while cellulase-less mutant is generated for the selective degradation of lignin over cellulose [96]. Even though biological pretreatment is energy-saving and requires no chemicals, it still suffers the main drawback such as low degradation rate and efficiency for lignin and hemicellulose [54,96].

Mechanical pretreatment is applied in the production of microfibrillated cellulose and nanofibrilated cellulose [82]. The obtained cellulose fibers break into smaller sizes through high pressure homogenizing or grinding processes. For instance, the cellulose fibers are consistently agitated with a high shear homogenizer for some time which results in nanofibrils [81]. Nevertheless, mechanical treatment is highly energy-intensive which leads to alternative methods such as enzymatic and acetylation treatments [82].

Lately, agricultural biomass is the most cited and it is a preferred substrate to isolate cellulose and its derivatives. The broad range of cellulose-based raw materials includes corn straw, OPEFB, rice husk, durian shell, wood, cotton, potato tubers, and soybean stock [43,51]. The average cellulose contents in wood, cotton, jute, flax, and ramie, are 40–50 wt%, 87–90 wt%, 60–65 wt%, 70–80 wt%, and 70–75 wt%, respectively [44]. Plant cellulose is mainly utilized in the textile, pulp and paper, packaging, and pharmaceutical industries for harvesting, processing, and handling [43]. Nevertheless, plant cellulose has disadvantages, such as being easily degraded at temperatures above 200 °C, and a high moisture content, which hinders the subsequent carbonization heat treatment [47]. Overall, the benefits of plant cellulose surpass its shortcomings and it is commonly chosen for mass production compared to bacterial cellulose due to lower costs [82].

### 3.2. Bacterial Cellulose

Bacterial cellulose is firstly reported as *Bacterium xylinum* by Brown in 1886 [104]. The bacterial cellulose is initially produced by bacterial strains named *Gluconacetobacter xylinus* (*G. xylinum*), previously known as *Acetobacter xylinum* (*A. xylinum*) [76]. Researcher Brown collected *G. xylinum* from a pellicle formed on the surface of beer [105]. *G. xylinum* can also be found in the fermentation of carbohydrates like rotten fruits and unpasteurized wine [106]. Other bacteria species such as *Agrobacterium*, *Pseudomonas*, *Rhizobium*, and *Sarcina*, are also able to produce bacterial cellulose [107]. Different bacteria species synthesize cellulose with various morphologies, structures, properties, and applications [108]. In comparison with other bacteria strains, *G. xylinum* is commonly used to produce bacterial cellulose commercially due to its relatively high productivity [109]. Regarding productivity, one bacterium is predicted to transform 108 glucose molecules per hour into cellulose [76].

Furthermore, *G. xylinum* is the most studied source and it will be used to demonstrate the proposed biochemical pathway from glucose to cellulose [107,110]. Cellulose is synthesized extracellularly, which is between the outer and cytoplasma membranes of *G. xylinum* [111]. The biosynthesis of bacterial cellulose requires cellulose synthase. The cellulose synthase is activated by the enzyme named cyclic diguanylmonophosphate (c-di-GMP) [107]. An individual *G. xylinum* cell could polymerize up to 200,000 glucose molecules per second into cellulose chains [112].

Looking into the formation of bacterial cellulose, an elongated cellulose chain is formed through the aggregation of about 6–8 glucose chains [111]. It is reported that there are approximately 50 to 80 pores arranged along the long axis of the *G. xylinum* cell [112]. These elongated cellulose chains or fibrils escape from the pores (diameter about 3 nm) on the surface of *G. xylinum* cell and are further assembled into ribbon-like microfibrils [106,111]. The bundles of ribbon-like microfibrils aggregate into a network of interwoven ribbons and finally form a bacterial cellulose membrane or pellicle [111]. Moreover, *G. xylinum* produces the pellicle on the surface of the liquid culture medium. The formed pellicle is a thick gel comprising ribbons of cellulose microfibrils and 99% water [107]. *G. xylinum* is an aerobic bacteria and it produces pellicle to protect itself from ultraviolet light [111]. The pellicle acts as a barrier to safeguard against other organisms and heavy-metal ions, and yet allows the diffusion of nutrients into the bacteria cell [105]. Therefore, the biosynthesis mechanism helps *G. xylinum* to reach the oxygen-rich surface for survival purposes.

The synthesis of bacterial cellulose focuses mainly on culturing methods and purification to promote cellulose microfibrillar growth and to eliminate the bacteria and other media, respectively [43]. Purification of bacterial cellulose normally serves the purpose of killing the bacteria and removing unwanted byproducts by using standard NaOH treatment [76]. The culture medium and culture conditions are the crucial factors in each type of the culturing method. In the culture medium, bacteria strain, carbon source, nitrogen source, nutrition, pH, and oxygen delivery affect the bacteria growth and properties of bacterial cellulose [108,113]. The optimum pH value of the culture medium ranges from 4 to 7 [76]. It is reported that bacteria perform efficiently in culture media containing rich carbon source and limited nitrogen source [113]. Meanwhile, the bacteria synthesize cellulose ribbons containing mixtures of I$\alpha$ and I$\beta$ lattices under different culture conditions including stirring, temperature, and additives. The I$\alpha$/I$\beta$ ratio and width of cellulose microfibrils can be changed without the presence of additives [76].

The common methods to prepare bacterial cellulose are static, agitated/shaking, and bioreactor cultures [108]. For static culture method, bacterial cellulose is aerobically synthesized in a thick cellulosic surface mat (pellicle) on the surface of the culture medium [112,114]. The overall yield of bacterial cellulose pellicle or hydrogel sheets depends on the surface area of the gas–liquid interface [115]. The formation and growth curve of bacterial cellulose is identical despite different culture media. The bacterial cellulose microfibers accumulate to form a hydrogel sheet structure. The bacteria strains help to form and grow bacterial cellulose hydrogel sheets along with the consumption of nutrients. Glucose and $CH_3COOH$ are the nutrients required for the bacteria strains [108]. When the bacterial cellulose gel sheet thickens, the oxygen supply is limited to the bacteria strains which inhibit the further synthesis of bacterial cellulose. As a result, the produced bacterial cellulose has high porosity. A long culture time and intense labor force cause low productivity of bacterial cellulose [116]. Static culture is the standard method to synthesize bacterial cellulose at lab-scale due to its simplicity and requires low shear force [108].

Agitated/shaking culture is introduced to promote bacteria cell growth [76]. Agitated/shaking culture provides shear force to the culture medium during the synthesis of bacterial cellulose. Upon shaking or agitation, forced aeration improves the respiration of bacteria to enhance the synthesis of bacterial cellulose [116]. Instead of forming hydrogel sheet structures, small irregular shaped (sphere-like) bacterial cellulose pellets are usually formed throughout the medium by continuously mixing with oxygen in the shaking culture method [117]. The size, geometry, and internal structure of the bacterial cellulose pellets are affected by the shaking or rotational speed. It is reported that hollow sphere-like bacterial cellulose pellets were obtained at 150 rpm whereas solid sphere-like bacterial cellulose pellets were obtained at 125 rpm [118]. The bacterial cellulose produced from agitated/shaking culture method has a lower degree of polymerization, crystallinity, and mechanical properties as compared to static culture [115]. In spite of increased oxygen delivery, agitated/shaking culture has lower productivity of bacterial cellulose than static culture due to the generation and accumulation of cellulose-negative mutants [116]. Only bacterial species (e.g., *G. xylinum*) that are resistant towards the mutant generation effect could result in higher productivity of bacterial cellulose [117].

Basic static and agitated/shaking cultures could not provide uniform mixing and oxygen delivery in the culture medium [117]. Their batch mode processes prevented the addition of supplementary nutrients which affects the mass production of bacterial cellulose. The bioreactor culture method uses specially designed reactors along with high-speed agitation. Small-sized bacterial cellulose granules are formed by rapid rotation speed and moving shafts within the bioreactors [117]. The examples of bioreactor are stirred tank bioreactor, airlift bioreactor, and rotating disc bioreactor. High energy consumption of stirred-tank bioreactors produce fibrous bacterial cellulose suspensions with a low degree of polymerization, low crystallinity, and low elastic modulus in comparison with bacterial cellulose hydrogel sheets, owing to high agitation and control of oxygen transfer [119]. The airlift bioreactor provides adequate oxygen supply to the culture medium and it requires significantly lower energy consumption than the stirred-tank bioreactor [120]. Rotating disc bioreactor produces a homogenous bacterial cellulose structure where the circular discs rotate to promote continuous interaction with air and liquid media [108]. The bacteria attach at the surface of the circular discs and absorb both the nutrients inside the culture medium and oxygen at the surface of the culture medium to synthesize bacterial cellulose [114].

The interesting properties of bacterial cellulose, such as outstanding mechanical properties, high moisture permeability, and biocompatibility, allows it to be used in different applications especially in the bioengineering field [121]. Bacterial cellulose fibers are used as reinforcing agents in composite due to their inert property [81]. The promising biocompatibility expanded the use of bacterial cellulose as a commercial wound care dressing material. Bacterial cellulose and its derivatives are also potential materials for scaffolds, drug-delivery systems, membrane, and filter materials [122]. Although bacterial cellulose has many advantages, it is still very expensive to produce in mass production [110]. The usage of various waste media and carbon sources produces bacterial cellulose with similar physicochemical properties as compared to commercial H–S media [108]. The promising waste materials allow the production of bacterial cellulose to be more economical and environmentally friendly. Hence, future research is set to develop commercial production of bacterial cellulose [122].

### 3.3. Comparison between Plant Cellulose and Bacterial Cellulose

Bacterial cellulose has an identical molecular structure to plant cellulose [119]. However, the chemical and physical properties of bacterial cellulose are different from plant cellulose. Bacterial cellulose has a higher purity which is pure or nearly 100% of cellulose content compared to plant cellulose (60–70%) [113]. The reason is that plant fibers normally contain polymers such as lignin, hemicellulose, and pectin, and also functional groups such as carbonyl and carboxyl, that are bonded with plant cellulose upon isolation and purification processes [106]. Other than the hydroxyl groups, bacterial cellulose is free of any other biopolymers or functional groups and eventually it does not require additional purification steps.

Cellulose fibrils in plant cellulose are formed within the plant cell wall matrix, whereas cellulose fibrils in bacterial cellulose are formed extracellularly and they are metabolically inert [112]. The extracellular synthesis and nanosized bacterial cellulose promote a stronger hydrogen bonding between cellulose fibrils than plant cellulose [81]. The degree of polymerization of bacterial cellulose and plant cellulose are 2000–6000 and 13,000–14,000, respectively [107]. The crystallinity of plant cellulose nanofibers ranges from 36 to 91% [123]. In comparison with plant cellulose, bacterial cellulose on average has a higher crystallinity of up to 90% with dominant I$\alpha$ crystal structures [124]. The higher thermal stability of bacterial cellulose is attributed to its high purity and crystallinity [113]. This allows the treated bacterial cellulose to have a maximum decomposition temperature of 350–355 °C [76].

Furthermore, the hydroxyl groups located on the surface of plant cellulose and bacterial cellulose contribute to their hydrophilicity properties. The larger surface area of bacterial cellulose imparts a higher liquid loading capacity than plant cellulose. Bacterial cellulose acts as a hydrogel and has higher water absorbing and holding capacities because it contains high water content, which is 90% and above [77,110]. The porosity of plant cellulose is improved by chemical treatment. The dense 3D network of fibrils assembled in the structure of bacterial cellulose forms porous sheets [114]. The random orientations in the 3D network are caused by the biosynthesis of bacterial cellulose [76]. In addition, both tensile strength (20–300 MPa) and Young's modulus of bacterial cellulose (sheet: 20,000 MPa; single fiber: 130,000 MPa) are more notable than that of plant cellulose [108]. With these properties, bacterial cellulose acts as a suitable thermal stabilizer in resins [113].

### 3.4. Other Cellulose Types

Cellulose can also be obtained from algae [125]. Algae are the fastest growing plant on Earth and they are branched into macroalgae (e.g., seaweeds) and microalgae [126]. With the unlimited and free sunlight from the sun, they grow rapidly by transforming solar energy into biomass effectively via photosynthesis. Macroalgae are also differentiated in colors based on their respective natural chlorophylls and pigments such as green seaweed (Chlorophyta), red seaweed (Rhodophyta), and brown seaweed (Phaeophyta) [127]. Cellulose is found in the cell walls of algae and it plays an important role as the building block of structural support in algae [128]. Other than cellulose, the cell walls of algae also comprise mannans, xylans, and sulfated glycans [126]. The functional groups of carboxyl, hydroxyl, amino, and sulfate can be found in these components [128]. They are responsible for the adsorption capability of algae in treating heavy metals [129].

The biosynthesis of cellulose mostly occurs at the plasma membrane of algae, except for those algae species that synthesize cellulose scales [130]. The linear and rosette-like terminal complexes are responsible for the biosynthesis, polymerization, and crystallization of cellulose [130]. Moreover, they play a role in assembling cellulose microfibrils [131]. For instance, *Valonia algae* contain large terminal complexes with nearly 10 catalytic sites to produce about 10 nm microfibrils [132].

The treatment methods for algal cellulose sources essentially incorporate culturing methods and purifications to eliminate the algae cell wall matrix [43]. Similar to plant cellulose and bacterial cellulose, NCC can be extracted from algae by utilizing acid hydrolysis, enzymatic hydrolysis, and mechanical treatments [133]. It was reported that the NCC produced from red algae waste through acid hydrolysis has high mechanical performance and good transparency [134]. Since enzymatic hydrolysis is more environmentally friendly and achieves a higher glucose yield than acid hydrolysis, it is more preferable to treat algal biomass for bioethanol production [126]. Algae and its derivatives have good biocompatibility and biodegradability. They are potential materials to construct hybrid and composite materials. In biomedical applications, algae-based polyesters are used as scaffolds and as controlled release of pharmaceutical agents [135].

Other than that, the animal source for cellulose is tunicate [136]. The name Tunicata is derived from a special integumentary tissue called a tunic, which encloses the whole epidermis of the animal cell [137]. The subphylum Tunicata is categorized into three classes which are *Ascidiacea*, *Thaliacea*, and *Appendicularia*. *Ascidiacea* (sea squirts) and *Thaliacea* contain tunics while *Appendicularia* does not contain tunic and yet produces cellulosic materials [136]. Tunic mainly consists of polysaccharides especially tunicin, in other words, tunicate cellulose, and proteins (e.g., collagen, pectin) [138]. The tunic is responsible for phagocytosis, pigmentation, colonial allorecognition, bioluminescence, photosymbiosis, innate immunity, chemical defense, tunic contraction, and impulse conduction in terms of biological functions [139].

Tunicate cellulose is chemically identical to both plant cellulose and bacterial cellulose. It is formed by the cellulose synthase which is found in the plasma membrane of the epidermal cells [136]. Besides cellulose biosynthesis, the cellulose synthase also contributes to the proper formation of tunic tissues and metamorphic events [140]. Similar to algae cellulose, cellulose is responsible for the skeletal structure in the tunic tissues. The hundreds of cellulose microfibril bundles are deposited in a multilayer pattern which is parallel to the epidermis surface [136]. The shape and dimensions of the cellulose microfibrils are influenced by their respective species. In the Ascidiacea class, mostly ascidians are reported to possess cellulose I microfibrils in the tunic tissues [137].

The synthesis of tunicate cellulose includes the isolation of mantel from the animal and the isolation of cellulose fibrils from the protein matrix [43]. In a simple procedure, the obtained tunicate tunic will be treated with acid hydrolysis, kraft cooking (involving alkaline treatment), and bleaching [136]. The purification of tunicate cellulose uses alkali solutions (e.g., NaOH and KOH) and acidic solutions (e.g., $CH_3COOH$, nitric acid) at elevated temperatures [136]. The tunicate cellulose also can be treated with strong acid hydrolysis to prepare NCC. The collected tunicate NCCs had lengths of 500–3000 nm, widths of 10–30 nm, and aspect ratios of 10–200 [141]. Thus, tunicate cellulose is another promising material for the preparation of NCC and composite film applications [142].

Tunicate cellulose possesses a high specific surface area (150–170 $m^2/g$), high crystallinity (95%), and a reactive surface containing hydroxyl groups which promote good mechanical properties [143]. Modified NCCs are widely used in the biomedical engineering field, such as in scaffolds and biomarkers. The tunicate NCCs with high aspect ratios are more difficult to detach from the cell surface compared to smaller cotton NCCs during the inhalation studies [144]. Furthermore, the tunicate NCC membranes fabricated by using the vacuum-assisted self-assembly method had high mechanical strength, distinguished pH- and temperature-stability, good cycling performance, and achieved highly efficient separation of oily water [145]. Therefore, tunicate species like *Ciona intestinalis* could be farmed at a large scale to produce commercial tunicate cellulose for the production of chemicals, materials, and biofuels [136].

## 4. Photocatalyst Nanomaterials

$TiO_2$, ZnO, g-$C_3N_4$, and graphene are the common semiconductor photocatalysts applied in the degradation of organic pollutants. In general, a semiconductor photocatalyst consists of a band gap that separates the valence band and conduction band. The band gap energy determines the applicability of a semiconductor in photocatalysis [146]. The semiconductor photocatalysts with wide band gap energies depend on the electron excitation by obtaining additional energy from ultraviolet (UV) light radiation. The electrons excited from the valence band to the conduction band induce charge separation. Hence, the electron–hole pairs are formed to engage in the redox reactions for the degradation of organic pollutants [147]. Semiconductor photocatalyst often faces the significant drawback of the rapid recombination of the electron–hole pairs [148]. This hinders the generation of hydroxyl radicals (•OH) which play a crucial part in photocatalytic degradation.

### *4.1. Titanium Dioxide ($TiO_2$)*

The application of $TiO_2$ electrode was first prepared in 1967 and it had demonstrated heterogeneous photocatalytic oxidation through the splitting of water molecules under UV light radiation [149]. To date, $TiO_2$ is the most studied semiconductor photocatalyst for the removal of dyes and phenolic compounds from wastewater [150]. This is because of its low toxicity, low cost, high chemical stability, and high thermal stability [151]. The high stability of $TiO_2$ allows it to stand out from other semiconductors such as gallium phosphide and cadmium sulfide, which generate toxic byproducts [152]. $TiO_2$ is applied in commercial applications and products (e.g., cosmetics, catalysts, desiccant) due to inert and long-term photostability properties [153].

Besides that, $TiO_2$ also possesses high UV absorption and superhydrophilicity which is important for the photocatalytic degradation of organic pollutants and solar fuel production [154]. $TiO_2$ is thermodynamically efficient versus normal hydrogen electrode (NHE) at pH 7 [155]. The photogenerated electrons demonstrate higher reduction strength due to the more negative conduction band potential of $TiO_2$ which is −0.5 V. Meanwhile, the generated holes demonstrate higher oxidation strength due to the more positive valence band potential of $TiO_2$ which is + 2.7 V.

Despite these potentials, there are some limitations of $TiO_2$ that affect the performance of $TiO_2$ in the decomposition of organic chemical compounds. Firstly, the large band gap of $TiO_2$ limits the application of $TiO_2$ in certain photocatalytic degradation processes. It can only absorb about 4–5% of the solar spectrum. Secondly, the fast recombination of photogenerated electron–hole pairs lowers the photocatalytic efficiency of $TiO_2$ [156].

Thirdly, the weak adsorption of organic pollutants on the surface of $TiO_2$ resulted in the poor affinity of $TiO_2$ and slower rate of photocatalytic degradation [150]. Fourthly, $TiO_2$ nanoparticles aggregate easily, owing to the large surface area-to-volume and surface change in certain media [157]. The aggregation of $TiO_2$ nanoparticles forms aggregates or clusters with sizes hundreds of times bigger than their initial sizes. These $TiO_2$ aggregates obstruct active sites from exposure to light radiation and eventually inhibit photocatalytic activity [150]. Lastly, the recovery of $TiO_2$ nanoparticles remains a revolving issue in wastewater remediation. To prevent the release of free nanoparticles into the water, $TiO_2$ is synthesized in thin films or immobilized on substrates [147].

Next, the band gap of a photocatalyst greatly influences its catalytic performance. Figure 4 illustrates the band gap energy of different photocatalysts. The redox potential level of adsorbate species and band gap energy govern the possibility and rate of charge transfer [158]. The conduction band energy of $TiO_2$ is located slightly higher than the reduction potential of oxygen molecules ($O_2$), which makes it easier for the electron migration from the conduction band of $TiO_2$ to $O_2$ [147]. Meanwhile, the valence band energy of $TiO_2$ is located lower than the oxidation potential of most electron donors. This allows the transfer of oxidative holes to the •OH radicals adsorbed on $TiO_2$ surface and eventually enhances the redox reactions of pollutant degradation.

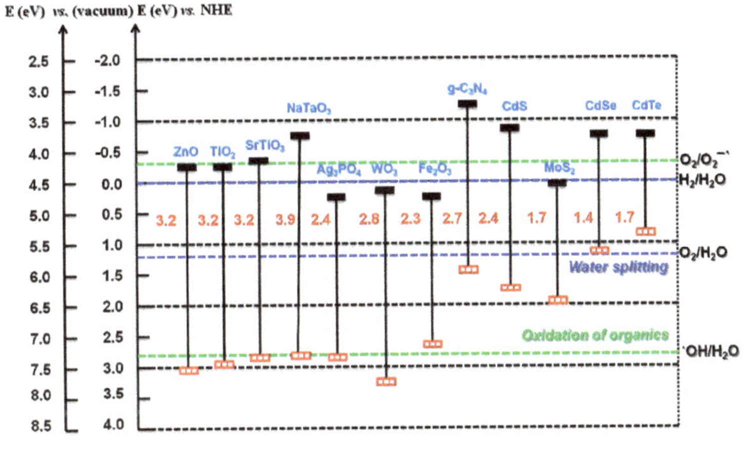


**Figure 4.** Band gap energy of various semiconductor photocatalysts. No special permission is required to reuse all or part of article published by MDPI [159].

$TiO_2$ is an n-type semiconductor, and it consists of three crystal structures which are anatase, brookite, and rutile as shown in Figure 5. In these crystal structures, the titanium atoms coordinate with six oxygen atoms to form $TiO_6$ octahedron units [151]. The 3d orbitals of titanium atoms form the lower part of the conduction band of $TiO_2$, while the overlapping of 2p orbitals of oxygen atoms forms the valence band of $TiO_2$ [155]. The band gap energies of anatase, rutile, and brookite are determined around 3.2 eV, 3.0 eV, and 3.1 eV, respectively [160]. The electron excitation under UV light radiation at wavelengths below 400 nm is necessary due to the large band gap energies of $TiO_2$ crystal structures [152]. These crystal structures or polymorphs influence the photocatalytic activity of $TiO_2$. The photocatalytic activity of a semiconductor is mainly dependent on the light absorption, reduction and oxidation rates, and electron–hole recombination rate [161].

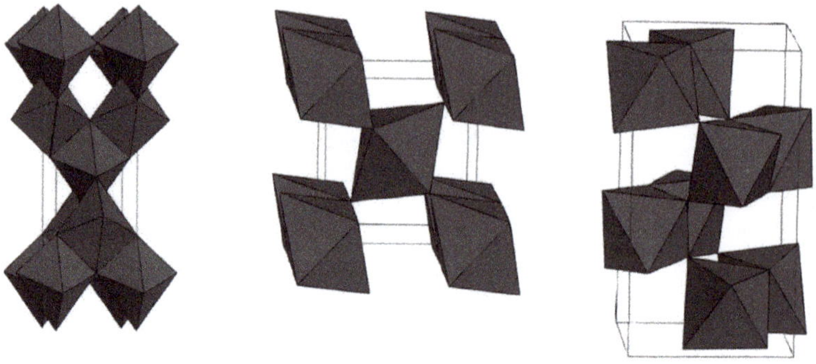


**Figure 5.** Crystal structures of $TiO_2$. Reproduced with permission from [152].

In the anatase crystal structure, the $TiO_6$ octahedras are shared or connected through their corners (vertices) [151]. In other words, the octahedras share four edges which results in the tetragonal structure of anatase [162]. In anatase $TiO_2$, the Ti−Ti lengths are longer and the Ti−O lengths are shorter compared to rutile $TiO_2$ [161]. The anatase crystal structure is normally found in solution-phase preparation methods of $TiO_2$ like sol-gel process [152,160]. Anatase is only stable at low temperatures. It is reported that anatase is thermodynamically stable in equivalent-sized $TiO_2$ nanoparticles with sizes smaller than 11 nm [161]. Upon, calcination, anatase will transform to rutile at temperatures above 600 °C.

Next, the defects on the $TiO_2$ affect the reduction activity. The defects tend to trap the electrons and lower the probability to recombine with holes. The depth of electron trap in anatase, rutile, and brookite are <0.1 eV, ~0.9 eV, and ~0.4 eV, respectively [163]. In comparison with brookite and rutile, anatase has the smallest depth of electron trap that indicates the presence of a larger number of free or shallowly trapped electrons. These electrons have higher reactivity than deeply trapped electrons. The lifetime of photogenerated electrons is prolonged. Consequently, anatase $TiO_2$ has the highest photocatalytic reduction.

Furthermore, anatase $TiO_2$ is known as an indirect band gap semiconductor. Looking into this, anatase has a longer lifetime of photogenerated electrons and holes as compared to brookite and rutile. Apart from this, the photogenerated electrons and holes in anatase can migrate easily from the innermost to the outermost surface of $TiO_2$ due to lighter effective mass [164]. Additionally, the dominant (101) and (001) facets contribute to the high photocatalytic activity of anatase $TiO_2$. The (001) facet contains rich under-bonded titanium atoms and a large Ti−O−Ti bond angle [155]. Anatase $TiO_2$ usually exhibits higher photocatalytic activity compared to brookite and rutile $TiO_2$ due to longer lifetime and lighter effective mass of photogenerated electrons and holes [164].

Rutile has octahedras sharing two edges to form a tetragonal crystalline structure which is identical to anatase [165]. In the rutile crystal structure, each octahedron is connected with eight similar octahedrons and this differentiates it from anatase where each octahedron is connected with ten similar octahedrons [155]. Rutile is thermodynamically stable in equivalent-sized $TiO_2$ nanoparticles with sizes larger than 35 nm [161]. Rutile is stable at high temperature. This is proven where anatase and brookite transform into rutile at temperatures above 600–700 °C [166].

As explained earlier, rutile has the largest depth of electron trap, and this signifies that more electrons are deeply trapped at the defects of $TiO_2$. The lifetime of holes is prolonged because the trapped electrons are unable to recombine with the holes [163]. Consequently, rutile $TiO_2$ has the highest photocatalytic oxidation due to the higher availability of electron-scavenger. In addition, the smaller band gap energy of rutile $TiO_2$ allows it to seize photons to produce electron and hole pairs that can be further used by anatase $TiO_2$ [150]. It was reported that the heterojunctions of anatase/rutile $TiO_2$ demonstrated higher photocatalytic activity than pure anatase or rutile $TiO_2$ [167]. This was due to their matched band levels that restrain the recombination of photogenerated charge carriers.

In the brookite crystal structure, titanium atoms are located at the center, while oxygen atoms are located at each corner [168]. The $TiO_6$ octahedra share three edges and form an orthorhombic crystalline structure [162]. Upon the precipitation in an acidic medium at low temperature, brookite is usually formed as the byproduct [152]. It is reported that brookite is thermodynamically stable in equivalent-sized $TiO_2$ nanoparticles with sizes between 11 and 35 nm [161]. Similar to anatase, brookite will also transform to rutile upon calcination at high temperatures because it is metastable.

Brookite is a direct band gap semiconductor [164]. Based on the exposed surfaces of (201) and (210) facets, it can become oxidative and reductive surface facets, respectively. The moderate depth of the electron trap in brookite $TiO_2$ resulted in both reactive electrons and holes. This explains the higher activity of brookite $TiO_2$ in some photocatalytic reactions [163]. However, the detailed behavior of photogenerated electrons and holes in brookite $TiO_2$ is not completely established yet [169]. Pure phase brookite is uncommon, difficult to prepare, and hence the discussion of brookite regarding its photocatalytic properties is limited.

### 4.2. *Zinc Oxide (ZnO)*

Besides of $TiO_2$, ZnO is also a favorable semiconductor photocatalyst. The biocompatibility, excellent physicochemical stability, low production cost, great photocatalytic property, and high photosensitivity of ZnO promote its utilization in various energy conversion and photocatalytic activities [170,171]. In addition, ZnO displays strong luminescence in the green-white spectrum region which is appropriate for phosphor applications [172]. ZnO is an n-type semiconductor that has a direct and wide band gap in the UV region. The band gap energies of ZnO at low temperature and room temperature are 3.44 eV and 3.37 eV, respectively [172]. ZnO possesses a higher free-exciton binding energy (60 meV) than gallium nitride (25 meV). This shows that the excitonic emission in ZnO can occur at temperatures higher than room temperature. The photocatalytic degradation of ZnO is identical to $TiO_2$ whereby the generated charge carriers produce free radicals to degrade organic pollutants under UV irradiation [173].

However, the application of pure ZnO is limited by some drawbacks. Firstly, ZnO requires expensive UV light for the wide band gap excitation. Secondly, the fast recombination of charge carriers in ZnO impedes the migration of charge carriers towards the outer surface of ZnO that leads to the retardation of degradation process [170]. Thirdly, the aggregation of ZnO particles reduces its dispersion and blocks the exposed active facets of ZnO [174]. Fourthly, ZnO faces photocorrosion under UV irradiation [175]. The ZnO powder dissolves in strong acidic and alkaline solutions, and eventually passivates to produce an inert outer layer of zinc hydroxide [174]. This determined that pure ZnO does not have sufficient activity under solar energy [170]. Lastly, the recovery of ZnO nanoparticles from post-treatment effluents is necessary owing to its toxicity. Upon respiratory and digestion uptake, the ZnO nanoparticles increase blood viscosity and heighten oxidative stress and cellular inflammatory response in the mammalian body [176].

ZnO is classified as an II-VI compound semiconductor whereby its ionicity remains at the cut-off point between covalent and ionic semiconductors [177]. Figure 6 shows that ZnO consists of three common crystal structures, which are rocksalt, zinc blende, and wurtzite. In these crystal structures, each anion is connected with four cations located at the edges of a tetrahedron, and vice versa [177]. The rocksalt structure of ZnO is considered rare and it can be obtained at high pressure [178]. Rocksalt has a cubic structure which is similar to zinc blende. The crystal structure of zinc blende is metastable. There are four atoms found in each unit cell of the zinc blende crystal structure. Each atom of group II is bonded with four atoms of group VI tetrahedrally and vice versa. It is noticeable that the tetrahedral coordination of zinc blende is identical to wurtzite.

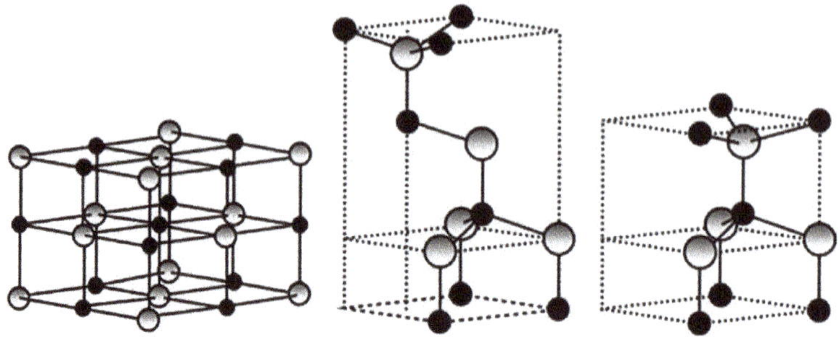


**Figure 6.** Crystal structures of ZnO: (**a**) rocksalt, (**b**) zinc blende, and (**c**) wurtzite. Reproduced with permission from [178].

Wurtzite is the most thermodynamically stable phase among the three crystal structures of ZnO. Wurtzite has a hexagonal structure at ambient conditions [178]. The stacking sequence of close-packed diatomic planes differentiates the crystal structure of wurtzite from zinc blende [177]. The crystal structure of zinc blende comprises triangularly arranged atoms in the close-packed (111) planes where the stacking order is in terms of "AaBbCcAaBbCc" along the (111) direction. On the other hand, the crystal structure of wurtzite comprised of triangularly arranged alternating biatomic close-packed (0001) planes where the stacking order is in terms of "AaBbAaBb" along the (0001) direction. The uppercase and lowercase letters indicate two distinct types of constituents. The low symmetry structure of wurtzite combines with a large electromechanical coupling to promote stronger piezoelectric and pyroelectric properties [172]. Thus, ZnO exhibits high piezoelectric constant, and it is commonly applied in sensors, transducers, and actuators.

### 4.3. *Graphitic Carbon Nitride (g-$C_3N_4$)*

In recent years, metal-free semiconductor photocatalysts received great attention due to their unique physicochemical properties. Carbon nitride semiconductors exist in different allotropes such as α-$C_3N_4$, β-$C_3N_4$, g-$C_3N_4$, and cubic-$C_3N_4$ [179]. Here, g-$C_3N_4$ is classified as the most stable allotrope under ambient conditions and this is proven with its popularity in photocatalytic applications [180]. This polymeric g-$C_3N_4$ has recently been reported as an easily available and simple photocatalyst for water splitting reactions without the presence of noble metals [181]. This outcome triggered the broad investigations of g-$C_3N_4$ in photocatalytic degradation of pollutants, carbon dioxide reduction, and hydrogen evolution [182]. The easy preparation, low cost, abundance in nature, and sustainable properties give rise to the widespread application of g-$C_3N_4$ [180].

Next, the tunable electronic band structure of g-$C_3N_4$ by nanomorphology or doping modifications improves its utilization in solar energy conversions (e.g., photoelectrochemical cells) [183]. During the photocatalytic process, the zero evolution of nitrogen gas ($N_2$) determined the strong covalent bonding in pure g-$C_3N_4$ and exhibited high chemical stability of g-$C_3N_4$ [182]. In comparison with $TiO_2$ and ZnO, g-$C_3N_4$ has a medium band gap of 2.7 eV, which enables it to absorb light at about 450–460 nm [184]. Under solar irradiation, the more negative conduction band potential (−1.3 V) and the more positive valence band potential (+1.4 V) of g-$C_3N_4$ versus NHE at pH 7 demonstrate its thermodynamic efficiency in participating redox reactions [155].

Although g-$C_3N_4$ contributes outstanding merits, it encounters some restrictions in photocatalytic activities. Firstly, the activity of g-$C_3N_4$ is often restricted by the fast recombination of charge carriers. Secondly, the small specific surface area (<10 $m^2/g$) of g-$C_3N_4$ gives rise to its limited light-harvesting capability [185]. The reason is that a larger surface area indicates more reactive sites and higher light-harvesting capability. Thirdly, g-$C_3N_4$ is well known for its poor crystallinity and surface defects, which resulting in low conductivity in photocatalytic activities [182]. The crystallinity of g-$C_3N_4$ can be enhanced via the ionothermal approach [186]. Fourthly, the bulk g-$C_3N_4$ produced from conventional methods comprises multiple layers of two-dimensional (2D) counterparts [187]. The low specific surface area and irregular morphology of bulk g-$C_3N_4$ restrict its photocatalytic power [187]. Furthermore, pristine g-$C_3N_4$ also possesses high exciton binding energy as compared to inorganic photocatalysts [186]. Lastly, the organic semiconductor g-$C_3N_4$ encounters charge transport issues owing to the presence of grain boundaries [179].

Looking into the crystal structure of g-$C_3N_4$, it is an n-type and indirect semiconductor that exhibits 2D and 3D graphitic carbon [188]. The pyridinic and graphitic nitrogen in these structures are enriched with nitrogen content. g-$C_3N_4$ is identified as a conjugated polymeric arrangement and is composed of s-triazine or tri-s-triazine monomers interconnected via tertiary amines as shown in Figure 7 [179]. The combination of aromatic s-triazine rings and conjugated 2D polymer of s-triazine construct the $\pi$-conjugated planar layers which are similar to graphite [183]. Based on the density functional theory, it is determined that the carbon atoms are the ideal sites for the proton reduction to hydrogen, whereas the nitrogen atoms are the ideal sites for the water oxidation to oxygen [189]. The lone pair of nitrogen and the $\pi$ bonding electronic states contribute to the stability of the lone pair state which results in the unique electronic structure of g-$C_3N_4$ [190].

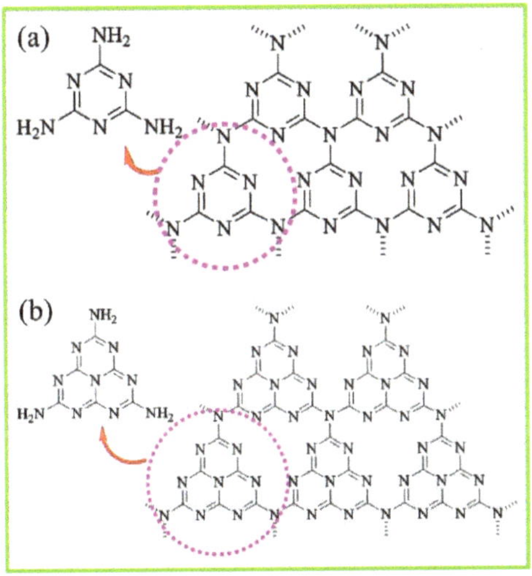


**Figure 7.** Structures of g-$C_3N_4$: (**a**) s-triazine based, and (**b**) tri-s-triazine based. Reproduced with permission from [184].

Moreover, g-$C_3N_4$ can withstand thermal heat up to 600 °C in the air [183]. The high thermal and chemical stability of g-$C_3N_4$ is contributed by the high degree of condensation and the structure of tri-s-triazine ring. This polymeric g-$C_3N_4$ can be readily synthesized by thermal condensation with low-cost nitrogen-rich precursors (e.g., melamine, cyanamide, urea) [189]. The strong covalent bonds are responsible for the honeycomb arrangements of the atoms in the layers while the weak Van der Waals forces are responsible for stacking between the 2D sheets [179]. The Van der Waals forces also give rise to the good chemical stability of g-$C_3N_4$ in common solvents (e.g., water, alcohols, diethyl ether) [183].

### *4.4. Graphene*

Geim and Novoselov were awarded the Nobel Prize for investigating the remarkable properties of graphene in 2010. The large theoretical specific surface area of graphene

(2630 $m^2/g$) gives rise to its strong adsorption capacity [191]. Graphene possesses good electron mobility that is 200 times higher than silica (1000 $cm^2/Vs$) [192]. The fast-moving charge carriers do not scatter around under the presence of metal impurities while traveling through thousands of interatomic distances. This shows that graphene is suitable to be applied in high-speed operations due to power saving properties [192].

Furthermore, the cost of graphene production is low and it interacts strongly with transitional metals [193]. Graphene acts as a conductive support and an alternative electron sink whereby it accepts, stores, and shuttles photogenerated electrons [191]. This corresponds to the high conductivity and promising work function of graphene. Eventually, graphene exhibits high electrical conductivity (2000 S/m), high thermal conductivity (5000 W/mK), and good environmental compatibility [191]. The largest surface area, fastest electron mobility, highest conductivity, and outstanding electronic capability of graphene make it stand out from other materials like carbon nanotubes, graphite, and common metals.

As graphene is the mediator of electron transport, it prolongs the lifetime of photogenerated charge carriers and improves both the extraction and separation of charges [191]. This further elevates the photocatalytic activity of graphene-based composites. In addition, graphene serves as a building platform for the epitaxial growth of semiconductor nanostructures [171]. It also impedes the aggregation of these nanostructures which results in the improvements of exposed surface area and photocatalytic performance. Thus, graphene and its derivatives are widely applied in areas involving optical electronics, energy conversions, energy storages, photocatalysis, and photosensitizers [194].

Nevertheless, some limitations interrupt the application of graphene. Firstly, graphene faces the common problem of rapid recombination of electron–hole pairs. Secondly, graphene is a semi-metallic material and it has a zero band gap [195]. The zero band gap is formed from the contact between the $\pi$ (bonding) and $\pi^*$ (antibonding) orbitals and the Brillouin zone corners. Alternatives like chemical doping could disrupt the lattice symmetry in graphene and eventually open a band gap with the help of foreign atoms. Thirdly, graphene-based powders tend to agglomerate easily and self-restack to form graphite [196]. The restacking phenomenon is irreversible and it is caused by the $\pi$–$\pi$ stacking and Van der Waals forces located between the graphene sheets [197]. This causes the reduction of exposed surface area and ionic pathways [198]. Fourthly, pristine graphene is insoluble and it does not disperse uniformly in common solvents [194,199]. Lastly, graphene and its derivatives also encounter the familiar issues of recovery and separation after pollutant treatments.

Graphene consists of a monolayer of $sp^2$-bonded carbon atoms [200]. It displays a hexagonal honeycomb lattice that is composed of two equivalent carbon sublattices [191]. Graphene is the basic structure of graphitic carbon allotropes as shown in Figure 8 [201]. The zero-dimensional fullerene is formed from wrapping a graphene sheet into a buckyball. The one-dimensional carbon nanotubes are formed from rolling graphene sheets into cylinder structures. The 3D graphite is formed from the stacking of graphene sheets as mentioned earlier. Monolayer and bi-layer graphene consists of one hole type and one electron type. The few layer graphene consists of three to nine graphene sheets whereas the multi-layer graphene is made up of ten or more graphene sheets [202]. The delocalized $\pi$ bonds govern the graphene sheet-to-sheet interactions [203]. They allow the delocalization of $\pi$ electrons along the basal plane.

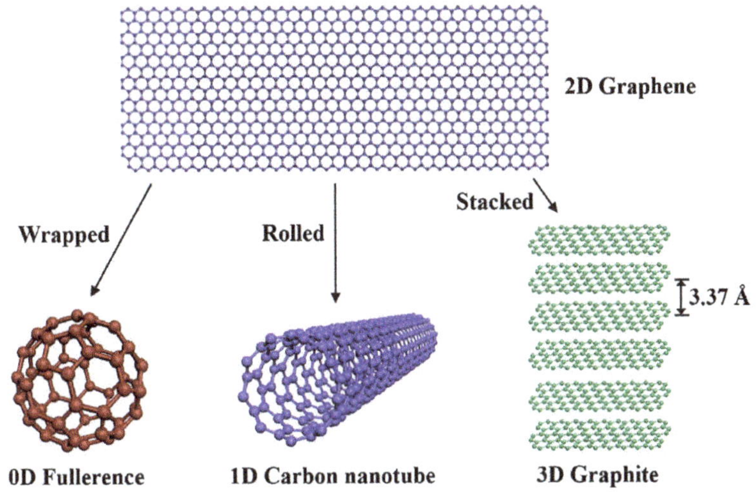


**Figure 8.** Structures of graphene and its graphitic carbon allotropes. Reproduced with permission from [201].

Based on the structure of graphene, the $\pi$ electrons found between the two adjacent carbons within its neighboring $2p_z$ orbitals are related to the delocalized $\pi$ and $\pi^*$ bands [191]. Here, the delocalized $\pi$ band forms the highest occupied valence band while the $\pi^*$ band forms the lowest unoccupied conduction band with their respective $\pi$ electrons. These two bands come into contact at the Dirac points, or, in other words, neutrality points. This indicated the semi-metallic properties and zero band gap of graphene. The $\pi$ bonds are also responsible for the electrical conductivity of graphene [204]. The extended $\pi$–$\pi$ conjugation contributes to the excellent strength of graphene. In terms of mechanical strength, graphene is reported with a Young's modulus of 1 TPa [200]. Moreover, the surface of graphene is enriched with active functional groups like ketonic and quinonic [204]. This allows the functionalization of graphene and graphene-based materials because the functional groups can link easily with foreign molecules which results in the enhancement of photocatalytic performance.

## 5. Synthesizing Method of $TiO_2$-Based Photocatalyst

Next, $TiO_2$ can be prepared by various methods such as hydrothermal synthesis, sol-gel synthesis and chemical vapor deposition synthesis. These methods offer benefits like controlling the structural phases and stoichiometry, regulating the particle size, surface morphology, and homogeneity, promoting production of high purity nanomaterials, and are cost effective [205]. In the early stages of most synthesis, amorphous $TiO_2$ with high surface area and porous structure is produced but these appealing properties tend to diminish after calcination at high temperatures where amorphous $TiO_2$ is transformed into crystalline $TiO_2$ [206]. Table 3 shows the photocatalytic degradation of pollutants using $TiO_2$ catalysts prepared from hydrothermal method, sol-gel method, and chemical vapor deposition method.

**Table 3.** Performance of $TiO_2$ prepared from different synthesis methods in pollutant treatments.

| Synthesis Method | Sample | Conditions | Pollutant Type | Treatment Time (min) | Degradation (%) | Ref |
|---|---|---|---|---|---|---|
| Hydrothermal | $TiO_2$ nanowires | (1) Titanium (IV) butoxide (TBT), ethanol, and 10 M NaOH within autoclave at 180 °C for 24 h<br>(2) Calcined at 650 °C for 2 h | 10 mg/L Rhodamine B | 60 | 100 | [207] |
| | $TiO_2$ nanorods | (1) Titanium (IV) isopropoxide (TTIP) and NaOH within autoclave at 180 °C for 24 h<br>(2) Annealed at 400 °C for 5 h | 10 mg/L Methyl Orange | 150 | 51 | [208] |
| | $TiO_2$ nanorod arrays | (1) TBT, HCl, and fluorine-doped tin oxide substrate within autoclave at 170 °C for 5 h<br>(2) Calcined at 450 °C for 2 h | 5 mg/L Bisphenol A | 180 | 49 | [209] |
| | $TiO_2$ nanotubes | (1) Titanium sulfate and urea within reaction vessel at 220 °C for 12 h<br>(2) Resultant $TiO_2$ powder and 10 M NaOH within Teflon at 150 °C for 12 h | 20 mg/L Tetracycline Hydro-chloride | 60 | 23 | [210] |
| | $TiO_2$ nanosheets | (1) TBT and hydrofluoric acid within autoclave at 180 °C for 24 h | 10 mg/L Rhodamine B | 80 | 36.5 | [211] |
| Sol-gel | $TiO_2$ | (1) Deionized water, ethanol, $CH_3COOH$, and sodium dodecyl sulfate stirred for 30 min<br>(2) Resultant mixture and TTIP stirred at 25 °C for 24 h followed by at 60 °C for 24 h<br>(3) Calcined at 450 °C for 4 h | 20 mg/L Methylene Blue | 30 | 99 | [212] |
| | $TiO_2$ | (1) Titanium (IV) chloride ($TiCl_4$) and $H_2SO_4$, followed by ammonia to achieve pH 7–8<br>(2) Annealed at 500 °C | 10 mg/L Thymol | 120 | 9.65 | [213] |
| | $TiO_2$ | (1) TBT, $CH_3COOH$, and water stirred at 80 °C<br>(2) Annealed at 500 °C for 5 h | 0.03 mg/L Methylene Blue | 120 | 96 | [214] |
| | $TiO_2$ | (1) $TiCl_4$, deionized double distilled water, and 30% ammonium hydroxide stand for 1 h | 20 mg/L Alizarin | 60 | 71 | [215] |
| | $TiO_2$ | (1) TTIP, isopropanol, acetylacetone, and water followed by HCl to achieve pH 3, the resultant mixture A stirred for 4 h and aged for 24 h<br>(2) Isopropanol and water followed by HCl to achieve pH 3, the resultant mixture B refluxed at 60 °C for 24 h and aged for 24 h<br>(3) Resultant mixture A and B stirred for 1 h, heated at 60 °C for 2 h, and followed by thermal treatment at 500 °C for 1 h | 50 mg/L Quinoline | 180 | 51 | [216] |

**Table 3.** *Cont.*

| Synthesis Method | Sample | Conditions | Pollutant Type | Treatment Time (min) | Degradation (%) | Ref |
|---|---|---|---|---|---|---|
| Chemical vapor deposition | $TiO_2$ | (1) Reactive sputtering using a titanium (Ti) target with gaseous mixture of argon (Ar) and $O_2$ (2) Resultant $TiO_2$ deposited on polyvinylidene difluoride membrane for 1 h | 4.1587 mg/L Methylene Blue | 240 | 92 | [217] |
| | $TiO_2$ | (1) TTIP as precursor and Ar as carrier gas (2) Resultant $TiO_2$ deposited on alumina balls at 500 °C for 1 h | 246.22 mg/L Nitro-benzene | 100 | 99 | [218] |
| | | | 20 mg/L Cimetidine | 180 | 98.2 | [219] |
| | $TiO_2$/clay | (1) TTIP as precursor and $N_2$ as carrier gas (2) Resultant $TiO_2$ deposited on ion-exchanged $Na^+$–clay at 600 °C for 5 h | 75 mg/L Methyl Green | 60 | 87.2 | [220] |
| | Zinc ferrite@ $TiO_2$ | (1) TTIP as precursor, Ar as carrier gas, and $O_2$ as oxidant gas (2) Resultant $TiO_2$ deposited on zinc ferrite nanofibers at 150 °C for 1 h (3) Calcined at 550 °C for 3 h | 20 mg/L Methylene Blue | 180 | 98 | [221] |
| | Sulfur (S)-doped $TiO_2$ | (1) TTIP as precursor and $N_2$ as carrier and purge gas (2) Resultant $TiO_2$ deposited on borosilicate glass substrate at 400 °C and 50 mbar (3) Resultant $TiO_2$ film and hydrogen-2v.% hydrogen sulfide at 50 °C for 60 min | 5 mg/L Methyl Orange | 300 | 72.1 | [222] |

### 5.1. Hydrothermal Synthesis

In general, hydrothermal synthesis is known as the crystallization above the room temperature and pressure through heterogeneous reactions in an aqueous/non-aqueous medium [223]. The hydrothermal synthesis is normally conducted within a sealed Teflon-lined stainless steel autoclave that can resist high temperature and pressure conditions. These conditions are applied to chemical substances that are hard to dissolve in solutions. It also can be assisted by ultrasonic irradiation, microwaves, and surface directing agents [224]. The crystallization in the hydrothermal synthesis involves two stages which are the nucleation and development of crystals [225].

In fact, hydrothermal synthesis produces a semiconductor, i.e., $TiO_2$ that exhibits defect-free nano $TiO_2$ crystals with large specific surface area, less particle agglomeration, narrow particle size distribution, anatase formation at temperature lower than 200 °C, and low energy consumption [206]. Experimental parameters such as hydrothermal temperature, time, pressure, solvent type, and titanium precursors can be controlled to modify the final product of $TiO_2$ including crystallinity and porosity. However, the application of $TiO_2$ to synthesize nanomaterials is limited, owing to the expansive sealed autoclave, inability to view the reaction in progress, and safety issues regarding the high temperature and pressure conditions [225].

In addition, the surface morphology of semiconductor photocatalysts could also affect their photocatalytic activities. For instance, Baral et al. [226] reported that the photocatalytic degradation efficiency of 20 mg/L dyes (Rhodamine B, Rhodamine—6G, Congo Red, Methyl Blue, and Methyl Orange) in the presence of α-manganese dioxide ($MnO_2$)

nanorods achieved 95–100% after 10 min under visible light irradiation (8 mW/cm$^2$). The high photocatalytic activity of α-$MnO_2$ nanorods was related to their one-dimensional morphology with a high aspect ratio and low photoluminescence intensity. The distinctive one-dimensional nanorod facilitated the separation of photogenerated charge carriers, as progressing charge separation along a single channel could lengthen the recombination time taken for electrons and holes.

Similar findings were reported on the performance of $TiO_2$ photocatalysts with different morphologies fabricated via hydrothermal synthesis as shown in Table 3. $TiO_2$ morphologies like nanoparticles, nanotubes, and nanosheets have been synthesized to test for environmental decontamination [223]. $TiO_2$ nanoparticles with small primary particle size have large specific surface area and high pore volume that can improve the adsorption of contaminants, light-harvesting ability, and photocatalytic degradation. It is reported that the base solution employed in hydrothermal synthesis affects the morphology of one-dimensional $TiO_2$ nanostructures. NaOH is commonly used as the solvent in hydrothermal synthesis to influence the crystalline phase and morphology of $TiO_2$ nanostructures [224].

### 5.2. Sol-Gel Synthesis

Sol-gel synthesis is commonly used to prepare semiconductor metal oxide nanomaterials like $TiO_2$ and ZnO. The sol-gel synthesis involves four processes which are hydrolysis, polycondensation, drying, and thermal decomposition [205]. In a typical sol-gel synthesis, the chemically active precursors are mixed uniformly in a liquid phase to undergo hydrolysis and condensation processes [227]. Here, water, alcohol, acid, and base can support the hydrolysis of the precursors [205]. A stable sol is formed after hydrolysis and condensation processes. The limiting factor of sol-gel synthesis is the necessary aging time for the catalyst. The range of aging time is 8–48 h [228]. Upon aging, a gel with 3D network structure is formed. After drying, thermal decomposition is carried out to remove the organic/inorganic precursors. In short, the sol-gel synthesis is focused on controllable hydrolysis, condensation of precursor type, and inorganic polymerization of the catalyst nanoparticles [205].

Next, the advantages of sol-gel synthesis are the formation of nanosized $TiO_2$ with high purity at low temperature, possible stoichiometry-control process, good homogeneity, and fabrication of composite [229]. It is also simple, economical, and does not require any special and expensive instruments [227,230]. Despite that, the sol-gel synthesis may produce amorphous or weakly crystalline semiconductor metal oxide nanomaterials owing to the low fabrication temperature [205]. This points to the need for thermal decomposition such as annealing and calcination for further crystallization. Thermal annealing may induce hard aggregation and inter-particle sintering within the catalyst [231].

### 5.3. Chemical Vapor Deposition Synthesis

Chemical vapor deposition is commonly applied to produce semiconductors, i.e., $TiO_2$ thin films. Generally, a thin and conformal solid film will be deposited on the surface of a substrate via chemical reactions of gaseous materials [232]. Here, single or multiple volatile precursors are introduced to the substrate at elevated temperature and pressure in an inert atmosphere [233]. The volatile precursors react or decompose onto the substrate to generate the ideal film thickness. The synthesis temperature of chemical vapor deposition method is usually applied at 200–1600 °C [232]. This method is still gaining momentum in the application of semiconductors, corrosion and wear-resistant coatings, monolithic components, etc. [234].

The chemical vapor deposition differs from physical vapor deposition where the films are formed under no chemical reactions in physical vapor deposition. It is reported that the durability, adhesion, and uniformity of film produced via chemical vapor deposition is better than physical vapor deposition [158]. Aging, drying, and reduction processes are also not required by applying chemical vapor deposition. Moreover, chemical vapor deposition gives advantages in terms of fabricating uniform and pure films, short fabrication time, allowing film formations on the inner pipe surfaces, and promoting good compatibility and adhesion [158,235]. Despite these advantages, drawbacks such as high deposition temperature, expensive vacuum systems, and safety issues with presence of corrosive gases hinder the application growth of chemical vapor deposition.

## 6. Fabrication and Performance of Cellulose Composite Catalysts

In general, the $TiO_2$ photocatalyst encounters the main problem of recombination of the photogenerated electrons and holes. This causes a reduction in quantum yield and wastage of energy [147]. The modification of $TiO_2$ photocatalyst with foreign atoms can alter the band gap and expand the adsorption range optically to improve the photocatalytic activity [236]. It is reported that the doping of $TiO_2$ with metal and non-metal elements can increase charge separation to overcome the recombination problem mentioned earlier. The doping of $TiO_2$ leads to a smaller photocatalyst loading required, lesser energy required, higher reusability of photocatalyst, shorter photocatalytic degradation time required, and higher photocatalytic activity [147].

The doping of $TiO_2$ with metal elements includes noble metals, transition metals, and rare earth metals. Metal doping, or in other words, cation doping, mainly contributes towards the downward shift of the conduction band [237]. The conduction band of $TiO_2$ contains the Ti 3d, 4s, and 4p orbitals whereby the Ti 3d orbitals govern the lower section of the conduction band. On the other hand, non-metal doping is an alternative approach to metal doping. Non-metal doping, or in other words, anion doping, mainly contributes towards the upward shift of the valence band. As a result, a new valence band is rebuilt to reduce the band gap. Other than doping, other modifications of photocatalyst involving the incorporation of cellulose and MOFs are further discussed later. Table 4 shows the photocatalytic degradation of dye by applying photocatalyst composites that incorporated cellulose, metal/non-metal doping, and MOFs.

**Table 4.** Performance of cellulose composites, cellulose/metal doped composites, cellulose/non-metal doped composites, and cellulose/MOF composites as photocatalysts in the degradation of organic dyes.

| Composite Type | Sample | Dye Type | Catalyst Loading (g/L) | Power (W) | Treatment Time (min) | Degradation (%) | Ref |
|---|---|---|---|---|---|---|---|
| Cellulose | ZnO/CNF | 5 mg/L Methylene Blue | 2 | 9 | 30 | 96 | [238] |
| | Chromium oxide/cellulose | 10 mg/L Crystal Violet | 0.1 | 25 | 40 | 99.65 | [239] |
| | Bismuth oxybromide/cellulose-derived carbon nanofibers | 10 mg/L Rhodamine B | 0.5 | 200 | 60 | 100 | [240] |
| | Beta-iron oxyhydroxide (β-FeOOH)/cellulose | 10 mg/L Methylene Blue | 1.3 | 300 | 30 | 99.89 | [241] |
| | Bacterial cellulose (BC)/polydopamine/$TiO_2$ | 20 mg/L Methylene Blue | 0.6 | 500 | 20 | 99.5 | [242] |
| | | 20 mg/L Methyl Orange | | | 30 | 95.1 | |
| | | 20 mg/L Rhodamine B | | | 60 | 100 | |

**Table 4.** *Cont.*

| Composite Type | Sample | Dye Type | Catalyst Loading (g/L) | Power (W) | Treatment Time (min) | Degradation (%) | Ref |
|---|---|---|---|---|---|---|---|
| Cellulose/metal doped | Copper (Cu)@cuprous oxide/reduced graphene oxide/cellulose | 10 mg/L Methyl Orange | 3 | 350 | 120 | 92.8 | [243] |
| | Aluminum-doped ZnO/cellulose | 10 mg/L Methyl Orange | 3 | 500 | 360 | 89.9 | [244] |
| | Cu-CNF/$TiO_2$ | 50 mg/L Reactive Brilliant Red K-2BP | 0.6 | 300 | 120 | 96.57 | [245] |
| | | 50 mg/L Cationic Red X-GRL | | | | 99.73 | |
| | Ag-cadmium selenide (CdSe)/graphene oxide@cellulose acetate | 5 mg/L Malachite Green | 4 | 300 | 25 | 97 | [246] |
| | CNF-Indium-doped Mo(O,S)$_2$ | 10 mg/L Methylene Blue | 1 | 150 | 30 | 100 | [247] |
| | | 10 mg/L Methyl Orange | | | 240 | 90 | |
| | | 10 mg/L Rhodamine B | | | 240 | 100 | |
| Cellulose/non-metal doped | Regenerated cellulose membrane-templated C-doped/core shell $TiO_2$ | 10 mg/L Methylene Blue | 0.05 | 300 | 120 | 90.1 | [248] |
| | Nitrogen (N)-doped BC/$TiO_2$ | 10 mg/L Methyl Blue | 0.5 | 300 | 30 | 100 | [249] |
| | | 10 mg/L Rhodamine B | 0.5 | | 35 | | |
| | | 20 mg/L Methyl Orange | 1 | | 15 | | |
| | N and S doped carbon dot CNF | 5 mg/L Methylene Blue | - | 1000 | 25 | 98 | [250] |
| | C–$TiO_2$/cellulose acetate | 20 mg/L Reactive Red-195 | 5 | 125 | 60 | 99.15 | [251] |
| Cellulose/MOF | Europium-MOF@viscose fabric | 20 mg/L Rhodamine B | - | 500 | 120 | 97 | [252] |
| | Phosphotungstic acid/zeolitic imidazolate framework(ZIF)-8@cellulose | 10 mg/L Methylene Blue | 0.6 | - | 30 | 99.8 | [253] |
| | Ag@silver chloride@Material Institute Lavoisier(MIL)-100(Fe)/cotton fabric | 20 mg/L Methylene Blue | 0.125–0.15 | 500 | 40 | 100 | [254] |
| | | 20 mg/L Rhodamine B | | | | | |
| | β-FeOOH@MIL-100(Fe)/cellulose/polyvinyl pyrrolidone | 20 mg/L Methylene Blue | 0.125 | 500 | 20 | 99.4 | [255] |

Cellulose contain various functional groups like hydroxyl, carboxyl, and amino groups [256]. This also beneficial for the photocatalytic degradation of organic pollutants to less harmful substances. The cellulose-based adsorbents are also effective in the removal of pollutants including metal ions, dyes and pesticides as shown in Table 5. Besides, cellulose fibers are known for their large surface area, porous structure, low dielectric permittivity, biodegradability, and strong tensile strength [257]. Chen et al. [258] reported that the cellulose hydrogel promotes uniform dispersion of $TiO_2$ nanoparticles and graphene oxide sheets while retaining the structure of the cellulose matrix. Cellulose could also act as a support to promote dye adsorption and delay the recombination of electron and holes that enhances the photocatalytic performance [259]. The significant findings on the reported cellulose-based catalyst composites were discussed as shown in Table 6.

**Table 5.** Performance of cellulose and cellulose/MOF adsorbents on the removal of pollutants.

| Composite Type | Sample | Adsorbent Dosage (g/L) | Pollutant Type | pH | Contact Time (min) | Adsorption Capacity (mg/g) | Ref |
|---|---|---|---|---|---|---|---|
| Cellulose | Cellulose-*g*-hydroxyethyl methacrylate-*co*-glycidyl methacrylate | 1 | 20 mg/L Malachite Green | 7 | 360 | 24.88 | [260] |
| | | | 20 mg/L Crystal Violet | | | 19.51 | |
| | Cellulose-*g*-2-acrylamido-2-methylpropane sulfonic acid-*co*-glycidyl methacrylate | 1 | 100 mg/L $Cu^{2+}$ | 6 | 120 | 78.247 | [261] |
| | | | 100 mg/L $Ni^{2+}$ | 5 | | 69.061 | |
| | Magnetite-functionalized NCCs/starch-g-(2-acrylamido-2methyl propane sulfonate-co-acrylic acid) | 1 | 1000 mg/L Crystal Violet | 9 | 120 | 2500 | [262] |
| | | | 1000 mg/L Methylene Blue | | | 1428.6 | |
| | Cadmium sulfide@silanized CNF | 1 | 9.597 mg/L Methylene Blue | 11 | 360 | 26.66 | [263] |
| | | | 10.5255 mg/L Safranin-O | 11 | | 17.857 | |
| | | | 28.0472 mg/L Chlorpyrifos | 3 | | 86.9565 | |
| | Dual cross-linked—alginate/treated biomass bead | 0.4 | 210 mg/L $Pb^{2+}$ | 5 | 120 | 206.75 | [264] |
| Cellulose/MOF | Cellulose acetate/MOF-derived porous carbon | 0.1 | 50 mg/L Methylene Blue | 11 | 360 | 41.36 | [265] |
| | MOF-199/cellulose/chitosan | 0.3 | 50 mg/L Methylene Blue | 7.5 | 1440 | 161.7 | [266] |
| | BC@ZIF-8 | 0.28 | 180 mg/L $UO_2^{2+}$ | 3 | 120 | 387.13 | [267] |
| | MOF-199@cellulose acetate | 0.8 | 20 mg/L Dimethoate | 7 | 360 | 321.9 | [268] |
| | Waste paper@polystyrene sulfonate@Cu-MOF | 1 | 150 mg/L $Li^{+}$ | 9 | 360 | 9.69 | [269] |

**Table 6.** Significant findings on the performance of cellulose-based photocatalysts in the degradation of pollutants.

| Composite Type | Sample | Significant Findings | Ref |
|---|---|---|---|
| Cellulose | ZnO/NCC | • The incorporation of NCCs increased the specific surface area of bare ZnO from 13.014 $m^2/g$ to 32.202 $m^2/g$.<br>• The photocatalytic degradation of Methylene Blue (10 mg/L) in the presence of ZnO and ZnO/NCC reached 65.87% and 88.62%, respectively, after 120 min under solar light irradiation. | [270] |
| | $g\text{-}C_3N_4\text{−}CNF$ | • The band gap energy of $g\text{-}C_3N_4\text{−}CNF$ nanocomposite foam (2.7 eV) was lower than bare $g\text{-}C_3N_4$ (2.9 eV).<br>• The photocatalytic degradation of Rhodamine B (5 mg/L) in the presence of $g\text{-}C_3N_4$ and $g\text{-}C_3N_4\text{−}CNF$ achieved 16% and 47%, respectively, after 6 h of visible light irradiation in the absence of agitation. | [271] |

**Table 6.** *Cont.*

| Composite Type | Sample | Significant Findings | Ref |
| --- | --- | --- | --- |
| | Cellulose/bismuth vanadate | • The addition of cellulose to bismuth vanadate reduced its crystallite size and band gap energy from 22.03 nm to 6.55 nm and from 2.93 eV to 2.44 eV, respectively.<br>• The photocatalytic degradation of Methyl Orange (3.2734 mg/L) in the presence of cellulose/bismuth vanadate reached 80% after 60 min under visible light irradiation (200 W). | [259] |
| Cellulose/ metal doped | Molybdenum-doped $TiO_2$ films templated by NCCs | • The specific surface area of NCCs-templated $TiO_2$ films doped with molybdenum (149 $m^2/g$) was larger than NCCs-templated $TiO_2$ films (103 $m^2/g$).<br>• The photocatalytic degradation of trichloroethylene (200 ppmv) in the presence of NCCs-templated $TiO_2$ film doped with 5 at% molybdenum achieved 100% after 30 min under UV irradiation (15 W). | [272] |
| | Iron (II, III) oxide($Fe_3O_4$)/ praseodymium-bismuth oxychloride (BiOCl)/cellulose | • The specific surface areas of $Fe_3O_4$/BiOCl, $Fe_3O_4$/praseodymium-BiOCl, and $Fe_3O_4$/praseodymium-BiOCl/cellulose were 3.5315 $m^2/g$, 4.2115 $m^2/g$, and 17.6303 $m^2/g$, respectively.<br>• The photocatalytic degradation of Rhodamine B (10 mg/L) in the presence of $Fe_3O_4$/praseodymium-BiOCl/cellulose accomplished 99.8% after 120 min under visible light irradiation (300 W). | [273] |
| | Ag/$TiO_2$@ cellulose-derived carbon beads | • The band gap energy of Ag/$TiO_2$@cellulose-derived carbon beads (2.21 eV) was lower than pure $TiO_2$ (3.08 eV).<br>• The photocatalytic degradation of ceftriaxone sodium (25 mg/L) in the presence of Ag/$TiO_2$@cellulose-derived carbon beads reached 91.92% after 270 min under visible light irradiation (500 W). | [274] |
| Cellulose/ non-metal doped | Lanthanum–N–$TiO_2$–cellulose/silicon dioxide ($SiO_2$) | • The band gap energy of $TiO_2$ was reduced from 3.59 eV to 2.81 eV.<br>• The photocatalytic degradation of crude oil solution (20 mg/L) in the presence of lanthanum–N–$TiO_2$–cellulose/$SiO_2$ achieved 92% after 180 min under sunlight. | [275] |
| | N-$TiO_2$/C | • The photocatalytic degradation of 4-nitrophenol (10 mg/L) in the presence of N-$TiO_2$/C reached 80% after 420 min under visible light irradiation (24 W).<br>• The effect of N doping improved the photocatalytic activity of $TiO_2$/C by 26%. | [276] |
| | TEMPO-oxidized cellulose-ZnO | • The specific surface area and pore volume of TEMPO-oxidized cellulose-ZnO (20.1 $m^2/g$ and 0.09 $cm^3/g$) were higher than bare ZnO (9.1 $m^2/g$ and 0.05 $cm^3/g$).<br>• The band gap energy of ZnO was reduced from 3.15 eV to 3.11 eV.<br>• The photocatalytic degradation of Methyl Orange (5 mg/L) in the presence of TEMPO-oxidized cellulose-ZnO obtained 95.4% after 120 min under UV irradiation (6 W).<br>• There was an improvement of 29.5% in the photocatalytic activity of TEMPO-oxidized cellulose-ZnO as compared to bare ZnO. | [277] |
| Cellulose/ MOF | $TiO_2$/ magnetic-MIL-101(chromium) | • The band gap energy of $TiO_2$/magnetic-MIL-101(chromium) (1.61 eV) was lower than pure $TiO_2$ (3.1 eV).<br>• The photocatalytic degradation of Acid Red 1 (20 mg/L) in the presence of $TiO_2$/magnetic-MIL-101(chromium) achieved 90% after 60 min under visible light irradiation (500 W). | [278] |

**Table 6.** *Cont.*

| Composite Type | Sample | Significant Findings | Ref |
|---|---|---|---|
| | (Ag & palladium (Pd))@MIL-125-$NH_2$@cellulose acetate | • The band gap energy of MIL-125-$NH_2$@cellulose acetate, Ag@MIL-125-$NH_2$@cellulose acetate, and Pd@MIL-125-$NH_2$@cellulose acetate films were 2.53, 2.38, and 1.99 eV, respectively.<br>• The photocatalytic reduction of 2-nitrophenol (25 mg/L) in the presence of Ag-doped and Pd-doped MIL-125-$NH_2$@cellulose acetate films reached 87.4% and 95.1%, respectively, after 120 min under visible light irradiation (12 W).<br>• The metal-doped MIL-125-$NH_2$@cellulose acetate films doubly increased the photocatalytic reduction of 2-nitrophenol in 120 min as compared to MIL-125-$NH_2$@cellulose acetate alone. | [279] |
| | Cellulose acetate@Ti-MIL-$NH_2$ | • The insertion of Ti-MIL-$NH_2$ within the cellulose acetate film increased the surface area from 408.9 $m^2/g$ to 632.3 $m^2/g$.<br>• The band gap energy of cellulose acetate@Ti-MIL-$NH_2$ (2.42 eV) was slightly lower than pure Ti-MIL-$NH_2$ (2.51 eV).<br>• The photocatalytic degradation of paracetamol (30 mg/L) in the presence of cellulose acetate@Ti-MIL-$NH_2$ achieved 78% after 30 min under visible light irradiation (12 W). | [280] |

### *6.1. Metal Doping*

The purpose of metal doping is to minimize the band gap energy and to redshift the absorption from the UV region. The doping of metal ions induces an impurity level (intermediate energy level) that enhances the visible light absorption by taking the role of electron donor or acceptor [237]. Looking into this, metal doping introduces defects in the lattice structure of $TiO_2$ or modifies the lattice degree and further prolongs the lifetime of $TiO_2$ [281]. The metal ions can be easily doped into the lattice structure of $TiO_2$ due to their close similarity in ionic radius [147]. According to previous research, the modification of photocatalysts (e.g., $TiO_2$) with foreign atoms, i.e., metal or non-metal ion doping, can alter the band gap by expanding the light absorption range optically. The minimum energy required to excite an electron from the valence band to the conduction band is known as the band gap energy [282]. A lower band gap energy is advantageous for visible light absorption and photocatalytic activity [283]. When the band gap energy of photocatalysts is reduced, lesser energy from light irradiation is required to excite the electrons from the valence band to the conduction band of photocatalysts. Subsequently, this improves the redox reactions and enhances the photocatalytic activity of photocatalysts.

Despite these advantages, the required amount of metals must be minimized owing to their exceedingly high price [236]. Metal doping triggers localized d states in the band gap of $TiO_2$ that becomes the recombination centers of electron–hole pairs [237]. When there are too many traps for charge carriers positioned on the surface of the catalyst bulk, the mobility of charge carriers will be lowered and they are more likely to recombine before reaching the surface [284]. This is often found in $TiO_2$ photocatalysts loaded with metal content higher than 5 wt% [285]. Noble metals such as Ag and gold (Au) are toxic in nature which makes it necessary to use them in small amounts. Metal-doped $TiO_2$ photocatalysts also encounter photocorrosion and thermal instability upon the photocatalytic degradation of organic pollutants [286].

Noble metals such as Ag, Au, platinum, and Pd, enhance the performance of $TiO_2$ through charge transferring, electron trapping, and reduction of band gap energy [147]. Upon doping, it is preferable for the electrons to move from $TiO_2$ to the noble metals and the movement tends to continue until the Fermi energy of both $TiO_2$ and noble metals reaches an equilibrium state. A potential barrier is generated as a result of the distortion of band structure between $TiO_2$ and the noble metals. This makes it difficult for electrons to move from $TiO_2$ to the noble metal. As a consequence, there is an accumulation of negative electrons at the interface of $TiO_2$. Looking into this, a positive charge layer is generated below the $TiO_2$ surface to maintain electrical neutrality and this leads to the bending of a conduction band called Schottky barrier [284]. This barrier traps electrons effectively to prevent the electrons from flowing back to $TiO_2$. The performance of noble-metal-doped $TiO_2$ can be improved via surface plasmon resonance of noble metals, surface adsorption of $TiO_2$ with organic compounds, and adjustment of Fermi level between $TiO_2$ and noble metals to improve charge excitation and separation, and reduction of band gap energy [147].

Transition metals such as manganese, iron (Fe), and chromium, are also studied as metal dopants on $TiO_2$. Transition metals that exhibit various oxidation states are capable of trapping electrons and impede their recombination [284]. For example, Fe has three ion forms which are $Fe^{2+}$, $Fe^{3+}$, and $Fe^{4+}$. Nevertheless, it is reported that the transition metal is likely to cause thermal instability in anatase $TiO_2$. Rare earth metals including yttrium and scandium, are investigated as metal dopants on $TiO_2$ photocatalyst. They consist of incomplete 4f and unoccupied 5d orbitals that can be incorporated into $TiO_2$ [147]. The presence of rare earth metal−O−Ti bonds on the surface of $TiO_2$ photocatalyst impedes the growth of nanocrystalline $TiO_2$.

### 6.2. Non-Metal Doping

As stated earlier, the non-metal doping of $TiO_2$ shifts the valence band upwards. Unlike metal doping, non-metal doping does not affect the shifting of conduction band [237]. An intermediate energy level is formed between the conduction bands of $TiO_2$ and non-metal dopants. Non-metal doping inhibits the recombination of electron–hole pairs, improves the redox potential of $OH^\bullet$, and finally enhances the quantum efficiency of $TiO_2$ [147]. N, fluorine, S, carbon (C), boron, and phosphorus are the common non-metal dopants on the $TiO_2$ photocatalyst.

Currently, N is the most studied dopant for non-metal doping of $TiO_2$ due to good stability, low ionization energy, and comparable atomic size [281,286]. Typically, N can be easily incorporated into $TiO_2$ in terms of substitutional and interstitial. The oxygen vacancies are the defects that can trap electrons at the defect sites [284]. Similar to metal doping, N doping also acts as an electron trapping center to decrease the recombination rate of electron–hole pairs. However, the main drawback of non-metal doping is the reduction of non-metal dopants during the annealing process [287]. The reduction of non-metal dopant in the resultant doped $TiO_2$ photocatalyst lowers its performance. Co-doping of $TiO_2$ is investigated to overcome the single non-metal doped $TiO_2$ photocatalysts.

### 6.3. Metal–Organic Frameworks (MOFs)

MOFs are known as new porous coordination polymers [288]. They consist of metal ions/clusters linked by multi-dentate organic linkers [289]. Their advantages, including large surface area, tunable pore size and structures, and being rich with active sites have been applied in catalysis and wastewater treatments. MOF-199, MIL-100, MIL-125, and ZIF-8 are some of the popular MOFs. Due to the large surface area and diverse functionality, MOFs act as appealing adsorbents and can be modified to obtain higher selective binding affinities for the targeted adsorbates [290]. Table 5 shows the removal of pollutants by cellulose/MOF composites via adsorption process.

Moreover, the performance of MOFs photocatalysts are similar to common semiconductors ($TiO_2$ and ZnO) and they are widely used in organic pollutant degradation and Cr(VI) reduction [289]. The tunable metal ions/clusters and organic linkers act as antennas to harvest light and generate electron–hole pairs for photocatalysis. Nevertheless, the excited electrons from highest occupied molecular orbital will travel to lowest unoccupied molecular orbital during the photocatalytic process in the presence of MOFs [289]. This terminology is different from the common semiconductors whereby the excited electrons will travel from the valence band to the conduction band.

## 7. Challenges and Future Perspective

Agricultural wastes have some agronomic benefits over wood as a biomass resource including shorter growth cycles, high biomass yield, and high carbohydrate content [291]. Nevertheless, the average increase rate of agricultural waste generated is 5–10% annually [292]. This rapid rate contributes to negative environmental impacts. The biomass wastes are mostly left in fields for natural decomposition, discarded at landfills, or incinerated for cooking process, drying process, and charcoal production [293]. This could give rise to GHG emissions and air quality deterioration. The conversion of biomass wastes into value-added materials supports the purpose of a circular economy and mitigates the biomass disposal problem.

Next, agricultural biomass contains high cellulose content. This potential cellulose source has drawn the attention of researchers to fabricate and develop biomass-derived cellulose and its derivatives over the years. Cellulose and its derivatives are commonly applied in industries such as paper, textiles, pharmaceuticals, and water treatment [294]. The common challenge faced during cellulose extraction from biomass is biomass recalcitrance which involves the resistance of the lignocellulosic biomass structure towards microbial or enzymatic degradation [295]. The removal of lignin and hemicellulose is necessary to obtain cellulosic materials with high purity. However, the harsh chemical treatments on biomass could lead to the formation of partially degraded cellulose and the generation of toxic effluents [294]. A greener approach should be taken into consideration for the fabrication of biomass-derived cellulosic products. Chlorine-based compounds are widely used in the bleaching and synthesis stages, and yet the effluents containing chlorine give out harmful impacts to the environment [296]. Chlorine-free preparation routes for biomass-derived cellulose products should be emphasized for future commercial-scale applications.

Semiconductor photocatalysts such as $TiO_2$, ZnO, g-$C_3N_4$, and graphene are widely used in the degradation of organic pollutants due to their high chemical stability and high thermal stability. The bulk forms of metal oxides (e.g., $TiO_2$, ZnO) are safer as compared to their nanoparticulate forms because the dispersion of nanoparticles with high oxidation potential in aqueous medium could attribute to higher toxicity [297]. The interesting features of cellulose make it a suitable candidate to fabricate cellulose-semiconductor photocatalyst composites. Besides modifications involving metal/non-metal doping, the incorporation of cellulose could also address the drawbacks of semiconductor photocatalysts alone. Cellulose could promote the uniform dispersion of nanoparticles, increase specific surface area, improve adsorption ability, inhibit recombination of charge carriers, and reduce band gap energy. Biomass-derived cellulose are usually considered to be good candidates as host materials of nanomaterials because they can improve the stability, retain the special morphology, and control the growth of nanoparticles. The excellent properties in terms of electrical, magnetic, and optical in inorganic nanoparticles, can be preserved in the polymer matrix. These potential characteristics are beneficial for the photocatalytic degradation of organic pollutants.

In recent years, numerous novel cellulose-semiconductor photocatalyst composites are being reported with high catalytic performances in water treatment applications. This has proven the importance and practical feasibility of cellulose in the development of photocatalyst nanomaterials. Other than the reported degradation mechanism of cellulose-based photocatalysts, a thorough investigation of the photostability and long usage period of cellulose should be taken into account. In the case of photocatalytic degradation, the photocatalysts are exposed to light irradiation for a long time period. Post-treatment characterization studies on the cellulose-based photocatalysts are encouraged to observe and determine any signs of photodegradation or altered structure of cellulose within the composite. A better understanding of this crucial feature could help develop sustainable and long-lasting cellulose-semiconductor photocatalysts in future research works.

## 8. Conclusions

Biopolymers, such as cellulose and chitosan, have gained tremendous attraction in environmental and biological applications due to their abundance and multi-functionality. The appealing characteristics of biopolymers including large surface area, non-toxic, antimicrobial activity, and biocompatibility have led to the applications in food packaging, wastewater treatment, and biofuel. Since cellulose is the main component of biomass, the advancements in cellulose isolation methods have enabled the utilization of biomass-derived cellulose in environmental applications. The hydroxyl groups found in the chemical structure of cellulose are responsible for controlling the crystalline packing, physical properties, and chemical reactivity of cellulose. This is beneficial for the functionalization of cellulose materials with foreign substances.

The promising aspects of cellulose, including that they are low cost, abundant, and environmentally friendly, are suitable for the fabrication of cellulose hybrid composite photocatalysts. The incorporation of cellulose could overcome the limitations of semiconductor photocatalysts. The large specific surface area and porous structure of cellulose materials could promote dye adsorption on the surface of photocatalysts. The introduction of cellulose minimizes the aggregation of semiconductor particles, impedes the recombination of charge carriers, and reduces the band gap energy of semiconductor photocatalysts. The generation of large amounts of reactive radicals enhances the photocatalytic performance. These notable characteristics are favorable for the photocatalytic degradation of organic dye. In addition, the modifications such as doping, and MOFs also enhanced the catalytic performance of cellulose composite photocatalysts. The high performance and sustainability of cellulose hybrid composite photocatalysts have proven the advantage of cellulose and showed a more promising advancement in the development of photocatalysts.

Agricultural biomass has a high cellulose content, making it a potential cellulose source. However, biomass recalcitrance limits and hinders the production of biomass-derived cellulose. High purity cellulose can be extracted from biomass by highly toxic chemical processes but doing so would also partially degrade the cellulose and produce harmful effluent or any byproducts. Therefore, a more environmentally friendly strategy, such as preparation routes for cellulose obtained from biomass without the use of chlorine, could ensure the discharge of less hazardous effluents into the environment. On the other hand, the high feasibility and significance of cellulose in the development of various photocatalysts has been reported over the years. Nevertheless, research on the photostability and long usage period of cellulose is limited. These critical aspects can be investigated through post-treatment characterization studies on the cellulose-based photocatalysts. It also serves as supporting information for the future development of robust cellulose-semiconductor photocatalysts.

**Author Contributions:** Conceptualization, writing—review and editing, Y.L.P.; data curation, writing—original draft preparation, Y.D.C.; methodology, resources, visualization, S.L.; resources, W.C.C.; resources and validation, C.W.L.; writing—review and editing, A.Z.A. All authors have read and agreed to the published version of the manuscript.

**Funding:** This research was supported in part with Ministry of Education (MOE) Malaysia that provided the Fundamental Research Grant Scheme (FRGS/1/2022/TK0/UTAR/02/34), Kurita Asia Research Grant (21Pmy076) provided by Kurita Water and Environment Foundation and Universiti Tunku Abdul Rahman (UTAR) Research Fund (IPSR/RMC/UTARRF/2020-C2/P01).

**Institutional Review Board Statement:** Not applicable.

**Informed Consent Statement:** Not applicable.

**Data Availability Statement:** Not applicable.

**Acknowledgments:** The authors would like to thank the Ministry of Education (MOE) Malaysia that provided the Fundamental Research Grant Scheme, the Universiti Tunku Abdul Rahman (UTAR) Research Fund and Kurita Water and Environment Foundation (KWEF) that provided the Overseas Research Grant for the financial support.

**Conflicts of Interest:** The authors declare no conflict of interest.

## References


1. Shen, Y. A Review on Hydrothermal Carbonization of Biomass and Plastic Wastes to Energy Products. *Biomass Bioenergy* **2020**, *134*, 105479. [CrossRef]
2. Chua, S.Y.; Periasamy, L.A.; Goh, C.M.H.; Tan, Y.H.; Mubarak, N.M.; Kansedo, J.; Khalid, M.; Walvekar, R.; Abdullah, E.C. Biodiesel Synthesis Using Natural Solid Catalyst Derived from Biomass Waste—A Review. *J. Ind. Eng. Chem.* **2020**, *81*, 41–60. [CrossRef]
3. Shanmuga Priya, M.; Divya, P.; Rajalakshmi, R. A Review Status on Characterization and Electrochemical Behaviour of Biomass Derived Carbon Materials for Energy Storage Supercapacitors. *Sustain. Chem. Pharm.* **2020**, *16*, 100243. [CrossRef]
4. Osman, A.I.; Abdelkader, A.; Farrell, C.; Rooney, D.; Morgan, K. Reusing, Recycling and up-Cycling of Biomass: A Review of Practical and Kinetic Modelling Approaches. *Fuel Process. Technol.* **2019**, *192*, 179–202. [CrossRef]
5. Khan, T.A.; Saud, A.S.; Jamari, S.S.; Rahim, M.H.A.; Park, J.W.; Kim, H.J. Hydrothermal Carbonization of Lignocellulosic Biomass for Carbon Rich Material Preparation: A Review. *Biomass Bioenergy* **2019**, *130*, 105384. [CrossRef]
6. Phromphithak, S.; Onsree, T.; Tippayawong, N. Machine Learning Prediction of Cellulose-Rich Materials from Biomass Pretreatment with Ionic Liquid Solvents. *Bioresour. Technol.* **2021**, *323*, 124642. [CrossRef] [PubMed]
7. Pennells, J.; Cruickshank, A.; Chaléat, C.; Godwin, I.D.; Martin, D.J. Sorghum as a Novel Biomass for the Sustainable Production of Cellulose Nanofibers. *Ind. Crops Prod.* **2021**, *171*, 113917. [CrossRef]
8. Chairungsri, W.; Subkomkaew, A.; Kijjanapanich, P.; Chimupala, Y. Direct Dye Wastewater Photocatalysis Using Immobilized Titanium Dioxide on Fixed Substrate. *Chemosphere* **2022**, *286*, 131762. [CrossRef] [PubMed]
9. Zhang, X.; Wang, J.; Dong, X.X.; Lv, Y.K. Functionalized Metal-Organic Frameworks for Photocatalytic Degradation of Organic Pollutants in Environment. *Chemosphere* **2020**, *242*, 125144. [CrossRef]
10. Pandey, B.; Prajapati, Y.K.; Sheth, P.N. Recent Progress in Thermochemical Techniques to Produce Hydrogen Gas from Biomass: A State of the Art Review. *Int. J. Hydrogen Energy* **2019**, *44*, 25384–25415. [CrossRef]
11. Nunes, L.J.R.; Causer, T.P.; Ciolkosz, D. Biomass for Energy: A Review on Supply Chain Management Models. *Renew. Sustain. Energy Rev.* **2020**, *120*, 109658. [CrossRef]
12. Olivares-Marín, M.; Fernández, J.A.; Lázaro, M.J.; Fernández-González, C.; Macías-García, A.; Gómez-Serrano, V.; Stoeckli, F.; Centeno, T.A. Cherry Stones as Precursor of Activated Carbons for Supercapacitors. *Mater. Chem. Phys.* **2009**, *114*, 323–327. [CrossRef]
13. Chandra, R.; Takeuchi, H.; Hasegawa, T. Methane Production from Lignocellulosic Agricultural Crop Wastes: A Review in Context to Second Generation of Biofuel Production. *Renew. Sustain. Energy Rev.* **2012**, *16*, 1462–1476. [CrossRef]
14. Li, T.; Takkellapati, S. The Current and Emerging Sources of Technical Lignins and Their Applications. *Biofuels Bioprod. Biorefining* **2018**, *12*, 756–787. [CrossRef] [PubMed]
15. Hameed, S.; Sharma, A.; Pareek, V.; Wu, H.; Yu, Y. A Review on Biomass Pyrolysis Models: Kinetic, Network and Mechanistic Models. *Biomass Bioenergy* **2019**, *123*, 104–122. [CrossRef]
16. Kumar, V.; Yadav, S.K.; Kumar, J.; Ahluwalia, V. A Critical Review on Current Strategies and Trends Employed for Removal of Inhibitors and Toxic Materials Generated during Biomass Pretreatment. *Bioresour. Technol.* **2020**, *299*, 122633. [CrossRef] [PubMed]
17. Ren, J.; Liu, Y.L.; Zhao, X.Y.; Cao, J.P. Biomass Thermochemical Conversion: A Review on Tar Elimination from Biomass Catalytic Gasification. *J. Energy Inst.* **2020**, *93*, 1083–1098. [CrossRef]
18. Food and Agricultural Organization of the United Nations (FAO). Overview: Major trends and issues. In *FAO Yearbook of Fishery and Aquaculture Statistics 2017*; Food and Agricultural Organization of the United Nations (FAO): Rome, Italy, 2019; pp. 1–108. ISBN 978-92-5-131669-6.
19. Hamed, I.; Özogul, F.; Regenstein, J.M. Industrial Applications of Crustacean By-Products (Chitin, Chitosan, and Chitooligosaccharides): A Review. *Trends Food Sci. Technol.* **2016**, *48*, 40–50. [CrossRef]
20. Zhang, J.; Xia, W.; Liu, P.; Cheng, Q.; Tahirou, T.; Gu, W.; Li, B. Chitosan Modification and Pharmaceutical/Biomedical Applications. *Mar. Drugs* **2010**, *8*, 1962–1987. [CrossRef] [PubMed]
21. Bakshi, P.S.; Selvakumar, D.; Kadirvelu, K.; Kumar, N.S. Chitosan as an Environment Friendly Biomaterial—A Review on Recent Modifications and Applications. *Int. J. Biol. Macromol.* **2020**, *150*, 1072–1083. [CrossRef]
22. Xiong, X.; Yu, I.K.M.; Cao, L.; Tsang, D.C.W.; Zhang, S.; Ok, Y.S. A Review of Biochar-Based Catalysts for Chemical Synthesis, Biofuel Production, and Pollution Control. *Bioresour. Technol.* **2017**, *246*, 254–270. [CrossRef] [PubMed]
23. Abdullah, S.H.Y.S.; Hanapi, N.H.M.; Azid, A.; Umar, R.; Juahir, H.; Khatoon, H.; Endut, A. A Review of Biomass-Derived Heterogeneous Catalyst for a Sustainable Biodiesel Production. *Renew. Sustain. Energy Rev.* **2017**, *70*, 1040–1051. [CrossRef]
24. Liu, J.; Jiang, J.; Meng, Y.; Aihemaiti, A.; Xu, Y.; Xiang, H.; Gao, Y.; Chen, X. Preparation, Environmental Application and Prospect of Biochar-Supported Metal Nanoparticles: A Review. *J. Hazard. Mater.* **2020**, *388*, 122026. [CrossRef]
25. Tan, X.; Liu, Y.; Zeng, G.; Wang, X.; Hu, X.; Gu, Y.; Yang, Z. Application of Biochar for the Removal of Pollutants from Aqueous Solutions. *Chemosphere* **2015**, *125*, 70–85. [CrossRef] [PubMed]
26. Ahmad, M.; Rajapaksha, A.U.; Lim, J.E.; Zhang, M.; Bolan, N.; Mohan, D.; Vithanage, M.; Lee, S.S.; Ok, Y.S. Biochar as a Sorbent for Contaminant Management in Soil and Water: A Review. *Chemosphere* **2014**, *99*, 19–33. [CrossRef]

27. Qian, K.; Kumar, A.; Zhang, H.; Bellmer, D.; Huhnke, R. Recent Advances in Utilization of Biochar. *Renew. Sustain. Energy Rev.* **2015**, *42*, 1055–1064. [CrossRef]
28. Nor, N.M.; Lau, L.C.; Lee, K.T.; Mohamed, A.R. Synthesis of Activated Carbon from Lignocellulosic Biomass and Its Applications in Air Pollution Control—A Review. *J. Environ. Chem. Eng.* **2013**, *1*, 658–666. [CrossRef]
29. González-García, P. Activated Carbon from Lignocellulosics Precursors: A Review of the Synthesis Methods, Characterization Techniques and Applications. *Renew. Sustain. Energy Rev.* **2018**, *82*, 1393–1414. [CrossRef]
30. Saleem, J.; Bin Shahid, U.; Hijab, M.; Mackey, H.; McKay, G. Production and Applications of Activated Carbons as Adsorbents from Olive Stones. *Biomass Convers. Biorefinery* **2019**, *9*, 775–802. [CrossRef]
31. Azari, A.; Nabizadeh, R.; Nasseri, S.; Mahvi, A.H.; Mesdaghinia, A.R. Comprehensive Systematic Review and Meta-Analysis of Dyes Adsorption by Carbon-Based Adsorbent Materials: Classification and Analysis of Last Decade Studies. *Chemosphere* **2020**, *250*, 126238. [CrossRef]
32. Demirbas, A. Agricultural Based Activated Carbons for the Removal of Dyes from Aqueous Solutions: A Review. *J. Hazard. Mater.* **2009**, *167*, 1–9. [CrossRef]
33. Nayak, A.; Bhushan, B.; Gupta, V.; Sharma, P. Chemically Activated Carbon from Lignocellulosic Wastes for Heavy Metal Wastewater Remediation: Effect of Activation Conditions. *J. Colloid Interface Sci.* **2017**, *493*, 228–240. [CrossRef]
34. Shrestha, D.; Rajbhandari, A. The Effects of Different Activating Agents on the Physical and Electrochemical Properties of Activated Carbon Electrodes Fabricated from Wood-Dust of *Shorea robusta*. *Heliyon* **2021**, *7*, e07917. [CrossRef]
35. Gao, Y.; Yue, Q.; Gao, B.; Li, A. Insight into Activated Carbon from Different Kinds of Chemical Activating Agents: A Review. *Sci. Total Environ.* **2020**, *746*, 141094. [CrossRef] [PubMed]
36. Zhang, Y.; Song, X.; Xu, Y.; Shen, H.; Kong, X.; Xu, H. Utilization of Wheat Bran for Producing Activated Carbon with High Specific Surface Area via NaOH Activation Using Industrial Furnace. *J. Clean. Prod.* **2019**, *210*, 366–375. [CrossRef]
37. Xiao, W.; Garba, Z.N.; Sun, S.; Lawan, I.; Wang, L.; Lin, M.; Yuan, Z. Preparation and Evaluation of an Effective Activated Carbon from White Sugar for the Adsorption of Rhodamine B Dye. *J. Clean. Prod.* **2020**, *253*, 119989. [CrossRef]
38. Wickramaarachchi, W.A.M.K.P.; Minakshi, M.; Gao, X.; Dabare, R.; Wong, K.W. Hierarchical Porous Carbon from Mango Seed Husk for Electro-Chemical Energy Storage. *Chem. Eng. J. Adv.* **2021**, *8*, 100158. [CrossRef]
39. Wickramaarachchi, K.; Minakshi, M.; Aravindh, S.A.; Dabare, R.; Gao, X.; Jiang, Z.T.; Wong, K.W. Repurposing N-Doped Grape Marc for the Fabrication of Supercapacitors with Theoretical and Machine Learning Models. *Nanomaterials* **2022**, *12*, 1847. [CrossRef]
40. Arun, S.; Kiran, K.U.V.; Kumar, S.M.; Karnan, M.; Sathish, M.; Mayavan, S. Effect of Orange Peel Derived Activated Carbon as a Negative Additive for Lead-Acid Battery under High Rate Discharge Condition. *J. Energy Storage* **2021**, *34*, 102225. [CrossRef]
41. Dou, Y.; Liu, X.; Wang, X.; Yu, K.; Liang, C. Jute Fiber Based Micro-Mesoporous Carbon: A Biomass Derived Anode Material with High-Performance for Lithium-Ion Batteries. *Mater. Sci. Eng. B Solid-State Mater. Adv. Technol.* **2021**, *265*, 115015. [CrossRef]
42. Farooq, A.; Patoary, M.K.; Zhang, M.; Mussana, H.; Li, M.; Naeem, M.A.; Mushtaq, M.; Farooq, A.; Liu, L. Cellulose from Sources to Nanocellulose and an Overview of Synthesis and Properties of Nanocellulose/Zinc Oxide Nanocomposite Materials. *Int. J. Biol. Macromol.* **2020**, *154*, 1050–1073. [CrossRef]
43. Moon, R.J.; Martini, A.; Nairn, J.; Simonsen, J.; Youngblood, J. Cellulose Nanomaterials Review: Structure, Properties and Nanocomposites. *Chem. Soc. Rev.* **2011**, *40*, 3941–3994. [CrossRef] [PubMed]
44. Gopi, S.; Balakrishnan, P.; Chandradhara, D.; Poovathankandy, D.; Thomas, S. General Scenarios of Cellulose and Its Use in the Biomedical Field. *Mater. Today Chem.* **2019**, *13*, 59–78. [CrossRef]
45. Azizi Samir, M.A.S.; Alloin, F.; Dufresne, A. Review of Recent Research into Cellulosic Whiskers, Their Properties and Their Application in Nanocomposite Field. *Biomacromolecules* **2005**, *6*, 612–626. [CrossRef] [PubMed]
46. Habibi, Y.; Lucia, L.A.; Rojas, O.J. Cellulose Nanocrystals: Chemistry, Self-Assembly, and Applications. *Chem. Rev.* **2010**, *110*, 3479–3500. [CrossRef] [PubMed]
47. John, M.J.; Thomas, S. Biofibres and Biocomposites. *Carbohydr. Polym.* **2008**, *71*, 343–364. [CrossRef]
48. Nsor-Atindana, J.; Chen, M.; Goff, H.D.; Zhong, F.; Sharif, H.R.; Li, Y. Functionality and Nutritional Aspects of Microcrystalline Cellulose in Food. *Carbohydr. Polym.* **2017**, *172*, 159–174. [CrossRef] [PubMed]
49. De Souza Lima, M.M.; Borsali, R. Rodlike Cellulose Microcrystals: Structure, Properties, and Applications. *Macromol. Rapid Commun.* **2004**, *25*, 771–787. [CrossRef]
50. Rieland, J.M.; Love, B.J. Ionic Liquids: A Milestone on the Pathway to Greener Recycling of Cellulose from Biomass. *Resour. Conserv. Recycl.* **2020**, *155*, 104678. [CrossRef]
51. Zaman, A.; Huang, F.; Jiang, M.; Wei, W.; Zhou, Z. Preparation, Properties, and Applications of Natural Cellulosic Aerogels: A Review. *Energy Built Environ.* **2020**, *1*, 60–76. [CrossRef]
52. Li, Y.; Lin, M.; Davenport, J.W. Ab Initio Studies of Cellulose I: Crystal Structure, Intermolecular Forces, and Interactions with Water. *J. Phys. Chem. C* **2011**, *115*, 11533–11539. [CrossRef]
53. Ng, H.M.; Sin, L.T.; Tee, T.T.; Bee, S.T.; Hui, D.; Low, C.Y.; Rahmat, A.R. Extraction of Cellulose Nanocrystals from Plant Sources for Application as Reinforcing Agent in Polymers. *Compos. Part B Eng.* **2015**, *75*, 176–200. [CrossRef]
54. Kumar, R.; Sharma, R.K.; Singh, A.P. Cellulose Based Grafted Biosorbents—Journey from Lignocellulose Biomass to Toxic Metal Ions Sorption Applications—A Review. *J. Mol. Liq.* **2017**, *232*, 62–93. [CrossRef]

55. Brinchi, L.; Cotana, F.; Fortunati, E.; Kenny, J.M. Production of Nanocrystalline Cellulose from Lignocellulosic Biomass: Technology and Applications. *Carbohydr. Polym.* **2013**, *94*, 154–169. [CrossRef]
56. Driemeier, C. Nanostructure of lignocellulose and its importance for biomass conversion into chemicals and biofuels. In *Advances of Basic Science for Second Generation Bioethanol from Sugarcane*; Buckeridge, M.S., De Souza, A.P., Eds.; Springer International Publishing: Cham, Switzerland, 2017; pp. 21–38. ISBN 978-3-319-49824-9.
57. Jones, D.; Ormondroyd, G.O.; Curling, S.F.; Popescu, C.-M.; Popescu, M.-C. Chemical compositions of natural fibres. In *Advanced High Strength Natural Fibre Composites in Construction*; Fan, M., Fu, F., Eds.; Elsevier: Amsterdam, The Netherlands, 2017; pp. 23–58. ISBN 9780081004302.
58. Siqueira, G.; Bras, J.; Dufresne, A. Cellulosic Bionanocomposites: A Review of Preparation, Properties and Applications. *Polymers* **2010**, *2*, 728–765. [CrossRef]
59. Illum, L.; Jabbal-Gill, I.; Hinchcliffe, M.; Fisher, A.N.; Davis, S.S. Chitosan as a Novel Nasal Delivery System for Vaccines. *Adv. Drug Deliv. Rev.* **2001**, *51*, 81–96. [CrossRef]
60. Wei, S.; Ching, Y.C.; Chuah, C.H. Synthesis of Chitosan Aerogels as Promising Carriers for Drug Delivery: A Review. *Carbohydr. Polym.* **2020**, *231*, 115744. [CrossRef] [PubMed]
61. Sampath, U.G.T.M.; Ching, Y.C.; Chuah, C.H.; Singh, R.; Lin, P.C. Preparation and Characterization of Nanocellulose Reinforced Semi-Interpenetrating Polymer Network of Chitosan Hydrogel. *Cellulose* **2017**, *24*, 2215–2228. [CrossRef]
62. Elgadir, M.A.; Uddin, M.S.; Ferdosh, S.; Adam, A.; Chowdhury, A.J.K.; Sarker, M.Z.I. Impact of Chitosan Composites and Chitosan Nanoparticle Composites on Various Drug Delivery Systems: A Review. *J. Food Drug Anal.* **2015**, *23*, 619–629. [CrossRef]
63. Arvanitoyannis, I.S.; Nakayama, A.; Aiba, S. ichi Chitosan and Gelatin Based Edible Films: State Diagrams, Mechanical and Permeation Properties. *Carbohydr. Polym.* **1998**, *37*, 371–382. [CrossRef]
64. Schmitz, C.; Auza, L.G.; Koberidze, D.; Rasche, S.; Fischer, R.; Bortesi, L. Conversion of Chitin to Defined Chitosan Oligomers: Current Status and Future Prospects. *Mar. Drugs* **2019**, *17*, 452. [CrossRef] [PubMed]
65. Andrady, A.L.; Xu, P. Elastic Behavior of Chitosan Films. *J. Polym. Sci. Part B Polym. Phys.* **1997**, *35*, 517–521. [CrossRef]
66. Jardine, A.; Sayed, S. Challenges in the Valorisation of Chitinous Biomass within the Biorefinery Concept. *Curr. Opin. Green Sustain. Chem.* **2016**, *2*, 34–39. [CrossRef]
67. Chaudhary, S.; Kumar, S.; Kumar, V.; Sharma, R. Chitosan Nanoemulsions as Advanced Edible Coatings for Fruits and Vegetables: Composition, Fabrication and Developments in Last Decade. *Int. J. Biol. Macromol.* **2020**, *152*, 154–170. [CrossRef] [PubMed]
68. Younes, I.; Rinaudo, M. Chitin and Chitosan Preparation from Marine Sources. Structure, Properties and Applications. *Mar. Drugs* **2015**, *13*, 1133–1174. [CrossRef] [PubMed]
69. Sánchez-Machado, D.I.; López-Cervantes, J.; Correa-Murrieta, M.A.; Sánchez-Duarte, R.G.; Cruz-Flores, P.; de la Mora-López, G.S. Chitosan. In *Nonvitamin and Nonmineral Nutritional Supplements*; Elsevier Inc.: Amsterdam, The Netherlands, 2019; pp. 485–493. ISBN 9780128124918.
70. Xu, Y.; Bajaj, M.; Schneider, R.; Grage, S.L.; Ulrich, A.S.; Winter, J.; Gallert, C. Transformation of the Matrix Structure of Shrimp Shells during Bacterial Deproteination and Demineralization. *Microb. Cell Fact.* **2013**, *12*, 90. [CrossRef] [PubMed]
71. Huang, G.; Liu, Y.; Chen, L. Chitosan and Its Derivatives as Vehicles for Drug Delivery. *Drug Deliv.* **2017**, *24*, 108–113. [CrossRef]
72. Ramkumar, R.; Minakshi, M. Fabrication of Ultrathin $CoMoO_4$ Nanosheets Modified with Chitosan and Their Improved Performance in Energy Storage Device. *Dalton Trans.* **2015**, *44*, 6158–6168. [CrossRef] [PubMed]
73. Ramkumar, R.; Minakshi, M. A Biopolymer Gel-Decorated Cobalt Molybdate Nanowafer: Effective Graft Polymer Cross-Linked with an Organic Acid for Better Energy Storage. *New J. Chem.* **2016**, *40*, 2863–2877. [CrossRef]
74. Ioelovich, M. Cellulose as a Nanostructured Polymer: A Short Review. *BioResources* **2008**, *3*, 1403–1418. [CrossRef]
75. Abdul Khalil, H.P.S.; Davoudpour, Y.; Islam, M.N.; Mustapha, A.; Sudesh, K.; Dungani, R.; Jawaid, M. Production and Modification of Nanofibrillated Cellulose Using Various Mechanical Processes: A Review. *Carbohydr. Polym.* **2014**, *99*, 649–665. [CrossRef] [PubMed]
76. Foresti, M.L.; Vázquez, A.; Boury, B. Applications of Bacterial Cellulose as Precursor of Carbon and Composites with Metal Oxide, Metal Sulfide and Metal Nanoparticles: A Review of Recent Advances. *Carbohydr. Polym.* **2017**, *157*, 447–467. [CrossRef]
77. Klemm, D.; Heublein, B.; Fink, H.P.; Bohn, A. Cellulose: Fascinating Biopolymer and Sustainable Raw Material. *Angew. Chem. Int. Ed.* **2005**, *44*, 3358–3393. [CrossRef] [PubMed]
78. Carpenter, A.W.; de Lannoy, C.F.; Wiesner, M.R. Cellulose Nanomaterials in Water Treatment Technologies. *Environ. Sci. Technol.* **2015**, *49*, 5277–5287. [CrossRef] [PubMed]
79. Kumar, R.; Strezov, V.; Weldekidan, H.; He, J.; Singh, S.; Kan, T.; Dastjerdi, B. Lignocellulose Biomass Pyrolysis for Bio-Oil Production: A Review of Biomass Pre-Treatment Methods for Production of Drop-in Fuels. *Renew. Sustain. Energy Rev.* **2020**, *123*, 109763. [CrossRef]
80. Jin, Z.; Katsumata, K.S.; Lam, T.B.T.; Iiyama, K. Covalent Linkages between Cellulose and Lignin in Cell Walls of Coniferous and Nonconiferous Woods. *Biopolymers* **2006**, *83*, 103–110. [CrossRef] [PubMed]
81. Bhat, A.H.; Khan, I.; Usmani, M.A.; Umapathi, R.; Al-Kindy, S.M.Z. Cellulose an Ageless Renewable Green Nanomaterial for Medical Applications: An Overview of Ionic Liquids in Extraction, Separation and Dissolution of Cellulose. *Int. J. Biol. Macromol.* **2019**, *129*, 750–777. [CrossRef]
82. Peng, B.; Yao, Z.; Wang, X.; Crombeen, M.; Sweeney, D.G.; Tam, K.C. Cellulose-Based Materials in Wastewater Treatment of Petroleum Industry. *Green Energy Environ.* **2020**, *5*, 37–49. [CrossRef]

83. Sharma, A.; Thakur, M.; Bhattacharya, M.; Mandal, T.; Goswami, S. Commercial Application of Cellulose Nano-Composites—A Review. *Biotechnol. Rep.* **2019**, *21*, e00316. [CrossRef] [PubMed]
84. Garba, Z.N.; Lawan, I.; Zhou, W.; Zhang, M.; Wang, L.; Yuan, Z. Microcrystalline Cellulose (MCC) Based Materials as Emerging Adsorbents for the Removal of Dyes and Heavy Metals—A Review. *Sci. Total Environ.* **2020**, *717*, 135070. [CrossRef]
85. Pachuau, L.; Dutta, R.S.; Hauzel, L.; Devi, T.B.; Deka, D. Evaluation of Novel Microcrystalline Cellulose from *Ensete glaucum* (Roxb.) Cheesman Biomass as Sustainable Drug Delivery Biomaterial. *Carbohydr. Polym.* **2019**, *206*, 336–343. [CrossRef]
86. Ren, H.; Shen, J.; Pei, J.; Wang, Z.; Peng, Z.; Fu, S.; Zheng, Y. Characteristic Microcrystalline Cellulose Extracted by Combined Acid and Enzyme Hydrolysis of Sweet Sorghum. *Cellulose* **2019**, *26*, 8367–8381. [CrossRef]
87. Baruah, J.; Deka, R.C.; Kalita, E. Greener Production of Microcrystalline Cellulose (MCC) from *Saccharum spontaneum* (Kans Grass): Statistical Optimization. *Int. J. Biol. Macromol.* **2020**, *154*, 672–682. [CrossRef] [PubMed]
88. Abu-Thabit, N.Y.; Judeh, A.A.; Hakeem, A.S.; Ul-Hamid, A.; Umar, Y.; Ahmad, A. Isolation and Characterization of Microcrystalline Cellulose from Date Seeds (*Phoenix dactylifera* L.). *Int. J. Biol. Macromol.* **2020**, *155*, 730–739. [CrossRef] [PubMed]
89. Fouad, H.; Kian, L.K.; Jawaid, M.; Alotaibi, M.D.; Alothman, O.Y.; Hashem, M. Characterization of Microcrystalline Cellulose Isolated from Conocarpus Fiber. *Polymers* **2020**, *12*, 2926. [CrossRef]
90. Islam, M.S.; Kao, N.; Bhattacharya, S.N.; Gupta, R.; Choi, H.J. Potential Aspect of Rice Husk Biomass in Australia for Nanocrystalline Cellulose Production. *Chin. J. Chem. Eng.* **2018**, *26*, 465–476. [CrossRef]
91. Hastuti, N.; Kanomata, K.; Kitaoka, T. Hydrochloric Acid Hydrolysis of Pulps from Oil Palm Empty Fruit Bunches to Produce Cellulose Nanocrystals. *J. Polym. Environ.* **2018**, *26*, 3698–3709. [CrossRef]
92. Trilokesh, C.; Uppuluri, K.B. Isolation and Characterization of Cellulose Nanocrystals from Jackfruit Peel. *Sci. Rep.* **2019**, *9*, 16709. [CrossRef]
93. Kian, L.K.; Saba, N.; Jawaid, M.; Alothman, O.Y.; Fouad, H. Properties and Characteristics of Nanocrystalline Cellulose Isolated from Olive Fiber. *Carbohydr. Polym.* **2020**, *241*, 116423. [CrossRef]
94. Nang An, V.; Chi Nhan, H.T.; Tap, T.D.; Van, T.T.T.; Van Viet, P.; Van Hieu, L. Extraction of High Crystalline Nanocellulose from Biorenewable Sources of Vietnamese Agricultural Wastes. *J. Polym. Environ.* **2020**, *28*, 1465–1474. [CrossRef]
95. Mishra, S.; Kharkar, P.S.; Pethe, A.M. Biomass and Waste Materials as Potential Sources of Nanocrystalline Cellulose: Comparative Review of Preparation Methods (2016–Till Date). *Carbohydr. Polym.* **2019**, *207*, 418–427. [CrossRef]
96. Sarkar, N.; Ghosh, S.K.; Bannerjee, S.; Aikat, K. Bioethanol Production from Agricultural Wastes: An Overview. *Renew. Energy* **2012**, *37*, 19–27. [CrossRef]
97. Sharma, N.; Bhardwaj, N.K.; Singh, R.B.P. Environmental Issues of Pulp Bleaching and Prospects of Peracetic Acid Pulp Bleaching: A Review. *J. Clean. Prod.* **2020**, *256*, 120338. [CrossRef]
98. Dilamian, M.; Noroozi, B. A Combined Homogenization-High Intensity Ultrasonication Process for Individualizaion of Cellulose Micro-Nano Fibers from Rice Straw. *Cellulose* **2019**, *26*, 5831–5849. [CrossRef]
99. Kumneadklang, S.; O-Thong, S.; Larpkiattaworn, S. Characterization of Cellulose Fiber Isolated from Oil Palm Frond Biomass. In *Materials Today: Proceedings of the First Materials Research Society of Thailand International Conference, Chiang Mai, Thailand, 31 October–3 November 2017*; Elsevier Ltd.: Amsterdam, The Netherlands, 2019; Volume 17, pp. 1995–2001.
100. Candido, R.G.; Gonçalves, A.R. Evaluation of Two Different Applications for Cellulose Isolated from Sugarcane Bagasse in a Biorefinery Concept. *Ind. Crops Prod.* **2019**, *142*, 111616. [CrossRef]
101. Qasim, U.; Ali, Z.; Nazir, M.S.; Ul Hassan, S.; Rafiq, S.; Jamil, F.; Al-Muhtaseb, A.H.; Ali, M.; Khan Niazi, M.B.; Ahmad, N.M.; et al. Isolation of Cellulose from Wheat Straw Using Alkaline Hydrogen Peroxide and Acidified Sodium Chlorite Treatments: Comparison of Yield and Properties. *Adv. Polym. Technol.* **2020**, *2020*, 9765950. [CrossRef]
102. Syafri, E.; Jamaluddin; Sari, N.H.; Mahardika, M.; Amanda, P.; Ilyas, R.A. Isolation and Characterization of Cellulose Nanofibers from *Agave gigantea* by Chemical-Mechanical Treatment. *Int. J. Biol. Macromol.* **2022**, *200*, 25–33. [CrossRef] [PubMed]
103. Pasma, S.A.; Daik, R.; Maskat, M.Y.; Hassan, O. Application of Box-Behnken Design in Optimization of Glucose Production from Oil Palm Empty Fruit Bunch Cellulose. *Int. J. Polym. Sci.* **2013**, *2013*, 104502. [CrossRef]
104. Brown, A.J. XLIII.—On an Acetic Ferment Which Forms Cellulose. *J. Chem. Soc. Trans.* **1886**, *49*, 432–439. [CrossRef]
105. Iguchi, M.; Yamanaka, S.; Budhiono, A. Bacterial Cellulose—A Masterpiece of Nature's Arts. *J. Mater. Sci.* **2000**, *35*, 261–270. [CrossRef]
106. Gatenholm, P.; Klemm, D. Bacterial Nanocellulose as a Renewable Material for Biomedical Applications. *MRS Bull.* **2010**, *35*, 208–213. [CrossRef]
107. Jonas, R.; Farah, L.F. Production and Application of Microbial Cellulose. *Polym. Degrad. Stab.* **1998**, *59*, 101–106. [CrossRef]
108. Wang, J.; Tavakoli, J.; Tang, Y. Bacterial Cellulose Production, Properties and Applications with Different Culture Methods—A Review. *Carbohydr. Polym.* **2019**, *219*, 63–76. [CrossRef] [PubMed]
109. Lin, S.P.; Loira Calvar, I.; Catchmark, J.M.; Liu, J.R.; Demirci, A.; Cheng, K.C. Biosynthesis, Production and Applications of Bacterial Cellulose. *Cellulose* **2013**, *20*, 2191–2219. [CrossRef]
110. Mohammadkazemi, F.; Azin, M.; Ashori, A. Production of Bacterial Cellulose Using Different Carbon Sources and Culture Media. *Carbohydr. Polym.* **2015**, *117*, 518–523. [CrossRef] [PubMed]
111. Klemm, D.; Schumann, D.; Udhardt, U.; Marsch, S. Bacterial Synthesized Cellulose—Artificial Blood Vessels for Microsurgery. *Prog. Polym. Sci.* **2001**, *26*, 1561–1603. [CrossRef]

112. Ross, P.; Mayer, R.; Benziman, M. Cellulose Biosynthesis and Function in Bacteria. *Microbiol. Mol. Biol. Rev.* **1991**, *55*, 35–58. [CrossRef]
113. Qiu, K.; Netravali, A.N. A Review of Fabrication and Applications of Bacterial Cellulose Based Nanocomposites. *Polym. Rev.* **2014**, *54*, 598–626. [CrossRef]
114. Islam, M.U.; Ullah, M.W.; Khan, S.; Shah, N.; Park, J.K. Strategies for Cost-Effective and Enhanced Production of Bacterial Cellulose. *Int. J. Biol. Macromol.* **2017**, *102*, 1166–1173. [CrossRef]
115. Pang, M.; Huang, Y.; Meng, F.; Zhuang, Y.; Liu, H.; Du, M.; Ma, Q.; Wang, Q.; Chen, Z.; Chen, L.; et al. Application of Bacterial Cellulose in Skin and Bone Tissue Engineering. *Eur. Polym. J.* **2020**, *122*, 109365. [CrossRef]
116. Chawla, P.R.; Bajaj, I.B.; Survase, S.A.; Singhal, R.S. Microbial Cellulose: Fermentative Production and Applications. *Food Technol. Biotechnol.* **2009**, *47*, 107–124.
117. Ul-Islam, M.; Khan, S.; Ullah, M.W.; Park, J.K. Bacterial Cellulose Composites: Synthetic Strategies and Multiple Applications in Bio-Medical and Electro-Conductive Fields. *Biotechnol. J.* **2015**, *10*, 1847–1861. [CrossRef]
118. Hu, Y.; Catchmark, J.M. Formation and Characterization of Spherelike Bacterial Cellulose Particles Produced by *Acetobacter xylinum* JCM 9730 Strain. *Biomacromolecules* **2010**, *11*, 1727–1734. [CrossRef] [PubMed]
119. Choi, S.M.; Shin, E.J. The Nanofication and Functionalization of Bacterial Cellulose and Its Applications. *Nanomaterials* **2020**, *10*, 406. [CrossRef]
120. Chao, Y.; Sugano, Y.; Kouda, T.; Yoshinaga, F.; Shoda, M. Production of Bacterial Cellulose by *Acetobacter xylinum* with an Air-Lift Reactor. *Biotechnol. Tech.* **1997**, *11*, 829–832. [CrossRef]
121. Wang, B.; Lv, X.; Chen, S.; Li, Z.; Sun, X.; Feng, C.; Wang, H.; Xu, Y. In Vitro Biodegradability of Bacterial Cellulose by Cellulase in Simulated Body Fluid and Compatibility in Vivo. *Cellulose* **2016**, *23*, 3187–3198. [CrossRef]
122. Römling, U.; Galperin, M.Y. Bacterial Cellulose Biosynthesis: Diversity of Operons, Subunits, Products, and Functions. *Trends Microbiol.* **2015**, *23*, 545–557. [CrossRef] [PubMed]
123. Jonoobi, M.; Oladi, R.; Davoudpour, Y.; Oksman, K.; Dufresne, A.; Hamzeh, Y.; Davoodi, R. Different Preparation Methods and Properties of Nanostructured Cellulose from Various Natural Resources and Residues: A Review. *Cellulose* **2015**, *22*, 935–969. [CrossRef]
124. Torres, F.G.; Arroyo, J.J.; Troncoso, O.P. Bacterial Cellulose Nanocomposites: An All-Nano Type of Material. *Mater. Sci. Eng. C* **2019**, *98*, 1277–1293. [CrossRef]
125. Khoo, C.G.; Dasan, Y.K.; Lam, M.K.; Lee, K.T. Algae Biorefinery: Review on a Broad Spectrum of Downstream Processes and Products. *Bioresour. Technol.* **2019**, *292*, 121964. [CrossRef]
126. Sirajunnisa, A.R.; Surendhiran, D. Algae—A Quintessential and Positive Resource of Bioethanol Production: A Comprehensive Review. *Renew. Sustain. Energy Rev.* **2016**, *66*, 248–267. [CrossRef]
127. Zheng, L.X.; Chen, X.Q.; Cheong, K.L. Current Trends in Marine Algae Polysaccharides: The Digestive Tract, Microbial Catabolism, and Prebiotic Potential. *Int. J. Biol. Macromol.* **2020**, *151*, 344–354. [CrossRef] [PubMed]
128. Rangabhashiyam, S.; Balasubramanian, P. Characteristics, Performances, Equilibrium and Kinetic Modeling Aspects of Heavy Metal Removal Using Algae. *Bioresour. Technol. Rep.* **2019**, *5*, 261–279. [CrossRef]
129. Cheng, S.Y.; Show, P.L.; Lau, B.F.; Chang, J.S.; Ling, T.C. New Prospects for Modified Algae in Heavy Metal Adsorption. *Trends Biotechnol.* **2019**, *37*, 1255–1268. [CrossRef] [PubMed]
130. Tsekos, I. The Sites of Cellulose Synthesis in Algae: Diversity and Evolution of Cellulose-synthesizing Enzyme Complexes. *J. Phycol.* **1999**, *35*, 635–655. [CrossRef]
131. Itoh, T. Cellulose Synthesizing Complexes in Some Giant Marine Algae. *J. Cell Sci.* **1990**, *95*, 309–319. [CrossRef]
132. Rånby, B.; Rambo, C.R. Natural cellulose fibers and membranes: Biosynthesis. In *Reference Module in Materials Science and Materials Engineering*; Elsevier: Amsterdam, The Netherlands, 2017; pp. 1–7. ISBN 9780128035818.
133. Sampath Udeni Gunathilake, T.M.; Ching, Y.C.; Chuah, C.H.; Rahman, N.A.; Liou, N.S. Recent Advances in Celluloses and Their Hybrids for Stimuli-Responsive Drug Delivery. *Int. J. Biol. Macromol.* **2020**, *158*, 670–688. [CrossRef] [PubMed]
134. Mu, R.; Hong, X.; Ni, Y.; Li, Y.; Pang, J.; Wang, Q.; Xiao, J.; Zheng, Y. Recent Trends and Applications of Cellulose Nanocrystals in Food Industry. *Trends Food Sci. Technol.* **2019**, *93*, 136–144. [CrossRef]
135. Syrpas, M.; Venskutonis, P.R. Algae for the production of bio-based products. In *Biobased Products and Industries*; Galanakis, C.M., Ed.; Elsevier Inc.: Amsterdam, The Netherlands, 2020; pp. 203–243. ISBN 9780128184936.
136. Zhao, Y.; Li, J. Excellent Chemical and Material Cellulose from Tunicates: Diversity in Cellulose Production Yield and Chemical and Morphological Structures from Different Tunicate Species. *Cellulose* **2014**, *21*, 3427–3441. [CrossRef]
137. Hirose, E.; Kimura, S.; Itoh, T.; Nishikawa, J. Tunic Morphology and Cellulosic Components of Pyrosomas, Doliolids, and Salps (Thaliacea, Urochordata). *Biol. Bull.* **1999**, *196*, 113–120. [CrossRef] [PubMed]
138. Bauermeister, A.; Branco, P.C.; Furtado, L.C.; Jimenez, P.C.; Costa-Lotufo, L.V.; da Cruz Lotufo, T.M. Tunicates: A Model Organism to Investigate the Effects of Associated-Microbiota on the Production of Pharmaceuticals. *Drug Discov. Today Dis. Model.* **2018**, *28*, 13–20. [CrossRef]
139. Hirose, E. Ascidian Tunic Cells: Morphology and Functional Diversity of Free Cells outside the Epidermis. *Invertebr. Biol.* **2009**, *128*, 83–96. [CrossRef]

140. Sasakura, Y.; Nakashima, K.; Awazu, S.; Matsuoka, T.; Nakayama, A.; Azuma, J.I.; Satoh, N. Transposon-Mediated Insertional Mutagenesis Revealed the Functions of Animal Cellulose Synthase in the Ascidian *Ciona intestinalis*. In Proceedings of the National Academy of Sciences, Washington, DC, USA, 1 March 2005; Volume 102, pp. 15134–15139.
141. Calvino, C.; Macke, N.; Kato, R.; Rowan, S.J. Development, Processing and Applications of Bio-Sourced Cellulose Nanocrystal Composites. *Prog. Polym. Sci.* **2020**, *103*, 101221. [CrossRef]
142. Zhao, Y.; Zhang, Y.; Lindström, M.E.; Li, J. Tunicate Cellulose Nanocrystals: Preparation, Neat Films and Nanocomposite Films with Glucomannans. *Carbohydr. Polym.* **2015**, *117*, 286–296. [CrossRef] [PubMed]
143. Ruiz, M.M.; Cavaillé, J.Y.; Dufresne, A.; Graillat, C.; Gérard, J. New Waterborne Epoxy Coatings Based on Cellulose Nanofillers. *Macromol. Symp.* **2001**, *169*, 211–222. [CrossRef]
144. Grishkewich, N.; Mohammed, N.; Tang, J.; Tam, K.C. Recent Advances in the Application of Cellulose Nanocrystals. *Curr. Opin. Colloid Interface Sci.* **2017**, *29*, 32–45. [CrossRef]
145. Wang, Z.; Li, N.; Zong, L.; Zhang, J. Recent Advances in Vacuum Assisted Self-Assembly of Cellulose Nanocrystals. *Curr. Opin. Solid State Mater. Sci.* **2019**, *23*, 142–148. [CrossRef]
146. Sharma, S.; Dutta, V.; Singh, P.; Raizada, P.; Rahmani-Sani, A.; Hosseini-Bandegharaei, A.; Thakur, V.K. Carbon Quantum Dot Supported Semiconductor Photocatalysts for Efficient Degradation of Organic Pollutants in Water: A Review. *J. Clean. Prod.* **2019**, *228*, 755–769. [CrossRef]
147. Al-Mamun, M.R.; Kader, S.; Islam, M.S.; Khan, M.Z.H. Photocatalytic Activity Improvement and Application of UV-$TiO_2$ Photocatalysis in Textile Wastewater Treatment: A Review. *J. Environ. Chem. Eng.* **2019**, *7*, 103248. [CrossRef]
148. Onkani, S.P.; Diagboya, P.N.; Mtunzi, F.M.; Klink, M.J.; Olu-Owolabi, B.I.; Pakade, V. Comparative Study of the Photocatalytic Degradation of 2–Chlorophenol under UV Irradiation Using Pristine and Ag-Doped Species of $TiO_2$, ZnO and ZnS Photocatalysts. *J. Environ. Manag.* **2020**, *260*, 110145. [CrossRef]
149. Fujishima, A.; Kohayakawa, K.; Honda, K. Hydrogen Production under Sunlight with an Electrochemical Photocell. *J. Electrochem. Soc.* **1975**, *122*, 1487–1489. [CrossRef]
150. Chen, D.; Cheng, Y.; Zhou, N.; Chen, P.; Wang, Y.; Li, K.; Huo, S.; Cheng, P.; Peng, P.; Zhang, R.; et al. Photocatalytic Degradation of Organic Pollutants Using $TiO_2$-Based Photocatalysts: A Review. *J. Clean. Prod.* **2020**, *268*, 121725. [CrossRef]
151. Pelaez, M.; Nolan, N.T.; Pillai, S.C.; Seery, M.K.; Falaras, P.; Kontos, A.G.; Dunlop, P.S.M.; Hamilton, J.W.J.; Byrne, J.A.; O'Shea, K.; et al. A Review on the Visible Light Active Titanium Dioxide Photocatalysts for Environmental Applications. *Appl. Catal. B Environ.* **2012**, *125*, 331–349. [CrossRef]
152. Khaki, M.R.D.; Shafeeyan, M.S.; Raman, A.A.A.; Daud, W.M.A.W. Application of Doped Photocatalysts for Organic Pollutant Degradation—A Review. *J. Environ. Manag.* **2017**, *198*, 78–94. [CrossRef] [PubMed]
153. Kamat, P.V. $TiO_2$ Nanostructures: Recent Physical Chemistry Advances. *J. Phys. Chem. C* **2012**, *116*, 11849–11851. [CrossRef]
154. Ibhadon, A.O.; Fitzpatrick, P. Heterogeneous Photocatalysis: Recent Advances and Applications. *Catalysts* **2013**, *3*, 189–218. [CrossRef]
155. Acharya, R.; Parida, K. A Review on $TiO_2$/g-$C_3N_4$ Visible-Light-Responsive Photocatalysts for Sustainable Energy Generation and Environmental Remediation. *J. Environ. Chem. Eng.* **2020**, *8*, 103896. [CrossRef]
156. Kusiak-Nejman, E.; Morawski, A.W. $TiO_2$/Graphene-Based Nanocomposites for Water Treatment: A Brief Overview of Charge Carrier Transfer, Antimicrobial and Photocatalytic Performance. *Appl. Catal. B Environ.* **2019**, *253*, 179–186. [CrossRef]
157. Xu, F. Review of Analytical Studies on $TiO_2$ Nanoparticles and Particle Aggregation, Coagulation, Flocculation, Sedimentation, Stabilization. *Chemosphere* **2018**, *212*, 662–677. [CrossRef]
158. Ola, O.; Maroto-Valer, M.M. Review of Material Design and Reactor Engineering on $TiO_2$ Photocatalysis for $CO_2$ Reduction. *J. Photochem. Photobiol. C Photochem. Rev.* **2015**, *24*, 16–42. [CrossRef]
159. Mun, S.J.; Park, S.J. Graphitic Carbon Nitride Materials for Photocatalytic Hydrogen Production via Water Splitting: A Short Review. *Catalysts* **2019**, *9*, 805. [CrossRef]
160. Reyes-Coronado, D.; Rodríguez-Gattorno, G.; Espinosa-Pesqueira, M.E.; Cab, C.; de Coss, R.; Oskam, G. Phase-Pure $TiO_2$ Nanoparticles: Anatase, Brookite and Rutile. *Nanotechnology* **2008**, *19*, 145605. [CrossRef] [PubMed]
161. Chen, X.; Mao, S.S. Titanium Dioxide Nanomaterials: Synthesis, Properties, Modifications and Applications. *Chem. Rev.* **2007**, *107*, 2891–2959. [CrossRef] [PubMed]
162. Di Mo, S.; Ching, W.Y. Electronic and Optical Properties of Three Phases of Titanium Dioxide: Rutile, Anatase, and Brookite. *Phys. Rev. B* **1995**, *51*, 13023–13032. [CrossRef]
163. Yamakata, A.; Vequizo, J.J.M. Curious Behaviors of Photogenerated Electrons and Holes at the Defects on Anatase, Rutile, and Brookite $TiO_2$ Powders: A Review. *J. Photochem. Photobiol. C Photochem. Rev.* **2019**, *40*, 234–243. [CrossRef]
164. Zhang, J.; Zhou, P.; Liu, J.; Yu, J. New Understanding of the Difference of Photocatalytic Activity among Anatase, Rutile and Brookite $TiO_2$. *Phys. Chem. Chem. Phys.* **2014**, *16*, 20382–20386. [CrossRef]
165. Rani, M.; Shanker, U. Green synthesis of $TiO_2$ and its photocatalytic activity. In *Handbook of Smart Photocatalytic Materials: Fundamentals, Fabrications and Water Resources Applications*; Hussain, C.M., Mishra, A.K., Eds.; Elsevier: Amsterdam, The Netherlands, 2020; pp. 11–61. ISBN 9780128190517.
166. Haggerty, J.E.S.; Schelhas, L.T.; Kitchaev, D.A.; Mangum, J.S.; Garten, L.M.; Sun, W.; Stone, K.H.; Perkins, J.D.; Toney, M.F.; Ceder, G.; et al. High-Fraction Brookite Films from Amorphous Precursors. *Sci. Rep.* **2017**, *7*, 15232. [CrossRef]

167. Zhou, T.; Chen, S.; Li, L.; Wang, J.; Zhang, Y.; Li, J.; Bai, J.; Xia, L.; Xu, Q.; Rahim, M.; et al. Carbon Quantum Dots Modified Anatase/Rutile $TiO_2$ Photoanode with Dramatically Enhanced Photoelectrochemical Performance. *Appl. Catal. B Environ.* **2020**, *269*, 118776. [CrossRef]
168. Di Paola, A.; Bellardita, M.; Palmisano, L. Brookite, the Least Known $TiO_2$ Photocatalyst. *Catalysts* **2013**, *3*, 36–73. [CrossRef]
169. Vequizo, J.J.M.; Matsunaga, H.; Ishiku, T.; Kamimura, S.; Ohno, T.; Yamakata, A. Trapping-Induced Enhancement of Photocatalytic Activity on Brookite $TiO_2$ Powders: Comparison with Anatase and Rutile $TiO_2$ Powders. *ACS Catal.* **2017**, *7*, 2644–2651. [CrossRef]
170. Pirhashemi, M.; Habibi-Yangjeh, A.; Rahim Pouran, S. Review on the Criteria Anticipated for the Fabrication of Highly Efficient ZnO-Based Visible-Light-Driven Photocatalysts. *J. Ind. Eng. Chem.* **2018**, *62*, 1–25. [CrossRef]
171. Raizada, P.; Sudhaik, A.; Singh, P. Photocatalytic Water Decontamination Using Graphene and ZnO Coupled Photocatalysts: A Review. *Mater. Sci. Energy Technol.* **2019**, *2*, 509–525. [CrossRef]
172. Janotti, A.; Van De Walle, C.G. Fundamentals of Zinc Oxide as a Semiconductor. *Rep. Prog. Phys.* **2009**, *72*, 126501. [CrossRef]
173. Zhu, D.; Zhou, Q. Action and Mechanism of Semiconductor Photocatalysis on Degradation of Organic Pollutants in Water Treatment: A Review. *Environ. Nanotechnol. Monit. Manag.* **2019**, *12*, 100255. [CrossRef]
174. Kumar, S.G.; Rao, K.S.R.K. Zinc Oxide Based Photocatalysis: Tailoring Surface-Bulk Structure and Related Interfacial Charge Carrier Dynamics for Better Environmental Applications. *RSC Adv.* **2015**, *5*, 3306–3351. [CrossRef]
175. Tekin, D.; Kiziltas, H.; Ungan, H. Kinetic Evaluation of ZnO/$TiO_2$ Thin Film Photocatalyst in Photocatalytic Degradation of Orange G. *J. Mol. Liq.* **2020**, *306*, 112905. [CrossRef]
176. Hou, J.; Wu, Y.; Li, X.; Wei, B.; Li, S.; Wang, X. Toxic Effects of Different Types of Zinc Oxide Nanoparticles on Algae, Plants, Invertebrates, Vertebrates and Microorganisms. *Chemosphere* **2018**, *193*, 852–860. [CrossRef]
177. Özgür, Ü.; Alivov, Y.I.; Liu, C.; Teke, A.; Reshchikov, M.A.; Doğan, S.; Avrutin, V.; Cho, S.-J.; Morkoç, H. A Comprehensive Review of ZnO Materials and Devices. *J. Appl. Phys.* **2005**, *98*, 041301. [CrossRef]
178. Lee, K.M.; Lai, C.W.; Ngai, K.S.; Juan, J.C. Recent Developments of Zinc Oxide Based Photocatalyst in Water Treatment Technology: A Review. *Water Res.* **2016**, *88*, 428–448. [CrossRef]
179. Naseri, A.; Samadi, M.; Pourjavadi, A.; Moshfegh, A.Z.; Ramakrishna, S. Graphitic Carbon Nitride (g-$C_3N_4$)-Based Photocatalysts for Solar Hydrogen Generation: Recent Advances and Future Development Directions. *J. Mater. Chem. A* **2017**, *5*, 23406–23433. [CrossRef]
180. Lam, S.M.; Sin, J.C.; Mohamed, A.R. A Review on Photocatalytic Application of G-$C_3N_4$/Semiconductor (CNS) Nanocomposites towards the Erasure of Dyeing Wastewater. *Mater. Sci. Semicond. Process.* **2016**, *47*, 62–84. [CrossRef]
181. Wang, X.; Maeda, K.; Thomas, A.; Takanabe, K.; Xin, G.; Carlsson, J.M.; Domen, K.; Antonietti, M. A Metal-Free Polymeric Photocatalyst for Hydrogen Production from Water under Visible Light. *Nat. Mater.* **2009**, *8*, 76–80. [CrossRef] [PubMed]
182. Mishra, A.; Mehta, A.; Basu, S.; Shetti, N.P.; Reddy, K.R.; Aminabhavi, T.M. Graphitic Carbon Nitride (g–$C_3N_4$)–Based Metal-Free Photocatalysts for Water Splitting: A Review. *Carbon N. Y.* **2019**, *149*, 693–721. [CrossRef]
183. Wang, Y.; Wang, X.; Antonietti, M. Polymeric Graphitic Carbon Nitride as a Heterogeneous Organocatalyst: From Photochemistry to Multipurpose Catalysis to Sustainable Chemistry. *Angew. Chem. Int. Ed.* **2012**, *51*, 68–89. [CrossRef]
184. Zhang, S.; Gu, P.; Ma, R.; Luo, C.; Wen, T.; Zhao, G.; Cheng, W.; Wang, X. Recent Developments in Fabrication and Structure Regulation of Visible-Light-Driven g-$C_3N_4$-Based Photocatalysts towards Water Purification: A Critical Review. *Catal. Today* **2019**, *335*, 65–77. [CrossRef]
185. Mamba, G.; Mishra, A.K. Graphitic Carbon Nitride (g-$C_3N_4$) Nanocomposites: A New and Exciting Generation of Visible Light Driven Photocatalysts for Environmental Pollution Remediation. *Appl. Catal. B Environ.* **2016**, *198*, 347–377. [CrossRef]
186. Zheng, Y.; Lin, L.; Wang, B.; Wang, X. Graphitic Carbon Nitride Polymers toward Sustainable Photoredox Catalysis. *Angew. Chem. Int. Ed.* **2015**, *54*, 12868–12884. [CrossRef]
187. Huang, D.; Li, Z.; Zeng, G.; Zhou, C.; Xue, W.; Gong, X.; Yan, X.; Chen, S.; Wang, W.; Cheng, M. Megamerger in Photocatalytic Field: 2D g-$C_3N_4$ Nanosheets Serve as Support of 0D Nanomaterials for Improving Photocatalytic Performance. *Appl. Catal. B Environ.* **2019**, *240*, 153–173. [CrossRef]
188. Prasad, C.; Tang, H.; Liu, Q.; Bahadur, I.; Karlapudi, S.; Jiang, Y. A Latest Overview on Photocatalytic Application of G-$C_3N_4$ Based Nanostructured Materials for Hydrogen Production. *Int. J. Hydrogen Energy* **2020**, *45*, 337–379. [CrossRef]
189. Xu, Y.; Kraft, M.; Xu, R. Metal-Free Carbonaceous Electrocatalysts and Photocatalysts for Water Splitting. *Chem. Soc. Rev.* **2016**, *45*, 3039–3052. [CrossRef] [PubMed]
190. Dong, G.; Zhang, Y.; Pan, Q.; Qiu, J. A Fantastic Graphitic Carbon Nitride (g-$C_3N_4$) Material: Electronic Structure, Photocatalytic and Photoelectronic Properties. *J. Photochem. Photobiol. C Photochem. Rev.* **2014**, *20*, 33–50. [CrossRef]
191. Li, X.; Yu, J.; Wageh, S.; Al-Ghamdi, A.A.; Xie, J. Graphene in Photocatalysis: A Review. *Small* **2016**, *12*, 6640–6696. [CrossRef]
192. Prasad, C.; Liu, Q.; Tang, H.; Yuvaraja, G.; Long, J.; Rammohan, A.; Zyryanov, G.V. An Overview of Graphene Oxide Supported Semiconductors Based Photocatalysts: Properties, Synthesis and Photocatalytic Applications. *J. Mol. Liq.* **2020**, *297*, 111826. [CrossRef]
193. Kahng, S.; Yoo, H.; Kim, J.H. Recent Advances in Earth-Abundant Photocatalyst Materials for Solar $H_2$ Production. *Adv. Powder Technol.* **2020**, *31*, 11–28. [CrossRef]
194. Bai, H.; Li, C.; Shi, G. Functional Composite Materials Based on Chemically Converted Graphene. *Adv. Mater.* **2011**, *23*, 1089–1115. [CrossRef] [PubMed]

195. Rosman, N.N.; Mohamad Yunus, R.; Jeffery Minggu, L.; Arifin, K.; Salehmin, M.N.I.; Mohamed, M.A.; Kassim, M.B. Photocatalytic Properties of Two-Dimensional Graphene and Layered Transition-Metal Dichalcogenides Based Photocatalyst for Photoelectrochemical Hydrogen Generation: An Overview. *Int. J. Hydrogen Energy* **2018**, *43*, 18925–18945. [CrossRef]
196. Bano, Z.; Mazari, S.A.; Saeed, R.M.Y.; Majeed, M.A.; Xia, M.; Memon, A.Q.; Abro, R.; Wang, F. Water Decontamination by 3D Graphene Based Materials: A Review. *J. Water Process Eng.* **2020**, *36*, 101404. [CrossRef]
197. Nikokavoura, A.; Trapalis, C. Graphene and G-$C_3N_4$ Based Photocatalysts for $NO_x$ Removal: A Review. *Appl. Surf. Sci.* **2018**, *430*, 18–52. [CrossRef]
198. Pottathara, Y.B.; Tiyyagura, H.R.; Ahmad, Z.; Sadasivuni, K.K. Graphene Based Aerogels: Fundamentals and Applications as Supercapacitors. *J. Energy Storage* **2020**, *30*, 101549. [CrossRef]
199. Ji, X.; Xu, Y.; Zhang, W.; Cui, L.; Liu, J. Review of Functionalization, Structure and Properties of Graphene/Polymer Composite Fibers. *Compos. Part A Appl. Sci. Manuf.* **2016**, *87*, 29–45. [CrossRef]
200. Xiang, Q.; Yu, J.; Jaroniec, M. Graphene-Based Semiconductor Photocatalysts. *Chem. Soc. Rev.* **2012**, *41*, 782–796. [CrossRef] [PubMed]
201. Zhao, S.; Zhao, Z.; Yang, Z.; Ke, L.L.; Kitipornchai, S.; Yang, J. Functionally Graded Graphene Reinforced Composite Structures: A Review. *Eng. Struct.* **2020**, *210*, 110339. [CrossRef]
202. Sonia, F.J.; Aslam, M.; Mukhopadhyay, A. Understanding the Processing-Structure-Performance Relationship of Graphene and Its Variants as Anode Material for Li-Ion Batteries: A Critical Review. *Carbon N. Y.* **2020**, *156*, 130–165. [CrossRef]
203. Tarelho, J.P.G.; Soares dos Santos, M.P.; Ferreira, J.A.F.; Ramos, A.; Kopyl, S.; Kim, S.O.; Hong, S.; Kholkin, A. Graphene-Based Materials and Structures for Energy Harvesting with Fluids—A Review. *Mater. Today* **2018**, *21*, 1019–1041. [CrossRef]
204. Mehmood, A.; Mubarak, N.M.; Khalid, M.; Walvekar, R.; Abdullah, E.C.; Siddiqui, M.T.H.; Baloch, H.A.; Nizamuddin, S.; Mazari, S. Graphene Based Nanomaterials for Strain Sensor Application—A Review. *J. Environ. Chem. Eng.* **2020**, *8*, 103743. [CrossRef]
205. Verma, R.; Gangwar, J.; Srivastava, A.K. Multiphase $TiO_2$ Nanostructures: A Review of Efficient Synthesis, Growth Mechanism, Probing Capabilities, and Applications in Bio-Safety and Health. *RSC Adv.* **2017**, *7*, 44199–44224. [CrossRef]
206. Mamaghani, A.H.; Haghighat, F.; Lee, C.S. Systematic Variation of Preparation Time, Temperature, and Pressure in Hydrothermal Synthesis of Macro-/Mesoporous $TiO_2$ for Photocatalytic Air Treatment. *J. Photochem. Photobiol. A Chem.* **2019**, *378*, 156–170. [CrossRef]
207. Sun, Y.; Gao, Y.; Zeng, J.; Guo, J.; Wang, H. Enhancing Visible-Light Photocatalytic Activity of Ag-$TiO_2$ Nanowire Composites by One-Step Hydrothermal Process. *Mater. Lett.* **2020**, *279*, 128506. [CrossRef]
208. Santhi, K.; Navaneethan, M.; Harish, S.; Ponnusamy, S.; Muthamizhchelvan, C. Synthesis and Characterization of $TiO_2$ Nanorods by Hydrothermal Method with Different PH Conditions and Their Photocatalytic Activity. *Appl. Surf. Sci.* **2020**, *500*, 144058. [CrossRef]
209. Wang, H.; Qi, F.; Chen, X.; Guo, H.; Cui, W. Enhanced Photoelectrocatalytic Degradation by $TiO_2$ Nano-Arrays Decorated with Two-Dimensional Ultra-Thin P3HT Nanosheets. *Mater. Lett.* **2021**, *302*, 130432. [CrossRef]
210. Chen, M.; Sun, T.; Zhao, W.; Yang, X.; Chang, W.; Qian, X.; Yang, Q.; Chen, Z. In Situ Growth of Metallic 1T-$MoS_2$ on $TiO_2$ Nanotubes with Improved Photocatalytic Performance. *ACS Omega* **2021**, *6*, 12787–12793. [CrossRef]
211. Zhang, Q.; Wang, Y.; Zhu, X.; Liu, X.; Li, H. 1T and 2H Mixed Phase $MoS_2$ Nanobelts Coupled with $Ti^{3+}$ Self-Doped $TiO_2$ Nanosheets for Enhanced Photocatalytic Degradation of RhB under Visible Light. *Appl. Surf. Sci.* **2021**, *556*, 149768. [CrossRef]
212. Estrada-Flores, S.; Martínez-Luévanos, A.; Perez-Berumen, C.M.; García-Cerda, L.A.; Flores-Guia, T.E. Relationship between Morphology, Porosity, and the Photocatalytic Activity of $TiO_2$ Obtained by Sol–Gel Method Assisted with Ionic and Nonionic Surfactants. *Bol. Soc. Esp. Ceram. Vidr.* **2020**, *59*, 209–218. [CrossRef]
213. Abraham, C.; Devi, L.G. One-Pot Facile Sol-Gel Synthesis of W, N, C and S Doped $TiO_2$ and Its Application in the Photocatalytic Degradation of Thymol under the Solar Light Irradiation: Reaction Kinetics and Degradation Mechanism. *J. Phys. Chem. Solids* **2020**, *141*, 109350. [CrossRef]
214. Abbad, S.; Guergouri, K.; Gazaout, S.; Djebabra, S.; Zertal, A.; Barille, R.; Zaabat, M. Effect of Silver Doping on the Photocatalytic Activity of $TiO_2$ Nanopowders Synthesized by the Sol-Gel Route. *J. Environ. Chem. Eng.* **2020**, *8*, 103718. [CrossRef]
215. Pragathiswaran, C.; Smitha, C.; Abbubakkar, B.M.; Govindhan, P.; Krishnan, N.A. Synthesis and Characterization of $TiO_2$/ZnO-Ag Nanocomposite for Photocatalytic Degradation of Dyes and Anti-Microbial Activity. In *Materials Today: Proceedings of the International Conference on Advances in Materials Research, Tamil Nadu, India, 27–28 May 2021*; Elsevier Ltd.: Amsterdam, The Netherlands, 2021; Volume 45, pp. 3357–3364.
216. Jesus, M.A.M.L.; Ferreira, A.M.; Lima, L.F.S.; Batista, G.F.; Mambrini, R.V.; Mohallem, N.D.S. Micro-Mesoporous $TiO_2$/$SiO_2$ Nanocomposites: Sol-Gel Synthesis, Characterization, and Enhanced Photodegradation of Quinoline. *Ceram. Int.* **2021**, *47*, 23844–23850. [CrossRef]
217. De Filpo, G.; Pantuso, E.; Armentano, K.; Formoso, P.; Di Profio, G.; Poerio, T.; Fontananova, E.; Meringolo, C.; Mashin, A.I.; Nicoletta, F.P. Chemical Vapor Deposition of Photocatalyst Nanoparticles on PVDF Membranes for Advanced Oxidation Processes. *Membranes* **2018**, *8*, 35. [CrossRef]
218. Jeong, S.; Lee, H.; Park, H.; Jeon, K.J.; Park, Y.K.; Jung, S.C. Rapid Photocatalytic Degradation of Nitrobenzene under the Simultaneous Illumination of UV and Microwave Radiation Fields with a $TiO_2$ Ball Catalyst. *Catal. Today* **2018**, *307*, 65–72. [CrossRef]

219. Park, Y.K.; Ha, H.H.; Yu, Y.H.; Kim, B.J.; Bang, H.J.; Lee, H.; Jung, S.C. The Photocatalytic Destruction of Cimetidine Using Microwave-Assisted $TiO_2$ Photocatalysts Hybrid System. *J. Hazard. Mater.* **2020**, *391*, 122568. [CrossRef] [PubMed]
220. Hadjltaief, H.B.; Ben Zina, M.; Galvez, M.E.; Da Costa, P. Photocatalytic Degradation of Methyl Green Dye in Aqueous Solution over Natural Clay-Supported ZnO-$TiO_2$ Catalysts. *J. Photochem. Photobiol. A Chem.* **2016**, *315*, 25–33. [CrossRef]
221. Nada, A.A.; Nasr, M.; Viter, R.; Miele, P.; Roualdes, S.; Bechelany, M. Mesoporous $ZnFe_2O_4$@$TiO_2$ Nanofibers Prepared by Electrospinning Coupled to PECVD as Highly Performing Photocatalytic Materials. *J. Phys. Chem. C* **2017**, *121*, 24669–24677. [CrossRef]
222. Bento, R.T.; Correa, O.V.; Pillis, M.F. On the Surface Chemistry and the Reuse of Sulfur-Doped $TiO_2$ Films as Photocatalysts. *Mater. Chem. Phys.* **2021**, *261*, 124231. [CrossRef]
223. Mamaghani, A.H.; Haghighat, F.; Lee, C.S. Hydrothermal/Solvothermal Synthesis and Treatment of $TiO_2$ for Photocatalytic Degradation of Air Pollutants: Preparation, Characterization, Properties, and Performance. *Chemosphere* **2019**, *219*, 804–825. [CrossRef] [PubMed]
224. Gomathi Thanga Keerthana, B.; Solaiyammal, T.; Muniyappan, S.; Murugakoothan, P. Hydrothermal Synthesis and Characterization of $TiO_2$ Nanostructures Prepared Using Different Solvents. *Mater. Lett.* **2018**, *220*, 20–23. [CrossRef]
225. Ismael, M. A Review and Recent Advances in Solar-to-Hydrogen Energy Conversion Based on Photocatalytic Water Splitting over Doped-$TiO_2$ Nanoparticles. *Sol. Energy* **2020**, *211*, 522–546. [CrossRef]
226. Baral, A.; Das, D.P.; Minakshi, M.; Ghosh, M.K.; Padhi, D.K. Probing Environmental Remediation of RhB Organic Dye Using $\alpha$-$MnO_2$ under Visible- Light Irradiation: Structural, Photocatalytic and Mineralization Studies. *ChemistrySelect* **2016**, *1*, 4277–4285. [CrossRef]
227. Yu, J.; Lei, J.; Wang, L.; Zhang, J.; Liu, Y. $TiO_2$ Inverse Opal Photonic Crystals: Synthesis, Modification, and Applications—A Review. *J. Alloy. Compd.* **2018**, *769*, 740–757. [CrossRef]
228. Gomes, J.; Lincho, J.; Domingues, E.; Quinta-Ferreira, R.M.; Martins, R.C. N-$TiO_2$ Photocatalysts: A Review of Their Characteristics and Capacity for Emerging Contaminants Removal. *Water* **2019**, *11*, 373. [CrossRef]
229. Akpan, U.G.; Hameed, B.H. The Advancements in Sol-Gel Method of Doped-$TiO_2$ Photocatalysts. *Appl. Catal. A Gen.* **2010**, *375*, 1–11. [CrossRef]
230. Singh, R.; Dutta, S. A Review on $H_2$ Production through Photocatalytic Reactions Using $TiO_2$/$TiO_2$-Assisted Catalysts. *Fuel* **2018**, *220*, 607–620. [CrossRef]
231. MacWan, D.P.; Dave, P.N.; Chaturvedi, S. A Review on Nano-$TiO_2$ Sol-Gel Type Syntheses and Its Applications. *J. Mater. Sci.* **2011**, *46*, 3669–3686. [CrossRef]
232. Wang, Y.; He, Y.; Lai, Q.; Fan, M. Review of the Progress in Preparing Nano $TiO_2$: An Important Environmental Engineering Material. *J. Environ. Sci.* **2014**, *26*, 2139–2177. [CrossRef]
233. Zhang, Y.; Xiong, X.; Han, Y.; Zhang, X.; Shen, F.; Deng, S.; Xiao, H.; Yang, X.; Yang, G.; Peng, H. Photoelectrocatalytic Degradation of Recalcitrant Organic Pollutants Using $TiO_2$ Film Electrodes: An Overview. *Chemosphere* **2012**, *88*, 145–154. [CrossRef] [PubMed]
234. Shakeel Ahmad, M.; Pandey, A.K.; Abd Rahim, N. Advancements in the Development of $TiO_2$ Photoanodes and Its Fabrication Methods for Dye Sensitized Solar Cell (DSSC) Applications. A Review. *Renew. Sustain. Energy Rev.* **2017**, *77*, 89–108. [CrossRef]
235. Varshney, G.; Kanel, S.R.; Kempisty, D.M.; Varshney, V.; Agrawal, A.; Sahle-Demessie, E.; Varma, R.S.; Nadagouda, M.N. Nanoscale $TiO_2$ Films and Their Application in Remediation of Organic Pollutants. *Coord. Chem. Rev.* **2016**, *306*, 43–64. [CrossRef]
236. Do, H.H.; Nguyen, D.L.T.; Nguyen, X.C.; Le, T.H.; Nguyen, T.P.; Trinh, Q.T.; Ahn, S.H.; Vo, D.V.N.; Kim, S.Y.; Van Le, Q. Recent Progress in $TiO_2$-Based Photocatalysts for Hydrogen Evolution Reaction: A Review. *Arab. J. Chem.* **2020**, *13*, 3653–3671. [CrossRef]
237. Patil, S.B.; Basavarajappa, P.S.; Ganganagappa, N.; Jyothi, M.S.; Raghu, A.V.; Reddy, K.R. Recent Advances in Non-Metals-Doped $TiO_2$ Nanostructured Photocatalysts for Visible-Light Driven Hydrogen Production, $CO_2$ Reduction and Air Purification. *Int. J. Hydrogen Energy* **2019**, *44*, 13022–13039. [CrossRef]
238. Dehghani, M.; Nadeem, H.; Raghuwanshi, V.S.; Mahdavi, H.; Banaszak Holl, M.M.; Batchelor, W. ZnO/Cellulose Nanofiber Composites for Sustainable Sunlight-Driven Dye Degradation. *ACS Appl. Nano Mater.* **2020**, *3*, 10284–10295. [CrossRef]
239. Lu, M.; Cui, Y.; Zhao, S.; Fakhri, A. $Cr_2O_3$/Cellulose Hybrid Nanocomposites with Unique Properties: Facile Synthesis, Photocatalytic, Bactericidal and Antioxidant Application. *J. Photochem. Photobiol. B Biol.* **2020**, *205*, 111842. [CrossRef] [PubMed]
240. Gan, L.; Geng, A.; Song, C.; Xu, L.; Wang, L.; Fang, X.; Han, S.; Cui, J.; Mei, C. Simultaneous Removal of Rhodamine B and Cr(VI) from Water Using Cellulose Carbon Nanofiber Incorporated with Bismuth Oxybromide: The Effect of Cellulose Pyrolysis Temperature on Photocatalytic Performance. *Environ. Res.* **2020**, *185*, 109414. [CrossRef] [PubMed]
241. Wang, J.; Li, X.; Cheng, Q.; Lv, F.; Chang, C.; Zhang, L. Construction of $\beta$-FeOOH@tunicate Cellulose Nanocomposite Hydrogels and Their Highly Efficient Photocatalytic Properties. *Carbohydr. Polym.* **2020**, *229*, 115470. [CrossRef]
242. Yang, L.; Chen, C.; Hu, Y.; Wei, F.; Cui, J.; Zhao, Y.; Xu, X.; Chen, X.; Sun, D. Three-Dimensional Bacterial Cellulose/Polydopamine/$TiO_2$ Nanocomposite Membrane with Enhanced Adsorption and Photocatalytic Degradation for Dyes under Ultraviolet-Visible Irradiation. *J. Colloid Interface Sci.* **2020**, *562*, 21–28. [CrossRef] [PubMed]
243. Du, X.; Wang, Z.; Pan, J.; Gong, W.; Liao, Q.; Liu, L.; Yao, J. High Photocatalytic Activity of Cu@$Cu_2O$/RGO/Cellulose Hybrid Aerogels as Reusable Catalysts with Enhanced Mass and Electron Transfer. *React. Funct. Polym.* **2019**, *138*, 79–87. [CrossRef]
244. Li, H.; Zhang, L.; Lu, H.; Ma, J.; Zhou, X.; Wang, Z.; Yi, C. Macro-/Nanoporous Al-Doped ZnO/Cellulose Composites Based on Tunable Cellulose Fiber Sizes for Enhancing Photocatalytic Properties. *Carbohydr. Polym.* **2020**, *250*, 116873. [CrossRef]

245. Li, Y.; Pan, Y.; Zhang, B.; Liu, R. Adsorption and Photocatalytic Activity of Cu-Doped Cellulose Nanofibers/Nano-Titanium Dioxide for Different Types of Dyes. *Water Sci. Technol.* **2020**, *82*, 1665–1675. [CrossRef] [PubMed]
246. Ahmed, M.K.; Shalan, A.E.; Afifi, M.; El-Desoky, M.M.; Lanceros-Méndez, S. Silver-Doped Cadmium Selenide/Graphene Oxide-Filled Cellulose Acetate Nanocomposites for Photocatalytic Degradation of Malachite Green toward Wastewater Treatment. *ACS Omega* **2021**, *6*, 23129–23138. [CrossRef]
247. Abdullah, H.; Hsu, C.N.; Shuwanto, H.; Gultom, N.S.; Kebede, W.L.; Wu, C.M.; Lai, C.C.; Murakami, R.I.; Hirota, M.; Nakagaito, A.N.; et al. Immobilization of Cross-Linked In-Doped $Mo(O,S)_2$ on Cellulose Nanofiber for Effective Organic-Compound Degradation under Visible Light Illumination. *Prog. Nat. Sci. Mater. Int.* **2021**, *31*, 404–413. [CrossRef]
248. Mohamed, M.A.; Wan Salleh, W.N.; Jaafar, J.; Rosmi, M.S.; Hir, Z.A.M.; Abd Mutalib, M.; Ismail, A.F.; Tanemura, M. Carbon as Amorphous Shell and Interstitial Dopant in Mesoporous Rutile $TiO_2$: Bio-Template Assisted Sol-Gel Synthesis and Photocatalytic Activity. *Appl. Surf. Sci.* **2017**, *393*, 46–59. [CrossRef]
249. Liu, F.; Sun, Y.; Gu, J.; Gao, Q.; Sun, D.; Zhang, X.; Pan, B.; Qian, J. Highly Efficient Photodegradation of Various Organic Pollutants in Water: Rational Structural Design of Photocatalyst via Thiol-Ene Click Reaction. *Chem. Eng. J.* **2020**, *381*, 122631. [CrossRef]
250. Ahn, J.; Pak, S.; Kim, H. Synergetic Effect of Carbon Dot at Cellulose Nanofiber for Sustainable Metal-Free Photocatalyst. *Cellulose* **2021**, *28*, 11261–11274. [CrossRef]
251. Pham, X.N.; Pham, D.T.; Ngo, H.S.; Nguyen, M.B.; Doan, H.V. Characterization and Application of C–$TiO_2$ Doped Cellulose Acetate Nanocomposite Film for Removal of Reactive Red-195. *Chem. Eng. Commun.* **2021**, *208*, 304–317. [CrossRef]
252. Emam, H.E.; Abdelhamid, H.N.; Abdelhameed, R.M. Self-Cleaned Photoluminescent Viscose Fabric Incorporated Lanthanide-Organic Framework (Ln-MOF). *Dye Pigments* **2018**, *159*, 491–498. [CrossRef]
253. Wen, J.; Liu, H.; Zheng, Y.; Wu, Y.; Gao, J. A Novel of PTA/ZIF-8@cellulose Aerogel Composite Materials for Efficient Photocatalytic Degradation of Organic Dyes in Water. *Z. Anorg. Allg. Chem.* **2020**, *646*, 444–450. [CrossRef]
254. Lu, W.; Duan, C.; Liu, C.; Zhang, Y.; Meng, X.; Dai, L.; Wang, W.; Yu, H.; Ni, Y. A Self-Cleaning and Photocatalytic Cellulose-Fiber-Supported "Ag@AgCl@MOF-Cloth" Membrane for Complex Wastewater Remediation. *Carbohydr. Polym.* **2020**, *247*, 116691. [CrossRef] [PubMed]
255. Lu, W.; Duan, C.; Zhang, Y.; Gao, K.; Dai, L.; Shen, M.; Wang, W.; Wang, J.; Ni, Y. Cellulose-Based Electrospun Nanofiber Membrane with Core-Sheath Structure and Robust Photocatalytic Activity for Simultaneous and Efficient Oil Emulsions Separation, Dye Degradation and Cr(VI) Reduction. *Carbohydr. Polym.* **2021**, *258*, 117676. [CrossRef] [PubMed]
256. Zhao, Y.; Wang, Y.; Xiao, G.; Su, H. Fabrication of Biomaterial/$TiO_2$ Composite Photocatalysts for the Selective Removal of Trace Environmental Pollutants. *Chin. J. Chem. Eng.* **2019**, *27*, 1416–1428. [CrossRef]
257. Rajagopal, S.; Paramasivam, B.; Muniyasamy, K. Photocatalytic Removal of Cationic and Anionic Dyes in the Textile Wastewater by $H_2O_2$ Assisted $TiO_2$ and Micro-Cellulose Composites. *Sep. Purif. Technol.* **2020**, *252*, 117444. [CrossRef]
258. Chen, Y.; Xiang, Z.; Wang, D.; Kang, J.; Qi, H. Effective Photocatalytic Degradation and Physical Adsorption of Methylene Blue Using Cellulose/GO/$TiO_2$ Hydrogels. *RSC Adv.* **2020**, *10*, 23936–23943. [CrossRef]
259. Tavker, N.; Gaur, U.; Sharma, M. Cellulose Supported Bismuth Vanadate Nanocomposite for Effective Removal of Organic Pollutant. *J. Environ. Chem. Eng.* **2020**, *8*, 104027. [CrossRef]
260. Sharma, R.K.; Kumar, R. Functionalized Cellulose with Hydroxyethyl Methacrylate and Glycidyl Methacrylate for Metal Ions and Dye Adsorption Applications. *Int. J. Biol. Macromol.* **2019**, *134*, 704–721. [CrossRef]
261. Sharma, R.K.; Kumar, R.; Singh, A.P. Metal Ions and Organic Dyes Sorption Applications of Cellulose Grafted with Binary Vinyl Monomers. *Sep. Purif. Technol.* **2019**, *209*, 684–697. [CrossRef]
262. Moharrami, P.; Motamedi, E. Application of Cellulose Nanocrystals Prepared from Agricultural Wastes for Synthesis of Starch-Based Hydrogel Nanocomposites: Efficient and Selective Nanoadsorbent for Removal of Cationic Dyes from Water. *Bioresour. Technol.* **2020**, *313*, 123661. [CrossRef] [PubMed]
263. Komal; Gupta, K.; Kumar, V.; Tikoo, K.B.; Kaushik, A.; Singhal, S. Encrustation of Cadmium Sulfide Nanoparticles into the Matrix of Biomass Derived Silanized Cellulose Nanofibers for Adsorptive Detoxification of Pesticide and Textile Waste. *Chem. Eng. J.* **2020**, *385*, 123700. [CrossRef]
264. Iravani Mohammadabadi, S.; Javanbakht, V. Fabrication of Dual Cross-Linked Spherical Treated Waste Biomass/Alginate Adsorbent and Its Potential for Efficient Removal of Lead Ions from Aqueous Solutions. *Ind. Crops Prod.* **2021**, *168*, 113575. [CrossRef]
265. Tahazadeh, S.; Karimi, H.; Mohammadi, T.; Emrooz, H.B.M.; Tofighy, M.A. Fabrication of Biodegradable Cellulose Acetate/MOF-Derived Porous Carbon Nanocomposite Adsorbent for Methylene Blue Removal from Aqueous Solutions. *J. Solid State Chem.* **2021**, *299*, 122180. [CrossRef]
266. Liu, Q.; Yu, H.; Zeng, F.; Li, X.; Sun, J.; Li, C.; Lin, H.; Su, Z. HKUST-1 Modified Ultrastability Cellulose/Chitosan Composite Aerogel for Highly Efficient Removal of Methylene Blue. *Carbohydr. Polym.* **2021**, *255*, 117402. [CrossRef] [PubMed]
267. Zhou, Q.; Chen, J.J.; Jin, B.; Chu, S.; Peng, R. Modification of ZIF-8 on Bacterial Cellulose for an Efficient Selective Capture of U(VI). *Cellulose* **2021**, *28*, 5241–5256. [CrossRef]
268. Abdelhameed, R.M.; Abdel-Gawad, H.; Emam, H.E. Macroporous Cu-MOF@cellulose Acetate Membrane Serviceable in Selective Removal of Dimethoate Pesticide from Wastewater. *J. Environ. Chem. Eng.* **2021**, *9*, 105121. [CrossRef]

269. Bian, W.; Chen, J.; Chen, Y.; Xu, W.; Jia, J. A Novel Waste Paper Cellulose-Based Cu-MOF Hybrid Material Threaded by PSS for Lithium Extraction with High Adsorption Capacity and Selectivity. *Cellulose* **2021**, *28*, 3041–3054. [CrossRef]
270. Modi, S.; Fulekar, M.H. Synthesis and Characterization of Zinc Oxide Nanoparticles and Zinc Oxide/Cellulose Nanocrystals Nanocomposite for Photocatalytic Degradation of Methylene Blue Dye under Solar Light Irradiation. *Nanotechnol. Environ. Eng.* **2020**, *5*, 18. [CrossRef]
271. Anusuyadevi, P.R.; Riazanova, A.V.; Hedenqvist, M.S.; Svagan, A.J. Floating Photocatalysts for Effluent Refinement Based on Stable Pickering Cellulose Foams and Graphitic Carbon Nitride (g-$C_3N_4$). *ACS Omega* **2020**, *5*, 22411–22419. [CrossRef] [PubMed]
272. Yoon, Y.H.; Lee, S.Y.; Gwon, J.G.; Cho, H.J.; Wu, Q.; Kim, Y.H.; Lee, W.H. Photocatalytic Performance of Highly Transparent and Mesoporous Molybdenum-Doped Titania Films Fabricated by Templating Cellulose Nanocrystals. *Ceram. Int.* **2018**, *44*, 16647–16653. [CrossRef]
273. Yang, Q.; Zhai, Y.; Li, X.; Li, H. Synthesis of $Fe_3O_4$/Pr-BiOCl/Luffa Composites with Enhanced Visible Light Photoactivity for Organic Dyes Degradation. *Mater. Res. Bull.* **2018**, *106*, 409–417. [CrossRef]
274. Yang, J.; Luo, X. Ag-Doped $TiO_2$ Immobilized Cellulose-Derived Carbon Beads: One-Pot Preparation, Photocatalytic Degradation Performance and Mechanism of Ceftriaxone Sodium. *Appl. Surf. Sci.* **2021**, *542*, 148724. [CrossRef]
275. Zhao, J.; Chen, X.; Zhou, Y.; Tian, H.; Guo, Q.; Hu, X. Efficient Removal of Oil Pollutant via Simultaneous Adsorption and Photocatalysis Using La–N–$TiO_2$–Cellulose/$SiO_2$ Difunctional Aerogel Composite. *Res. Chem. Intermed.* **2020**, *46*, 1805–1822. [CrossRef]
276. RanguMagar, A.B.; Chhetri, B.P.; Parameswaran-Thankam, A.; Watanabe, F.; Sinha, A.; Kim, J.W.; Saini, V.; Biris, A.S.; Ghosh, A. Nanocrystalline Cellulose-Derived Doped Carbonaceous Material for Rapid Mineralization of Nitrophenols under Visible Light. *ACS Omega* **2018**, *3*, 8111–8121. [CrossRef]
277. Xiao, H.; Shan, Y.; Zhang, W.; Huang, L.; Chen, L.; Ni, Y.; Boury, B.; Wu, H. C-Nanocoated ZnO by TEMPO-Oxidized Cellulose Templating for Improved Photocatalytic Performance. *Carbohydr. Polym.* **2020**, *235*, 115958. [CrossRef]
278. Zhang, C.; Guo, D.; Shen, T.; Hou, X.; Zhu, M.; Liu, S.; Hu, Q. Titanium Dioxide/Magnetic Metal-Organic Framework Preparation for Organic Pollutants Removal from Water under Visible Light. *Colloids Surf. A Physicochem. Eng. Asp.* **2020**, *589*, 124484. [CrossRef]
279. Abdelhameed, R.M.; El-Shahat, M.; Emam, H.E. Employable Metal (Ag & Pd)@MIL-125-$NH_2$@cellulose Acetate Film for Visible-Light Driven Photocatalysis for Reduction of Nitro-Aromatics. *Carbohydr. Polym.* **2020**, *247*, 116695. [CrossRef]
280. Emam, H.E.; El-Shahat, M.; Abdelhameed, R.M. Observable Removal of Pharmaceutical Residues by Highly Porous Photoactive Cellulose Acetate@MIL-MOF Film. *J. Hazard. Mater.* **2021**, *414*, 125509. [CrossRef]
281. Xiu, Z.; Guo, M.; Zhao, T.; Pan, K.; Xing, Z.; Li, Z.; Zhou, W. Recent Advances in $Ti^{3+}$ Self-Doped Nanostructured $TiO_2$ Visible Light Photocatalysts for Environmental and Energy Applications. *Chem. Eng. J.* **2020**, *382*, 123011. [CrossRef]
282. Makuła, P.; Pacia, M.; Macyk, W. How to Correctly Determine the Band Gap Energy of Modified Semiconductor Photocatalysts Based on UV-Vis Spectra. *J. Phys. Chem. Lett.* **2018**, *9*, 6814–6817. [CrossRef]
283. Zhu, D.; Zhou, Q. Nitrogen Doped G-$C_3N_4$ with the Extremely Narrow Band Gap for Excellent Photocatalytic Activities under Visible Light. *Appl. Catal. B Environ.* **2021**, *281*, 119474. [CrossRef]
284. Abdullah, H.; Khan, M.M.R.; Ong, H.R.; Yaakob, Z. Modified $TiO_2$ Photocatalyst for $CO_2$ Photocatalytic Reduction: An Overview. *J. $CO_2$ Util.* **2017**, *22*, 15–32. [CrossRef]
285. Teh, C.M.; Mohamed, A.R. Roles of Titanium Dioxide and Ion-Doped Titanium Dioxide on Photocatalytic Degradation of Organic Pollutants (Phenolic Compounds and Dyes) in Aqueous Solutions: A Review. *J. Alloy. Compd.* **2011**, *509*, 1648–1660. [CrossRef]
286. Basavarajappa, P.S.; Patil, S.B.; Ganganagappa, N.; Reddy, K.R.; Raghu, A.V.; Reddy, C.V. Recent Progress in Metal-Doped $TiO_2$, Non-Metal Doped/Codoped $TiO_2$ and $TiO_2$ Nanostructured Hybrids for Enhanced Photocatalysis. *Int. J. Hydrogen Energy* **2020**, *45*, 7764–7778. [CrossRef]
287. Thambiliyagodage, C. Activity Enhanced $TiO_2$ Nanomaterials for Photodegradation of Dyes—A Review. *Environ. Nanotechnol. Monit. Manag.* **2021**, *16*, 100592. [CrossRef]
288. Wu, Z.; Yuan, X.; Zhang, J.; Wang, H.; Jiang, L.; Zeng, G. Photocatalytic Decontamination of Wastewater Containing Organic Dyes by Metal–Organic Frameworks and Their Derivatives. *ChemCatChem* **2016**, *9*, 41–64. [CrossRef]
289. Wang, C.C.; Wang, X.; Liu, W. The Synthesis Strategies and Photocatalytic Performances of $TiO_2$/MOFs Composites: A State-of-the-Art Review. *Chem. Eng. J.* **2020**, *391*, 123601. [CrossRef]
290. Bobbitt, N.S.; Mendonca, M.L.; Howarth, A.J.; Islamoglu, T.; Hupp, J.T.; Farha, O.K.; Snurr, R.Q. Metal-Organic Frameworks for the Removal of Toxic Industrial Chemicals and Chemical Warfare Agents. *Chem. Soc. Rev.* **2017**, *46*, 3357–3385. [CrossRef] [PubMed]
291. Pennells, J.; Godwin, I.D.; Amiralian, N.; Martin, D.J. Trends in the Production of Cellulose Nanofibers from Non-Wood Sources. *Cellulose* **2020**, *27*, 575–593. [CrossRef]
292. Debnath, B.; Haldar, D.; Purkait, M.K. A Critical Review on the Techniques Used for the Synthesis and Applications of Crystalline Cellulose Derived from Agricultural Wastes and Forest Residues. *Carbohydr. Polym.* **2021**, *273*, 118537. [CrossRef]
293. Zhou, C.; Wang, Y. Recent Progress in the Conversion of Biomass Wastes into Functional Materials for Value-Added Applications. *Sci. Technol. Adv. Mater.* **2020**, *21*, 787–804. [CrossRef] [PubMed]
294. Baghel, R.S.; Reddy, C.R.K.; Singh, R.P. Seaweed-Based Cellulose: Applications, and Future Perspectives. *Carbohydr. Polym.* **2021**, *267*, 118241. [CrossRef]

295. Hamawand, I.; Seneweera, S.; Kumarasinghe, P.; Bundschuh, J. Nanoparticle Technology for Separation of Cellulose, Hemicellulose and Lignin Nanoparticles from Lignocellulose Biomass: A Short Review. *Nano-Struct. Nano-Objects* **2020**, *24*, 100601. [CrossRef]
296. Hafemann, E.; Battisti, R.; Bresolin, D.; Marangoni, C.; Machado, R.A.F. Enhancing Chlorine-Free Purification Routes of Rice Husk Biomass Waste to Obtain Cellulose Nanocrystals. *Waste Biomass Valorization* **2020**, *11*, 6595–6611. [CrossRef]
297. Oyewo, O.A.; Elemike, E.E.; Onwudiwe, D.C.; Onyango, M.S. Metal Oxide-Cellulose Nanocomposites for the Removal of Toxic Metals and Dyes from Wastewater. *Int. J. Biol. Macromol.* **2020**, *164*, 2477–2496. [CrossRef] [PubMed]

polymers

MDPI

*Review*

# Biocomposite Materials Based on Poly(3-hydroxybutyrate) and Chitosan: A Review

**Yuliya Zhuikova, Vsevolod Zhuikov * and Valery Varlamov**

Research Center of Biotechnology of the Russian Academy of Sciences 33, Bld. 2 Leninsky Ave, Moscow 119071, Russia
* Correspondence: vsevolod1905@yandex.ru; Tel.: +7-915-320-73-80

**Abstract:** One of the important directions in the development of modern medical devices is the search and creation of new materials, both synthetic and natural, which can be more effective in their properties than previously used materials. Traditional materials such as metals, ceramics, and synthetic polymers used in medicine have certain drawbacks, such as insufficient biocompatibility and the emergence of an immune response from the body. Natural biopolymers have found applications in various fields of biology and medicine because they demonstrate a wide range of biological activity, biodegradability, and accessibility. This review first described the properties of the two most promising biopolymers belonging to the classes of polyhydroxyalkanoates and polysaccharides—polyhydroxybutyrate and chitosan. However, homopolymers also have some disadvantages, overcome which becomes possible by creating polymer composites. The article presents the existing methods of creating a composite of two polymers: copolymerization, electrospinning, and different ways of mixing, with a description of the properties of the resulting compositions. The development of polymer composites is a promising field of material sciences, which allows, based on the combination of existing substances, to develop of materials with significantly improved properties or to modify of the properties of each of their constituent components.

**Keywords:** biopolymers; chitosan; polyhydroxyalkanoates (PHA); poly(3-hydroxybutyrate) (PHB); electrospinning; composites; blends; biomedicine

**Citation:** Zhuikova, Y.; Zhuikov, V.; Varlamov, V. Biocomposite Materials Based on Poly(3-hydroxybutyrate) and Chitosan: A Review. *Polymers* **2022**, *14*, 5549. https://doi.org/10.3390/polym14245549

Academic Editors: Alberto Romero García and Łukasz Klapiszewski

Received: 25 October 2022
Accepted: 16 December 2022
Published: 18 December 2022



## 1. Introduction

Biopolymers are macromolecular compounds that are part of living organisms and are products of their vital activity.

Biopolymers are obtained from such biological sources as insects, crustaceans, and various microorganisms. Significant interest in biopolymers is associated both with the global attention to environmental pollution and with the presence of the critical advantages of biological polymers over synthetic ones. This eventually led to the commercialization of biopolymers and related products in various fields of biomedicine, ecology, and bioengineering [1–4]. The fact that biopolymers' breakdown byproducts are not harmful to the environment is one of their distinguishing characteristics [5].

Based on the fact that biopolymers are obtained naturally from the ecosystem, they have higher economic value and biodegradability [6]. Biopolymers can be both natural (synthesized in living organisms) and synthetic (synthesized under artificial conditions) (Figure 1). Polyesters are complex organic compounds with repeating ester bonds in their composition. Polyesters produced by microorganisms are biodegradable. However, most synthetic polyesters are not biodegradable. Polyhydroxyalkanoates, polylactide, polyglycolide, and polycaprolactone are the most famous polyesters. Polyesters are widely used in the production of drug delivery systems and the manufacture of materials for biomedical applications [7]. Polysaccharides are long-chain polymeric carbohydrates consisting of monosaccharide units bound together by glycosidic bonds. This carbohydrate can react with water (hydrolysis), which produces constituent sugars (monomeric saccharides) [8–10]. Cellulose and chitin are the most famous polysaccharides.

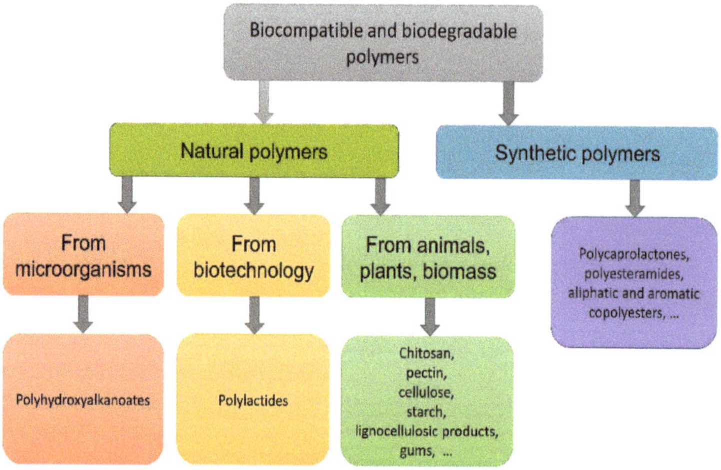


**Figure 1.** Classification of biodegradable polymers. Adapted from [11].

This review is devoted to the features of obtaining and studying the properties of materials based on polysaccharides and polyester. As noted above, polymers of natural and microbiological origin are considered promising alternatives to synthetic biopolymers in fields ranging from packaging materials to biomedicine. However, their use is often limited by certain disadvantages of each class of biopolymers. For example, high sensitivity to water and low mechanical and thermal stability are the main limitations for the industrial application of chitosan [12,13]. In addition, high solubility reduces its barrier properties, which can lead to the complete solubility of chitosan in food products. Blending chitosan with a hydrophobic polymer can minimize solubility problems [14].

The creation of copolymers and composites is intended to eliminate some of the shortcomings of homopolymers, such as poor mechanical characteristics, low strength, and insufficient biodegradability. Overcoming disadvantages is achieved by synthesizing copolymers or creating composites with other materials, both natural and synthetic [7,15–17].

Typically, composites are made by different methods, such as molding, extrusion, solvent casting, electrospinning, and intercalation [18,19]. The choice of a composite manufacturing method plays a decisive role in the mechanical and physical properties of polymeric biocomposites. For example, biocomposites based on banana fiber and polylactic acid have been made using three different processing techniques, namely direct injection molding (DIM), extrusion injection molding (EIM), and extrusion compression molding (ECM), to obtain better properties [20]. Biopolymer properties are also crucial for understanding the structural complexity of biocomposites [20–22].

The main problem faced by researchers of polyhydroxyalkanoates and chitosan composites is related to their different natures. The review presents data on the use of different methods of composite fabrication and investigations of their properties.

## 2. Biomaterials Based on Chitosan and PHB

### *2.1. Chitosan Characteristics and Applications*

Chitosan is a deacetylated derivative of chitin, mainly obtained from the exoskeletons of crustaceans [23–26]. Alternative sources of chitin are insect cuticles and fungal cell walls [27–33]. Chitosan is a linear heteropolysaccharide consisting of alternating units of glucosamine and N-acetyl-D-glucosamine connected by β-(1→4) glycosidic bonds, which have cationic properties: at pH $\leq$ 6.5, it has a positive charge (Figure 2).

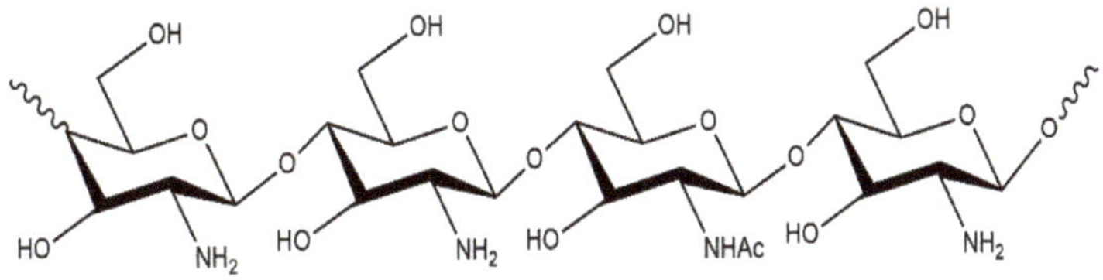


**Figure 2.** Chitosan structure.

Chitosan dissolves in aqueous solutions of acids. Chitosan salts (formate, lactate, citrate, acetate, ascorbate, etc.) are soluble in water [27]. Chitosan is characterized by such parameters as the degree of deacetylation and molecular weight [34], as well as the polydispersity index. These parameters affect many of the physicochemical and biological properties of chitosan, such as solubility, hydrophilicity, crystallinity, and cell affinity. At a pH value below 4, the amino groups of chitosan are protonated, which leads to electrostatic repulsion between the charged groups and swelling of the polymer. Free amino groups form intermolecular hydrogen bonds with the oxygen of neighboring chains. Amino groups make chitosan a cationic polyelectrolyte (pKa $\approx$ 6.5), one of the few in nature. Chitosan is protonated when dissolved in an aqueous acidic medium at pH < 6.5, but when dissolved, it has a high positive charge on the -NH3 + groups and the resulting soluble polysaccharide is positively charged. Chitosan aggregates polyanionic compounds. Chitosan has functional groups in its structure, which can be subjected to structural modification by chemical, radiation, and enzymatic methods in order to acquire new properties [34,35].

Solutions of chitosan with a high molecular weight have a high viscosity, which may limit the use of chitosan in some industries. Therefore, the process of depolymerization is widespread, with the formation of low molecular weight chitosan and oligosaccharides [36,37], which have a wide application potential [38]. Thus, chitosan with a molecular weight of 10 kDa has high antibacterial activity [39], increases disease resistance of plants [40], and demonstrates antioxidant activity [41]. Studies have shown that oligosaccharides [42,43] have lower viscosity, good absorbability, and solubility in water and under physiological conditions. Chitooligosaccharides demonstrated antimicrobial activity against a wide range of gram-positive (*M. luteus*, *S. mutans*, *S. faecalis*, *S. epidermis*, *S. aureus*, *B. subtilis*, *B. cereus*, *L. plantarum*, *B. bifidum*) and gram-negative (*E. coli*, *V. parahaemolyticus*, *V. Vulnificus*, *S. typhimurium*, *P. aeruginosa*) bacteria, as well as fungi (*Saccharomyces cerevisiae*, *Aspergillus niger*, *Candida*) [44–46]. Some researchers also report the effectiveness of oligosaccharides as antitumor agents, as well as providing an anti-inflammatory effect [47]. One of the areas of chitosan research is the development of new drug delivery systems, including those that have an antitumor effect. Thus, a stearic acid and chitooligosaccharide derivative was suggested as a carrier for anticancer agent intracellular transport and were effective against cells from breast cancer, liver cancer, and lung cancer [48].

The degradation of chitosan-based materials in the body leads to the formation of non-toxic amino sugars such as glucosamine or N-acetylglucosamine, which are completely absorbed by the human body. This allows considering it a promising candidate for a wide range of biomedical applications [49]. A significant number of studies have shown that chitosan has antibacterial properties [50–52], fungicidal [53], antitumor [54], absorbent, mucoadhesive[55,56], wound healing [57,58], hemostatic [59] properties.

Due to its polycationic properties, chitosan can interact with the surface of the cell membrane and be used as a material for bone tissue regeneration [60]. Chen et al. [61] developed a polymer stent made from chitosan films crosslinked with genipin to improve mechanical properties. The results showed not only improvement in mechanical properties but also reendothelialization of the implanted vascular stent.

The mechanical properties of chitosan have been widely studied. The authors [62] conclude that the ultimate tensile strength of chitosan fibrils is in the range of 121.5–308 MPa and Young's modulus is 7.9–22.7 GPa. The mechanical properties of chitosan fibers are determined by the molecular weight and degree of deacetylation, the solvent used, and the source of production. Furthermore, when the relative humidity is reduced from 93% to 11%, Young's tensile modulus of chitosan films increases from 10.9 ± 1.2 GPa to 18.8 ± 1.5 GPa [63]. The micromechanical properties of chitosan are also investigated using nanoindentation [64], with Young's modulus ranging from 1 to 3 GPa, which correlates with the results of [65].

The study [66] is devoted to the study of chitosan films stored under controlled conditions and the change of their properties over time. It is demonstrated that the films undergo significant changes in properties during storage due to changes in the structure associated with the Maillard reaction. The rearrangement of polymer chains during storage caused structural changes, changes in mechanical properties, changes in resistance to ultraviolet and visible light, and changes in hydrophobicity. Thus, during storage from 0 to 90 days, the tensile strength of low molecular weight chitosan changed from 55 to 76 MPa and that of high molecular weight chitosan from 61 to 76 MPa. Elongation at break values decreased. This means that the films became stiffer and less tensile, which can be explained by the loss of bound water. The authors conclude, however, that the functional properties of the chitosan films remained acceptable even after 90 days of storage. The use of natural or synthetic plasticizers [67] can improve the properties of chitosan films with retention for up to 10 months. Currently, chitosan is used to create products such as films, coatings, hydro- and cryogels, micro- and nanoparticles, and matrices used for the fabrication of wound healing agents [68,69], targeted delivery of medicines [70–73] and for ophthalmology [69,74,75]. Chitosan has the characteristics that are necessary for an ideal contact lens, such as optical transparency, mechanical stability, sufficient optical correction, gas permeability, wettability, and immunological compatibility [76,77]. It's a well-known immunomodulator [78]. Chitosan exhibits antitumor properties by inhibiting the growth of tumor cells due to the proliferation of cytolytic T-lymphocytes [44]. There are studies reporting its antiviral activity [79,80].

Mucoadhesive properties of chitosan are known, which appear due to electrostatic interactions with negatively charged epithelial membranes [81]. Thus, an improvement in the efficiency of drug delivery through the mucous membranes under the influence of chitosan was found, as well as prospects for its use in ophthalmology. [82]. A scientific study [83] describes biocompatible systems of chitosan nanoparticles with medicinal substances included in them, which can be retained on the ocular surface for a long time, penetrate the blood-ocular barrier, concentrate inside the eye, and also have a pronounced therapeutic effect.

Bioabsorbable suture materials based on chitosan are made. In work [84], a quaternized chitosan derivative was used for application to the Vicryl surface. The resulting material exhibited antibacterial activity and good cytocompatibility with human fibroblast cells. A significant number of studies report the ability of chitosan and its derivatives to accelerate wound healing [85] by modulating the secretion of enzymes and cytokines. Such materials are available in the form of electrospun fibers [86], hydrogels [87,88], membranes [89,90], films [91], 3D scaffolds [92], and sponges [85].

One of the primary issues with materials based on unmodified chitosan is that they have poor mechanical qualities. This issue is resolved by functionalizing chitosan to produce chitosan derivatives, as well as by creating composites and mixes based on chitosan and other materials.

### 2.2. *Poly(3-hydroxybutyrate): Preparation, Structure, Applications*

Poly(3-hydroxybutyrate) (PHB) is one of the best-known polyhydroxyalkanoates (PHA). Polyhydroxyalkanoates are a family of biodegradable polyethers typically produced microbiologically using the highly efficient producer strain *Azotobacter chroococcum* 7B (Figure 3).

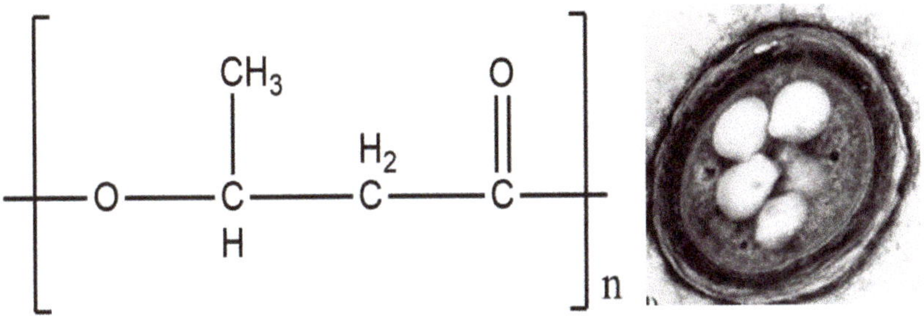

**Figure 3.** The formula of poly(3-hydroxybutyrate) and bacterial cell morphology of strain producer *A. chroococcum* during growth, reprinted with permission from [93].

PHAs are widely used in biomedicine [94] to create suture threads, wound healing materials, orthopedic pins, stents, devices for targeted tissue repair/regeneration, joint cartilage, bone implants, nerve fiber connections, etc. Although PHB has several advantages, it also has a few limitations that restrict its biological uses, including high crystallinity and brittleness.

A variety of microorganisms use polyhydroxyalkanoates as an intracellular energy source and carbon source [95]. Only prokaryotes can synthesize this type of polymer. Some PHA-synthesizing microorganisms are presented in Table 1. Most microorganisms are capable of accumulating PHA from 30% to 80% of the cell's dry weight.

**Table 1.** Some representatives of prokaryotes that are capable of synthesizing PHA as an intracellular energy source.

| | |
|---|---|
| *Acinetobacter* | [96] |
| *Azotobacter* | [93,97–99] |
| *Bacillus* | [100,101] |
| *Clostridium* | [102] |
| *Escherichia* | [103] |
| *Halobacterium* | [104] |
| *Methylobacterium* | [105,106] |
| *Micrococcus* | [107] |
| *Nitrobacter* | [108] |
| *Parapedobacter* | [109] |
| *Pseudomonas* | [110,111] |
| *Rhizobium* | [112] |
| *Streptomyces* | [113,114] |

Poly(3-hydroxybutyrate) is a linear, isotactic polymer that can have a high molecular weight of up to about 3000 kDa. The conformational structure of PHB is a right-handed helix with a double helical axis. Helical conformation is stabilized by carbonyl/methyl interactions and is one of the few exceptions in nature in which its formation and stability do not depend on hydrogen bonds.

The physical and mechanical properties of PHAs, such as stiffness, brittleness, melting point, glass transition point, or resistance to organic solvents, can vary significantly, depending on the monomer composition [115]. For example, increasing the content of 3-hydroxyvalerate (HV) residues in the poly(3-hydroxybutyrate-co-3-hydroxyvalerate) (PHBV) copolymer leads to a decrease in the melting point from 180 °C for pure PHA to ~75 °C for the copolymer with 30–40 mol % HV content. There is also evidence that PHA isolated from P. oleovorans is able to dissolve in acetone, while the homopolymer of PHB shows very low solubility in acetone.

The crystalline structure of PHB is usually in the form of lamellae. The lamellae thickness varies from 4 to 10 nm depending on the molecular weight, solvent, and crystallization temperature [116,117].

The single crystal structure of PHB is a monolamellar system [118]. However, such products as films are usually multilamellar systems that assemble into multi-oriented lamellar crystals. In 3D-structures, such as scaffolds, PHB chains usually form spherulites [119]. In spherulites, lamellar PHB crystals grow radially stacked. Because of the tendency of lamellar crystals to twist, PHB spherulites usually have a banded texture. The periodicity and regularity of such twisted structures depend on both the molecular weight and the crystallization temperature of the polymer. The growth kinetics of PHB spherulites was investigated at various crystallization temperatures. At around 90 °C, the maximum crystallization volume was observed. The overall crystallization rate of PHB is maximal in the temperature range of 50–60 °C [108].

The molecular weight of poly(3-hydroxybutyrate) synthesized by wild-type bacteria ranges from $1 \times 10^4$ to $3 \times 10^6$ g/mol with a degree of polydispersity ~2 [93,98,119]. The glass transition temperature of PHB is ~4 °C, while the melting point is ~180 °C [120,121]. A bifurcated peak of melting temperature is also sometimes observed in homopolymers. This phenomenon can be explained by the presence of crystallites of different degrees of perfection, which can include the thermal prehistory of the sample and the broad molecular weight distribution. The densities of the crystalline and amorphous components of PHB are 1.26 and 1.18 g/cm$^3$, respectively. The mechanical properties of PHB, such as Young's modulus (~3.5 GPa), and tensile strength (~43 MPa), are similar to those of isotactic polypropylene. However, the elongation at break (5%) is much lower than that of polypropylene (400%). Consequently, PHB is a stiffer and more brittle plastic compared to polypropylene.

## 3. Formation of Composites Based on Chitosan and Poly(3-hydroxybutyrate)

Composites are multicomponent materials, usually consisting of a base (matrix) reinforced with fillers. The combination of dissimilar substances leads to the creation of a new material, the properties of which differ from the properties of each of the components. The creation of composites with hydrophilic and biocompatible polymers is one of the strategies for changing the surface properties and improving the biocompatibility of polyhydroxyalkanoates [122].

Composite materials based on PHB with the addition of chitosan will be more hydrophilic than materials based on pure PHB. This means that they will be more biocompatible and promising for creating tissue-engineered biomedical structures. It is also assumed that mixing chitosan with PHB will be able to provide a variety of primary amino groups for further modifications and thus diversify the possibilities of using such a blended material [123]. The bioactivity, osteoinductive, and osteoconductive capabilities of mixed compositions can be improved by including hydroxyapatite in their structure [124–127].

The coating for bioceramic scaffolds, consisting of Chitosan and PHB, with the addition of multi-walled carbon nanotubes, made it possible to obtain thermostable materials with good mechanical properties for tissue engineering [128]. Previously, the authors demonstrated that, in addition to PHB, coating nano-bioglass/$TiO_2$ scaffolds with chitosan made it possible to achieve a decrease in the contact angle from 84° to 42°, compressive strength of the scaffolds increased from 0.01 MPa for nBG/n$TiO_2$ to 0.19 MPa for nBG/n$TiO_2$ with PHB/Cs, improve biological activity, and cell proliferation [129]. However, such coatings were rapidly destroyed and tended to swell.

Chitin and chitosan have also been used to improve the mechanical and thermal properties of PHB and its copolymers. Chemical modification of chitin was carried out to increase the hydrophobicity (increase in contact angle from 33° for chitin and up to 70° for composite). The crystallization temperature is 22 °C lower in the composite compared to pure PHBV. The melting temperature of PHBV increased from 154.5 °C to 165 °C for composite. [130]. Films based on PHBV/acetylated chitin nanocrystals were produced by solution casting [131]. Studies of mechanical properties showed that the ultimate tensile strength and Young's modulus of PHBV are improved by about 24% and 43%, respectively. The contact angle increased from 31° for chitin to 68° for PHBV/chitin, and the Tc of composites is 5 °C higher than that of PHBV [131].

In addition, the melting technique cannot be applied because chitosan has a high melting point at which PHB will decompose. Additionally, the melting approach cannot be used because PHB would start to break down before chitosan melts due to its high melting point [132].

Using a micro compounder, mixtures were prepared as described in the article [133] at a temperature of 175 °C and a screw speed of 100 rpm. Initially, PHB was melted; then chitosan (5, 10, 20, 30, and 40 wt %) was introduced into the mixer. The granular samples were then subjected to injection molding. When studying the thermal properties of the obtained composites, it was shown that the inclusion of chitosan in PHB increases the glass transition temperature with a decrease in the melting temperature and crystallinity. The addition of 10 wt.% chitosan reduced the percentage of crystallinity of the chitosan/PHB composition. At the same time, a further increase in the concentration of chitosan did not cause significant changes. This is because the presence of very rigid chitosan molecules surrounding the PHB molecules made the PHB molecules in the composites inflexible and induced insufficient crystallization compared to pure PHB. Adding more chitosan reduced the thermal stability of the composites.

### 3.1. Copolymerization

A large number of studies confirm that chitosan, as well as its derivatives, has an antibacterial effect due to its ability to disrupt the normal functions of the bacterial cell membrane due to the reaction between the positive charges of chitosan and the negatively charged bacterial cell walls [134]. Usually, chitosan exhibits its antimicrobial activity in an acidic environment due to its poor solubility and the absence of protonated amino groups at pH 6.5 and above, which limits its use. One approach to overcoming some of the disadvantages of primary chitosan is to modify it with suitable functional groups. Graft copolymerization is a promising method for producing new types of hybrid materials based on chitosan with improved properties, thereby broadening its application in biomedicine and environmental protection [135,136]. This approach allows the formation of functional derivatives by covalent binding of the grafted molecule to the main chain of chitosan (Figure 4) [137].

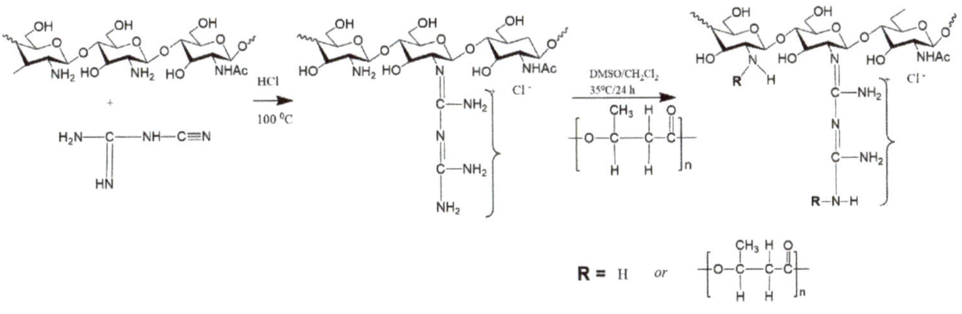


**Figure 4.** ChG−grafted PHB copolymer. Adapted from [137].

Polyhydroxyalkanoates can be used to modify chitosan. Thus, in the study by Arslan et al. [138], poly-(3-hydroxyoctanoate), as well as PHBV and linoleic acid, were grafted to chitosan via condensation reactions between carboxyl and amino groups, while the percentage of grafted polyester varied from 7 to 52% wt., depending on the molecular weight. It has been shown that the grafting percentage depends on the molecular weight, the structure of the grafted PHAs (steric effect), and, finally, the solubility of the polymers in the polymerization medium. So, the solubility of the grafted chitosan-g-PHA copolymer can be controlled by adjusting the grafting percentage.

In another study, chitosan derivatives formed viscous solutions in water. Although the original polymer is hydrophobic, its graft derivative exhibits amphiphilic behavior in which the degree of solubility is controlled by the percentage of graft. As a result, PHB-g-chitosan copolymers are obtained, which have strong elastic films with a low melting point. Due to their biocompatibility, amphiphilic behavior, and antimicrobial activity (Table 2), polymer grafts have good potential for medical applications such as tissue engineering and drug delivery systems [139].

**Table 2.** Antimicrobial activity of ChG-g-PHB Adapted from [139].

| Sample | Inhibition Zone (mm) | | |
|---|---|---|---|
| | *S. pneumonia* | *E. coli* | *A. fumigatus* |
| ChG-g-PHB | 21.30 ± 2.10 | 21.80 ± 2.10 | 19.40 ± 1.5 |
| Ampicillin | 23.80 ± 0.20 | - | - |
| Gentamicin | - | 19.90 ±0.30 | - |
| Amphotericin | - | - | 23.70 ± 0.10 |

The compatibility of chitosan and polyhydroxyalkanoates can be increased by functionalizing polymer chains by grafting functional groups or comonomers [140]. Vernaez et al. used maleic anhydride to modify PHBV using a Brabender measuring mixer. PHBV with a cross-linking agent concentration of 10% wt., similarly mixed with chitosan powder. It is concluded that functionalization with maleic anhydride effectively increases the compatibility of the polyester matrix with chitosan particles.

In work by Hu et al., acrylic acid and chitosan, onto which hyaluronic acid was immobilized, were sutured to ozone-treated PHBV membranes [141]. These PHBV/acrylic acid membranes, which were esterified to chitosan or chitooligosaccharides to increase hydrophilicity, had antibacterial properties and improved fibroblast cell attachment.

### 3.2. Electrospinning

The electrospinning process has shown significant potential for creating fibrous scaffolds that can mimic the structure of the extracellular matrix in natural tissues (Figure 5) [142,143]. Electrospun fiber scaffolds have improved structural properties, including greater surface area and higher porosity with interconnected pores that promote cell growth and nutrient exchange [144]. Many natural and synthetic polymers, including PHB and chitosan, have been investigated for the manufacture of fibrous materials [145]. In this study, PHB was successfully blended with chitosan using TFA as a co-solvent, and the blended solution was electrospun to fabricate fibrous scaffolds for cartilage engineering. This study showed that the addition of chitosan could increase the hydrophilicity and weight loss rate (and percentage) for PHB scaffolds while maintaining the mechanical properties in a suitable range. The results obtained indicate the great potential of fibrous scaffolds made from a mixture of PHB/chitosan for further additional studies in vitro and in vivo.

The method of fiber formation using electrospinning was used in [146]. The aim of this study was to determine the effect of a polyhydroxybutyrate/chitosan/nanobioglass nanofiber scaffold fabricated by electrospinning on the proliferation and differentiation of stem cells obtained from human exfoliated deciduous teeth into odontoblast-like cells. Trifluoroacetic acid was used as a solvent; first, poly(3-hydroxybutyrate) was dissolved, and then 15 wt % chitosan and 10 wt % bioglass nanoparticles were added. Then the solution was homogenized, and electrospinning was performed. According to the results, due to the presence of chitosan and bioglass nanoparticles, the resulting matrix had good cell viability and uniform properties and was suitable for pulp capping (Figure 6).

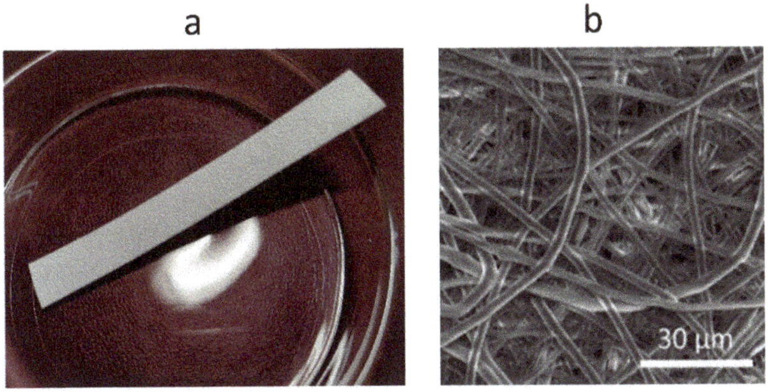


**Figure 5.** (**a**)—PHB film obtained by electrospinning; (**b**)—SEM image of PHB film, adapted with permission from [143].

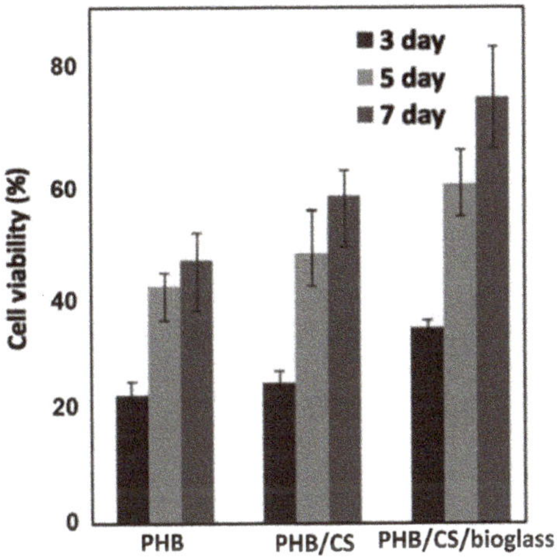


**Figure 6.** Cell viability, adapted from [146].

Peripheral nerve conductors can be constructed based on biodegradable polymers. Thus, in the article by Zhou et al., biocomposite was constructed using electrospinning based on chitosan and PHB [147]. In this case, the addition of chitosan made it possible to improve wettability. The results of TGA and DSC showed that PHB/chitosan polymers could be mixed in one phase with trifluoroacetic solvent in all compositions. Furthermore, a decrease in decomposition temperature (from 286.9 to 229.9 °C) and crystallinity (from 81.0% to 52.1%) with increasing chitosan content was demonstrated. The authors conclude that PHB/chitosan fibers with different percentages can be used as medical materials with a controlled degradation rate.

Electrospinning was used to create chitosan/PHBV nanofibers [148] with various PHBV/chitosan ratios (4:1 and 2:3, respectively) as cytocompatible degradable scaffolds. Such materials ensured the proliferation and differentiation of L929 fibroblast cells, as well as a controlled rate of biodegradation [149]. PHBV/chitosan at a ratio of 4:1 demonstrated the best cell proliferation (53 ± 9% compared to PHBV/C [2:3] (32 ± 9%)) and wound healing rate (smaller wound areas (26 ± 11%), compared to control (59 ± 17%).) in vivo.

In another study, the authors developed a material based on PHB-chitosan with TFA as a solvent [150]. Aluminum oxide nanowires were added as a reinforcing phase, and

3D scaffolds were obtained by the electrospinning method. Chitosan addition reduces the tensile strength from 2.81 ± 0.15 MPa to 0.89 ± 0.26 MPa and the modulus from 126.3 ± 22.2 to 44.6 ± 0.2 MPa. The alumina provided optimum mechanical properties (11.18 ± 1.24 MPa) and increased surface roughness (492.6 ± 67 nm against 346.2 ± 23 nm for PHB-CTS ) as well as good biological properties in vitro.

Thus, composite electrospun matrices are promising materials, primarily for biomedicine. This is due to the possibility of creating a material with certain predetermined characteristics and the ease of its functionalization and modification. However here, it is important to note one problem, which is usually characteristic of materials that are difficult to mix under normal conditions. Namely, rather little is known about the homogeneity and distribution of components in such mixed materials, i.e., whether the components are combined into a single fiber or are independent networks [151]. These issues require further investigation in the creation of electroformed PHB-chitosan composites.

### *3.3. Blending Polymer Solutions in Different Solvents*

Blending is a simple and effective way to obtain biomaterials with the desired properties. However, blends are often developed from the same class of biomaterials. At the same time, insufficient attention is paid to the creation of mixtures based on hydrophilic biopolymers and hydrophobic polyesters. The creation of composites and mixtures based on chitosan and PHB is associated with certain difficulties, which include selecting the optimal solvent suitable for chitosan and PHB. As mentioned above, new types of biomaterials based on PHB, and chitosan can be obtained using various mixing methods (Figure 7).

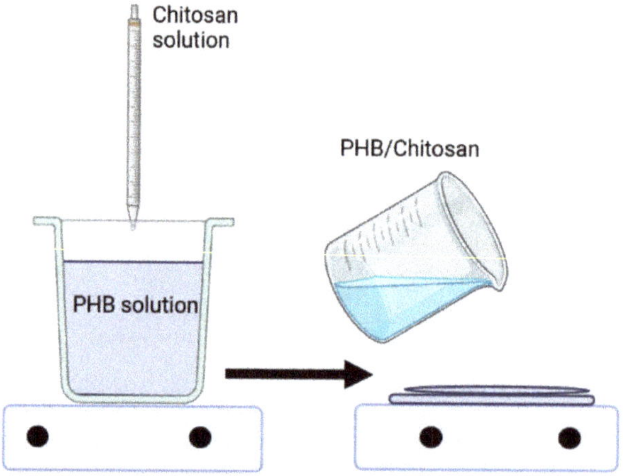


**Figure 7.** A method of casting polymer solutions.

It is expected that PHB/chitosan mixtures will have good biocompatibility and high biodegradability, so they can be used as biomedical materials. The authors [152] note that the properties of mixtures of chitosan and PHA depend on the method of their preparation:

1. A method of casting polymer solutions. By casting from a PHB/chitosan mixture solution, compositions with different polymer ratios can be prepared, which means that the mixtures will have the properties of both components [153]. The main advantage of the method of mixing by casting from polymer solutions is its simplicity. Thus, to prepare composites, chitosan and PHB were dissolved in 1,1,1,3,3,3-hexafluoro-2-propanol, and the resulting mixture was poured onto a Teflon cup [154]. The introduction of chitosan reduces the crystallization of PHB from 65% to ~35% for PHB/CS 50:50 due to the harsh environment of chitosan and the formation of intermolecular hydrogen bonds between the components.

2. Blending Precipitation method [155]. In this case, PHB or its copolymers were dissolved in DMSO and mixed with a solution of chitosan in acetic acid with DMSO. Then the mixture was precipitated in excess of acetone. The precipitate was filtered and dried. The pre-deposition mixing method can be used as an effective way to improve the plasticity of chitosan by manipulating the production conditions and composition [156].

It was shown that hydrogen bonds formed between the components of the mixture, and the addition of chitosan led to a decrease in the degree of PHB crystallinity. Similar results were obtained by Khasanah et al. [157] in the study of PHB/chitin films. The crystallinity of PHB decreased (22% for PHB/CS 50:50) in mixtures compared to the pure polymer (77%), and new intermolecular hydrogen bonds formed between the CO groups of PHB, and the O–H, N–H groups of chitin. At the same time, a decrease in the degree of crystallinity makes it possible to improve the physical properties of PHB, which means expanding the potential for its practical application [154,158].

Propylene carbonate and acetic acid were used as solvents, respectively, for PHB and chitosan and further mixing of polymer solutions [159]. This made it possible to avoid the use of toxic fluorinated co-solvents. The same volumes of solutions with different concentrations of polymers were used. After mixing, acetone was added to precipitate the polymers, which were then washed, frozen, and lyophilized. As a result, composite scaffolds (chitosan-PHB-hydroxyapatite) were obtained.

3. The method of casting in the emulsion. Chitosan dissolved in acetic acid was mixed with a solution of PHB in chloroform. After casting into films, the mixture was neutralized with 0.5 M sodium hydroxide, washed with water, and dried. In this case, the chitosan solution should be sufficiently viscous, and the amount of PHB should not exceed 30% so that PHB drops dispersed in the chitosan matrix retain their stability [160]. 3T3 fibroblasts were used to study cytocompatibility. Fibroblasts were attached to the films and had a normal spreading morphology. Moreover, experiments on adhesion and cell proliferation showed that PHB/chitosan films had better cytocompatibility compared to pure chitosan films, probably due to better PHB biocompatibility and higher surface roughness of the films from the mixture [152].

In work, Ivantsova and co-authors [161] created a new biodegradable polymer composition for the controlled release of drugs based on PHB and chitosan with different percentages of components (10–90% wt.). In this case, PHB was used as a biodegradable component, and chitosan was added to improve the physical properties and the possibility of further modification due to the presence of reactive amino groups in the composition. The components were mixed using a Brabender laboratory mixer at 150 °C for 10 min, and the resulting powder was hot pressed at 160 °C.

The authors [162] formed films based on PHA with the inclusion of various polysaccharides (chitosan, pectin, hyaluronic acid) by mixing equal amounts of polymer solutions. PHB and PHBV were dissolved in chloroform, and chitosan was dissolved in 0.1 N acetic acid at pH 5.5. The solutions were filtered and mixed. When the solvent was evaporated, the films were washed with distilled water. The resulting porous films had the flexibility and plasticity characteristic of PHB and good biocompatibility with HaCaT cells (Table 3). However, there was a slight inhibition of cell proliferation on materials containing chitosan.

**Table 3.** Comparison of the properties of PHB and PHB/Cs adapted from [162].

| Sample | Young's Modulus (MPa) | Elongation at Break (%) | Tensile Strength (MPa) | Contact Angle (Degrees) | Proliferation of HaCaT Cells (% vs. Control) |
|---|---|---|---|---|---|
| PHB | 1640 | 1.4 | 12.9 | 84 | 22 ± 19 |
| PHB/Cs | 334 | 1.6 | 3.3 | 93 | 90 ± 10 |

According to Wenling et al., chitosan/PHB films were prepared by the emulsification/casting/evaporation method [160]. A solution of chitosan in acetic acid and a solution of PHB in chloroform were prepared separately. Scanning electron microscopy showed that PHB microspheres were formed and captured by chitosan matrices, which made the film surface rough. With an increase in the PHB content, the film surface roughness increased, and the swelling coefficient of the PHB/chitosan films decreased (9.7 ± 5.0% for PHB and 119.3 ± 4.3% for Cs/PHB 70:30). This can be explained by the hydrophobicity of PHB. This result indicates that the swelling capacity of chitosan films can be reduced by adding PHB. The stress-strain curves of the blended films were almost linear and similar to the stress-strain curves of pure chitosan films, which meant that all blended films were resilient and strong. With an increase in the amount of PHB, the elastic modulus of the films decreased (8.7 ± 1.6 MPa for Cs/PHB 90:10 to 4.9 ± 0.6 MPa for Cs/PHB 70:30), and the elongation at break increased (43.3 ± 5.9% (Cs/PHB 90:10) to (82.9 ± 9.3% CS/PHB 70:30). In addition, such films demonstrated higher tensile strength compared to chitosan films (4.0 ± 1.1 MPa for Cs/PHB 90:10 to 3.4 ± 0.5 MPa for Cs/PHB 70:30)).

Iordansky et al. investigated various issues of creating mixed compositions based on PHB and chitosan for targeted drug transport using rifampicin as an example [163,164]. At the same time, methods were proposed for creating emulsions with different percentages of biopolymers by various methods [165,166] using solutions of PHB in chloroform and chitosan in 1% acetic acid. Using SEM and DSC, the composition was shown to separate into two immiscible phases, especially in the PHB concentration range of 50–60%, when the separation of the components was observed, and the PHB globules were embedded in the chitosan matrix. The authors conclude that the formation of a heterophasic immiscible structure is not a disadvantage in the development of biodegradable materials due to improved bioavailability and the potential ability to control the rate of degradation of this polymer system.

### 3.4. *Using the Same Solvent for Both Polymers*

Ikejima et al. [154] reported the development of biodegradable polyester/polysaccharide blend films made from bacterial poly(3-hydroxybutyrate) with chitin and chitosan. The crystallinity of such mixtures decreased with an increase in the number of polysaccharides from ~65% to ~35%, which was shown by DSC. Using the FTIR method, this trend was confirmed since, with an increase in the number of polysaccharides, the intensity of the absorption bands of the carbonyl sites changed. It was found that the suppression of PHB crystallization occurs more strongly when mixing chitosan compared with chitin. It was found that PHB in mixtures (by 13C NMR-spectroscopy) is immersed in a "glassy" medium of the polysaccharide. The resonances of chitosan in mixtures were significantly broadened compared to those of chitin. This was explained by the formation of hydrogen bonds between the carbonyl groups of PHB and the amide -NH groups of chitin and chitosan. Crystallization and environmental biodegradability have been shown for mixtures of PHB and chitosan.

In another work by T. Ikejima and Y. Inoue, 1,1,1,3,3,3-hexafluoro-2-propanol (HFIP) was used as the solvent [167]. The films were made by casting from a solution. Dynamic mechanical, and thermal analysis showed that the thermal transition temperatures of PHB amorphous regions were similar to those of pure PHB for PHB/chitin and PHB/chitosan blends (~15 °C). It is also indicated that the studied films of mixtures were biodegradable in water.

Measurement of the contact angle of scaffolds from a mixture of PHB/chitosan dissolved in TFA showed that with an increase in the content of chitosan in the mixture, the contact angle decreases, and the hydrophilicity of the mixture increases accordingly. Moreover, with an increase in the content of chitosan, the porosity of the scaffolds increases. The rate of degradation also increases. It was shown that during 14 weeks of decomposition in a buffer solution, scaffolds containing 40% chitosan lost 40% of their mass, while pure PHB lost about 10% [132]. In addition, the authors found that an increase in the content of chitosan increased the contribution of the amorphous fraction. This result indicates the suppression of PHB crystallization when mixed with chitosan. An in vitro degradability study showed that the rate of degradation of mixed scaffolds is higher than that of pure PHB scaffolds, and the dissolution of chitosan can neutralize the acidity of PHB degradation products. Based on the obtained results, the authors conclude that the prospects for using the developed scaffolds in the field of bone and cartilage tissue engineering are good.

Cheung et al. [168] also demonstrated a decrease in the degree of crystallinity of PHB and PHBV when mixed with chitosan (Figure 8).

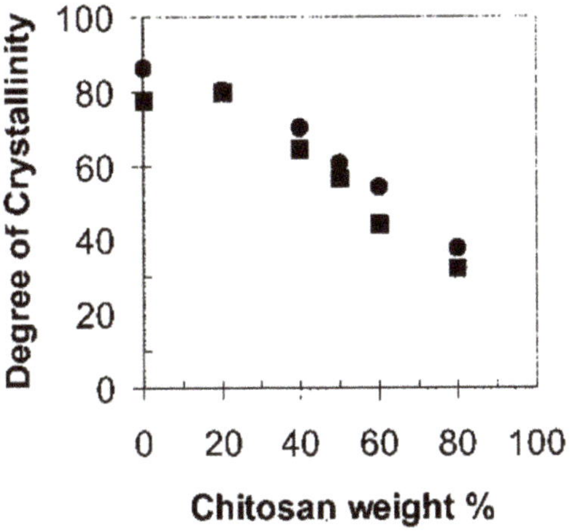


**Figure 8.** The degree of crystallinity of PHB (•) and P(HB-co-HV) (■), adapted from [168].

There is also a decrease in the melting point of the mixture compared to pure PHB. In this work, films based on chitosan/PHB and chitosan/PHBV were obtained by casting from a solution of 1,1,1,3,3,3-hexafluoro-2-propanol with the addition of 1% acetic acid. At the same time, the polymers were dissolved separately at concentrations of 10 g/l and mixed in different weight ratios (100/0, 80/20, 60/40, 50/50, 40/60, 20/80, and 0/100). Then solutions were slowly evaporated at an ambient temperature for 24 h. The authors conclude that there is an intermolecular interaction between chitosan and PHB or PHBV in all compositions.

It should be noted that TFA and 1,1,1,3,3,3-hexafluoro-2-propanol are the most commonly used solvents for preparing compositions based on PHB and chitosan. They are highly toxic and quite expensive. Therefore, the search and study of other methods for obtaining mixtures based on chitosan and PHB remain relevant.

The researchers [169] used PHB and chitosan for the synthesis of gold nanoparticles because chitosan has a positive charge under acidic conditions and, as a result, can form conjugates with negatively charged polymers such as PHB. At the same time, a solution of chitosan in acetic acid was prepared and treated with ultrasound, and a weighted portion of PHB was added to it in a ratio of 4.16:1 under optimal conditions. The resulting conjugate was completely soluble, with chitosan acting as a reducing agent in it and PHB as a stabilizer. Previously, a similar method for the preparation of amide conjugates of partially depolymerized PHB and chitosan was described in [170] due to the interaction of terminal carboxyl groups of PHB with amino groups of chitosan and water-soluble conjugates were obtained.

In a number of papers, glacial acetic acid has been proposed as a solvent for polyhydroxybutyrate. Choonut et al. [171] heated a sample of PHB in glacial acetic acid with constant stirring to 118 °C until the sample was completely dissolved. Then, films were prepared by casting on a heated 120 °C Petri dish to accelerate the evaporation of the solvent. Similarly, PHB films were obtained in [172], and it was shown that the choice of film preparation temperature and the solvent evaporation rate is important in terms of such material properties as the degree of crystallinity, surface roughness, transparency, and mechanical characteristics. Thus, films obtained by casting at high temperatures had a lower tensile strength and deformation, a higher degree of crystallinity (78% against 60.5%

in chloroform) and were also more transparent and smoother. The authors explained the increase in the crystallinity of the material by the fact that at a higher temperature in acetic acid, more thermal energy was available for the formation and growth of crystal structures and by the fact that chloroform is a more compatible solvent for PHB.

In their study Mukheem et al. obtained and characterized the antimicrobial properties of composite films based on PHB and chitosan and incorporated boron nitride nanoparticles [173]. PHB was dissolved in glacial acetic acid at a temperature of 118 °C. Then the solution was cooled, and nanoparticles were added at various mass ratios. A solution of chitosan in 1% acetic acid was added dropwise (at a ratio of PHB and chitosan 10:1, respectively). The resulting mixture was sonicated. Films were formed by casting on a glass plate heated to 80 °C.

### *3.5. Summary of the above Methods*

Based on the work described above, creating new composites from two biopolymers of different natures is a non-trivial problem. Table 4 summarizes the main methods used to make composites of both PHB and chitosan.

**Table 4.** Advantages and disadvantages of methods for creating composites.

| Methods | References | Advantages | Disadvantages |
|---|---|---|---|
| Extrusion, melt functionalization | [140,174,175] | There is no need to use expensive and toxic solvents. | Melting temperatures of PHB and chitosan are different. Thermal degradation of polymers and reduction in molecular weight can occur. Expensive specific equipment is required. |
| Copolymerization | [137–140] | Covalent bonding ensures the creation of a single branched structure. | Difficulty in synthesis, selection of optimal conditions is necessary, in some cases it is not easy to determine the degree of grafting |
| Electrospinning | [143–150] | Creating products with a variety of structures (controlled fiber size, porosity) | Requires expensive equipment. Difficult to manufacture. Limited choice of solvents |
| Blending in different solutions | [152–161] | The relative ease of creating the material. No specific equipment is required. | It is necessary to carefully select the mixing conditions, the ratio of polymers due to the problem of stability |
| Blending in a common solution | [132,154,167–173] | Relative stability, simplicity of manufacturing | The most widely used common solvents are expensive and toxic. |

Each method is not universal; each has its own advantages and disadvantages. For example, the future product will be expensive and difficult to manufacture, limiting its economic viability. Inconsistencies in physical properties (melting point) make it difficult to use melt extrusion methods. The use of some solvents, such as trifluoroacetic acid, is limited by their toxicity. However, the ease of making a composite by mixing it in a common solution makes it one of the most commonly used methods. Among all solvents, acetic acid occupies a separate place. It is possible to dissolve both PHB and chitosan in acetic acid. It is cheap and non-toxic, which is an advantage over other solvents.

## 4. Prospects for Applications of PHB/Chitosan Composite Materials

Most of the options for the practical application of materials based on both chitosan and poly(3-hydroxybutyrate) are associated with the biomedical field. Composite materials based on them also have prospects for application in medicine. For example, hydrogels for injection with the possibility of therapeutic loading were presented [176]. Gels based on PHB were obtained, and chitosan was added using the solvent-exchange method, which made it possible to load chitosan successfully into the PHB matrix. PHB was dissolved in chloroform at 95 °C; tetrahydrofuran was added and then immersed in methanol. Alcohol gels were extracted by soaking them in a solution of chitosan with acetic acid, which led to the formation of hydrogels. The hydrogels were lyophilized. Gels, with the addition of chitosan, showed more elastic behavior, improved compressive strength, and increased wettability. The increased hydrophilicity of the gel network may result in more efficient rebinding with water after the degradation of composite hydrogels compared to pure PHB. Drug-containing PHB/chitosan hydrogels showed excellent injectability and stability and also showed pH-driven release. The release of doxorubicin was significantly accelerated by lowering the pH of the medium from 7.46 to 4 due to the repulsion between positively charged chitosan and doxorubicin. This paves the way for the future development of a controlled drug release system in typical cancer microenvironments that are characterized by low acidity.

PHB and chitosan were used to create reservoir-type composite microparticles that could be used in the field of controlled drug delivery [177]. At the same time, PHB microspheres containing piroxicam or ketoprofen as model objects [178] were also included in chitosan matrices. The addition of chitosan crosslinked with glutaraldehyde at various concentrations makes it possible to control the release of drugs.

The potential uses of materials based on chitosan and PHB are not just limited to the biomedical industry. In the creation of biodegradable plastics, natural polymers are thought to be promising alternatives to synthetic polymers. In order to create new packaging materials that won't pollute the environment, their use can be justifiable. Ecosystem preservation and pollution reduction are both greatly improved by the use of biopolymers [179].

Due to chitosan's specific antibacterial activity, which demonstrates potent activity against bacteria, fungi, and yeast [180], its use in this field is justified. This makes it possible to create antimicrobial packaging materials [17]. In order to develop antimicrobial packaging, chitosan and its nanoforms have been widely used to reduce microbial growth in food products. Notably, they have been widely used to make edible coatings and edible films to increase the shelf life of food products [181]. A nanocomposite obtained by dispersing silk nanodisks in a chitosan matrix has been used as an edible coating to increase the shelf life of perishable food products [182].

Poly-3-hydroxybutyrate is used to make bioplastics because its characteristics are similar to those of typical petroleum-based polymers such as polypropylene (PP), polystyrene (PS), polyethylene (PE), and polyethylene terephthalate (PET) [183]. However, its application in the food packaging sector is underdeveloped due to the moderate barrier, thermal, and mechanical properties of this biopolymer [184]. Therefore, for this purpose, PHB is often combined with other materials, such as nanoparticles [185]. Chitosan has been used to modify PHB in the creation of food packaging [186]. Biohydrothermally synthesized ZnO-Ag nanocomposites were used as a filler to improve mechanical properties and impart additional antimicrobial properties. The resulting material showed excellent prospects for replacing non-degradable plastic for food packaging.

## 5. Conclusions

One of the strategies for improving the properties of biopolymers is the creation of new materials by obtaining biocomposites from polymers that differ in their properties. This review is devoted to the methods of creating and studying the properties and applications of composite materials based on polysaccharide chitosan and polyester PHB. Some of the disadvantages of the two classes of biomaterials, polyesters, and polysaccharides, can be overcome by blending them. For example, the incorporation of hydrophilic biopolymers such as chitosan into hydrophobic polyesters can provide functional groups for further modification, such as conjugation with growth factors, as well as reduce the crystallinity of the polymers and vary their biodegradability. In turn, the addition of polyesters to hydrophilic biopolymers can reduce their tendency to overswell. In this review, the features of obtaining biocomposites using different methods were considered: electrospinning, copolymerization, and the creation of mixed compositions by mixing polymer solutions in various solvents. Blending the two classes of biomaterials will lead to the development of some new biodegradable materials with improved properties for various applications.

**Author Contributions:** Conceptualization, Y.Z. and V.Z.; investigation, Y.Z. and V.Z., writing—original draft preparation, Y.Z. and V.Z.; writing—review and editing Y.Z. and V.Z.; supervision, V.V.; project administration, Y.Z.; funding acquisition, Y.Z. All authors have read and agreed to the published version of the manuscript.

**Funding:** This research was funded by Russian Science Foundation, grant number 22-73-00240 (https://rscf.ru/en/project/22-73-00240/ (Agreement of 1 August 2022).

**Institutional Review Board Statement:** Not applicable.

**Data Availability Statement:** Not applicable.

**Conflicts of Interest:** The authors declare no conflict of interest.

## References


1. Kabir, E.; Kaur, R.; Lee, J.; Kim, K.-H.; Kwon, E.E. Prospects of biopolymer technology as an alternative option for non-degradable plastics and sustainable management of plastic wastes. *J. Clean. Prod.* **2020**, *258*, 120536. [CrossRef]
2. Sivakanthan, S.; Rajendran, S.; Gamage, A.; Madhujith, T.; Mani, S. Antioxidant and antimicrobial applications of biopolymers: A review. *Food Res. Int.* **2020**, *136*, 109327. [CrossRef]
3. Pradhan, B.; Bharti, D.; Chakravarty, S.; Ray, S.S.; Voinova, V.V.; Bonartsev, A.P.; Pal, K. Internet of Things and Robotics in Transforming Current-Day Healthcare Services. *J. Health Eng.* **2021**, *2021*, 1–15. [CrossRef]
4. Bonartsev, A.P.; Bonartseva, G.A.; Reshetov, I.V.; Kirpichnikov, M.P.; Shaitan, K.V. Application of polyhydroxyalkanoates in medicine and the biological activity of natural poly(3-hydroxybutyrate). *Acta Nat.* **2019**, *11*, 4–16. [CrossRef]
5. Shrivastava, A.; Dondapati, S. Biodegradable composites based on biopolymers and natural bast fibres: A review. *Mater. Today Proc.* **2021**, *46*, 1420–1428. [CrossRef]
6. Udayakumar, G.P.; Muthusamy, S.; Selvaganesh, B.; Sivarajasekar, N.; Rambabu, K.; Banat, F.; Sivamani, S.; Sivakumar, N.; Hosseini-Bandegharaei, A.; Show, P.L. Biopolymers and composites: Properties, characterization and their applications in food, medical and pharmaceutical industries. *J. Environ. Chem. Eng.* **2021**, *9*, 105322. [CrossRef]
7. Aaliya, B.; Sunooj, K.V.; Lackner, M. Biopolymer composites: A review. *Int. J. Biobased Plast.* **2021**, *3*, 40–84. [CrossRef]
8. Kartik, A.; Akhil, D.; Lakshmi, D.; Gopinath, K.P.; Arun, J.; Sivaramakrishnan, R.; Pugazhendhi, A. A critical review on production of biopolymers from algae biomass and their applications. *Bioresour. Technol.* **2021**, *329*, 124868. [CrossRef]
9. Han, X.; Zheng, Y.; Munro, C.J.; Ji, Y.; Braunschweig, A.B. Carbohydrate nanotechnology: Hierarchical assembly using nature's other information carrying biopolymers. *Curr. Opin. Biotechnol.* **2015**, *34*, 41–47. [CrossRef]
10. Rong, S.Y.; Mubarak, N.; Tanjung, F.A. Structure-property relationship of cellulose nanowhiskers reinforced chitosan biocomposite films. *J. Environ. Chem. Eng.* **2017**, *5*, 6132–6136. [CrossRef]
11. Mehrpouya, M.; Vahabi, H.; Janbaz, S.; Darafsheh, A.; Mazur, T.R.; Ramakrishna, S. 4D printing of shape memory polylactic acid (PLA). *Polymer* **2021**, *230*, 124080. [CrossRef]
12. Kumari, S.V.G.; Pakshirajan, K.; Pugazhenthi, G. Recent advances and future prospects of cellulose, starch, chitosan, polylactic acid and polyhydroxyalkanoates for sustainable food packaging applications. *Int. J. Biol. Macromol.* **2022**, *221*, 163–182. [CrossRef]
13. Haghighi, H.; Licciardello, F.; Fava, P.; Siesler, H.W.; Pulvirenti, A. Recent advances on chitosan-based films for sustainable food packaging applications. *Food Packag. Shelf Life* **2020**, *26*, 100551. [CrossRef]
14. Claro, P.I.C.; Neto, A.R.S.; Bibbo, A.C.C.; Mattoso, L.H.C.; Bastos, M.S.R.; Marconcini, J.M. Biodegradable Blends with Potential Use in Packaging: A Comparison of PLA/Chitosan and PLA/Cellulose Acetate Films. *J. Polym. Environ.* **2016**, *24*, 363–371. [CrossRef]
15. Hazrati, K.; Sapuan, S.; Zuhri, M.; Jumaidin, R. Preparation and characterization of starch-based biocomposite films reinforced by Dioscorea hispida fibers. *J. Mater. Res. Technol.* **2021**, *15*, 1342–1355. [CrossRef]
16. Chang, I.; Lee, M.; Tran, A.T.P.; Lee, S.; Kwon, Y.-M.; Im, J.; Cho, G.-C. Review on biopolymer-based soil treatment (BPST) technology in geotechnical engineering practices. *Transp. Geotech.* **2020**, *24*, 100385. [CrossRef]
17. Priyadarshi, R.; Roy, S.; Ghosh, T.; Biswas, D.; Rhim, J.-W. Antimicrobial nanofillers reinforced biopolymer composite films for active food packaging applications—A review. *Sustain. Mater. Technol.* **2021**, *321*, e00353. [CrossRef]
18. Manu, T.; Nazmi, A.R.; Shahri, B.; Emerson, N.; Huber, T. Biocomposites: A review of materials and perception. *Mater. Today Commun.* **2022**, *31*, 103308. [CrossRef]
19. Shanmugam, V.; Mensah, R.A.; Försth, M.; Sas, G.; Restás, Á.; Addy, C.; Xu, Q.; Jiang, L.; Neisiany, R.E.; Singha, S.; et al. Circular economy in biocomposite development: State-of-the-art, challenges and emerging trends. *Compos. Part C Open Access* **2021**, *5*, 100138. [CrossRef]
20. Komal, U.K.; Lila, M.K.; Singh, I. PLA/banana fiber based sustainable biocomposites: A manufacturing perspective. *Compos. Part B Eng.* **2020**, *180*, 107535. [CrossRef]
21. Makvandi, P.; Ghomi, M.; Ashrafizadeh, M.; Tafazoli, A.; Agarwal, T.; Delfi, M.; Akhtari, J.; Zare, E.N.; Padil, V.V.; Zarrabi, A.; et al. A review on advances in graphene-derivative/polysaccharide bionanocomposites: Therapeutics, pharmacogenomics and toxicity. *Carbohydr. Polym.* **2020**, *250*, 116952. [CrossRef]

22. Yaashikaa, P.; Kumar, P.S.; Karishma, S. Review on biopolymers and composites—Evolving material as adsorbents in removal of environmental pollutants. *Environ. Res.* **2022**, *212*, 113114. [CrossRef]
23. Kaur, S.; Dhillon, G.S. Recent trends in biological extraction of chitin from marine shell wastes: A review. *Crit. Rev. Biotechnol.* **2015**, *35*, 44–61. [CrossRef]
24. Younes, I.; Rinaudo, M. Chitin and chitosan preparation from marine sources. Structure, properties and applications. *Mar. Drugs* **2015**, *13*, 1133–1174. [CrossRef]
25. Domard, A. A perspective on 30 years research on chitin and chitosan. *Carbohydr. Polym.* **2011**, *84*, 696–703. [CrossRef]
26. Tao, F.; Cheng, Y.; Shi, X.; Zheng, H.; Du, Y.; Xiang, W.; Deng, H. Applications of chitin and chitosan nanofibers in bone regenerative engineering. *Carbohydr. Polym.* **2020**, *230*, 115658. [CrossRef]
27. Bakshi, P.S.; Selvakumar, D.; Kadirvelu, K.; Kumar, N.S. Chitosan as an environment friendly biomaterial—A review on recent modifications and applications. *Int. J. Biol. Macromol.* **2020**, *150*, 1072–1083. [CrossRef]
28. Ghormade, V.; Pathan, E.K.; Deshpande, M.V. Can fungi compete with marine sources for chitosan production? *Int. J. Biol. Macromol.* **2017**, *104*, 1415–1421. [CrossRef]
29. Muzzarelli, R.; Ilari, P.; Tarsi, R.; Dubini, B.; Xia, W. Chitosan from *Absidia coerulea*. *Carbohydr. Polym.* **1994**, *25*, 45–50. [CrossRef]
30. Di Mario, F.; Rapanà, P.; Tomati, U.; Galli, E. Chitin and chitosan from Basidiomycetes. *Int. J. Biol. Macromol.* **2008**, *43*, 8–12. [CrossRef]
31. Cai, J.; Yang, J.; Du, Y.; Fan, L.; Qiu, Y.; Li, J.; Kennedy, J.F. Enzymatic preparation of chitosan from the waste *Aspergillus niger mycelium* of citric acid production plant. *Carbohydr. Polym.* **2006**, *64*, 151–157. [CrossRef]
32. Gonil, P.; Sajomsang, W. Applications of magnetic resonance spectroscopy to chitin from insect cuticles. *Int. J. Biol. Macromol.* **2012**, *51*, 514–522. [CrossRef]
33. Khayrova, A.; Lopatin, S.; Varlamov, V. Black Soldier Fly *Hermetia illucens* as a Novel Source of Chitin and Chitosan. *Int. J. Sci.* **2019**, *8*, 81–86. [CrossRef]
34. Svirshchevskaya, E. Correlation Analysis of Chitosan Physicochemical Parameters Determined by Different Methods. *Org. Med. Chem. Int. J.* **2017**, *1*, 74–82. [CrossRef]
35. Mourya, V.; Inamdar, N.N. Chitosan-modifications and applications: Opportunities galore. *React. Funct. Polym.* **2008**, *68*, 1013–1051. [CrossRef]
36. Sashiwa, H.; Shigemasa, Y.; Roy, R. Chemical modification of chitosan 8: Preparation of chitosan–dendrimer hybrids via short spacer. *Carbohydr. Polym.* **2002**, *47*, 191–199. [CrossRef]
37. Varlamov, V.P.; Il'Ina, A.V.; Shagdarova, B.T.; Lunkov, A.P.; Mysyakina, I.S. Chitin/Chitosan and Its Derivatives: Fundamental Problems and Practical Approaches. *Biochem. Moscow* **2020**, *85*, 154–176. [CrossRef]
38. Harish Prashanth, K.V.; Tharanathan, R.N. Chitin/chitosan: Modifications and their unlimited application potential—An overview. *Trends Food Sci. Technol.* **2007**, *18*, 117–131. [CrossRef]
39. Fei Liu, X.; Guan, Y.L.; Yang, D.Z.; Li, Z.; Yao, K. De Antibacterial action of chitosan and carboxymethylated chitosan. *J. Appl. Polym. Sci.* **2000**, *79*, 1324–1335. [CrossRef]
40. Vasyukova, N.I.; Zinov'eva, S.V.; Il'inskaya, L.I.; Perekhod, E.A.; Chalenko, G.I.; Gerasimova, N.G.; Il'ina, A.V.; Varlamov, V.P.; Ozeretskovskaya, O.L. Modulation of Plant Resistance to Diseases by Water-Soluble Chitosan. *Appl. Biochem. Microbiol.* **2001**, *371*, 103–109. [CrossRef]
41. Laokuldilok, T.; Potivas, T.; Kanha, N.; Surawang, S.; Seesuriyachan, P.; Wangtueai, S.; Phimolsiripol, Y.; Regenstein, J.M. Physicochemical, antioxidant, and antimicrobial properties of chitooligosaccharides produced using three different enzyme treatments. *Food Biosci.* **2017**, *18*, 28–33. [CrossRef]
42. Naveed, M.; Phil, L.; Sohail, M.; Hasnat, M.; Baig, M.M.F.A.; Ihsan, A.U.; Shumzaid, M.; Kakar, M.U.; Khan, T.M.; Akabar, M.D.; et al. Chitosan oligosaccharide (COS): An overview. *Int. J. Biol. Macromol.* **2019**, *129*, 827–843. [CrossRef]
43. Yuan, X.; Zheng, J.; Jiao, S.; Cheng, G.; Feng, C.; Du, Y.; Liu, H. A review on the preparation of chitosan oligosaccharides and application to human health, animal husbandry and agricultural production. *Carbohydr. Polym.* **2019**, *220*, 60–70. [CrossRef] [PubMed]
44. Kim, S.-K.; Rajapakse, N. Enzymatic production and biological activities of chitosan oligosaccharides (COS): A review. *Carbohydr. Polym.* **2005**, *62*, 357–368. [CrossRef]
45. Lee, S.-H.; Park, J.-S.; Kim, S.-K.; Ahn, C.-B.; Je, J.-Y. Chitooligosaccharides suppress the level of protein expression and acetylcholinesterase activity induced by Aβ25–35 in PC12 cells. *Bioorg. Med. Chem. Lett.* **2009**, *19*, 860–862. [CrossRef] [PubMed]
46. Rahman, H.; Hjeljord, L.G.; Aam, B.B.; Sørlie, M.; Tronsmo, A. Antifungal effect of chito-oligosaccharides with different degrees of polymerization. *Eur. J. Plant Pathol.* **2015**, *141*, 147–158. [CrossRef]
47. Azuma, K.; Osaki, T.; Minami, S.; Okamoto, Y. Anticancer and Anti-Inflammatory Properties of Chitin and Chitosan Oligosaccharides. *J. Funct. Biomater.* **2015**, *6*, 33–49. [CrossRef] [PubMed]
48. Huang, X.; Zhang, X.; Wang, X.; Wang, C.; Tang, B. Microenvironment of alginate-based microcapsules for cell culture and tissue engineering. *J. Biosci. Bioeng.* **2012**, *114*, 640–647. [CrossRef]
49. Frank, L.A.; Onzi, G.R.; Morawski, A.S.; Pohlmann, A.R.; Guterres, S.S.; Contri, R.V. Chitosan as a coating material for nanoparticles intended for biomedical applications. *React. Funct. Polym.* **2020**, *147*, 104459. [CrossRef]
50. Moon, C.; Seo, D.-J.; Song, Y.-S.; Jung, W.-J. Antibacterial activity of various chitosan forms against *Xanthomonas axonopodis* pv. glycines. *Int. J. Biol. Macromol.* **2020**, *156*, 1600–1605. [CrossRef]

51. Chudinova, Y.V.; Shagdarova, B.T.; Il'ina, A.V.; Varlamov, V.P. Antibacterial effect of peptide conjugates with a quaternized chitosan derivative and its estimation by the method of atomic force microscopy. *Appl. Biochem. Microbiol.* **2016**, *52*, 496–501. [CrossRef]
52. Verlee, A.; Mincke, S.; Stevens, C.V. Recent developments in antibacterial and antifungal chitosan and its derivatives. *Carbohydr. Polym.* **2017**, *164*, 268–283. [CrossRef]
53. Qin, Y.; Li, P.; Guo, Z. Cationic chitosan derivatives as potential antifungals: A review of structural optimization and applications. *Carbohydr. Polym.* **2020**, *236*, 116002. [CrossRef] [PubMed]
54. Chakrabarti, A.; Talukdar, D.; Pal, A.; Ray, M. Immunomodulation of macrophages by methylglyoxal conjugated with chitosan nanoparticles against Sarcoma-180 tumor in mice. *Cell. Immunol.* **2014**, *287*, 27–35. [CrossRef]
55. Cesari, A.; Fabiano, A.; Piras, A.M.; Zambito, Y.; Uccello-Barretta, G.; Balzano, F. Binding and mucoadhesion of sulfurated derivatives of quaternary ammonium-chitosans and their nanoaggregates: An NMR investigation. *J. Pharm. Biomed. Anal.* **2020**, *177*, 112852. [CrossRef]
56. Badhe, R.V.; Nanda, R.K.; Chejara, D.R.; Choonara, Y.E.; Kumar, P.; du Toit, L.C.; Pillay, V. Microwave-assisted facile synthesis of a new tri-block chitosan conjugate with improved mucoadhesion. *Carbohydr. Polym.* **2015**, *130*, 213–221. [CrossRef] [PubMed]
57. Augustine, R.; Rehman, S.R.U.; Ahmed, R.; Zahid, A.A.; Sharifi, M.; Falahati, M.; Hasan, A. Electrospun chitosan membranes containing bioactive and therapeutic agents for enhanced wound healing. *Int. J. Biol. Macromol.* **2020**, *156*, 153–170. [CrossRef] [PubMed]
58. Singh, R.; Shitiz, K.; Singh, A. Chitin and chitosan: Biopolymers for wound management. *Int. Wound J.* **2017**, *14*, 1276–1289. [CrossRef]
59. Wang, L.; Hao, F.; Tian, S.; Dong, H.; Nie, J.; Ma, G. Targeting polysaccharides such as chitosan, cellulose, alginate and starch for designing hemostatic dressings. *Carbohydr. Polym.* **2022**, *291*, 119574. [CrossRef]
60. Rodríguez-Vázquez, M.; Vega-Ruiz, B.; Ramos-Zúñiga, R.; Saldaña-Koppel, D.A.; Quiñones-Olvera, L.F. Chitosan and Its Potential Use as a Scaffold for Tissue Engineering in Regenerative Medicine. *Biomed. Res. Int.* **2015**, *2015*, 821279. [CrossRef]
61. Chen, M.-C.; Liu, C.-T.; Tsai, H.-W.; Lai, W.-Y.; Chang, Y.; Sung, H.-W. Mechanical properties, drug eluting characteristics and in vivo performance of a genipin-crosslinked chitosan polymeric stent. *Biomaterials* **2009**, *30*, 5560–5571. [CrossRef] [PubMed]
62. Perrin, N.; Mohammadkhani, G.; Moghadam, F.H.; Delattre, C.; Zamani, A. Biocompatible fibers from fungal and shrimp chitosans for suture application. *Curr. Res. Biotechnol.* **2022**, *4*, 530–536. [CrossRef]
63. Zhang, Q.; Chen, Y.; Wei, P.; Zhong, Y.; Chen, C.; Cai, J. Extremely strong and tough chitosan films mediated by unique hydrated chitosan crystal structures. *Mater. Today* **2021**, *51*, 27–38. [CrossRef]
64. Luna, R.; Touhami, F.; Uddin, M.J.; Touhami, A. Effect of temperature and pH on nanostructural and nanomechanical properties of chitosan films. *Surf. Interfaces* **2022**, *29*, 101706. [CrossRef]
65. Lewandowska, K.; Sionkowska, A.; Kaczmarek, B.; Furtos, G. Mechanical and Morphological Studies of Chitosan/Clay Composites. *Mol. Cryst. Liq. Cryst.* **2014**, *590*, 193–198. [CrossRef]
66. Leceta, I.; Peñalba, M.; Arana, P.; Guerrero, P.; De La Caba, K. Ageing of chitosan films: Effect of storage time on structure and optical, barrier and mechanical properties. *Eur. Polym. J.* **2015**, *66*, 170–179. [CrossRef]
67. Janik, W.; Ledniowska, K.; Nowotarski, M.; Kudła, S.; Knapczyk-Korczak, J.; Stachewicz, U.; Nowakowska-Bogdan, E.; Sabura, E.; Nosal-Kovalenko, H.; Turczyn, R.; et al. Chitosan-based films with alternative eco-friendly plasticizers: Preparation, physicochemical properties and stability. *Carbohydr. Polym.* **2023**, *301*, 120277. [CrossRef]
68. Movaffagh, J.; Khatib, M.; Bazzaz, B.S.F.; Taherzadeh, Z.; Hashemi, M.; Moghaddam, A.S.; Tabatabaee, S.A.; Azizzadeh, M.; Jirofti, N. Evaluation of wound-healing efficiency of a functional Chitosan/Aloe vera hydrogel on the improvement of re-epithelialization in full thickness wound model of rat. *J. Tissue Viability* **2022**, *31*, 649–656. [CrossRef]
69. Shagdarova, B.; Konovalova, M.; Zhuikova, Y.; Lunkov, A.; Zhuikov, V.; Khaydapova, D.; Il'Ina, A.; Svirshchevskaya, E.; Varlamov, V. Collagen/Chitosan Gels Cross-Linked with Genipin for Wound Healing in Mice with Induced Diabetes. *Materials* **2021**, *15*, 15. [CrossRef]
70. Li, X.; Hetjens, L.; Wolter, N.; Li, H.; Shi, X.; Pich, A. Charge-reversible and biodegradable chitosan-based microgels for lysozyme-triggered release of vancomycin. *J. Adv. Res.* **2022**. [CrossRef]
71. Huang, Y.; Huang, Z.; Liu, H.; Zhang, X.; Cai, Q.; Yang, X. Photoluminescent biodegradable polyorganophosphazene: A promising scaffold material for in vivo application to promote bone regeneration. *Bioact. Mater.* **2020**, *5*, 102–109. [CrossRef] [PubMed]
72. Kedir, W.M.; Abdi, G.F.; Goro, M.M.; Tolesa, L.D. Pharmaceutical and drug delivery applications of chitosan biopolymer and its modified nanocomposite: A review. *Heliyon* **2022**, *8*, e10196. [CrossRef] [PubMed]
73. Singha, I.; Basu, A. Chitosan based injectable hydrogels for smart drug delivery applications. *Sensors Int.* **2022**, *3*, 100168. [CrossRef]
74. Popova, E.V.; Tikhomirova, V.E.; Beznos, O.V.; Chesnokova, N.B.; Grigoriev, Y.V.; Klyachko, N.L.; Kost, O.A. Chitosan-covered calcium phosphate particles as a drug vehicle for delivery to the eye. *Nanomed. Nanotechnol. Biol. Med.* **2022**, *40*, 102493. [CrossRef] [PubMed]
75. Drozd, N.N.; Lunkov, A.P.; Shagdarova, B.T.; Zhuikova, Y.V.; Il'ina, A.V.; Varlamov, V.P. Chitosan/heparin layer-by-layer coatings for improving thromboresistance of polyurethane. *Surf. Interfaces* **2022**, *28*, 101674. [CrossRef]

76. Yang, C.; Wang, M.; Wang, W.; Liu, H.; Deng, H.; Du, Y.; Shi, X. Electrodeposition induced covalent cross-linking of chitosan for electrofabrication of hydrogel contact lenses. *Carbohydr. Polym.* **2022**, *292*, 119678. [CrossRef]
77. Lin, X.; Liu, J.; Zhou, F.; Ou, Y.; Rong, J.; Zhao, J. Poly(2-hydroxyethyl methacrylate-co-quaternary ammonium salt chitosan) hydrogel: A potential contact lens material with tear protein deposition resistance and antimicrobial activity. *Biomater. Adv.* **2022**, *136*, 212787. [CrossRef] [PubMed]
78. Chang, S.-H.; Wu, G.-J.; Wu, C.-H.; Huang, C.-H.; Tsai, G.-J. Oral administration with chitosan hydrolytic products modulates mitogen-induced and antigen-specific immune responses in BALB/c mice. *Int. J. Biol. Macromol.* **2019**, *131*, 158–166. [CrossRef]
79. Chirkov, S.N. The Antiviral Activity of Chitosan (Review). *Appl. Biochem. Microbiol.* **2002**, *38*, 1–8. [CrossRef]
80. Zhang, S.; Zhang, Q.; Chen, J.; Dong, H.; Cui, A.; Sun, L.; Wang, N.; Li, J.; Qu, Z. Cost-effective chitosan thermal bonded nonwovens serving as an anti-viral inhibitor layer in face mask. *Mater. Lett.* **2022**, *318*, 132203. [CrossRef]
81. Pardeshi, C.V.; Belgamwar, V.S. Controlled synthesis of N,N,N-trimethyl chitosan for modulated bioadhesion and nasal membrane permeability. *Int. J. Biol. Macromol.* **2016**, *82*, 933–944. [CrossRef] [PubMed]
82. Başaran, E.; Yenilmez, E.; Berkman, M.S.; Büyükköroğlu, G.; Yazan, Y. Chitosan nanoparticles for ocular delivery of cyclosporine A. *J. Microencapsul.* **2014**, *31*, 49–57. [CrossRef] [PubMed]
83. Li, N.; Zhuang, C.; Wang, M.; Sun, X.; Nie, S.; Pan, W. Liposome coated with low molecular weight chitosan and its potential use in ocular drug delivery. *Int. J. Pharm.* **2009**, *379*, 131–138. [CrossRef]
84. Yang, Y.; Yang, S.-B.; Wang, Y.-G.; Zhang, S.-H.; Yu, Z.-F.; Tang, T.-T. Bacterial inhibition potential of quaternised chitosan-coated VICRYL absorbable suture: An in vitro and in vivo study. *J. Orthop. Transl.* **2017**, *8*, 49–61. [CrossRef] [PubMed]
85. Jayakumar, R.; Prabaharan, M.; Kumar, P.T.S.; Nair, S.V.; Tamura, H. Biomaterials based on chitin and chitosan in wound dressing applications. *Biotechnol. Adv.* **2011**, *29*, 322–337. [CrossRef]
86. Zhou, Y.; Yang, D.; Chen, X.; Xu, Q.; Lu, F.; Nie, J. Electrospun water-soluble carboxyethyl chitosan/poly(vinyl alcohol) nanofibrous membrane as potential wound dressing for skin regeneration. *Biomacromolecules* **2008**, *9*, 349–354. [CrossRef]
87. Ribeiro, M.P.; Espiga, A.; Silva, D.; Baptista, P.; Henriques, J.; Ferreira, C.; Silva, J.C.; Borges, J.P.; Pires, E.; Chaves, P.; et al. Development of a new chitosan hydrogel for wound dressing. *Wound Repair Regen.* **2009**, *17*, 817–824. [CrossRef]
88. Boucard, N.; Viton, C.; Agay, D.; Mari, E.; Roger, T.; Chancerelle, Y.; Domard, A. The use of physical hydrogels of chitosan for skin regeneration following third-degree burns. *Biomaterials* **2007**, *28*, 3478–3488. [CrossRef]
89. Mi, F.-L.; Shyu, S.-S.; Wu, Y.-B.; Lee, S.-T.; Shyong, J.-Y.; Huang, R.-N. Fabrication and characterization of a sponge-like asymmetric chitosan membrane as a wound dressing. *Biomaterials* **2000**, *22*, 165–173. [CrossRef]
90. Silva, S.S.; Luna, S.M.; Gomes, M.E.; Benesch, J.; Pashkuleva, I.; Mano, J.F.; Reis, R.L. Plasma surface modification of chitosan membranes: Characterization and preliminary cell response studies. *Macromol. Biosci.* **2008**, *8*, 568–576. [CrossRef]
91. Zhang, X.; Yang, D.; Nie, J. Chitosan/polyethylene glycol diacrylate films as potential wound dressing material. *Int. J. Biol. Macromol.* **2008**, *43*, 456–462. [CrossRef] [PubMed]
92. Zhang, Y.; He, H.; Gao, W.-J.; Lu, S.-Y.; Liu, Y.; Gu, H.-Y. Rapid adhesion and proliferation of keratinocytes on the gold colloid/chitosan film scaffold. *Mater. Sci. Eng. C* **2009**, *29*, 908–912. [CrossRef]
93. Bonartsev, A.P.; Zharkova, I.I.; Yakovlev, S.G.; Myshkina, V.L.; Mahina, T.K.; Voinova, V.V.; Zernov, A.L.; Zhuikov, V.A.; Akoulina, E.A.; Ivanova, E.V.; et al. Biosynthesis of poly(3-hydroxybutyrate) copolymers by Azotobacter chroococcum 7B: A precursor feeding strategy. *Prep. Biochem. Biotechnol.* **2017**, *47*, 173–184. [CrossRef] [PubMed]
94. Jose, A.A.; Hazeena, S.H.; Lakshmi, N.M.; B, A.K.; Madhavan, A.; Sirohi, R.; Tarafdar, A.; Sindhu, R.; Awasthi, M.K.; Pandey, A.; et al. Bacterial biopolymers: From production to applications in biomedicine. *Sustain. Chem. Pharm.* **2022**, *25*, 100582. [CrossRef]
95. Kourmentza, C.; Plácido, J.; Venetsaneas, N.; Burniol-Figols, A.; Varrone, C.; Gavala, H.N.; Reis, M.A.M. Recent Advances and Challenges towards Sustainable Polyhydroxyalkanoate (PHA) Production. *Bioengineering* **2017**, *4*, 55. [CrossRef] [PubMed]
96. Anburajan, P.; Kumar, A.N.; Sabapathy, P.C.; Kim, G.-B.; Cayetano, R.D.; Yoon, J.-J.; Kumar, G.; Kim, S.-H. Polyhydroxy butyrate production by Acinetobacter junii BP25, Aeromonas hydrophila ATCC 7966, and their co-culture using a feast and famine strategy. *Bioresour. Technol.* **2019**, *293*, 122062. [CrossRef]
97. García, A.; Segura, D.; Espín, G.; Galindo, E.; Castillo, T.; Peña, C. High production of poly-β-hydroxybutyrate (PHB) by an *Azotobacter vinelandii* mutant altered in PHB regulation using a fed-batch fermentation process. *Biochem. Eng. J.* **2014**, *82*, 117–123. [CrossRef]
98. Myshkina, V.L.; Nikolaeva, D.A.; Makhina, T.K.; Bonartsev, A.; Bonartseva, G. Effect of growth conditions on the molecular weight of poly-3-hydroxybutyrate produced by *Azotobacter chroococcum* 7B. *Appl. Biochem. Microbiol.* **2008**, *44*, 482–486. [CrossRef]
99. Bonartsev, A.; Yakovlev, S.; Boskhomdzhiev, A.; Zharkova, I.; Bagrov, D.; Myshkina, V.; Mahina, T.; Kharitonova, E.; Samsonova, O.; Zernov, A.; et al. The Terpolymer Produced by *Azotobacter Chroococcum* 7B: Effect of Surface Properties on Cell Attachment. *PLoS ONE* **2013**, *8*, e57200. [CrossRef]
100. Lee, H.-J.; Kim, S.-G.; Cho, D.-H.; Bhatia, S.K.; Gurav, R.; Yang, S.-Y.; Yang, J.; Jeon, J.-M.; Yoon, J.-J.; Choi, K.-Y.; et al. Finding of novel lactate utilizing *Bacillus* sp. YHY22 and its evaluation for polyhydroxybutyrate (PHB) production. *Int. J. Biol. Macromol.* **2022**, *201*, 653–661. [CrossRef]
101. Mohammed, S.; Panda, A.N.; Ray, L. An investigation for recovery of polyhydroxyalkanoates (PHA) from Bacillus sp. BPPI-14 and *Bacillus* sp. BPPI-19 isolated from plastic waste landfill. *Int. J. Biol. Macromol.* **2019**, *134*, 1085–1096. [CrossRef] [PubMed]

102. Lemgruber, R.D.S.P.; Valgepea, K.; Tappel, R.; Behrendorff, J.B.; Palfreyman, R.W.; Plan, M.; Hodson, M.P.; Simpson, S.D.; Nielsen, L.K.; Köpke, M.; et al. Systems-level engineering and characterisation of *Clostridium autoethanogenum* through heterologous production of poly-3-hydroxybutyrate (PHB). *Metab. Eng.* **2019**, *53*, 14–23. [CrossRef] [PubMed]
103. Jung, H.-R.; Choi, T.-R.; Han, Y.H.; Park, Y.-L.; Park, J.Y.; Song, H.-S.; Yang, S.-Y.; Bhatia, S.K.; Gurav, R.; Park, H.; et al. Production of blue-colored polyhydroxybutyrate (PHB) by one-pot production and coextraction of indigo and PHB from recombinant *Escherichia coli*. *Dye. Pigment.* **2020**, *173*, 107889. [CrossRef]
104. Kirk, R.G.; Ginzburg, M. Ultrastructure of two species of halobacterium. *J. Ultrastruct. Res.* **1972**, *41*, 80–94. [CrossRef]
105. Tyagi, B.; Gupta, B.; Khatak, D.; Meena, R.; Thakur, I.S. Genomic analysis, simultaneous production, and process optimization of extracellular polymeric substances and polyhydroxyalkanoates by *Methylobacterium* sp. ISTM1 by utilizing molasses. *Bioresour. Technol.* **2022**, *354*, 127204. [CrossRef]
106. Höfer, P.; Vermette, P.; Groleau, D. Production and characterization of polyhydroxyalkanoates by recombinant *Methylobacterium extorquens*: Combining desirable thermal properties with functionality. *Biochem. Eng. J.* **2011**, *54*, 26–33. [CrossRef]
107. Mohanrasu, K.; Rao, R.G.R.; Dinesh, G.; Zhang, K.; Sudhakar, M.; Pugazhendhi, A.; Jeyakanthan, J.; Ponnuchamy, K.; Govarthanan, M.; Arun, A. Production and characterization of biodegradable polyhydroxybutyrate by *Micrococcus luteus* isolated from marine environment. *Int. J. Biol. Macromol.* **2021**, *186*, 125–134. [CrossRef]
108. Zhang, M.; Zhu, C.; Gao, J.; Fan, Y.; He, L.; He, C.; Wu, J. Deep-level nutrient removal and denitrifying phosphorus removal (DPR) potential assessment in a continuous two-sludge system treating low-strength wastewater: The transition from nitration to nitritation. *Sci. Total Environ.* **2020**, *744*, 140940. [CrossRef]
109. Tyagi, B.; Takkar, S.; Meena, R.; Thakur, I.S. Production of polyhydroxybutyrate (PHB) by *Parapedobacter* sp. ISTM3 isolated from Mawsmai cave utilizing molasses as carbon source. *Environ. Technol. Innov.* **2021**, *24*, 101854. [CrossRef]
110. Nitschke, M.; Costa, S.G.; Contiero, J. Rhamnolipids and PHAs: Recent reports on Pseudomonas-derived molecules of increasing industrial interest. *Process. Biochem.* **2011**, *46*, 621–630. [CrossRef]
111. Aloui, H.; Khomlaem, C.; Torres, C.A.; Freitas, F.; Reis, M.A.; Kim, B.S. Enhanced co-production of medium-chain-length polyhydroxyalkanoates and phenazines from crude glycerol by high cell density cultivation of *Pseudomonas chlororaphis* in membrane bioreactor. *Int. J. Biol. Macromol.* **2022**, *211*, 545–555. [CrossRef] [PubMed]
112. Lakshman, K.; Rastogi, N.; Shamala, T. Simultaneous and comparative assessment of parent and mutant strain of *Rhizobium meliloti* for nutrient limitation and enhanced polyhydroxyalkanoate (PHA) production using optimization studies. *Process Biochem.* **2004**, *39*, 1977–1983. [CrossRef]
113. Ramachander, T.; Rawal, S. PHB synthase from *Streptomyces aureofaciens* NRRL 2209. *FEMS Microbiol. Lett.* **2005**, *242*, 13–18. [CrossRef] [PubMed]
114. Krishnan, S.; Chinnadurai, G.S.; Perumal, P. Polyhydroxybutyrate by *Streptomyces* sp.: Production and characterization. *Int. J. Biol. Macromol.* **2017**, *104*, 1165–1171. [CrossRef] [PubMed]
115. Brandl, H.; Gross, R.A.; Lenz, R.W.; Fuller, R.C. Plastics from Bacteria and for Bacteria: Poly(β-hydroxyalkanoates) as Natural, Biocompatible, and Biodegradable Polyesters. In *Microbial Bioproducts*; Springer: Berlin/Heidelberg, Germany, 1990; Volume 41, pp. 77–93.
116. Birley, C.; Briddon, J.; Sykes, K.E.; Barker, P.A.; Organ, S.J.; Barham, P.J. Morphology of single crystals of poly (hydroxybutyrate) and copolymers of hydroxybuty rate and hydroxyvalerate. *J. Mater. Sci.* **1995**, *30*, 633–638. [CrossRef]
117. Zhuikov, V.A.; Bonartsev, A.P.; Zharkova, I.; Bykova, G.S.; Taraskin, N.Y.; Kireynov, A.V.; Kopitsyna, M.N.; Bonartseva, G.A.; Shaitan, K.V. Effect of Poly(ethylene glycol) on the Ultrastructure and Physicochemical Properties of the Poly(3-hydroxybutyrate). *Macromol. Symp.* **2017**, *375*, 1600189. [CrossRef]
118. Zhuikov, V.A.; Bonartsev, A.P.; Bagrov, D.; Yakovlev, S.; Myshkina, V.L.; Makhina, T.K.; Bessonov, I.V.; Kopitsyna, M.N.; Morozov, A.S.; Rusakov, A.A.; et al. Mechanics and surface ultrastructure changes of poly(3-hydroxybutyrate) films during enzymatic degradation in pancreatic lipase solution. *Mol. Cryst. Liq. Cryst.* **2017**, *648*, 236–243. [CrossRef]
119. Barham, P.J.; Keller, A.; Otun, E.L.; Holmes, P.A. Crystallization and morphology of a bacterial thermoplastic: Poly-3-hydroxybutyrate. *J. Mater. Sci.* **1984**, *19*, 2781–2794. [CrossRef]
120. Vakhrusheva, A.; Endzhievskaya, S.; Zhuikov, V.; Nekrasova, T.; Parshina, E.; Ovsiannikova, N.; Popov, V.; Bagrov, D.; Minin, A.; Sokolova, O.S. The role of vimentin in directional migration of rat fibroblasts. *Cytoskeleton* **2019**, *76*, 467–476. [CrossRef]
121. Zhuikov, V.A.; Zhuikova, Y.V.; Makhina, T.K.; Myshkina, V.L.; Rusakov, A.; Useinov, A.; Voinova, V.V.; Bonartseva, G.A.; Berlin, A.A.; Bonartsev, A.P.; et al. Comparative Structure-Property Characterization of Poly(3-Hydroxybutyrate-Co-3-Hydroxyvalerate)s Films under Hydrolytic and Enzymatic Degradation: Finding a Transition Point in 3-Hydroxyvalerate Content. *Polymers* **2020**, *12*, 728. [CrossRef]
122. Li, X.; Liu, K.L.; Wang, M.; Wong, S.Y.; Tjiu, W.C.; He, C.; Goh, S.H.; Li, J. Improving hydrophilicity, mechanical properties and biocompatibility of poly[(R)-3-hydroxybutyrate-co-(R)-3-hydroxyvalerate] through blending with poly[(R)-3-hydroxybutyrate]-alt-poly(ethylene oxide). *Acta Biomater.* **2009**, *5*, 2002–2012. [CrossRef] [PubMed]
123. Shih, W.-J.; Chen, Y.-H.; Shih, C.-J.; Hon, M.-H.; Wang, M.-C. Structural and morphological studies on poly(3-hydroxybutyrate acid) (PHB)/chitosan drug releasing microspheres prepared by both single and double emulsion processes. *J. Alloys Compd.* **2007**, *434–435*, 826–829. [CrossRef]
124. Degli Esposti, M.; Chiellini, F.; Bondioli, F.; Morselli, D.; Fabbri, P. Highly porous PHB-based bioactive scaffolds for bone tissue engineering by in situ synthesis of hydroxyapatite. *Mater. Sci. Eng. C* **2019**, *100*, 286–296. [CrossRef] [PubMed]

125. Cavalcante, M.D.P.; de Menezes, L.R.; Rodrigues, E.J.D.R.; Tavares, M.I.B. In vitro characterization of a biocompatible composite based on poly(3-hydroxybutyrate)/hydroxyapatite nanoparticles as a potential scaffold for tissue engineering. *J. Mech. Behav. Biomed. Mater.* **2022**, *128*, 105138. [CrossRef] [PubMed]
126. Akoulina, E.A.; Demianova, I.V.; Zharkova, I.I.; Voinova, V.V.; Zhuikov, V.A.; Khaydapova, D.D.; Chesnokova, D.V.; Menshikh, K.A.; Dudun, A.A.; Makhina, T.K.; et al. Growth of Mesenchymal Stem Cells on Poly(3-Hydroxybutyrate) Scaffolds Loaded with Simvastatin. *Bull. Exp. Biol. Med.* **2021**, *171*, 172–177. [CrossRef]
127. Volkov, A.V.; Muraev, A.A.; Zharkova, I.I.; Voinova, V.V.; Akoulina, E.A.; Zhuikov, V.A.; Khaydapova, D.D.; Chesnokova, D.V.; Menshikh, K.A.; Dudun, A.A.; et al. Poly(3-hydroxybutyrate)/hydroxyapatite/alginate scaffolds seeded with mesenchymal stem cells enhance the regeneration of critical-sized bone defect. *Mater. Sci. Eng. C* **2020**, *114*, 110991. [CrossRef]
128. Parvizifard, M.; Karbasi, S. Physical, mechanical and biological performance of PHB-Chitosan/MWCNTs nanocomposite coating deposited on bioglass based scaffold: Potential application in bone tissue engineering. *Int. J. Biol. Macromol.* **2020**, *152*, 645–662. [CrossRef]
129. Parvizifard, M.; Karbasi, S.; Salehi, H.; Bakhtiari, S.S.E. Evaluation of physical, mechanical and biological properties of bioglass/titania scaffold coated with poly (3-hydroxybutyrate)-chitosan for bone tissue engineering applications. *Mater. Technol.* **2019**, *35*, 75–91. [CrossRef]
130. Wang, J.; Wang, Z.; Li, J.; Wang, B.; Liu, J.; Chen, P.; Miao, M.; Gu, Q. Chitin nanocrystals grafted with poly(3-hydroxybutyrate-co-3-hydroxyvalerate) and their effects on thermal behavior of PHBV. *Carbohydr. Polym.* **2012**, *87*, 784–789. [CrossRef]
131. Wang, B.; Li, J.; Zhang, J.; Li, H.; Chen, P.; Gu, Q.; Wang, Z. Thermo-mechanical properties of the composite made of poly (3-hydroxybutyrate-co-3-hydroxyvalerate) and acetylated chitin nanocrystals. *Carbohydr. Polym.* **2013**, *95*, 100–106. [CrossRef]
132. Karbasi, S.; Khorasani, S.N.; Ebrahimi, S.; Khalili, S.; Fekrat, F.; Sadeghi, D. Preparation and characterization of poly (hydroxy butyrate)/chitosan blend scaffolds for tissue engineering applications. *Adv. Biomed. Res.* **2016**, *5*, 177. [CrossRef] [PubMed]
133. Rajan, R.; Sreekumar, P.A.; Joseph, K.; Skrifvars, M. Thermal and mechanical properties of chitosan reinforced polyhydroxybutyrate composites. *J. Appl. Polym. Sci.* **2012**, *124*, 3357–3362. [CrossRef]
134. Holappa, J.; Hjálmarsdóttir, M.; Másson, M.; Rúnarsson, Ö.; Asplund, T.; Soininen, P.; Nevalainen, T.; Järvinen, T. Antimicrobial activity of chitosan N-betainates. *Carbohydr. Polym.* **2006**, *65*, 114–118. [CrossRef]
135. Makuška, R.; Gorochovceva, N. Regioselective grafting of poly(ethylene glycol) onto chitosan through C-6 position of glucosamine units. *Carbohydr. Polym.* **2006**, *64*, 319–327. [CrossRef]
136. Li, G.; Zhuang, Y.; Mu, Q.; Wang, M.; Fang, Y. Preparation, characterization and aggregation behavior of amphiphilic chitosan derivative having poly (l-lactic acid) side chains. *Carbohydr. Polym.* **2008**, *72*, 60–66. [CrossRef]
137. Salama, H.E.; Saad, G.R.; Sabaa, M.W. Synthesis, characterization and antimicrobial activity of biguanidinylated chitosan- g -poly[( R )-3-hydroxybutyrate]. *Int. J. Biol. Macromol.* **2017**, *101*, 438–447. [CrossRef]
138. Arslan, H.; Hazer, B.; Yoon, S.C. Grafting of poly(3-hydroxyalkanoate) and linoleic acid onto chitosan. *J. Appl. Polym. Sci.* **2007**, *103*, 81–89. [CrossRef]
139. Salama, H.E.; Aziz, M.S.A.; Saad, G.R. Thermal properties, crystallization and antimicrobial activity of chitosan biguanidine grafted poly(3-hydroxybutyrate) containing silver nanoparticles. *Int. J. Biol. Macromol.* **2018**, *111*, 19–27. [CrossRef]
140. Vernaez, O.; Neubert, K.J.; Kopitzky, R.; Kabasci, S. Compatibility of Chitosan in Polymer Blends by Chemical Modification of Bio-based Polyesters. *Polymers* **2019**, *11*, 1939. [CrossRef]
141. Hu, S.-G.; Jou, C.-H.; Yang, M. Protein adsorption, fibroblast activity and antibacterial properties of poly(3-hydroxybutyric acid-co-3-hydroxyvaleric acid) grafted with chitosan and chitooligosaccharide after immobilized with hyaluronic acid. *Biomaterials* **2003**, *24*, 2685–2693. [CrossRef]
142. Loeffler, A.P.; Ma, P.X. Bioinspired Nanomaterials for Tissue Engineering. In *Nanotechnologies for the Life Sciences*; Wiley-VCH Verlag GmbH & Co. KGaA: Weinheim, Germany, 2012.
143. Surmenev, R.A.; Ivanov, A.N.; Cecilia, A.; Baumbach, T.; Chernozem, R.V.; Mathur, S.; Surmeneva, M.A. Electrospun composites of poly-3-hydroxybutyrate reinforced with conductive fillers for in vivo bone regeneration. *Open Ceram.* **2022**, *9*, 100237. [CrossRef]
144. Matsumoto, H.; Tanioka, A. Functionality in Electrospun Nanofibrous Membranes Based on Fiber's Size, Surface Area, and Molecular Orientation. *Membranes* **2011**, *1*, 249–264. [CrossRef] [PubMed]
145. Sadeghi, D.; Karbasi, S.; Razavi, S.; Mohammadi, S.; Shokrgozar, M.A.; Bonakdar, S. Electrospun poly(hydroxybutyrate)/chitosan blend fibrous scaffolds for cartilage tissue engineering. *J. Appl. Polym. Sci.* **2016**, *133*, 44171. [CrossRef]
146. Khoroushi, M.; Foroughi, M.R.; Karbasi, S.; Hashemibeni, B.; Khademi, A.A. Effect of Polyhydroxybutyrate/Chitosan/Bioglass nanofiber scaffold on proliferation and differentiation of stem cells from human exfoliated deciduous teeth into odontoblast-like cells. *Mater. Sci. Eng. C* **2018**, *89*, 128–139. [CrossRef] [PubMed]
147. Zhou, Y.; Li, Y.; Li, D.; Yin, Y.; Zhou, F. Electrospun PHB/Chitosan Composite Fibrous Membrane and Its Degradation Behaviours in Different pH Conditions. *J. Funct. Biomater.* **2022**, *13*, 58. [CrossRef] [PubMed]
148. Veleirinho, B.; Ribeiro-do-Valle, R.M.; Lopes-da-Silva, J.A. Processing conditions and characterization of novel electrospun poly (3-hydroxybutyrate-co-hydroxyvalerate)/chitosan blend fibers. *Mater. Lett.* **2011**, *65*, 2216–2219. [CrossRef]
149. Veleirinho, B.; Coelho, D.S.; Dias, P.F.; Maraschin, M.; Ribeiro-do-Valle, R.M.; Lopes-da-Silva, J.A. Nanofibrous poly(3-hydroxybutyrate-co-3-hydroxyvalerate)/chitosan scaffolds for skin regeneration. *Int. J. Biol. Macromol.* **2012**, *51*, 343–350. [CrossRef]

150. Toloue, E.B.; Karbasi, S.; Salehi, H.; Rafienia, M. Potential of an electrospun composite scaffold of poly (3-hydroxybutyrate)-chitosan/alumina nanowires in bone tissue engineering applications. *Mater. Sci. Eng. C* **2019**, *99*, 1075–1091. [CrossRef]
151. Pavlova, E.; Nikishin, I.; Bogdanova, A.; Klinov, D.; Bagrov, D. The miscibility and spatial distribution of the components in electrospun polymer–protein mats. *RSC Adv.* **2020**, *10*, 4672–4680. [CrossRef]
152. Chen, C.; Dong, L. Biodegradable Blends Based on Microbial Poly(3-Hydroxybutyrate) and Natural Chitosan. In *Biodegradable Polymer Blends and Composites from Renewable Resources*; John Wiley & Sons, Inc.: Hoboken, NJ, USA, 2009; pp. 227–237.
153. Ignatova, M.; Manolova, N.; Rashkov, I.; Markova, N. Quaternized chitosan/κ-carrageenan/caffeic acid–coated poly(3-hydroxybutyrate) fibrous materials: Preparation, antibacterial and antioxidant activity. *Int. J. Pharm.* **2016**, *513*, 528–537. [CrossRef]
154. Ikejima, T.; Yagi, K.; Inoue, Y. Thermal properties and crystallization behavior of poly(3-hydroxybutyric acid) in blends with chitin and chitosan. *Macromol. Chem. Phys.* **1999**, *200*, 413–421. [CrossRef]
155. Chen, C.; Zhou, X.; Zhuang, Y.; Dong, L. Thermal behavior and intermolecular interactions in blends of poly(3-hydroxybutyrate) and maleated poly(3-hydroxybutyrate) with chitosan. *J. Polym. Sci. Part B Polym. Phys.* **2005**, *43*, 35–47. [CrossRef]
156. Kolhe, P.; Kannan, R.M. Improvement in Ductility of Chitosan through Blending and Copolymerization with PEG: FTIR Investigation of Molecular Interactions. *Biomacromolecules* **2002**, *4*, 173–180. [CrossRef] [PubMed]
157. Khasanah; Reddy, K.R.; Sato, H.; Takahashi, I.; Ozaki, Y. Intermolecular hydrogen bondings in the poly(3-hydroxybutyrate) and chitin blends: Their effects on the crystallization behavior and crystal structure of poly(3-hydroxybutyrate). *Polymer* **2015**, *75*, 141–150. [CrossRef]
158. Suttiwijitpukdee, N.; Sato, H.; Zhang, J.; Hashimoto, T.; Ozaki, Y. Intermolecular interactions and crystallization behaviors of biodegradable polymer blends between poly (3-hydroxybutyrate) and cellulose acetate butyrate studied by DSC, FT-IR, and WAXD. *Polymer* **2011**, *52*, 461–471. [CrossRef]
159. Medvecky, L. Microstructure and Properties of Polyhydroxybutyrate-Chitosan-Nanohydroxyapatite Composite Scaffolds. *Sci. World J.* **2012**, *2012*, 537973. [CrossRef] [PubMed]
160. Cao, W.; Wang, A.; Jing, D.; Gong, Y.; Zhao, N.; Zhang, X. Novel biodegradable films and scaffolds of chitosan blended with poly(3-hydroxybutyrate). *J. Biomater. Sci. Polym. Ed.* **2005**, *16*, 1379–1394. [CrossRef]
161. Ivantsova, E.L.; Iordanskii, A.L.; Kosenko, R.Y.; Rogovina, S.Z.; Grachev, A.V.; Prut, É.V. Poly(3-hydroxybutyrate)-chitosan: A new biodegradable composition for prolonged delivery of biologically active substances. *Pharm. Chem. J.* **2011**, *45*, 51–55. [CrossRef]
162. Peschel, G.; Dahse, H.-M.; Konrad, A.; Wieland, G.D.; Mueller, P.-J.; Martin, D.P.; Roth, M. Growth of keratinocytes on porous films of poly(3-hydroxybutyrate) and poly(4-hydroxybutyrate) blended with hyaluronic acid and chitosan. *J. Biomed. Mater. Res. Part A* **2008**, *85A*, 1072–1081. [CrossRef]
163. Iordanskii, A.L.; Ivantsova, E.L.; Kosenko, R.Y.; Zernova, Y.N.; Rogovina, S.Z.; Filatova, A.G.; Gumargalieva, K.Z.; Rusanova, S.N.; Stoyanov, O.V.; Zaikov, G.E. Diffusion and structural characteristics of compositions based on polyhydroxybutyrate and chitosan for directed transport of medicinal substances. part 2. *Bull. Kazan Technol. Univ.* **2013**, *16*, 162–164. (In Russian)
164. Iordanskii, A.L.; Ivantsova, E.L.; Kosenko, R.Y.; Zernova, Y.N.; Rogovina, S.Z.; Filatova, A.G.; Gumargalieva, K.Z.; Rusanova, S.N.; Stoyanov, O.V.; Zaikov, G.E. Diffusion and structural characteristics of compositions based on polyhydroxybutyrate and chitosan for directed transport of medicinal substances. part 3. *Bull. Kazan Technol. Univ.* **2014**, *17*, 145–148. (In Russian)
165. Karpova, S.G.; Jordan, A.L.; Olkhov, A.A.; Povov, A.A.; Lomakin, S.M.; Shilkina, N.S.; Zaikov, G.E. Effect of rolling on the structure of fibrous materials based on poly(3-hydroxybutyrate) with chitosan obtained by electroforming. *Bull. Univ. Technol.* **2015**, *18*, 109–115. (In Russian)
166. Karpova, S.G.; Iordanskii, A.L.; Klenina, N.S.; Popov, A.A.; Lomakin, S.M.; Shilkina, N.G.; Rebrov, A.V. Changes in the structural parameters and molecular dynamics of polyhydroxybutyrate-chitosan mixed compositions under external influences. *Russ. J. Phys. Chem. B* **2013**, *7*, 225–231. [CrossRef]
167. Ikejima, T.; Inoue, Y. Crystallization behavior and environmental biodegradability of the blend films of poly(3-hydroxybutyric acid) with chitin and chitosan. *Carbohydr. Polym.* **2000**, *41*, 351–356. [CrossRef]
168. Cheung, M.K.; Wan, K.P.; Yu, P.H. Miscibility and morphology of chiral semicrystalline poly-(R)-(3-hydroxybutyrate)/chitosan and poly-(R)-(3-hydroxybutyrate-co-3-hydroxyvalerate)/chitosan blends studied with DSC,1HT1 andT1? CRAMPS. *J. Appl. Polym. Sci.* **2002**, *86*, 1253–1258. [CrossRef]
169. Silvestri, D.; Wacławek, S.; Sobel, B.; Torres-Mendieta, R.; Novotný, V.; Nguyen, N.H.A.; Ševců, A.; Padil, V.V.T.; Müllerová, J.; Stuchlík, M.; et al. A poly(3-hydroxybutyrate)–chitosan polymer conjugate for the synthesis of safer gold nanoparticles and their applications. *Green Chem.* **2018**, *20*, 4975–4982. [CrossRef]
170. Yalpani, M.; Marchessault, R.H.; Morin, F.G.; Monasterios, C.J. Synthesis of poly(3-hydroxyalkanoate) (PHA) conjugates: PHA-carbohydrate and PHA-synthetic polymer conjugates. *Macromolecules* **1991**, *24*, 6046–6049. [CrossRef]
171. Choonut, A.; Prasertsan, P.; Klomklao, S.; Sangkharak, K. An Environmentally Friendly Process for Textile Wastewater Treatment with a Medium-Chain-Length Polyhydroxyalkanoate Film. *J. Polym. Environ.* **2021**, *29*, 3335–3346. [CrossRef]
172. Anbukarasu, P.; Sauvageau, D.; Elias, A. Tuning the properties of polyhydroxybutyrate films using acetic acid via solvent casting. *Sci. Rep.* **2015**, *5*, 17884. [CrossRef]
173. Mukheem, A.; Shahabuddin, S.; Akbar, N.; Miskon, A.; Sarih, N.M.; Sudesh, K.; Khan, N.A.; Saidur, R.; Sridewi, N. Boron nitride doped polyhydroxyalkanoate/chitosan nanocomposite for antibacterial and biological applications. *Nanomaterials* **2019**, *9*, 645. [CrossRef]

174. Zaccone, M.; Patel, M.K.; De Brauwer, L.; Nair, R.; Montalbano, M.L.; Monti, M.; Oksman, K. Influence of Chitin Nanocrystals on the Crystallinity and Mechanical Properties of Poly(hydroxybutyrate) Biopolymer. *Polymers* **2022**, *14*, 562. [CrossRef] [PubMed]
175. Patel, M.K.; Zaccone, M.; De Brauwer, L.; Nair, R.; Monti, M.; Martinez-Nogues, V.; Frache, A.; Oksman, K. Improvement of Poly(lactic acid)-Poly(hydroxy butyrate) Blend Properties for Use in Food Packaging: Processing, Structure Relationships. *Polymers* **2022**, *14*, 5104. [CrossRef] [PubMed]
176. Kang, J.; Yun, S.I. Chitosan-reinforced PHB hydrogel and aerogel monoliths fabricated by phase separation with the solvent-exchange method. *Carbohydr. Polym.* **2022**, *284*, 119184. [CrossRef]
177. Bazzo, G.; Lemos-Senna, E.; Pires, A. Poly(3-hydroxybutyrate)/chitosan/ketoprofen or piroxicam composite microparticles: Preparation and controlled drug release evaluation. *Carbohydr. Polym.* **2009**, *77*, 839–844. [CrossRef]
178. Bazzo, G.C.; Lemos-Senna, E.; Gonçalves, M.C.; Pires, A.T.N. Effect of preparation conditions on morphology, drug content and release profiles of poly(hydroxybutyrate) microparticles containing piroxicam. *J. Braz. Chem. Soc.* **2008**, *19*, 914–921. [CrossRef]
179. Hall, A.R.; Geoghegan, M. Polymers and biopolymers at interfaces. *Rep. Prog. Phys.* **2018**, *81*, 036601. [CrossRef]
180. Garavand, F.; Cacciotti, I.; Vahedikia, N.; Rehman, A.; Tarhan, Ö.; Akbari-Alavijeh, S.; Shaddel, R.; Rashidinejad, A.; Nejatian, M.; Jafarzadeh, S.; et al. A comprehensive review on the nanocomposites loaded with chitosan nanoparticles for food packaging. *Crit. Rev. Food Sci. Nutr.* **2020**, *62*, 1383–1416. [CrossRef]
181. Boura-Theodoridou, O.; Giannakas, A.; Katapodis, P.; Stamatis, H.; Ladavos, A.; Barkoula, N.-M. Performance of ZnO/chitosan nanocomposite films for antimicrobial packaging applications as a function of NaOH treatment and glycerol/PVOH blending. *Food Packag. Shelf Life* **2020**, *23*, 100456. [CrossRef]
182. Ghosh, T.; Mondal, K.; Giri, B.S.; Katiyar, V. Silk nanodisc based edible chitosan nanocomposite coating for fresh produces: A candidate with superior thermal, hydrophobic, optical, mechanical and food properties. *Food Chem.* **2021**, *360*, 130048. [CrossRef]
183. Serafim, L.S.; Lemos, P.C.; Oliveira, R.; Reis, M.A.M. Optimization of polyhydroxybutyrate production by mixed cultures submitted to aerobic dynamic feeding conditions. *Biotechnol. Bioeng.* **2004**, *87*, 145–160. [CrossRef]
184. Xu, P.; Yang, W.; Niu, D.; Yu, M.; Du, M.; Dong, W.; Chen, M.; Jan Lemstra, P.; Ma, P. Multifunctional and robust polyhydroxyalkanoate nanocomposites with superior gas barrier, heat resistant and inherent antibacterial performances. *Chem. Eng. J.* **2020**, *382*, 122864. [CrossRef]
185. Manikandan, N.A.; Pakshirajan, K.; Pugazhenthi, G. Preparation and characterization of environmentally safe and highly biodegradable microbial polyhydroxybutyrate (PHB) based graphene nanocomposites for potential food packaging applications. *Int. J. Biol. Macromol.* **2020**, *154*, 866–877. [CrossRef] [PubMed]
186. Zare, M.; Namratha, K.; Ilyas, S.; Hezam, A.; Mathur, S.; Byrappa, K. Smart Fortified PHBV-CS Biopolymer with ZnO–Ag Nanocomposites for Enhanced Shelf Life of Food Packaging. *ACS Appl. Mater. Interfaces* **2019**, *11*, 48309–48320. [CrossRef] [PubMed]

*polymers*

MDPI

*Article*

# Bioconversion of Used Transformer Oil into Polyhydroxyalkanoates by *Acinetobacter* sp. Strain AAAID-1.5

**Shehu Idris [1,2], Rashidah Abdul Rahim [1], Ahmad Nazri Saidin [3] and Amirul Al-Ashraf Abdullah [1,4,*]**

1 School of Biological Sciences, Universiti Sains Malaysia, Gelugor 11800, Malaysia
2 Department of Microbiology, Kaduna State University, P.M.B. 2339, Kaduna 800283, Nigeria
3 TNB Research Sdn Bhd, Kajang 43000, Malaysia
4 Centre for Chemical Biology, Universiti Sains Malaysia, Bayan Lepas 11900, Malaysia
* Correspondence: amirul@usm.my; Tel.: +60-1-3439-3662

**Abstract:** In this research, the utilisation of used transformer oil (UTO) as carbon feedstock for the production of polyhydroxyalkanoate (PHA) was targeted; with a view to reducing the environmental challenges associated with the disposal of the used oil and provision of an alternative to non-biodegradable synthetic plastic. *Acinetobacter* sp. strain AAAID-1.5 is a PHA-producing bacterium recently isolated from a soil sample collected in Penang, Malaysia. The PHA-producing capability of this bacterium was assessed through laboratory experiments in a shake flask biosynthesis under controlled culture conditions. The effect of some biosynthesis factors on growth and polyhydroxyalkanoate (PHA) accumulation was also investigated, the structural composition of the PHA produced by the organism was established, and the characteristics of the polymer were determined using standard analytical methods. The results indicated that the bacteria could effectively utilise UTO and produce PHA up to 34% of its cell dry weight. Analysis of the effect of some biosynthesis factors revealed that the concentration of carbon substrate, incubation time, the concentration of yeast extract and utilisation of additional carbon substrates could influence the growth and polymer accumulation in the test organism. Manipulation of culture conditions resulted in an enhanced accumulation of the PHA. The data obtained from GC-MS and NMR analyses indicated that the PHA produced might have been composed of 3-hydroxyoctadecanoate and 3-hydroxyhexadecanoate as the major monomers. The physicochemical analysis of a sample of the polymer revealed an amorphous elastomer with average molecular weight and polydispersity index (PDI) of 110 kDa and 2.01, respectively. The melting and thermal degradation temperatures were 88 °C and 268 °C, respectively. The findings of this work indicated that used transformer oil could be used as an alternative carbon substrate for PHA biosynthesis. Also, *Acinetobacter* sp. strain AAAID-1.5 could serve as an effective agent in the bioconversion of waste oils, especially UTO, to produce biodegradable plastics. These may undoubtedly provide a foundation for further exploration of UTO as an alternative carbon substrate in the biosynthesis of specific polyhydroxyalkanoates.

**Keywords:** bioconversion; used transformer oil; polyhydroxyalkanoates; *Acinetobacter* sp.; bioplastics

**Citation:** Idris, S.; Rahim, R.A.; Saidin, A.N.; Abdullah, A.A.-A. Bioconversion of Used Transformer Oil into Polyhydroxyalkanoates by *Acinetobacter* sp. Strain AAAID-1.5. *Polymers* **2023**, *15*, 97. https://doi.org/10.3390/polym15010097

Academic Editors: Rosane Michele Duarte Soares and Vsevolod Aleksandrovich Zhuikov

Received: 23 November 2022
Revised: 19 December 2022
Accepted: 20 December 2022
Published: 26 December 2022



## 1. Introduction

Synthetic plastics have become important commodities that have improved the quality of human life, replacing packaging materials like glass and paper [1]. The interest in the development of biodegradable plastics is largely generated from the problems associated with conventional plastics in a global environment [2]. Typical petroleum-derived plastics are non-biodegradable and mostly gather or aggregate around our environment, a problem that calls for great concern among communities, waste management agencies, and policymakers. Undoubtedly, plastic wastes pose a serious problem to landscape, marine animals, and wildlife and have since become an "environmental eye sore" [3]. Managing plastics wastes has become a global concern. Although it is hard to completely stop the use of petroleum-based plastics owing to their versatile utility, it is possible to substitute or reduce their usage with better alternatives by promoting the production and application of biodegradable polymers that have similar material properties [4]. Predictably, the plastics industry may witness a paradigm shift in the current century; this may ultimately bring about a change from an all-petroleum-based industrial economy to one that encompasses a relatively broader base of materials that include but not limited to fermentation byproducts [5]. The over-dependence on conventional plastics brings about waste accumulation and greenhouse gas emissions.

Consequently, recent technologies are directed towards developing bio-green materials that exert insignificant environmental side effects [6]. Petroleum-based mineral oils have been used in electrical transformers, primarily for insulating purposes. The oil serves the function of cooling the transformers. However, the long-term usage of transformer oil results in changes in its physical and chemical characteristics, which makes it unfit for cooling and insulating purposes. Thus, after being used up, the disposal of used transformer oil (UTO) from electrical power stations, as well as a large number of electrical transformers located in populated areas and shopping centres throughout the world, is becoming increasingly complex; this is so because it could contaminate waterways and soil if serious spills happen, this problem necessitates the need for providing an immediate solution [7]. Coincidentally, there is a continuous search for cheaper substrates to produce biodegradable plastics to reduce the high production cost that remains a big challenge to commercialising these eco-friendly alternative products. Consequently, several attempts are being made to produce bioplastics using waste oils. For instance, waste glycerol [8,9], palm oil [10], crude glycerol [11] and waste frying oil [12].

Polyhydroxyalkanoates (PHAs) represent a versatile group of prokaryotic reserve materials that display high potential for application in numerous fields of the plastic market, partly due to their plastic-like properties [13]. PHA are polyesters synthesised and accumulated by a number of taxonomically different microorganisms under nutrient-limited conditions, particularly when a carbon source is readily available. These biomolecules serve as carbon and energy storage materials; their presence has also been established to enhance resistance to various stress conditions [14]. For instance, recent research demonstrated that the biological role of PHAs goes beyond their storage function since their presence in cytoplasm increases the stress resistance of microbes [15]. PHAs have recently gained so much attention, especially in research institutions and industry. Undoubtedly, they are valuable materials with unique and desirable features; these important biological molecules attract attention as "green" alternatives to petrochemical plastics. However, The major disadvantage of these important biopolymers is their high production cost [1]. PHAs are mostly classified based on the number of carbon atoms in their respective monomers. Short chain length polyhydroxyalkanoates (*scl*-PHAs) have 3-5 carbon, medium chain length polyhydroxyalkanoates (*mcl*-PHAs) have 6-14 carbon, while the long chain length polyhydroxyalkanoates (lcl-PHAs) have 15 and above carbon atoms [3,16–18]. Among these classes of PHAs, *lcl*-PHAs is relatively the least explored, and this is evident in the limited number of published works on this important class of PHAs. In comparison with the short-chain-length PHAs, the medium-chain-length PHAs (*mcl*-PHAs) are relatively less abundant and frequently produced by bacterial species belonging to the genus *Pseudomonas* [19].

PHA-accumulating microorganisms can be isolated from diverse ecological niches like water sediments, sludge, rhizosphere, marine region, and coastal water body sediments [20]. Such habitats are often rich in organic nutrients and poor in other nutrients to meet the metabolic requirements of the starving microbial population, especially PHA-accumulating ones [21]. It has been established through extensive research that several PHA-producing microbes synthesise PHA toward the end of the log phase of their growth cycle [4]. It is also found that in later stages of their life cycle, they often use it as a carbonosomes [22]. Evidently, the presence of PHA inclusions within the microbial cytosol has served and will continue to serve as a chemotaxonomic signature for detecting various microbial isolates [4,23]. Numerous screening procedures have been developed for the detection of microorganisms that accumulate PHAs. Staining techniques involving the use of Nile red [24], Sudan black [25], and Nile blue [26] have, over the years, been used. Although these staining techniques cannot distinguish PHA inclusions from other lipids inclusion, a number of PHA-producing organisms have been successfully detected using this method, especially when complemented with other methods for confirmation [27]. Although analytical and molecular approaches are, in most cases, relatively time-consuming; and, therefore, difficult to apply in throughput screening, they are considered more reliable.

Despite the fact that PHA and other biodegradable polymers offer numerous environmental advantages such as composability, biodegradability, and biocompatibility, as well as the ease with which they become embedded in the natural carbon cycle, the main obstacle to commercialisation is their production cost. Consequently, the PHAs market penetration still lags behind the high expectations of the scientific community [28]. However, there is no doubt that PHAs biosynthesis and its related technologies are creating an industrial value chain ranging from materials, fermentation, and energy to medical fields [29]. PHAs can be applied in a number of ways that include, among others: packaging films, bags and containers, feminine hygiene products, surgical pins, sutures, swabs, wound dressing and staples, biodegradable carriers of drugs, medicines, insecticides, herbicides, or fertilisers, especially for long term dosage, replacements of bones and plates. In addition, PHAs are also applicable as starting materials for chiral compounds [30]. The relatively high production cost and the public awareness of the negative environmental impact of fluid mineral fuels-related products have been part of the driving force toward the search for novel raw materials with a 'green agenda' [31]. It is based on this background that the attempt was made to convert the UTO into PHAs with a view to providing an alternative to non-biodegradable plastics as well as reducing the production cost of PHAs.

## 2. Materials and Methods

### 2.1. Isolation of Organism and Growth Conditions

The bacterium used in this work was isolated from a soil sample obtained around Sungai Pinang Malaysia (Lat. 5°8′57″ Lon. 100°24′38″) [32]. The mineral salt medium (MSM) used for enrichment cultivation contains gram per litre of $K_2HPO_4$ (5.8); $KH_2PO_4$ (3.7); $(NH_4)_2SO_4$ (1.1); $MgSO_4{\cdot}7H_2O$ (0.2); bacteriological agar (15). Other components include 1.0 mL of micro-element solution containing of $FeSO_4{\cdot}7H_2O$ (2.78); $CaCl_2{\cdot}2H_2O$ (1.67); $MnCl_2{\cdot}4H_2O$ (1.98); $CoSO_4{\cdot}7H_2O$ (2.81); $CuCl_2{\cdot}2H_2O$ (0.17); $ZnSO_4{\cdot}2H_2O$ (0.29); in 0.1M, HCl [33]. The MSM was supplemented with 4%($v/v$) emulsified UTO (The emulsification was done in 1:1 ratio with Tween 80) as the sole carbon source. The enrichment culture was set up by inoculating 1 g of the soil sample into a 250 mL conical flask containing 100 mL of the medium and incubated at 30 °C for 48 h with an agitation speed of 200 rpm. Serial dilutions of the enrichment culture were made, after which 0.1 mL aliquot was taken from a selected dilution tube and inoculated on MSM agar (containing UTO as the sole carbon source) using the spread plate technique. The plates were incubated at 30 °C for 3–5 days. Pure colonies were sub-cultured and maintained in 20% glycerol at −20 °C.

### 2.2. Screening for PHA Accumulation

The stock culture of the isolated bacteria was activated in 30 mL nutrient-rich broth (10 g/L peptone, 2 g/L yeast extract and 10 g/L Lab Lemco powder). It was incubated at 30 °C for 18 h with an agitation speed of 150 rpm. After that, the culture was serially diluted using distilled water. An aliquot was then inoculated on solid MSM containing five(5) μg/mL Nile red [24]. The plates were incubated for 48 h at 30 °C. Colonies that formed were replicated onto a fresh MSM agar plate. The original plates were then exposed to ultraviolet illumination (320 nm) to identify PHA-accumulating organisms. Colonies that exhibit pink fluorescence were tentatively considered PHAs producers. Fluorescence microscopy was also used to further identify the PHA-accumulating isolates; in this approach, about 1 mL of 72 h old culture grown in MSM was transferred into 2.5 mL Eppendorf tube;

after which 50 μL of Nile red solution (5 μg/mL) and 0.5 mL distilled water were added to the cell suspension and vortex-mixed immediately. The cell suspension was kept at room temperature for an hour and then centrifuged at 1000 rpm for 5 min. The supernatant was discarded, and the pellet was washed twice with distilled water. Another 0.5 mL of distilled water was then added to the pellet and mixed. Subsequently, 10 μL of the stained cell suspension was placed onto a clean glass slide and covered with a glass slip. The edge of the glass slip was sealed using cutex. Finally, the prepared slides were observed using a fluorescence microscope (Olympus BX53, Olympus Optical Co., Ltd., Tokyo, Japan) equipped with an Olympus DP72 camera. The prepared slide was observed under X100 UV compatible objective to detect the presence of PHAs granules. The granules normally excite Nile red and produce red fluorescence [34].

PHA Granules Visualization

Visualisation of the PHAs granules within the bacterial cell was accomplished using transmission electron microscopy (TEM) of the thin section of the bacterial cells embedded in Spur's resin. Prior to resin block preparation, the test isolate was grown in a PHA production medium for 72 h at 30 °C. The culture was then harvested, and the cell pellet was washed twice with distilled water and fixed with McDowell-Trump fixative for one hour [35]. The fixed cells were subsequently treated with 1%(*v*/*v*) Osmium tetroxide for 1hr and suspended in 3% agar, followed by sequential dehydration in ethanol [50%, 75%, 95%, 100% (*v*/*v*); 30min each and finally with 100% acetone for 10 min]. The dehydrated cells were then embedded in low-viscosity Spur's resin and cured overnight at 60 °C in the oven [36]. Ultrathin sections (1 μm thickness) were cut using a microtome (Power Tome PC-RMC Product, Boeckeler Instrument Inc., Tucson, AZ, USA). The thin section was placed onto the copper grids and stained sequentially with uranyl acetate and lead citrate solution for 15 min. Lastly, the thin section was observed at an acceleration voltage of 120 kV in the transmission electron microscope (Philip CM 12/STEM and JLM-2000FX11).

### 2.3. *PHA Biosynthesis in Shake Flask*

The PHA production was accomplished in a 250 mL Erlenmeyer flask containing 50 mL MSM supplemented with 2%(*v*/*v*) used transformer oil. The pH of the medium was adjusted to 6.8. the culture was incubated for 72 h at 30 °C, 200 rpm. After incubation, the culture broth was then centrifuged at 10,000× *g* for 10min. The cells pellet obtained was washed twice and freeze-dried in a freeze drier machine (LABCONCO, Kansas, USA). The cell dry weight (CDW) in g/L was estimated using an established method as described elsewhere [37]. The PHA content was subsequently analysed using gas chromatography (GC). The GC analysis was achieved using GC system (Shimadzu 2014, Kyoto, Japan) equipped with Flame ionization detector (FID) and fused silica capillary column (SPB-1 30 m × 0.25 mm × 0.25 mm). A 2 μL of the methyl ester obtained via methanolysis of the freeze-dried cells was injected into the system. An initial column temperature of 50 °C which was ramped to a final temperature of 250 °C in a continuous step of 5 °C/min was adopted with a total analysis time was set at 26.33 min.

### 2.4. *Effect of Some Biosynthesis Factors on Growth and PHA Accumulation*

The effect of some variable biosynthesis factors, such as concentration of carbon source (UTO), incubation time, and concentration of yeast extract, on the cell's growth and polymer accumulation was assessed. All the experiments were carried out in three replicates, from which mean values were calculated accordingly.

#### 2.4.1. Effect of Carbon Source Concentration

A varying concentration of waste transformer oil ranging from 0.5, 1.0, 1.5, 2.0, and 2.5%(*v*/*v*) was used in order to assess the effect of the carbon source concentration on the organism's growth and polymer accumulation. In the experimental setup, a centrifuged and washed cell inoculum (0.06 g/L) from the nutrient-rich broth was cultured in 50 mL

MSM at 30 °C for 72 h with an agitation speed of 200 rpm. At the end of incubation, the growth in terms of cell dry weight (CDW) in g/L was deduced from the standard growth curve after measurement of the optical density (OD) at 540 nm. The biomass was recovered by centrifugation and then freeze-dried. The PHA content was determined using GC analysis.

#### 2.4.2. Effect of Incubation Time

The MSM was inoculated with 0.06 g/L inoculum of the bacteria and incubated at 30 °C with an agitation speed of 200 rpm. As the biosynthesis progresses, the sampling was made at a regular time interval of 12 h within a period of 120 h. Growth was monitored using spectrophotometry, and the CDW was determined by comparing the OD reading with the standard curve. The biomass was recovered by centrifugation and freeze-dried; PHA content analysis was subsequently carried out using GC techniques.

#### 2.4.3. Effect of Yeast Extract Concentration

The effect of yeast extract on the growth and PHA accumulation of the bacteria was assessed in the basal MSM medium supplemented with yeast extract at varying concentrations (0.5, 1.0, 1.5, 2.0, and 2.5 g/L). An inoculum (0.06 g/L) was cultured in 50 mL of the medium at 30 °C for 72 h with an agitation speed of 200 rpm. The bacterial growth through comparison of the OD with the standard growth curve. PHA content was later determined via GC analysis.

#### 2.4.4. Effect of Co-Carbon Substrates

In order to explore the metabolic capability of the test organism in terms of PHA synthesis, the effect of the combination of the main carbon substrate (UTO) with other additional/co-carbon sources was assessed. A varying concentration of UTO was used with a fixed amount of either oleic acid or palm oil (PO). In each case, 0.06 g/L inoculum was cultured in 50 mL MSM at 30 °C for 72 h with an agitation speed of 200 rpm. Growth was monitored using spectrophotometry, and the cell dry weight was subsequently determined. The biomass was recovered and freeze-dried, after which the PHA content was determined using Gas chromatography of the derivatised methyl ester.

### *2.5. Analytical Methods*

#### 2.5.1. Polymer Extraction

The PHA was extracted using the solvent extraction method. Briefly, 200 mL of chloroform ($CHCl_3$) was added to 1g of the dried cells in a Schott Duran bottle. The bottle was then tightly covered, and the cell suspension was stirred continuously using a magnetic stirrer for 48 h at ambient temperature. The cell debris was filtered using Whatman's number 1 filter, and the solvent containing dissolve polymer was then concentrated to a volume of 10–15 mL at 60 °C using a rotary evaporator (Eyela, N-1000, Tokyo, Japan). The polymer in the concentrated filtrate was precipitated/recrystallised by dropwise addition of the filtrate into ten (10) volumes of well-stirred, chilled methanol. The suspension of the polymer was allowed to settle, after which the methanol was decanted, and the traces of the methanol were allowed to evaporate, leaving the sticky polymer for purification and subsequent analysis. Purification was achieved by dissolving the recovered polymer in chloroform, followed by concentrating the solution and the precipitation as done during the initial extraction [38].

#### 2.5.2. GC/GC-MS Analysis

The PHAs quantification was carried out through GC analysis using caprylic methyl ester (CME) as an internal standard. About 20 mg of freeze-dried cells were weighed into screw-cap tubes and subjected to acid methanolysis at 100 °C for 140 min. The methanolysis solution contains 85% (*v*/*v*) methanol and 15% (*v*/*v*) sulphuric acid [39]. The derivatised methyl esters were subsequently analysed by GC (Shimadzu GC-2014,

Kyoto Japan) equipped with a capillary column SPB-1 (30 m length, 0.25 mm internal diameter and 25 µm film thickness; Supelco, Bellefonte, PA, USA) connected to a flame ionisation detector (FID). Nitrogen gas was utilised as carrier gas. Sample injection was made by autoinjector (Shimadzu AOC-20i) connected to the GC system. The injector and detector temperatures were set at 260 and 280 °C, respectively. The column temperature was ramped up from 60 to 280 °C at 5 °C/min with a total analysis time of 26.33 min. Identification of the monomer components of the polymer was achieved by GC coupled with mass spectroscopy.

#### 2.5.3. FT-IR Analysis

The FT-IR analysis was carried out using Fourier-transform infrared spectroscope (FT-IR-8300, Shimadzu, Kyoto, Japan). The sample casting was done on KBr pellets; the infrared spectrum was recorded between 400 and 4500 1/cm and 45 scans [40].

#### 2.5.4. NMR Analysis

The PHA recovered from the test organism was subjected to both $^1$H and $^{13}$C NMR using an FTNMR spectrometer (Acend$^{TM}$ 500; Bruker, Switzerland). were carried out. Briefly, 5 mg of the polymer was dissolved in 2 mL deuterated chloroform ($CDCl_3$) containing 0.03% (*v*/*v*) tetramethylsilane (TMS) as a reference standard. The solution was then filtered using a polytetrafluoroethylene (PTFE) filter (11807–25; Sartorius, Goettingen, Germany). The $^1$H spectrum was acquired at 500 MHz, 25 °C, with a sampling pulse of 3sec against TMS [41]. The $^{13}$C spectrum was also measured at 125.7 MHz, 27 °C with a sampling pulse of 1seconds. Chemical shifts were referenced to the residual proton peak of the deuterated chloroform at 7.26 ppm and to the carbon peak at 77 ppm [42].

#### 2.5.5. Size Exclusion Chromatography (SEC)

The molecular weight (Mw) data of the polymer were obtained using size exclusion chromatography system (Lachrom Merck–Hitachi, Darmstadt, Germany) equipped with a refractive index detector. The sample solution (1% *w*/*v*) was prepared in chloroform and filtered through a 0.45 µm pore size Sartorius membranes filter. Approximately 20 µL was injected with the flow rate set at 1.0 mL/min. The columns used were placed in series with exclusion limits of $10^6$, $10^5$, $10^4$ and $10^3$ Da. Chloroform with narrow polystyrene polydispersity (~1.1) was used for the calibration curve. The calculations were accomplished using clarity chromatography software version 8.0 [40].

#### 2.5.6. Thermogravimetry Analysis (TGA)

The thermogravimetry analysis was carried out using a TGA analyser (Netzsch TG 209, Selb, Germany). About 10 mg of pure polymer was used, and the sample was heated at a rate of 10 °C per min from 48.9 °C temperature to 898.6 °C in a nitrogen atmosphere [43].

#### 2.5.7. Differential Scanning Calorimetry (DSC)

About 15 mg of the pure polymer was measured and kept at 25 °C for 5 min. The sample was then heated to 125 °C through an incremental rate of 10 °C per minute to suppress the memory effect; the sample was kept at this temperature (125 °C) for 5 min and followed by cooling down to −100 °C at a rate of 20 °C per min and kept for 5 min before final heating to 350 °C at the rate of 10 °C per min. Melting temperature (*Tm*) was determined and recorded accordingly from endothermic peaks in the initial heating scan.

## 3. Results

### *3.1. Strain Characterisation and Screening for PHA Production*

Nile red staining was applied in the microscopic observation for possible accumulation of PHA within the cell by the test organism. Brightly orange inclusions were observed under a fluorescent microscope; this indicates the possible presence of PHA granules. Further confirmation was made via transmission electron microscopy in which PHA granules were detected within the bacterial cells. This isolate was found to be a Gram-negative coccobacillus and designated as *Acinetobacter* sp. strain AAAID-1.5 with gene accession number MZ411700 [32]. The organism was used in the PHA production in shake flasks via a batch process biosynthesis with UTO as the sole carbon source.

### 3.2. PHA Biosynthesis in Shake Flask

The polymer biosynthesis in the shake flask experiment revealed that the bacterium had accumulated PHAs up to 34% of its dry weight (Table 1). From the table, it can also be seen that the polymer produced by the organism is composed of 3-hydroxyhexadecanoate and 3-hydroxyoctadecanoate monomers with mole fractions of 15 and 85 mol%, respectively.

**Table 1.** PHA biosynthesis by *Acinetobacter* sp. strain AAAID-1.5 using used transformer oil as the sole carbon source.

| Biosynthesis Factors | CDW (g/L) [a] | PHA Content (wt.%) [b] | PHA (g/L) | RCDW (g/L) | Monomer Composition (mol%) [b] | |
|---|---|---|---|---|---|---|
| | | | | | 3HHD | 3HOD |
| 2% ($v/v$) UTO 200 rpm, 30 °C, 72 h | 2.10 ± 0.04 | 34 ± 1 | 0.72 ± 0.24 | 1.38 ± 0.29 | 15 ± 1 | 85 ± 1 |

**Key**: CDW: Cell dry weight, RCDW = Residual cell dry weight, [a]: Cells were harvested after 72 h of incubation. Values are means ± SD of three experimental replications, [b]: Values are mean ± SD of three experimental replicates calculated from GC analysis, 3HHD: 3-hydroxyhexadecanoate, 3HOD: 3-hydroxyoctadecanoate.

### 3.3. Effect of Carbon Source Concentration

Varying concentrations of carbon source with a fixed (1.1 g/L) supply of nitrogen source was tested in shake flask biosynthesis. The influence of increasing concentration of the carbon source on the growth and PHA production is shown in Table 2. The results showed that the PHA concentration increased from 0.37 ± 0.03 g/L at 0.5%($v/v$) concentration of the carbon source to the highest value of 0.72 ± 0.24 g/L at 2.0%($v/v$) concentration of the same carbon source. Similarly, the growth showed an increasing trend with the increase in carbon source concentration. The molar fractions of the two monomeric components were roughly in a 1:4 ratio for 3HHD and 3HOD, respectively.

**Table 2.** Effect of concentration of the carbon source on growth and PHA biosynthesis by *Acinetobacter* sp. strain AAAID-1.5.

| Concentration of UTO (%$v/v$) | CDW (g/L) [a] | PHA Content (wt.%) [b] | PHA Concentration (g/L) | RCDW (g/L) | Monomer Composition (mol%) [b] | |
|---|---|---|---|---|---|---|
| | | | | | 3HHD | 3HOD |
| 0.5 | 0.94 ± 0.03 | 39 ± 2 | 0.37 ± 0.03 | 0.56 ± 0.01 | 17 ± 0 | 83 ± 0 |
| 1.0 | 1.43 ± 0.10 | 36 ± 1 | 0.51 ± 0.03 | 0.92 ± 0.07 | 14 ± 0 | 86 ± 0 |
| 1.5 | 1.58 ± 0.24 | 36 ± 2 | 0.56 ± 0.06 | 1.01 ± 0.18 | 15 ± 0 | 85 ± 0 |
| 2.0 | 2.10 ± 0.04 | 34 ± 4 | 0.72 ± 0.24 | 1.38 ± 0.29 | 15 ± 1 | 85 ± 1 |
| 2.5 | 2.11 ± 0.09 | 33 ± 6 | 0.69 ± 0.13 | 1.41 ± 0.15 | 20 ± 2 | 80 ± 2 |

**Key**: CDW: Cell dry weight, RCDW = Residual cell dry weight, [a]: Cells were harvested after 72 h of incubation. Values are means ± SD of three experimental replications, [b]: Values are mean ± SD of three experimental replicates calculated from GC analysis, 3HHD: 3-hydroxyhexadecanoate, 3HOD: 3-hydroxyoctadecanoate.

### 3.4. Effect of Incubation Time

The results obtained from this analysis are presented in Table 3. From the results, it is evident that PHA accumulation began in the early phase of the growth cycle. The PHA concentration almost doubled from 0.57 ± 0.05 g/L at 24 h to 0.96 ± 0.07 g/L at 48 h. The increase continued steadily and reached a maximum value of 1.19 ± 0.08 g/L at 72 h.

**Table 3.** Effect of incubation time on the growth and PHA biosynthesis by *Acinetobacter* sp. strain AAAID-1.5.

| Incubation time (h) | CDW (g/L) [a] | PHA Content (wt.%) [b] | PHA Concentration (g/L) | RDCW (g/L) | Monomer Composition (mol%) [b] | |
|---|---|---|---|---|---|---|
| | | | | | 3HHD | 3HOD |
| 24 | 1.90 ± 0.09 | 30 ± 1 | 0.57 ± 0.05 | 1.33 ± 0.03 | 24 ± 0 | 76 ± 0 |
| 48 | 2.46 ± 0.04 | 39 ± 3 | 0.96 ± 0.07 | 1.49 ± 0.07 | 17 ± 3 | 83 ± 3 |
| 72 | 2.51 ± 0.08 | 48 ± 2 | 1.19 ± 0.08 | 1.32 ± 0.02 | 17 ± 3 | 83 ± 3 |
| 96 | 2.34 ± 0.07 | 50 ± 2 | 1.17 ± 0.08 | 1.17 ± 0.04 | 16 ± 5 | 84 ± 5 |
| 120 | 1.90 ± 0.09 | 50 ± 2 | 0.95 ± 0.06 | 0.95 ± 0.06 | 19 ± 0 | 81 ± 0 |

**Key**: CDW: Cell dry weight, RCDW = Residual cell dry weight, [a]: Cells were harvested after 72hr of incubation. Values are means ± SD of three experimental replications, [b]: Values are mean ± SD of three experimental replicates calculated from GC analysis, 3HHD: 3-hydroxyhexadecanoate, 3HOD: 3-hydroxyoctadecanoate.

### 3.5. Effect of Yeast Extract

The effect of yeast extract as a supplement to the biosynthesis medium on bacterial growth and PHA accumulation was investigated. A control experiment was set without addition of yeast extract. The results are summarised in Table 4. It was observed that the addition of a lower concentration of yeast extract was more favourable to the polymer accumulation. The highest PHA concentration (1.25 ± 0.07 g/L) was observed at a yeast extract concentration of 1.0 g/L. Conversely, the biomass concentration increased steadily from 2.87 ± 0.04 g/L at a yeast extract concentration of 0.5 g/L to a maximum value of 3.46 ± 0.06 g/L at a concentration of 2.0 g/L. The 3HHD to 3HOD monomer ratio did not show wider variation across the range of the yeast extract concentration tested.

**Table 4.** Effect of yeast extract concentration on the growth and PHA biosynthesis by *Acinetobacter* sp. strain AAAID-1.5.

| Conc. of YE (g/L) | CDW (g/L) [a] | PHA Content (wt.%) [b] | PHA Concentration (g/L) | RCDW (g/L) | Monomer Composition (mol%) [b] | |
|---|---|---|---|---|---|---|
| | | | | | 3HHD | 3HOD |
| 0.5 | 2.87 ± 0.04 | 40 ± 2 | 1.15 ± 0.06 | 1.73 ± 0.05 | 17 ± 1 | 83 ± 1 |
| 1.0 | 3.36 ± 0.05 | 37 ± 2 | 1.25 ± 0.07 | 2.11 ± 0.11 | 19 ± 3 | 81 ± 3 |
| 1.5 | 3.63 ± 0.08 | 34 ± 1 | 1.23 ± 0.02 | 2.41 ± 0.11 | 17 ± 2 | 83 ± 2 |
| 2.0 | 3.46 ± 0.06 | 34 ± 4 | 1.17 ± 0.15 | 2.30 ± 0.10 | 18 ± 0 | 82 ± 0 |
| 2.5 | 3.42 ± 0.05 | 31 ± 1 | 1.08 ± 0.04 | 2.35 ± 0.10 | 18 ± 0 | 82 ± 0 |
| Control | 2.10 ± 0.04 | 34 ± 1 | 0.72 ± 0.24 | 1.38 ± 0.29 | 15 ± 1 | 85 ± 1 |

**Key**: YE: Yeast extract, CDW: Cell dry weight, RCDW = Residual cell dry weight, [a]: Cells were harvested after 72 h of incubation. Values are means ± SD of three experimental replications, [b]: Values are mean ± SD of three experimental replicates calculated from GC analysis, 3HHD: 3-hydroxyhexadecanoate, 3HOD: 3-hydroxyoctadecanoate.

### 3.6. Effect of Additional/Co-carbon Substrates

The effect of Co-carbon substrates on growth and polymer accumulation was studied, in which the main carbon source (UTO) was combined with either of the two additional carbon substrates. The palm oil used as one of the additional substrates composed of 0.8% arachidic acid (C20:0), 0.3% lauric acid (C12:0), 0.2% linolenic acid (C18:3), 10.4% linoleic acid (C18:2), 47.7% oleic acid (C18:1), 3.7% stearic acid (C18:0), 36.3% palmitic acid (C16:0), 0.2% palmitoleic acid (C16:1) and 0.8% myristic acid (C14:0). The PHA accumulation by an organism during growth on different combination carbon sources (C1, C2, C3, and C4) at 30 °C for 72 h is presented in Table 5. The control experiment was set with the UTO as the sole carbon substrate. From the results, it was observed that the use of either of the two additional carbon substrates (palm oil and oleic acid) had a profound influence on both the PHA accumulation and the molar concentration of the monomeric components of the PHA produced. For instance, up to 3.02 g/L was recorded in the case of C4; this represents a 200% increase compared to the control experiment.

**Table 5.** Effect of additional/Co-carbon substrates on the growth and PHA biosynthesis by *Acinetobacter* sp. strain AAAID-1.5.

| Combination of Carbon Source | CDW (g/L) [a] | PHA Content (wt.%) [b] | PHA Concentration (g/L) | RCDW (g/L) | Monomer Composition (mol%) [b] | |
|---|---|---|---|---|---|---|
| | | | | | 3HHD | 3HOD |
| C1 | 1.91 ± 0.05 | 45 ± 2 | 0.85 ± 0.03 | 1.06 ± 0.07 | 29 ± 2 | 71 ± 2 |
| C2 | 2.10 ± 0.08 | 58 ± 7 | 1.21 ± 0.10 | 0.90 ± 0.17 | 33 ± 2 | 67 ± 2 |
| C3 | 3.54 ± 0.13 | 67 ± 9 | 2.37 ± 0.39 | 1.17 ± 0.30 | 10 ± 1 | 90 ± 1 |
| C4 | 3.85 ± 0.14 | 78 ± 4 | 3.02 ± 0.27 | 0.83 ± 0.14 | 09 ± 0 | 91 ± 0 |
| Control | 2.10 ± 0.04 | 34 ± 1 | 0.72 ± 0.24 | 1.38 ± 0.29 | 15 ± 1 | 85 ± 1 |

**Key**: CDW: Cell dry weight, RCDW = Residual cell dry weight, [a]: Cells were harvested after 72 h of incubation. Values are means ± SD of three experimental replications, [b]: Values are mean ± SD of three experimental replicates calculated from GC analysis, 3HHD: 3-hydroxyhexadecanoate, 3HOD: 3-hydroxyoctadecanoate, **C1**: 2% *v/v* UTO + 0.74% *v/v* palm oil, C2: 1% *v/v* UTO + 0.74 palm oil, C3: 2% *v/v* UTO + 0.74% *v/v* oleic acid, C4: 1% *v/v* UTO + 0.74% *v/v* oleic acid.

## *3.7. PHA Structure and Characterisation*

The identification of monomer components of the PHA was achieved using GC-MS analysis. The spectral matching by the GC-MS system compared to the NIST08 standard reference library revealed the presence of 3-hydroxyhexadecanoate and 3-hydroxyoctadecanoate as the major monomer constituents. The FT-IR spectrum obtained from the analysis of the PHA extracted from the bacteria is shown in Figure 1; the sharp bending seen at 1722.67 $cm^{-1}$ could be attributable to the carbonyl stretching of the polyester. The bends at 2922.82 and 2858.09 $cm^{-1}$ signify the presence of the alkyl (-$CH_3$) group, while the one at 1458.33 $cm^{-1}$ could be due to asymmetric bending of -$CH_3$ or -$CH_2$ bending. The bending at 1397.01 $cm^{-1}$ is considered to have stemmed from the symmetric bending of -$CH_3$. The C-O stretching was represented by the bending at 1278.38 $cm^{-1}$. Absorption spectra in the range of 800 to 1200 $cm^{-1}$ signalled the existence of carbon-to-carbon stretches of an alkane group, while the band at about 721 $cm^{-1}$ is typical of -$CH_2$ side chains.

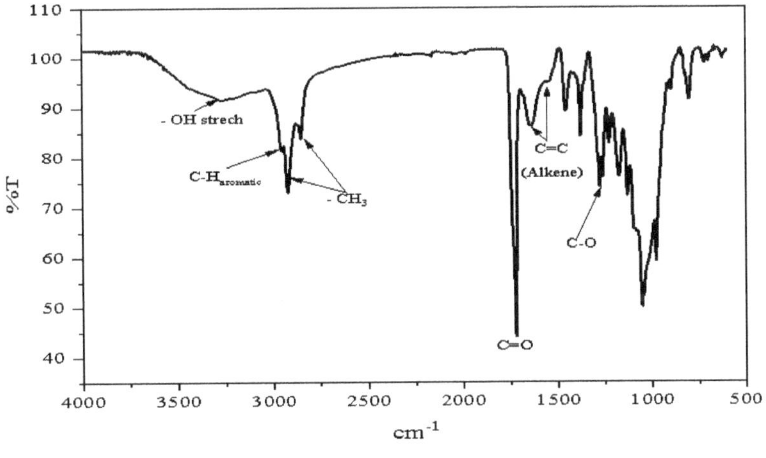


**Figure 1.** FT-IR spectrum of PHA produced *Acinetobacter* sp. strain AAAID-1.5 grown in MSM containing used transformer oil as carbon source.

Figures 2 and 3 show the spectra of the $^1H$ and $^{13}C$ NMR analyses. The chemical shifts were recorded in ppm relative to the signal of deuterated chloroform as an internal reference. In Figure 2, the singlet peak at δ = 5.2 ppm was assigned to the methine proton of the *β*-carbon, whereas the triplet peaks at δ = 2.6 ppm were deemed to be for methylene proton of the alpha carbon. The peak at 0.9 ppm was assigned to the methyl protons (-$CH_3$) of the terminal carbon of the side chain, while those at 1.3 and 1.5 ppm were considered to have emanated from methylene protons, the peak at zero ppm was considered to be due to hydrogen atoms of the tetramethylsilane (I = TMS). The proposed PHA structure is shown in the figure.

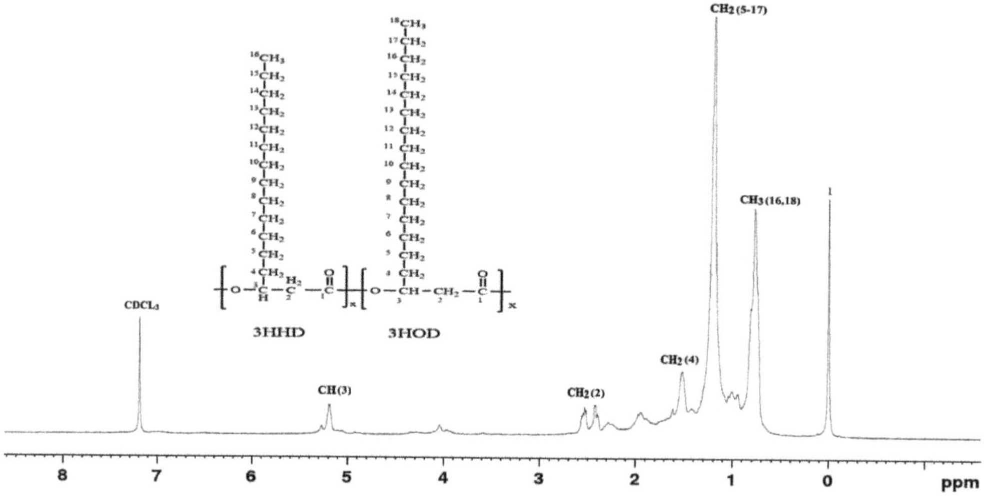


**Figure 2.** The [1]H NMR spectrum of the PHA produced by *Acinetobacter* sp. strain AAAID-1.5 in a biosynthesis medium containing used transformer oil as a carbon source.

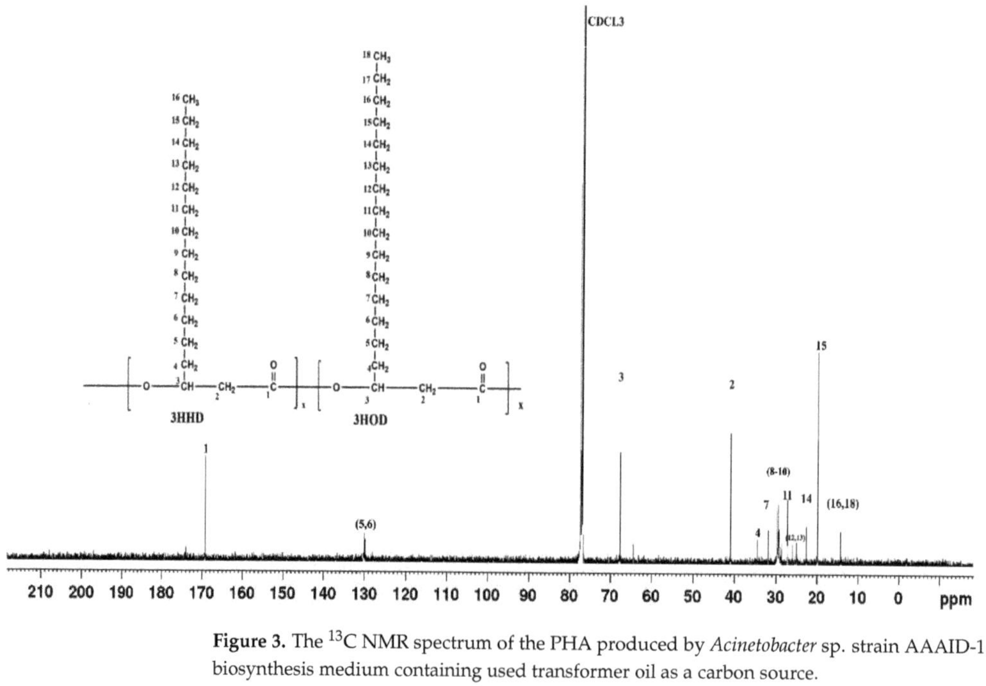


**Figure 3.** The [13]C NMR spectrum of the PHA produced by *Acinetobacter* sp. strain AAAID-1.5 in a biosynthesis medium containing used transformer oil as a carbon source.

In Figure 3, it can be observed from the chemical shift that the carbonyl groups (C=O) of the PHA backbone resonated at 169.14 ppm. The signals at 19.76 to 67.62 ppm correspond to methyl, methylene, and methine of the monomers.

In Table 6, some characteristics of the polymer produced by the bacterium at 4%($v/v$) emulsified UTO, 30 °C, agitation speed of 200 rpm and incubation time of 72 h are presented. The results show that the polymer had an average molecular weight of 110.45 kDa and a polydispersity index of 2.01. The melting temperature was approximately 88 °C, while the highest degradation temperature was 268 °C.

**Table 6.** Some characteristic features of the PHA produced by *Acinetobacter* sp. Strain AAAID-1.5.

| Parameter | Value |
|---|---|
| $M_w$ (kDa) | 110.45 |
| $M_n$ (kDa) | 55.059 |
| PDI | 2.01 |
| $T_m$ (°C) | 88 |
| $T_d$ (°C) | 268 |

## 4. Discussion

The successful isolation of *Acinetobacter* sp. strain AAAID-1.5 from a soil sample and its characterisation as a potent PHA-producing organism indicate that natural environments, especially soil, remain the reservoirs of effective biological agents that can be used to accomplish certain biotechnological processes. The ability of this bacterium to utilise the UTO as the sole carbon substrate to produce PHA has underscored its potential as a 'seed' for the valorisation of waste oil, particularly UTO, in bioplastic industries. A wide variety of wild-type bacteria, both Gram-negative and Gram-positive, such as *Pseudomonas*, *Cupriavidus*, *Bacillus* etc., have been implicated with the biosynthesis of PHA [44,45]. Similarly, *Acinetobacter* sp. was previously implicated with the ability to accumulate PHA, especially medium chain length PHAs [46]. Furthermore, a recent study described a species of *Acinetobacter* capable of utilising waste transformer oil as a sole carbon source [47,48]. The utilisation of waste streams for the biosynthesis of value-added products such as PHA is considered a cost-efficient strategy and can help address waste disposal problems [49,50]. Although some bacteria have been reported to synthesise PHA as much as 90% of their cell dry weight [6], A PHA accumulation of up to 34% of CDW was achieved by *Acinetobacter* sp. strain AAAID-1.5 in the shake flask experiment, and this indicated that the organism is capable of converting the UTO into PHA; to a reasonable extent that can be further explored for possible commercialisation. These findings substantiated a previous report in which a certain strain of *Acinetobacter* was reported to have accumulated PHA to the extent of approximately 25% CDW during biosynthesis in a medium containing various oil substrates [46]. Transformer oil and the wastes generated from there are a complex mixture of hydrocarbons, thus, the biosynthesis of PHA from such complex mixture was made possible probably due the ability of the bacterium to degrade the oil components using hydrolytic enzymes into chemical intermediates that can easily be channelled into the PHA biosynthetic pathway. For instance, *Acinetobacter* was reported to have the capacity of converting naphthalene (a major component of transformer oil) into acetyl-CoA via salicylic acid [51]. Oil hydrolysis during bacterial growth can also produce free fatty acids, that may undergo $\beta$-oxidation to generate precursors for PHA biosynthesis [52]. On the other hand, the limited growth shown by the organism in the first shake flask experiment presented in Table 1 where 2.1 g/L CDW was recorded; could be due to the presence of chemical components in the UTO that are resistant to microbial attack which might have caused toxic effect and slowed down the growth.

Regarding the effect of the carbon substrate concentration on bacterial growth and PHA accumulation, as presented in Table 2, it was observed that the results indicated that the increase in the concentration of the carbon source had a profound effect on the growth and PHA accumulation by the bacterium. PHA biosynthesis is normally induced by unbalanced growth conditions in the presence of excess carbon sources while essential nutrients such as nitrogen, oxygen, sulphur, and phosphorus are in limited supply [53]. The slight decline in PHA accumulation at carbon source concentration beyond 2%v/v might result from an inhibitory effect of some toxic chemical constituents of the carbon source that are hard to degrade. For instance, high concentrations of polyaromatic hydrocarbons (PAHs) may be toxic and can hinder metabolic activity. The ratio of molar concentration 3HHD to 3HOD was roughly 1:8 across the various concentrations tested, which could reflect the carbon composition of the bacterium's degradation products derived from the UTO during biosynthesis, especially if fatty acids are the intermediate products. Occasionally, PHAs are synthesised with monomers equal in the number of carbon atoms to the fatty acid substrate [54].

The influence of incubation time on the PHA accumulation by the bacterium in this work showed an increasing trend until a maximum concentration of 1.19 $\pm$ 0.08 g/L was achieved at 72 h, as shown in Table 3. The decline observed beyond 72 h might be linked to the exhaustion of nutrients and accumulation of toxic metabolic waste in the late stationary phase of the growth cycle, as previously asserted elsewhere [55]. A significant accumulation of biosynthesis products during the late exponential phase is a typical feature of growth-linked biosynthesis [56,57]. Thus, the PHA accumulation pattern exhibited by *Acinetobacter* sp. strain AAAID-1.5 could be considered a growth-linked biosynthesis. Microbially produced polymers may vary as a function of the time at which the biomass is harvested [58]. Interestingly, the molar fractions of the monomer units of the PHA (3HHD and 3HOD) did not show a wider variation throughout the incubation time; this suggests that the molar concentration of the monomeric components of the polymer might be independent of the incubation time in this case.

With respect to the effect of yeast extract as a supplement to the biosynthesis medium on bacterial growth and PHA accumulation, the results, as presented in Table 4, the findings revealed that the addition of this supplement had improved the growth and PHA biosynthesis. This is evident when comparing the experimental control results with the other set of experiments in which the biosynthesis medium was supplemented with various concentrations of yeast extract. For instance, PHA accumulation increased from 0.72 g/L in the control experiment to 1.25 g/L in an experiment where yeast extract was added at a concentration of 1 g/L. this represents roughly a 42% increase. However, the ratio of the two monomer components of the polymer was roughly the same across different yeast extract concentration regimes. Yeast extract has been reported in a number of reports to have improved microbial growth and biosynthesis of some target products. Specifically, it was reported to have promoted PHA production and bacterial growth in recent studies by many authors [59–61]. The effect of yeast extract could be linked to its minerals and vitamin content which are critical to enzyme functioning and can ultimately facilitate the metabolic process.

The effect of additional carbon substrates in which either palm oil or oleic acid was added to the biosynthesis medium that already contained UTO showed that manipulating the carbon source could boost the PHA accumulation by the test organism. The fatty acid composition of the PO utilised as one of the additional carbon substrates is 0.8% arachidic acid (C20:0), 0.3% lauric acid (C12:0), 10.4% linoleic acid (C18:2), 0.2% linolenic acid (C18:3), 47.7% oleic acid (C18:1), 3.7% stearic acid (C18:0), 36.3% palmitic acid (C16:0), 0.2% palmitoleic acid (C16:1) and 0.8% myristic acid (C14:0). All the four different combinations (C1, C2, C3, and C4) of carbon substrate have produced a relative higher PHA content compared to the control experiment in which only UTO was used. This suggests that the organism might have exhibited a preference for carbon substrate. The free fatty acid is assumed to be more easily utilised by the bacterium than the complex mixture of hydrocarbons in UTO that may require degradation before usage. Palm oil and other organic oils, such as plant oils, are good feedstocks that can serve as excellent carbon sources for PHA biosynthesis [62]. Generally, oils are attractive carbon substrates for the sustainable production of PHAs, and they have been producing higher yields for both biomass and polymer synthesis [10]. PHA biosynthetic pathways are intricately linked to the organism's central metabolic pathways such as glycolysis, de-novo fatty acids synthesis, $\beta$-oxidation, Krebs cycle, serine pathway, amino acid catabolism and Calvin cycle [63–65]. Therefore, the higher PHA content achieved upon the addition of the PO or oleic acid implied that the $\beta$-oxidation could have been a key pathway employed by *Acinetobacter* sp. strain AAAID-1.5. The improved PHA production upon the addition of oleic acid or PO has substantiated their influence on the biosynthesis mcl-PHAs in particular, as well as their attractive feature in PHA biosynthesis in general. For instance, PO was reported to have the ability to minimise the toxic effects of certain carbon substrates on bacterial cells [66]. Thus, the usage of PO in combination with UTO is assumed to have abated the suspected toxic effect of some constituents of the UTO.

With regards to the characterisation of the PHA in terms of physicochemical properties, the polymer produced appeared yellowish, amorphous, and sticky, with typical features of elastomeric polyester. The FT-IR spectrum recorded in the range of 600 to 4000 cm$^{-1}$ showed spectral bending of a typical PHA that is comparable with the one previously reported by Cruz and his associates [67]. Interestingly, the sample of the PHA analysed had expressed substantial peaks similarity at various points compared to the spectrum obtained from *mcl*-PHA as reported elsewhere [41]. Furthermore, the data obtained from the GC-MS and NMR analyses of the synthesised polymer revealed that the PHA might have been composed of 3HHD and 3HOD as the major monomer constituents. This is comparable with a similar finding extensively described elsewhere [32]. In Table 6, the molecular weight and thermal properties of the polymer produced by *Acinetobacter* sp. strain AAAID-1.5 are presented. The number average molecular weight ($M_n$) determined via Size exclusion chromatography in chloroform relative to polystyrene standard was 55.059 kDa. The average molecular weight and polydispersity index (PDI) were 110 kDa and 2.01 respectively. Whereas the melting temperature was about 88 °C and the highest degradation temperature was 268 °C. According to a previous report, *mcl*-PHA produced by *Acinetobacter* sp. ASC1 using crude glycerol as carbon source presents $M_n$ = 47 kDa, $M_w$ = 88 kDa and PDI = 1.9 [46]. The PHA produced in the present work presented a relatively higher $M_w$, $M_n$ and PDI values, this might be due to longer alkyl side chain of the polymer produced in this study. Polydispersity index was determined by $M_w/M_n$ as calculated in SEC. The PDI value (2.01) falls within the range of a typical *mcl*-PHA. For instance, it was reported that the PDI of *mcl*-PHA ranges between 1.1 to 6.0 [68]. The thermal temperature of *mcl*-PHA ranges between 30–80 °C [69]. However, a slightly higher value (88 °C) recorded in this work might be a result of factors that influence the thermal properties of the polyester. Temperature characteristics are the most variable features of PHAs, and they are critical to processing this important polymer. Such characteristic features determine the conditions under which the polymer can be processed and the properties of the resulting products [70]. Being a medium-chain length PHA which is elastomeric, the polymer produced in this work could be applied as biomedical materials, especially in the production of surgical mesh, sutures, vein valves, cardiovascular patches etc. Likewise, it could be applied in the production of packaging materials as well as in agriculture in the production of agricultural nets, mulch and in the controlled delivery of biofertilizer. Medium-chain length PHAs are preferred in biomedical applications because of their low crystallinity and are relatively more flexible in addition to being biocompatible [71,72].

## 5. Conclusions

The potential application of used transformer oil as feedstocks for PHA biosynthesis was established. Our study demonstrates that *mcl*-PHA polymers could be synthesised by a wild strain of *Acinetobacter* sp. utilising UTO as the sole carbon substrate. Furthermore, it was established that certain growth and biosynthesis parameters could influence the polymer accumulation in biosynthesis under control conditions. The study also provided insight into the potential of UTO as an alternative carbon feed in the production of biodegradable plastics using a biotechnological approach.

**Author Contributions:** Conceptualization, A.A.-A.A.; methodology, formal analysis, S.I; resources, A.A.-A.A. and A.N.S.; writing—original draft preparation, S.I.; writing—review and editing, S.I. and A.A.-A.A.; supervision, R.A.R.; project administration, A.A.-A.A.; funding acquisition, A.A.-A.A. and A.N.S. All authors have read and agreed to the published version of the manuscript.

**Funding:** This research was funded by Universiti Sains Malaysia (1001/PBIOLOGI/8011067) and TNB Research Sdn Bhd (304/PBIOLOGI/6501224/T138).

**Data Availability Statement:** Not applicable.

**Acknowledgments:** The authors would like to thank the management of Universiti Sains Malaysia (U.S.M.) and the Ministry of Higher Education Malaysia (MOHE) for providing the enabling environment for the conduct of this study. The support of TNB Research Sdn Bhd is also acknowledged.

**Conflicts of Interest:** The authors declare no conflict of interest.

## References

1. Możejko-Ciesielska, J.; Kiewisz, R. Bacterial polyhydroxyalkanoates: Still fabulous? *Microbiol. Res.* **2016**, *192*, 271–282. [CrossRef] [PubMed]
2. Singh, A.K.; Mallick, N. Enhanced production of SCL-LCL-PHA co-polymer by sludge-isolated *Pseudomonas aeruginosa* MTCC 7925. *Lett. Appl. Microbiol.* **2008**, *46*, 350–357. [CrossRef] [PubMed]
3. Kalia, V.C.; Raizada, N.; Sonakya, V. Bioplastics. *J. Sci. Ind. Res.* **2000**, *59*, 433–445.
4. Mohapatra, S.; Maity, S.; Dash, H.R.; Das, S.; Pattnaik, S.; Rath, C.C.; Samantaray, D. Bacillus and biopolymer: Prospects and challenges. *Biochem. Biophys. Rep.* **2017**, *12*, 206–213. [CrossRef] [PubMed]
5. Rosseto, M.; Rigueto, C.V.; Krein, D.D.; Balbé, N.P.; Massuda, L.A.; Dettmer, A. Biodegradable Polymers: Opportunities and Challenges. *Org. Polym.* **2020**, *27*, 37–41. [CrossRef]
6. Chee, J.; Yoga, S.; Lau, N.; Ling, S.; Abed, R.M.M. Bacterially Produced Polyhydroxyalkanoate (PHA): Converting Renewable Resources into Bioplastics. *Curr. Res. Technol. Educ. Top. Appl. Microbiol. Microb. Biotechnol.* **2010**, *2*, 1395–1404.
7. Prasanna Raj Yadav, S.; Saravanan, C.G.; Vallinayagam, R.; Vedharaj, S.; Roberts, W.L. Fuel and engine characterization study of catalytically cracked waste transformer oil. *Energy Convers. Manag.* **2015**, *96*, 490–498. [CrossRef]
8. Cavalheiro, J.M.B.T.; de Almeida, M.C.M.D.; Grandfils, C.; da Fonseca, M.M.R. Poly(3-hydroxybutyrate) production by *Cupriavidus necator* using waste glycerol. *Process Biochem.* **2009**, *44*, 509–515. [CrossRef]
9. Teeka, J.; Imai, T.; Cheng, X.; Reungsang, A.; HIGu, T.; Yamamoto, K.; Sekine, M. Screening of PHA-Producing Bacteria Using Biodiesel-Derived Waste Glycerol as a Sole Carbon Source. *J. Water Environ. Technol.* **2010**, *8*, 373–381. [CrossRef]
10. Loo, C.Y.; Lee, W.H.; Tsuge, T.; Doi, Y.; Sudesh, K. Biosynthesis and characterization of poly(3-hydroxybutyrate-co-3- hydroxyhexanoate) from palm oil products in a *Wautersia eutropha* mutant. *Biotechnol. Lett.* **2005**, *27*, 1405–1410. [CrossRef]
11. Teeka, J.; Imai, T.; Reungsang, A.; Cheng, X.; Yuliani, E.; Thiantanankul, J.; Poomipuk, N.; Yamaguchi, J.; Jeenanong, A.; Higuchi, T.; et al. Characterization of polyhydroxyalkanoates (PHAs) biosynthesis by isolated *Novosphingobium* sp. THA-AIK7 using crude glycerol. *J. Ind. Microbiol. Biotechnol.* **2012**, *39*, 749–758. [CrossRef] [PubMed]
12. Fernández, D.; Rodríguez, E.; Bassas, M.; Viñas, M.; Solanas, A.M.; Llorens, J.; Marqués, A.M.; Manresa, A. Agro-industrial oily wastes as substrates for PHA production by the new strain *Pseudomonas aeruginosa* NCIB 40045: Effect of culture conditions. *Biochem. Eng. J.* **2005**, *26*, 159–167. [CrossRef]
13. Koller, M.; Rodríguez-Contreras, A. Techniques for tracing PHA-producing organisms and for qualitative and quantitative analysis of intra- and extracellular PHA. *Eng. Life Sci.* **2015**, *15*, 558–581. [CrossRef]
14. Mravec, F.; Obruca, S.; Krzyzanek, V.; Sedlacek, P.; Hrubanova, K.; Samek, O.; Kucera, D.; Benesova, P.; Nebesarova, J. Accumulation of PHA granules in *Cupriavidus necator* as seen by confocal fluorescence microscopy. *FEMS Microbiol. Lett.* **2016**, *363*, 1–7. [CrossRef] [PubMed]
15. Obruca, S.; Sedlacek, P.; Koller, M.; Kucera, D.; Pernicova, I. Involvement of polyhydroxyalkanoates in stress resistance of microbial cells: Biotechnological consequences and applications. *Biotechnol. Adv.* **2018**, *36*, 856–870. [CrossRef]
16. Kato, M.; Bao, H.J.; Kang, C.K.; Fukui, T.; Doi, Y. Production of a novel copolyester of 3-hydroxybutyric acid and medium-chain-length 3-hydroxyalkanoic acids by *Pseudomonas* sp. 61-3 from sugars. *Appl. Microbiol. Biotechnol.* **1996**, *45*, 363–370. [CrossRef]
17. Khanna, S.; Srivastava, A.K. Recent advances in microbial polyhydroxyalkanoates. *Process Biochem.* **2005**, *40*, 607–619. [CrossRef]
18. Liz, J.A.Z.E.; Jan-Roblero, J.; De La Serna, J.Z.D.; De León, A.V.P.; Hernández-Rodríguez, C. Degradation of polychlorinated biphenyl (PCB) by a consortium obtained from a contaminated soil composed of *Brevibacterium*, *Pandoraea* and *Ochrobactrum*. *World J. Microbiol. Biotechnol.* **2009**, *25*, 165–170. [CrossRef]
19. De Eugenio, L.I.; Escapa, I.F.; Morales, V.; Dinjaski, N.; Galán, B.; García, J.L.; Prieto, M.A. The turnover of medium-chain-length polyhydroxyalkanoates in *Pseudomonas putida* KT2442 and the fundamental role of PhaZ depolymerase for the metabolic balance. *Environ. Microbiol.* **2010**, *12*, 207–221. [CrossRef]

20. Mohapatra, S.; Mohanta, P.R.; Sarkar, B.; Daware, A.; Kumar, C.; Samantaray, D.P. Production of Polyhydroxyalkanoates (PHAs) by Bacillus Strain Isolated from Waste Water and Its Biochemical Characterization. *Proc. Natl. Acad. Sci. India Sect. B Biol. Sci.* **2017**, *87*, 459–466. [CrossRef]
21. Koller, M.; Gasser, I.; Schmid, F.; Berg, G. Linking ecology with economy: Insights into polyhydroxyalkanoate-producing microorganisms. *Eng. Life Sci.* **2011**, *11*, 222–237. [CrossRef]
22. Lee, S.Y. Plastic bacteria? Progress and prospects for polyhydroxyalkanoate production in bacteria. *Trends Biotechnol.* **1996**, *14*, 431–438. [CrossRef]
23. Nicolaus, B.; Lama, L.; Esposito, E.; Manca, M.C.; Improta, R.; Bellitti, M.R.; Duckworth, A.W.; Grant, W.D.; Gambacorta, A. *Haloarcula* spp able to biosynthesize exo- and endopolymers. *J. Ind. Microbiol. Biotechnol.* **1999**, *23*, 489–496. [CrossRef]
24. Spiekermann, P.; Rehm, B.H.A.; Kalscheuer, R.; Baumeister, D.; Steinbüchel, A. A sensitive, viable-colony staining method using Nile red for direct screening of bacteria that accumulate polyhydroxyalkanoic acids and other lipid storage compounds. *Arch. Microbiol.* **1999**, *171*, 73–80. [CrossRef] [PubMed]
25. Steinbiiehel, A.; Friind, C.; Jendrossek, D.; Schlegei, H.G. Unable to Derepress the Fermentative Alcohol Dehydrogenase. *Arch. Microbiol.* **1987**, *6*, 178–186. [CrossRef]
26. Ostle, A.G.; Holt, J.G. Nile blue A as a fluorescent stain for poly-β-hydroxybutyrate. *Appl. Environ. Microbiol.* **1982**, *44*, 238–241. [CrossRef]
27. Higuchi-Takeuchi, M.; Morisaki, K.; Numata, K. A screening method for the isolation of polyhydroxyalkanoate-producing purple non-sulfur photosynthetic bacteria from natural seawater. *Front. Microbiol.* **2016**, *7*, 1–7. [CrossRef]
28. Koller, M. Established and advanced approaches for recovery of microbial polyhydroxyalkanoate (PHA) biopolyesters from surrounding microbial biomass. *EuroBiotech J.* **2020**, *4*, 113–126. [CrossRef]
29. Chen, G.Q. A microbial polyhydroxyalkanoates (PHA) based bio- and materials industry. *Chem. Soc. Rev.* **2009**, *38*, 2434–2446. [CrossRef]
30. Sang Yup, L.; Lee, S.Y. Bacterial Polyb ydroxyalkanoates. *Biotechnol. Bioeng.* **1996**, *49*, 1–14. [CrossRef]
31. Keshavarz, T.; Roy, I. Polyhydroxyalkanoates: Bioplastics with a green agenda. *Curr. Opin. Microbiol.* **2010**, *13*, 321–326. [CrossRef] [PubMed]
32. Idris, S.; Rahim, R.A.; Amirul, A.A.A. Bioprospecting and Molecular Identification of Used Transformer Oil-Degrading Bacteria for Bioplastics Production. *Microorganisms* **2022**, *10*, 583. [CrossRef] [PubMed]
33. Amirul, A.A.; Syairah, S.N.; Yahya, A.R.M.; Azizan, M.N.M.; Majid, M.I.A. Synthesis of biodegradable polyesters by Gram negative bacterium isolated from Malaysian environment. *World J. Microbiol. Biotechnol.* **2008**, *24*, 1327–1332. [CrossRef]
34. López-Cortés, A.; Lanz-Landázuri, A.; García-Maldonado, J.Q. Screening and isolation of PHB-producing bacteria in a polluted marine microbial mat. *Microb. Ecol.* **2008**, *56*, 112–120. [CrossRef] [PubMed]
35. McDowell, E.M.; Trump, B.F. Histologic fixatives suitable for diagnostic light and electron microscopy. *Arch. Pathol. Lab. Med.* **1976**, *100*, 405–414.
36. Martin, C.; Sofla, A.Y.N. A method for bonding PDMS without using plasma. In *ASME International Mechanical Engineering Congress and Exposition*; ASME: New York, NY, USA, 2010; Volume 10, pp. 557–560. [CrossRef]
37. Du, G.; Chen, J.; Yu, J.; Lun, S. Continuous production of poly-3-hydroxybutyrate by *Ralstonia eutropha* in a two-stage culture system. *J. Biotechnol.* **2001**, *88*, 59–65. [CrossRef]
38. Ren, Q.; Sierro, N.; Kellerhals, M.; Kessler, B.; Witholt, B. Properties of engineered poly-3-hydroxyalkanoates produced in recombinant *Escherichia coli* strains. *Appl. Environ. Microbiol.* **2000**, *66*, 1311–1320. [CrossRef]
39. Braunegg, G.; Sonnleitner, B.; Lafferty, R.M. A rapid gas chromatographic method for the determination of poly-β-hydroxybutyric acid in microbial biomass. *Eur. J. Appl. Microbiol. Biotechnol.* **1978**, *6*, 29–37. [CrossRef]
40. Sánchez, R.J.; Schripsema, J.; Da Silva, L.F.; Taciro, M.K.; Pradella, J.G.C.; Gomez, J.G.C. Medium-chain-length polyhydroxyalkanoic acids (PHAmcl) produced by *Pseudomonas putida* IPT 046 from renewable sources. *Eur. Polym. J.* **2003**, *39*, 1385–1394. [CrossRef]
41. Ansari, N.F.; M. Annuar, M.S. Functionalization of medium-chain-length poly(3-hydroxyalkanoates) as amphiphilic material by graft copolymerization with glycerol 1,3-diglycerolate diacrylate and its mechanism. *J. Macromol. Sci. Part A Pure Appl. Chem.* **2018**, *55*, 66–74. [CrossRef]
42. Gross, R.A.; DeMello, C.; Lenz, R.W.; Brandi, H.; Fuller, R.C. Biosynthesis and Characterization of Poly(β-hydroxyalkanoates) Produced by *Pseudomonas oleovorans*. *Macromolecules* **1989**, *22*, 1106–1115. [CrossRef]
43. Guo, W.; Duan, J.; Geng, W.; Feng, J.; Wang, S.; Song, C. Comparison of medium-chain-length polyhydroxyalkanoates synthases from *Pseudomonas mendocina* NK-01 with the same substrate specificity. *Microbiol. Res.* **2013**, *168*, 231–237. [CrossRef] [PubMed]
44. Philip, S.; Keshavarz, T.; Roy, I. Polyhydroxyalkanoates: Biodegradable polymers with a range of applications. *J. Chem. Technol. Biotechnol.* **2007**, *82*, 233–247. [CrossRef]
45. Leong, Y.K.; Show, P.L.; Ooi, C.W.; Ling, T.C.; Lan, J.C.W. Current trends in polyhydroxyalkanoates (PHAs) biosynthesis: Insights from the recombinant *Escherichia coli*. *J. Biotechnol.* **2014**, *180*, 52–65. [CrossRef] [PubMed]
46. Muangwong, A.; Boontip, T.; Pachimsawat, J.; Napathorn, S.C. Medium chain length polyhydroxyalkanoates consisting primarily of unsaturated 3-hydroxy-5-cis-dodecanoate synthesized by newly isolated bacteria using crude glycerol. *Microb. Cell Fact.* **2016**, *15*, 1–17. [CrossRef]

47. De Sisto, A.; Fusella, E.; Urbina, H.; Leyn, V.; Naranjo, L. Molecular characterization of bacteria isolated from waste electrical transformer oil. *Moscow Univ. Chem. Bull.* **2008**, *63*, 120–125. [CrossRef]
48. Rojas-Avelizapa, N.G.; Rodríguez-Vázquez, R.; Enríquez-Villanueva, F.; Martínez-Cruz, J.; Poggi-Varaldo, H.M. Transformer oil degradation by an indigenous microflora isolated from a contaminated soil. *Resour. Conserv. Recycl.* **1999**, *27*, 15–26. [CrossRef]
49. Koller, M. A review on established and emerging fermentation schemes for microbial production of polyhydroxyalkanoate (PHA) biopolyesters. *Fermentation* **2018**, *4*, 30. [CrossRef]
50. Koller, M.; Bona, R.; Braunegg, G.; Hermann, C.; Horvat, P.; Kroutil, M.; Martinz, J.; Neto, J.; Pereira, L.; Varila, P. Production of polyhydroxyalkanoates from agricultural waste and surplus materials. *Biomacromolecules* **2005**, *6*, 561–565. [CrossRef]
51. Shuai, J.J.; Tian, Y.S.; Yao, Q.H.; Peng, R.H.; Xiong, F.; Xiong, A.S. Identification and analysis of polychlorinated biphenyls (PCBs)-biodegrading bacterial strains in shanghai. *Curr. Microbiol.* **2010**, *61*, 477–483. [CrossRef]
52. Ashby, R.D.; Solaiman, D.K.Y.; Foglia, T.A. Bacterial poly(hydroxyalkanoate) polymer production from the biodiesel co-product stream. *J. Polym. Environ.* **2004**, *12*, 105–112. [CrossRef]
53. Rai, R.; Keshavarz, T.; Roether, J.A.; Boccaccini, A.R.; Roy, I. Medium chain length polyhydroxyalkanoates, promising new biomedical materials for the future. *Mater. Sci. Eng. R Rep.* **2011**, *72*, 29–47. [CrossRef]
54. Tappel, R.C.; Wang, Q.; Nomura, C.T. Precise control of repeating unit composition in biodegradable poly(3-hydroxyalkanoate) polymers synthesized by *Escherichia coli*. *J. Biosci. Bioeng.* **2012**, *113*, 480–486. [CrossRef]
55. Valappil, S.P.; Misra, S.K.; Boccaccini, A.R.; Keshavarz, T.; Bucke, C.; Roy, I. Large-scale production and efficient recovery of PHB with desirable material properties, from the newly characterised *Bacillus cereus* SPV. *J. Biotechnol.* **2007**, *132*, 251–258. [CrossRef] [PubMed]
56. Asad-Ur-Rehman Aslam, A.; Masood, R.; Aftab, M.N.; Ajmal, R. Ikram-Ul-Haq: Production and characterization of a thermostable bioplastic (Poly-s-hydroxybutyrate) from *Bacillus cereus* NRRL-b-3711. *Pak. J. Bot.* **2016**, *48*, 349–356.
57. Flora, G.D.; Bhatt, K.; Tuteja, U. Optimization of culture conditions for poly A-Hydroxybutyrate production from isolated Bacillus species. *J. Cell Tissue Res.* **2010**, *10*, 2235.
58. Cromwick, A.M.; Foglia, T.; Lenz, R.W. The microbial production of poly(hydroxyalkanoates) from tallow. *Appl. Microbiol. Biotechnol.* **1996**, *46*, 464–469. [CrossRef]
59. Kim, T.J.; Lee, E.Y.; Kim, Y.J.; Cho, K.S.; Ryu, H.W. Degradation of polyaromatic hydrocarbons by *Burkholderia cepacia* 2A-12. *World J. Microbiol. Biotechnol.* **2003**, *19*, 411–417. [CrossRef]
60. Chen, J.; Li, W.; Zhang, Z.Z.; Tan, T.W.; Li, Z.J. Metabolic engineering of *Escherichia coli* for the synthesis of polyhydroxyalkanoates using acetate as a main carbon source. *Microb. Cell Fact.* **2018**, *17*, 1–12. [CrossRef]
61. Khanna, S.; Srivastava, A.K. Statistical media optimization studies for growth and PHB production by *Ralstonia eutropha*. *Process Biochem.* **2005**, *40*, 2173–2182. [CrossRef]
62. Akiyama, M.; Tsuge, T.; Doi, Y. Environmental life cycle comparison of polyhydroxyalkanoates produced from renewable carbon resources by bacterial fermentation. *Polym. Degrad. Stab.* **2003**, *80*, 183–194. [CrossRef]
63. Lu, J.; Tappel, R.C.; Nomura, C.T. Mini-review: Biosynthesis of poly(hydroxyalkanoates). *Polym. Rev.* **2009**, *49*, 226–248. [CrossRef]
64. Tan, G.Y.A.; Chen, C.L.; Li, L.; Ge, L.; Wang, L.; Razaad, I.M.N.; Li, Y.; Zhao, L.; Mo, Y.; Wang, J.Y. Start a research on biopolymer polyhydroxyalkanoate (PHA): A review. *Polymers* **2014**, *6*, 706–754. [CrossRef]
65. Madison, L.L.; Huisman, G.W. Metabolic Engineering of Poly(3-Hydroxyalkanoates): From DNA to Plastic. *Microbiol. Mol. Biol. Rev.* **1999**, *63*, 21–53. [CrossRef] [PubMed]
66. Loo, C.; Sudesh, K. Polyhydroxyalkanoates: Bio-based microbial plastics and their properties. *Malays. Polym. J.* **2007**, *2*, 31–57.
67. Cruz, M.V.; Freitas, F.; Paiva, A.; Mano, F.; Dionísio, M.; Ramos, A.M.; Reis, M.A.M. Valorization of fatty acids-containing wastes and byproducts into short- and medium-chain length polyhydroxyalkanoates. *New Biotechnol.* **2016**, *33*, 206–215. [CrossRef]
68. Chanprateep, S. Current trends in biodegradable polyhydroxyalkanoates. *J. Biosci. Bioeng.* **2010**, *110*, 621–632. [CrossRef]
69. Hany, R.; Hartmann, R.; Böhlen, C.; Brandenberger, S.; Kawada, J.; Löwe, C.; Zinn, M.; Witholt, B.; Marchessault, R.H. Chemical synthesis and characterization of POSS-functionalized poly[3-hydroxyalkanoates]. *Polymer (Guildf)* **2005**, *46*, 5025–5031. [CrossRef]
70. Volova, T.G.; Syrvacheva, D.A.; Zhila, N.O.; Sukovatiy, A.G. Synthesis of P(3HB-co-3HHx) copolymers containing high molar fraction of 3-hydroxyhexanoate monomer by *Cupriavidus eutrophus* B10646. *J. Chem. Technol. Biotechnol.* **2016**, *91*, 416–425. [CrossRef]
71. Pappalardo, F.; Fragalà, M.; Mineo, P.G.; Damigella, A.; Catara, A.F.; Palmeri, R.; Rescifina, A. Production of filmable medium-chain-length polyhydroxyalkanoates produced from glycerol by *Pseudomonas mediterranea*. *Int. J. Biol. Macromol.* **2014**, *65*, 89–96. [CrossRef]
72. Tang, H.J.; Neoh, S.Z.; Sudesh, K. A review on poly (3-hydroxybutyrate-*co*-[P(3HB-*co*-3HHx)] and genetic modi fi cations that affect its production. *Front. Bioeng. Biotechnol.* **2022**, *10*, 2234. [CrossRef]

*polymers*

MDPI

*Article*

# The Influence of Plasticizers and Accelerated Ageing on Biodegradation of PLA under Controlled Composting Conditions

**Pavel Brdlík *, Jan Novák, Martin Borůvka, Luboš Běhálek and Petr Lenfeld**

Faculty of Mechanical Engineering, Technical University of Liberec, Studentska 1402/2, 46117 Liberec, Czech Republic
* Correspondence: pavel.brdlik@tul.cz; Tel.: +420-485-353-335

**Abstract:** The overall performance of plasticizers on common mechanical and physical properties, as well as on the processability of polylactic acid (PLA) films, is well-explored. However, the influence of plasticizers on biodegradation is still in its infancy. In this study, the influence of natural-based dicarboxylic acid-based ester plasticizers (MC2178 and MC2192), acetyl tributyl citrate (ATBC Citroflex A4), and polyethylene glycol (PEG 400) on the biodegradation of extruded PLA films was evaluated. Furthermore, the influence of accelerated ageing on the performance properties and biodegradation of films was further investigated. The biodegradation of films was determined under controlled thermophilic composting conditions (ISO 14855-1). Apart from respirometry, an evaluation of the degree of disintegration, differential scanning calorimetry (DSC), thermogravimetric analysis (TGA), Fourier transform infrared spectroscopy (FT-IR), and scanning electron microscopy (SEM) of film surfaces was conducted. The influence of melt-processing with plasticizers has a significant effect on structural changes. Especially, the degree of crystallinity has been found to be a major factor which affects the biodegradation rate. The lowest biodegradation rates have been evaluated for films plasticized with PEG 400. These lower molecular weight plasticizers enhanced the crystallinity degrees of the PLA phase due to an increase in chain mobility. On the contrary, the highest biodegradation rate was found for films plasticized with MC2192, which has a higher molecular weight and evoked minimal structural changes of the PLA. From the evaluated results, it could also be stated that migration of plasticizers, physical ageing, and chain scission of films prompted by ageing significantly influenced both the mechanical and thermal properties, as well as the biodegradation rate. Therefore, the ageing of parts has to be taken into consideration for the proper evolution of the biodegradation of plasticized PLA and their applications.

**Keywords:** PLA films; plasticizers; biodegradation; ageing; composting

**Citation:** Brdlík, P.; Novák, J.; Borůvka, M.; Běhálek, L.; Lenfeld, P. The Influence of Plasticizers and Accelerated Ageing on Biodegradation of PLA under Controlled Composting Conditions. *Polymers* **2023**, *15*, 140. https://doi.org/10.3390/polym15010140

Academic Editor: Vsevolod Aleksandrovich Zhuikov

Received: 25 November 2022
Revised: 14 December 2022
Accepted: 19 December 2022
Published: 28 December 2022



## 1. Introduction

In 2020, more than 29 million tons of plastic post-consumer waste were collected in the EU27 + 3 [1]. Although compared to previous years, the ratio of recycling and energy recovery is increasing, still, more than 23% of plastic waste ends up in landfills [1]. If the world population and demand for plastics continues within the current trend, the plastics demand will achieve 1000 million tons by 2050 [2]. Therefore, it is evident that waste management and environmental pollution will continue to increase. Undoubtedly, the largest end-used market is the packing industry (40.5%), where the most used materials are polyethylene terephthalate (PET), polyethylene (PE), and polypropylene (PP). Despite being potentially recyclable, monomaterial flexible packaging solutions based on these petroleum-based polymers, in most cases, end up in energy recovery or landfills. The environmentally friendly solution could be the incorporation of biodegradable polymers that are made from renewable sources. Nevertheless, the amount of biodegradable biopolymers in this segment is still lower than 1% [1,3].

There are currently several commercially available biodegradable biopolymers on the market, such as polylactic acid (PLA), polyhydroxyalkanoates (PHA), polybutylene adipate terephthalate (PBAT), and starch-based blends. These materials represent nearly 60% of thermoplastic biopolymer production capacities [3]. Especially, PLA, due to its high transparency, gloss, scratch resistance, and relatively high strength and toughness, could be an interesting alternative to petroleum-based polymers in the packing industry and in agriculture and textile segments. However, PLA is further characterized by low ductility and resistance to fracture. In order to improve the utility properties of PLA, plasticizers are often added to the PLA. Plasticizers, most-often low molecular weight polymers or oligomers, evoke the enhancement of macromolecular distance that ensures that the intermolecular forces decrease and increase the mobility of the system. Consequently, the glass transition temperature is shifted to lower temperatures and brittleness is reduced. Phthalic acid has become the most-used class of plasticizers in the 21st century [4]. However, issues such as problematic degradation, migration, and negative impact on human health were a catalyst for the development of new non-toxic, environmentally friendly, and biodegradable green-plasticizers [4–6]. Plenty of researchers have focused on this issue in the last two decades. Emad et al. [7] evaluated the thermal stability, and mechanical and morphological properties of PLA plasticized with epoxidized palm oil. A remarkable increase in the ductility of PLA by the addition of epoxidized cottonseed oil (ESCO) was presented in the study of Verdi et al. [8]. Ljungberg et al. [9] presented a decrease in the storage modulus and thermal properties (melting and glass transition temperatures) of PLA plasticized with tributyl citrate (TBC). Maiza et al. [10] studied the plasticizing effect of triethyl citrate (TEC) and acetyl tributyl citrate (ATBC). Besides a decrease in the glass transition temperature, dynamic storage modulus, and thermal stability, no color changes in the PLA films were observed in this study. Special effort was given to analyzing the applicability of poly(ethylene glycol) (PEG) plasticizer in the past few decades [11–13]. Courgneau et al. [14] compared the mechanical and thermal properties of PLA plasticized with ATBC and PEG in the content range from 2.5 wt. % to 20 wt. %. The improvement of mechanical properties is not the only aspect that has to be taken into consideration for the evaluation applicability of plasticizers in the packaging segment. Other fundamental aspects are film barrier properties (oxygen and water permeability) and migration of plasticizers. Courgneau et al. [14] presented that ATBC plasticizer maintains its gas barrier properties and water vapour transmission up to 13 wt. %. However, the barrier properties decrease in PLA already at 9 wt. % content of PEG plasticizer. On the contrary, Sessini et al. [15] achieved increasing hydrophobicity by incorporating 10 wt. % of limonene oxide (LO) produced from the peel of citrus fruits. Another issue is the migration of plasticizers that results in changes in physical and mechanical properties and color changes in plasticized PLA [10]. Kodal et al. [16] declared, through investigation with scanning electron microscopy, that the migration of PEG 400 in the amount of 20 wt. % content occurred. Also, Tsou et al. [17] reported some migration tendencies of ATBC. On the contrary, in the study of Burgos et al. [18], changes in thermal, mechanical, structural, and barrier properties that could be linked to the migration of oligomeric lactic acid (OLA) plasticizer even at 25 wt. % have not been observed.

Regarding the previous summary, the influence of chemistry, polarity, content, and molecular weight of plasticizers on thermal, mechanical, and barrier properties, as well as migration, has been discussed many times. However, from the view of environmental aspects, one of the most important questions, "How the plasticizers influence biodegradation of PLA," is still not well explained. PLA is polyester, where hydrolysis and enzymatic/microbial activity are the dominant degradation mechanisms. The biodegradation process could, despite the influence of environmental characteristics such as pH, temperature, moisture, and atmosphere composition, occur from several weeks up to years. For the degradation of PLA, temperature is one of the most critical parameters. Itävaara et al. [19] and Kale et al. [20] reported that faster biodegradation of PLA is achieved in thermophilic conditions compared to mesophilic ones. The higher temperature of the medium evokes a decrease in intermolecular binding forces that cause a more straightforward hydrolysis reaction and the attachment of microbes/enzymes. Therefore, incorporating low molecular weight plasticizers into PLA could evoke faster biodegradation rates due to the increased macromolecular chain mobility. On the contrary, the increasing macromolecule chain mobility could stimulate changes in morphology structure and enhance the crystallinity degree. Kolstad et al. [21] reported very poor biodegradability of semicrystalline PLA compared to amorphous PLA. It has been reported that the amorphous regions are easily assimilated by microorganisms [22]. Consequently, the biodegradation rate of PLA could also decrease with the incorporation of plasticizers. Furthermore, the polar character and end-chain group of plasticizers influence oxygen permeability and hydrophobicity. Another critical factor which affects biodegradation is the chemical composition of plasticizers. Regarding the previous summary, it is obvious that the effect of plasticizers on biodegradation rate complicates many factors and cannot be simply predicted. Therefore, the current work is dedicated to investigating the influence of the most often used plasticizers, PEG and ATBC, on the biodegradation rate of PLA under controlled thermophilic composting (ISO 1455-1). Besides PEG and ATBC, new plasticizers, MC 2178 and MC2192, based on 100% biobased dicarboxylic acid esters, were investigated. The reason behind this was to study the determining influence of the molecular weight of plasticizers on the biodegradation rate. The influence of ageing on the utility properties of plasticized PLA is another underexplored area of this research environment. Therefore, the influence of accelerated ageing on the changes in mechanical and thermal properties, and also on biodegradation, was also investigated.

## 2. Materials and Methods

The commercial 100% biobased PLA Luminy L 130 was purchased from Total Energies Caorbion (Gorinchem, The Netherlands). Luminy L 130 is high heat (the melting point is 175 °C, glass transition temperature is 60 °C), medium flow PLA (melt flow index 23 g/10 min, ISO 1133-A) homopolymer with minimal 99% L-isomer stereochemical purity. PLA Luminy L 130 was plasticized with PEG 400 (Sigma-Aldrich, Taufkirchen, Germany), ATBC Citroflex A4 (Vertellus Holding LLC, Linz, Austria), dicarboxylic acid-based plasticizers MC 2178 and MC 2192 (Emery Oleochemicals GmbH, Dusseldorf, Germany). The characteristic properties of plasticizers are listed in Table 1. All the plasticizers have food contact approval.

**Table 1.** Properties of plasticizers.

| Plasicizers | Molecular Mass ($g \cdot mol^{-1}$) | Density ($kg/m^3$) | Viscosity (mPas) | Sources |
|---|---|---|---|---|
| PEG 400 | 380–420 | 1.125 (at 20 °C) | 30–45 (at 25 °C) | [23–25] |
| ATBC citroflex A4 | 402 | 1.048 (at 25 °C) | 53.7 (at 25 °C) | [25–27] |
| MC 2178 | 1250 | 1.03–1.07 (at 20 °C) | 650–750 (at 20 °C) | [28] |
| MC 2192 | 4236 | 1.04–1.10 (at 20 °C) | 4000–6000 (at 20 °C) | [29] |

### 2.1. Preparation of PLA Films

Before the production process, any moisture from PLA pellets was removed using the vacuum oven VD53 (Binder, Germany) at 80 °C for 12 h. Further, the PLA was plasticized in compounder Collin ZK 25 P (COLLIN Lab & Pilot Solutions GmbH, Maitenbeth, Germany) equipped with automatic volumetric DVL LIQUIDOSER (Moretto, Italy), which is designed especially for dosing liquid additives. The plasticizers are commonly used in the range from 5 to 20 wt. %. The lower concentration of plasticizer ensures a lower potential of migration and better miscibility with PLA. However, the improvement in flexibility and fracture resistance is lower. The influence on barrier properties will be lower than in higher concentrations. Consequently, a lower influence of plasticizer on biodegradability of PLA could be assumed. Therefore, the dosing of plasticizers to compounder was established to achieve a final concentration of 15 wt. % in PLA. The temperature profile from 145 °C to 165 °C and speed of 160 rpm was used for PLA compounding. The plasticized PLA was further transformed in an extrusion head set up to 180 °C and pelletizer running at

3000 rpm. Before film processing, another vacuum drying process for 12 h at 80 °C was incorporated for moisture removing. The PLA films were extruded on twin screw extruder MC 15 HT (Xplore, Netherlands) and equipped with flat film die (0.4 mm gap size) at the constant melt temperature of 185 °C and 100 rpm screw speed.

### 2.2. Accelerated Ageing

The ageing process was performed in climatic chamber SUN 3600 (Weiss Technik, Balingen, Germany) incorporated with two 4 kW metal halide (MH) lamps. The climatic chamber made it possible to use irradiation intensity from 400 to 1150 W/m$^2$ in the spectral region of radiation from 300 nm to 2450 nm. Consequently, simulation of solar radiation at ground level that is defined in standard IEC 60068-2-5 could be tested. The condition of ageing was used with respect to DIN 75 220 standard. After the conditioning process (25 °C, 50% relative humidity, 240 h) in the climatic chamber (Teseco, Kostelec nad Orlicí, Czech Republic), films were exposed to constant radiation intensity 1000 W/m$^2$ for 240 h. The recommended temperature of chamber 42 °C was due to the potential decrease in glass transition in plasticized PLA films decreased to 28 °C. The relative humidity was adjusted to 65%, an average value in the summer months in the Czech Republic [30].

### 2.3. Mechanical Properties

The influence of plasticizers and accelerated ageing process on mechanical properties were evaluated by determination of tensile modulus ($E_t$), tensile strength ($\sigma_m$), and nominal tensile strain at break ($\varepsilon_{tb}$). TiraTest tensile testing machine (Tira GmbH, Schalkau, Germany), which is equipped with KAF type of load cells (ranging from 0 to 1000 N, sensitivity 2.0 mV/V) and with WA-type series displacement transducers (max. linearity deviation: − 0.13% and nominal supply voltage: 80 mV/V), was used for this study. The PLA films were, according to ISO 527-3 standard, trimmed to a size of 15 ± 0.2 mm width and 160 ± 2 mm length. Before testing, samples were conditioned in the climatic chamber (Teseco, Kostelec nad Orlicí, Czech Republic) at temperature of 25 °C and 50% relative humidity for 240 h. According to the ISO 527 standard, load speed of 1 mm/min was used for determination of tensile modulus. For the tensile strength and strain at break, load speed of 5 mm/min was applied. The results were evaluated from 15 measurements.

### 2.4. Rheological Properties

The evaluation of rheological properties of conditioned samples was further incorporated for determination of accelerated ageing impact on the stability of plasticized PLA films. The changes in rheological properties were evaluated with melt flow rate (MVR) analysis according to ISO 1133-1 (190 °C/2.16 kg). The Ceast 7028 (Instron, Buckinghamshire, UK) melt flow tester from company Instron (Buckinghamshire, UK) was used.

### 2.5. Analysis of Biodegradability under Thermophilic Composting

The method adapted from the ISO 14855-1 standard was used to analyse the influence of plasticizers on the biodegradation kinetics of PLA films in compost thermophilic (58 °C) environment. This method is based on evaluation of carbon dioxide amount evolution during the microbial degradation. The released carbon dioxide was detected with spirometer ECHO (ECHO d.o.o., Slovenske Konjce, Slovenia). Regarding the standard, the 10 g films were trimmed to pieces sized about 1 cm × 1 cm and placed into 2.8 l cylindrical hermetic vessels that contained 150 g of compost. The commonly available compost from the company AGRO CS (Říkov, Czech Republic), with 5.2 pH (measured by the Volcraft PH-100ATC pH meter, Conrad Electronic s.r.o., Praha, Czech Republic ), was used in this study. Before the biodegradation experiment, 50% humidity content of compost was adjusted by the halogen moisture analyser Mettler Toledo™ HX204 (Mettler Toledo, Columbus, OH, USA) and pebbles or other foreign objects larger than 2 mm were removed from the compost. For the proper microbial activity control of the compost, vessels with microcrystalline cellulose (Sigma-Aldrich, Saint-Quentin-Falavier, France) were used. The vessels were shielded from the light. The vessels were opened once a week and the compost was stirred to ensure an even distribution of moisture. Each biodegradation analysis was performed in duplicity. The percentage of biodegradation was determined in accordance with the following equation:

$$D_t = \frac{(CO_2)_T - (CO_2)_B}{T_{hCO_2}} \cdot 100 \quad (1)$$

where $(CO_2)_T$ is the cumulative amount of carbon dioxide evolved in the composting vessel containing the test material, $(CO_2)_B$ is the mean cumulative amount of carbon dioxide evolved in the blank vessels, and $(T_{hCO_2})$ is the theoretical amount of carbon dioxide that can be produced by the test material (all in g/vessel).

The theoretical amount of carbon dioxide can be determined via the following equation:

$$T_{hCO_2} = M_{TOT} \cdot C_{TOT} \cdot \frac{44}{12} \quad (2)$$

where $M_{TOT}$ is the total number of dry solids in the test material introduced into the composting vessel at the start of the test (in g), $C_{TOT}$ is the proportion of total organic carbon in the total dry solids in the test material (in g/g), and 44 and 12 are the molecular mass of carbon dioxide and the atomic mass of carbon, respectively. The individual proportions of total organic carbon of PLA films are listed in Table 2.

**Table 2.** The individual proportions of total organic carbon of components.

| Sample Designation | Proportions (%) |
|---|---|
| PLA | 50 |
| ATBC citroflex A1 | 59.7 |
| PEG 400 | 60.0 |
| MC2178 | 58.5 |
| MC2192 | 58.5 |

With respect to changes in organic carbon content, there is no possible way to do any supplemental analysis during a spirometry test. Therefore, the parallel experiment was carried out. The plasticized PLA films (size of 100 mm × 40 mm) were exposed to the same compost at the same condition as was used in above-introduced thermophilic composting analysis. The films were subjected to evaluation of disintegration degree every 14 days, being removed and conditioned (25 °C, 240 h, 50% relatively humidity) in climatic chamber (Teseco, Czech Republic). However, after 28 days, the experiment was finished because of intensive disintegration of some plasticized PLA films. The structural, thermal, and chemical changes, as well as surface roughness, were evaluated with differential scanning calorimetry (DSC), furrier transform infrared spectroscopy (FT-IR), thermogravimetric analysis (TGA), and scanning electron microscopy (SEM). The results were evaluated from 3 measurements. Therefore, the standard deviation has not been specified, only average values were provided.

### 2.6. *Differential Scanning Calorimetry (DSC)*

Thermal properties and structural changes were evaluated in a calorimeter DSC 1/700 (Mettler Toledo, Greifensee, Switzerland). Samples of approximately 5 mg taken from cross-section of the PLA films were sealed in aluminium pan and placed into DSC chamber where the constant nitrogen flow of 50 mL/min was adjusted. The samples were heated in temperature profile from 0 °C to 200 °C with heating rate of 10 °C/min. The samples were kept isothermal for 180 s at 200 °C, then the cooling process at rate of 10 °C/min was initiated. The glass transition temperature ($T_g$), cold crystallisation temperatures and enthalpies ($T_{cc}$, $\Delta H_{cc}$), melting temperatures and enthalpies ($T_m$, $\Delta H_m$), and primary crystallisation temperatures and enthalpies ($T_c$, $\Delta H_c$) were evaluated. The degree of crystallinity ($X_C$) was determined through the following equation:

$$X_C = \frac{\Delta H_m - \Delta H_c - \Delta H_{cc}}{\Delta H_m^0 \Delta w_m} = \frac{\Delta H}{\Delta H_m^0 \Delta w_m} \cdot 100 \quad (3)$$

where $\Delta H^0{}_m$ is the melting enthalpy of 100% crystalline PLA (106 J/g), $w_m$ is the mass fraction of PLA in the composites, and the $\Delta H$ is the enthalpy balance.

### *2.7. Fourier Transform Infrared Spectroscopy (FT-IR)*

The chemical changes in the PLA films were analysed using an infrared spectrometer Nicolet iS10 (Thermo Scientific, Waltham, MA, USA) in Attenuated Total Reflectance (ATR) mode using diamond crystal. The FTIR-ATR spectra were recorded in the range of 400–4000 $cm^{-1}$ by averaging 64 scans and using a resolution of 2 $cm^{-1}$.

### *2.8. Thermogravimetric Analysis (TGA)*

Thermal stability was evaluated using TGA2 instrument (Mettler Toledo, Switzerland). The samples were prepared with the same principle as for DSC analysis. They were taken from the cross-section of films in weights of 5 ± 0.5 mg. Further, heating from 50 °C to 600 °C at the heating ramp of 10 °C/min in nitrogen atmosphere was performed and the decomposition temperature at 5% weight loss ($T_{5\%}$) and 50% weight loss ($T_{50\%}$) were evaluated.

### *2.9. Scanning Electron Microscopy (SEM) Analysis*

The mechanism of degradation (surface, bulk) and surface changes were observed with field emission scanning electron microscopy (FE-SEM). To this purpose, the microscope TESCAN MIRA 3 (Tescan, Brno, Czech Republic) instrument with an accelerated voltage of 3 kV was used. The test samples were, prior to analysis, coated with 1 nm of platinum using Q150R ES (Quorum Technologies, UK).

## 3. Results

### *3.1. Mechanical Properties*

The results of the mechanical properties of as-produced and aged plasticized PLA films are shown in Figures 1–3. The plasticizers low molecular weight allows them to occupy intermolecular spaces between polymer chains. Due to this, they cause a reduction in energy for molecular motion, and the chain mobility is increased [10]. The influence of plasticizer on mechanical and rheological properties of PLA films will reflect its molecular weight and chemical composition (interaction between plasticizer and macromolecular chains of PLA). Consequently, different impacts on mechanical properties, such as fracture resistance and ductility enhancement, could be expected from the incorporation of plasticizers with different molecular weights and chemical compositions. The addition of 15 wt. % ATBC and PEG plasticizers to PLA evoked a significant increase in elongation at break under uniaxial loading of films. The 82% (PLA/ATBC) and 57% (PLA/PEG) increase in elongation were observed when compared to neat PLA. Courgneau et al. [14] reported equal efficiency of PLA/ATBC and PLA/PEG films at lower concentrations than 13 wt. %. However, at higher plasticizer contents, a higher elongation was found for PLA/ATBC. Rapa et al. [31] and Gálvez et al. [32] also found higher plasticizing efficiency of ATBC compared to PEG plasticizers at a higher content. The elongation at break of 250% (PLA/ATBC) and 140% (PLA/PEG) was achieved by the incorporation of 20 wt. % of plasticizer. The molecular weight influence of PEG on PLA plasticizer (15 wt. % concentration) is presented in the study by Darie-Niţă et al. [12]. The higher molecular weight of PEG 2000 and PEG 4000 evoked lower ductility changes in PLA (elongation lower than 5%). The influence of molecular weight on mechanical properties could be confirmed by the results of dicarboxylic acid-based plasticizers MC 2178 and MC2192. Only slight changes in elongation were observed for MC 2192, which had the highest molecular weight among those used. On the contrary, a significant enhancement in elongation (47%) was found for the PLA plasticized with MC 2178, which has the same chemical composition and a lower molecular weight. The low plasticizing efficiency of MC 2192 further evoked only a low decrease in tensile strength and modulus. Significant decreases in both properties were observed for MC 2178. The 19% decrease in tensile strength and 34% decrease in tensile modulus were evaluated for PLA/MC 2178. On the other hand, the incorporation of ATBC and PEG plasticizer into the PLA evoked significant changes. The 41% decrease in tensile strength and 37% decrease in tensile modulus were evaluated for ATBC-plasticized films. PEG-plasticised PLA films showed an even higher decrease in tensile strength (68%) and modulus (82%). Similar results were reported by Courgneau et al. [14], Rapa et al. [31], Gálvez et al. [32], and Greco et al. [33].

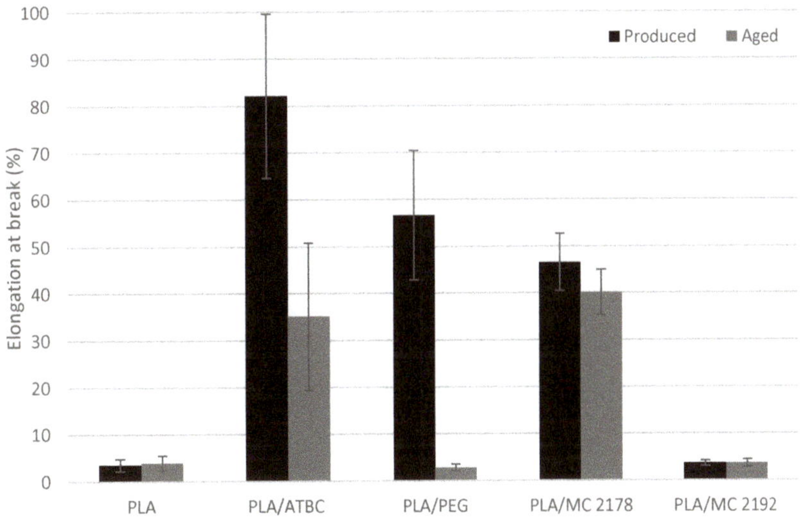


**Figure 1.** The evaluated results of elongation at break for as-produced and aged PLA films.

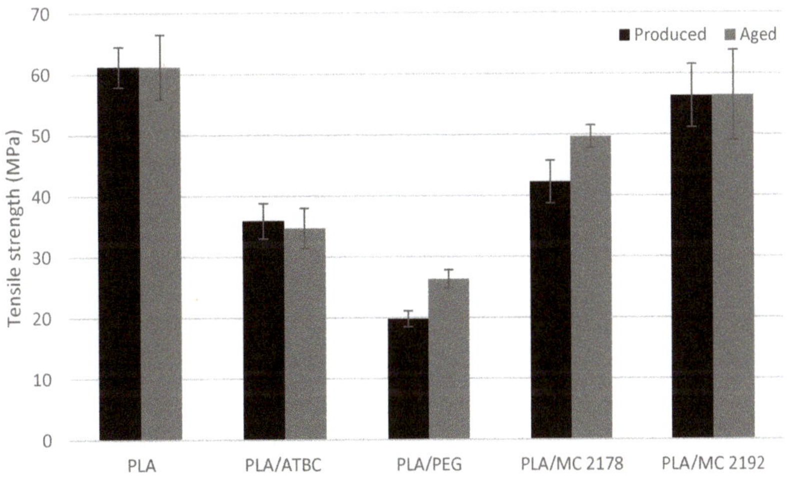


**Figure 2.** The evaluated results of tensile strength for as-produced and aged PLA films.

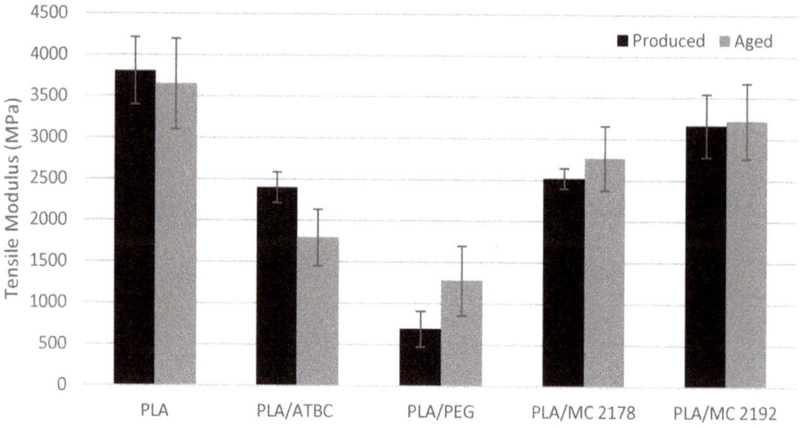


**Figure 3.** The evaluated results of tensile modulus for as-produced and aged PLA films.

The results of the mechanical properties of aged films were compared to as-produced ones, and no changes in neat PLA films could be found. However, the differences in results of as-produced and aged plasticized PLA films could be observed. The aged PLA films plasticized with ATBC showed a significant decrease in elongation (57%), a considerable decrease in tensile modulus (25%), and minimal changes in tensile strength (4%). This drop could be ascribed to structural changes, the degradation process (chain scission), and the migration of plasticizer. The chain scission evoked a reduction in intermolecular forces and increased the mobility of the system. Consequently, tensile strength and modulus decreased. Furthermore, the enhancement of brittleness, and decrease in elongation and viscosity, is characteristic of the degradation process. The migration of ATBC, due to the lower content of plasticizer, could cause the ductility and fracture resistance to decrease. On the contrary, tensile strength, modulus, and viscosity will increase [34]. Regarding the stated results, the degradation (chain scission) could be ascribed as the primary reason for the change in properties of aged PLA/ATBC films. This confirmed the result of Rapa et al. [31,35], where a low migration of ATBC at 20 wt. % in PLA was observed. The greatest changes in mechanical properties were found for aged PLA/PEG films. The elongation at break drop on the level of neat PLA. The tensile modulus and strength increased by around 85% and 33%, respectively. The migration of plasticizer could be ascribed as the major reason for these changes. A similar dependence was reported in several studies. Hu et al. [36,37] observed a significant increase in tensile modulus and strength, and a decrease in fracture strain of plasticized PLA with 30% of PEG after 720 h ageing under ambient conditions (23 °C, 50% RH). Also, the results of Kodal et al. [16] are in agreement with our conclusions. The changes in mechanical properties of aged, plasticized PLA (one year at the ambient condition) were considerable at a content of PEG higher than 5 wt. %. The influence of molecular weight on changes in mechanical properties is also presented in this study. Greater changes were achieved using plasticizers with lower molecular weight (PEG 400) than plasticizers with higher molecular weight (PEG 8000, PEG 35000). The aged PLA films with the dicarboxylic acid-based plasticizers did not show, in the higher molecular weight variant (MC2192), changes in tensile modulus, strength, or elongation at break. A slight increase in tensile modulus and strength, as well as a decrease in elongation, was noticed for dicarboxylic acid-based plasticizer with lower molecular weight (MC2178). However, the differences are, with respect to levels of standard deviations, very low. Consequently, the presumption of migration or macromolecular scission could not be declared.

### 3.2. Rheological Properties

The evaluated rheological properties (MVR) of as-produced and aged PLA films are introduced in Table 3. All the plasticizers showed, after production, enhancement in flowability of PLA. An enormous increase (221-fold enhancement of MVR) was observed in PLA films plasticized with PEG. PEG is a plasticizer with very low viscosity. On the contrary, the plasticized PLA films with dicarboxylic acid-based plasticizer MC 2192, which has the highest molecular weight among those used, showed relatively small changes (1.48 times enhancement of MVR). The MVR of PLA plasticized with the lower molecular weight variation (MC2178) was around 15% higher. The incorporation of ATBC plasticizer ensured 2.2 times enhancement in MVR of PLA.

**Table 3.** The evaluated results of Volume-Flow Rate (MVR) for as-produced and aged PLA films.

| MVR ($cm^3/min^{-1}$) | Sample Designation | | | | |
|---|---|---|---|---|---|
| | PLA | PLA/ ATBC | PLA/ PEG 400 | PLA/ MC 2178 | PLA/ MC 2192 |
| Produced | 13.9 ± 0.5 | 30.7 ± 0.1 | 3076 ± 216 | 23.9 ± 0.4 | 20.7 ± 0.1 |
| Aged | 14.5 ± 0.5 | 35.4 ± 0.1 | 1784 ± 43 | 32.0 ± 0.5 | 30.1 ± 0.4 |

No changes in MVR of as-produced neat PLA films and aged ones were noticed. As in mechanical properties, the significant changes in rheological properties (MVR) were evaluated for plasticized PLA films. Also, for rheological properties, the highest changes showed PLA films containing PEG plasticizer. The MVR decreased by 47% after ageing. This result confirmed our conclusion (Section 3.1), where a migration effect of PEG plasticizer was ascribed as a major reason for changes in mechanical properties. The lower content of plasticizer causes lower enhancement of chain mobility of PLA, which leads to a decrease in rheological properties (flowability). The aged PLA/ATBC films showed the opposite dependence. A 15% increase in MVR was found. The chain scission could be due to a decrease in intermolecular forces ascribed to this enhancement. Also, this result is in accordance with the previous conclusion where the chain scission was presumed as a major reason for property changes in aged PLA/ATBC films. Furthermore, a significant increase in MVR was found for aged PLA films plasticized with dicarboxylic acids (MC2178 and MC2192). Therefore, the chain scission is a very important effect that has significant influence on properties of these aged films.

### 3.3. Differential Scanning Calorimetry (DSC)

The results of the first non-isothermal heating of as-produced and aged PLA films are shown in Figure 4 and in Table 4. The glass transient temperatures were used to neglect the technological processing aspects evaluated from the second heating cycle. A shift in transient temperatures is evident from the estimated results of as-produced plasticized PLA films. The glass transition temperatures of PLA films plasticized with ATBC decreased compared to neat PLA film by about 18 °C, cold crystallization temperature by about 23 °C, and melt temperatures by about 3 °C. A shift of about 23 °C at the glass transition temperature, 25 °C of cold crystallization temperature, and about 6 °C of melt temperature was noticed for PLA/PEG films. Also, Rapa et al. [31], Galvez et al. [32], and Farah et al. [38] reported similar temperature changes for PLA/ATBC and PLA/PEG. In several studies [12,39], the lower shift of transient temperatures was evaluated for increasing molecular weight of the PEG plasticizer. The influence of molecular weight is evident if the transient temperatures of dicarboxylic acid-based plasticizers MC2178 and MC2192 are compared. The plasticizer MC2192, having the highest molecular weight among those used, showed lower changes in glass transition (about 7 °C) and no changes in melt temperatures. Nevertheless, a significant decrease in cold crystallization (about 21 °C) was evident from the evaluated results. The plasticizer MC2178, with the same chemical composition and a lower molecular weight, showed a significantly higher decrease in these values. The glass transition temperature decreased by about 20 °C, cold crystallization by 20 °C, and melt temperature by 2 °C. The appearance of a double peak at melting temperature was reported by Rapa et al. [31] and Greco et al. [33] for PLA films plasticized with PEG and ATBC. Also, in our previous study [40], we reported the presence of a bimodal peak for PLA films with 10 wt. % of ATBC. Wu et al. [41] ascribed the dual melting peak to the formation of different crystalline structures (α and α'crystals). However, only the single melting peaks were noticed for all produced films. This could be caused by a higher content of plasticizers (15 wt. %), which will evoke higher chain mobility enhancement during processing, by the used type of PLA. Luminy L130, which was used in this experiment, has a higher content of L-isomer (99% L-isomer stereochemical purity) and higher crystallization ability than Ingeo 3001D (95 wt. % of L- lactide), which was used in the previous experiment.

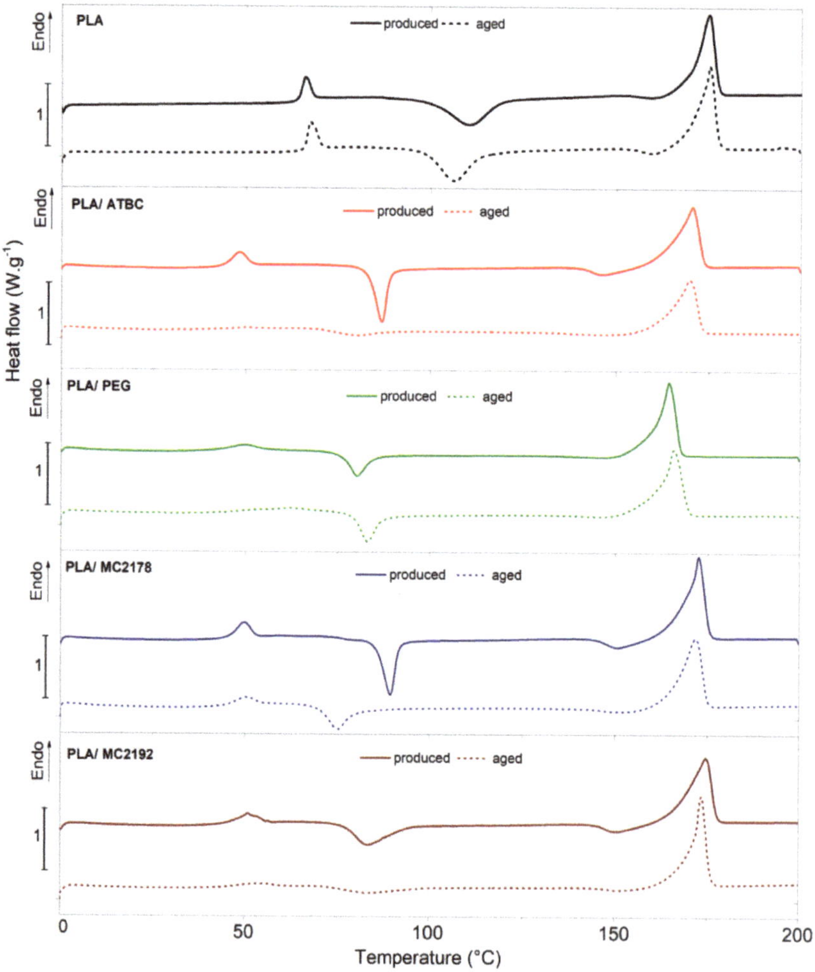


**Figure 4.** DSC curves of neat and plasticized PLA films after production and accelerated ageing.

**Table 4.** DSC data of neat and plasticized PLA films after production, accelerated ageing, and composting process.

| Sample Designation | | Exposition Time of Composting | $T_g$ (°C) | $T_{cc}$ (°C) | $\Delta H_{cc}$ (J/g) | $\Delta H_c$ (J/g) | $T_c$ (°C) | $T_m$ (°C) | $\Delta H_m$ (J/g) | $\Delta H$ (J/g) | $X_C$ (%) |
|---|---|---|---|---|---|---|---|---|---|---|---|
| PLA | Produced | Initial | 60.3 | 109.2 | 33.6 | 1.5 | 158.6 | 174.1 | 42.1 | 7.0 | 6.6 |
| | | 14 days | 59.8 | - | - | 1.4 | 161.9 | 175.4 | 51.3 | 49.9 | 47.0 |
| | | 28 days | 54.3 | - | - | - | - | 164.0 | 59.9 | 59.9 | 56.5 |
| | Aged | Initial | 60.1 | 106.3 | 33.2 | 2.3 | 159.4 | 174.8 | 41.7 | 5.4 | 5.1 |
| | | 14 days | 59.4 | - | - | 2.0 | 160.8 | 172.0 | 42.2 | 40.2 | 38.0 |
| | | 28 days | 54.5 | - | - | - | - | 165.5 | 59.7 | 59.7 | 56.3 |
| PLA/ ATBC | Produced | Initial | 41.7 | 86.4 | 23.8 | 6.4 | 146.6 | 171.3 | 42.7 | 12.5 | 13.9 |
| | | 14 days | 46.2 | - | - | 3.2 | 153.4 | 172.2 | 46.6 | 43.4 | - |
| | | 28 days | - | - | - | - | - | 170.4 | 49.4 | 49.4 | - |
| | Aged | Initial | 40.8 | 80.2 | 7.5 | 2.9 | 146.4 | 170.8 | 39.1 | 28.7 | - |
| | | 14 days | 44.9 | - | - | 2.3 | 160.8 | 172.1 | 41.7 | 39.4 | - |
| | | 28 days | - | - | - | - | - | 169.6 | 50.9 | 50.9 | - |
| PLA/ PEG | Produced | Initial | 37.6 | 84.1 | 13.4 | 1.7 | - | 167.9 | 41.3 | 26.26 | 29.5 |
| | | 14 days | 48.8 | - | - | - | - | 173.0 | 47.7 | 47.7 | - |
| | | 28 days | - | - | - | - | - | 172.0 | 48.0 | 48.0 | - |
| | Aged | Initial | 46.1 | 83.2 | 16.6 | 1.3 | - | 166.0 | 41.9 | 24.0 | - |
| | | 14 days | - | 93.9 | 6.9 | - | - | 168.8 | 50.9 | 44.0 | - |
| | | 28 days | - | - | - | - | - | 166.7 | 53.1 | 53.1 | - |
| PLA/ MC 2178 | Produced | Initial | 40.1 | 89.5 | 22.8 | 5.0 | 150.7 | 172.5 | 44.3 | 20.9 | 18.2 |
| | | 14 days | - | - | - | 1.7 | 159.9 | 176.3 | 46.2 | 44.4 | - |
| | | 28 days | - | - | - | - | - | 168.8 | 55.5 | 55.4 | - |
| | Aged | Initial | 43.2 | 75.6 | 17.1 | 3.4 | 153.6 | 172.5 | 41.2 | 20.7 | - |
| | | 14 days | - | - | - | - | - | 172.0 | 46.8 | 46.8 | - |
| | | 28 days | - | - | - | - | - | 169.6 | 55.9 | 55.8 | - |
| PLA/ MC2192 | Produced | Initial | 52.8 | 87.6 | 26.2 | 7.1 | 152.1 | 175.5 | 44.2 | 11.0 | 12.6 |
| | | 14 days | - | - | - | - | - | 163.4 | 59.2 | 59.2 | - |
| | | 28 days | - | - | - | - | - | 156.7 | 63.8 | 63.8 | - |
| | Aged | Initial | 51.5 | 80.6 | 26.4 | 5.9 | 151.1 | 174.4 | 45.5 | 13.2 | - |
| | | 14 days | - | - | - | - | - | 161.2 | 55.5 | 55.5 | - |
| | | 28 days | - | - | - | - | - | 156.8 | 60.6 | 60.6 | - |

From the evaluated results, it is evident that all plasticizers evoked, due to an increase in chain mobility, further structural changes. The highest crystallinity degree enhancements (crystallinity degree about 30%) were found for PLA/PEG films. On the contrary, the lowest enhancement of crystallinity (crystallinity degree of about 13%) was achieved for the highest molecular weight dicarboxylic acid-based plasticizer MC 2192. Also, PLA/ATBC films showed only a minor enhancement in crystallinity degree (about 14%). The dicarboxylic acid-based plasticizer MC2178 evoked, due to lower molecular weight (when compared to MC2192), greater improvement of PLA crystallinity (about 18% crystallinity degree).

No significant changes in transient temperatures between as-produced and aged neat PLA films have been found. Further, no change in crystallinity degree has been observed. Also, aged PLA films plasticized with ATBC did not show any considerable shift in transient temperature that made it possible to make a valuable statement. Nevertheless, significant changes in structure could be assumed for PLA/ATBC films after ageing. These structural changes were not evaluated via the degree of crystallinity but by the enthalpy changes due to the migration of plasticizer that could change weight content in PLA films. The cold crystallization and primary melt crystallization enthalpy markedly decreased in PLA/ATBC films after ageing. Consequently, the increase in enthalpy balance ΔH (Equation (3)) and enhancement of crystallinity degree could be assumed. In contrast, aged PLA films plasticized with PEG did not show any changes in enthalpies. Therefore, any valuable structure changes are not predicted. This could be ascribed to the high crystallization kinetics of PLA/PEG films, which already evoked a high level of structure order (degree of crystallinity) after production. The lower crystallization kinetics of PLA/ATBC films were stimulated by the supplied energy in the form of radiation during ageing and caused further structure order changes. Also, for as-produced PLA films that contain dicarboxylic acid-based plasticizer MC 2192, the lower crystallization kinetic (lower degree of crystallinity) was evaluated. However, due to higher molecular weight, only small changes in enthalpies were evaluated after accelerated ageing. Consequently, low structural changes could be assumed. Even lower changes in enthalpies were evaluated for MC 2178 plasticizer (lower molecular weight than MC 2192). This could be ascribed, as for PLA/PEG films, to the higher crystallization kinetics of PLA/MC2178 films during processing. If the glass transition temperatures of aged PLA films plasticized with PEG and MC2178 are compared to as-produced ones, the significant shifts are obvious. The glass transient temperatures increased by about 9 °C for PLA/PEG films and about 3 °C for PLA/MC 2178 films. The increased glass transient temperature is probably caused by the migration of plasticizers. The lower content of plasticizers in PLA will cause increased intramolecular bonding forces. On the contrary, any valuable shifting of glass transition was not evaluated for PLA/MC2192 films, as well as for PLA/ATBC films. Consequently, a low migration assumption could be declared.

The as-produced and aged neat PLA film exposed to thermophilic composting showed similar results (Table 4, Figures S1 and S2). After the first 14 days of composting, changes in transient temperatures had not been observed. However, the conditions in the thermophilic compost environment evoked the elimination of cold crystallization enthalpy, enhancement of melting enthalpy, and, consequently, enhancement of crystallization degree. Also, from the visual appearance of the films, it was evident that the thermophilic composting environment caused significant structural changes (loss of transparency). Further, 14 days of composting caused about a 6 °C decrease in the glass transient temperature, about 10 °C decreases in melt temperature, the elimination of primary crystallization enthalpy, and another enhancement in crystallinity degree. With regard to the increase in crystallinity degree, it can be stated that the further enhancement of crystallinity could be related to the degradation of the amorphous phase due to the hydrolysis. The degradation of amorphous parts (shortening of macromolecular chains within the amorphous region) could be ascribed to the reason for the decrease in transient temperatures. Jimenez et al. [42] reported that hydrolytic chain cleavage proceeds preferentially in the amorphous regions and leads, consequently, to an increase in crystallinity degree. The as-produced and aged PLA films plasticized with ATBC showed a slight increase in glass transient temperature (about 4 °C) after the first 14 days of composting. This phenomenon could be related to the migration and release of ATBC plasticizer in the environment. Similar to neat PLA films, the elimination of cold crystallization enthalpy and obvious increase in enthalpy balance (ΔH) was evaluated for as-produced and aged PLA/ATBC films. This increase could be ascribed to the aforementioned structure order changes, degradation of amorphous phase, or the migration of plasticizers. Further, 14 days of composting of as-produced and aged PLA/ATBC films evoked another enhancement of enthalpy balance (ΔH). However, due to decreasing content in the amorphous phase, it was not possible to evaluate glass transient temperature. The as-produced PLA/PEG films showed even higher shifting of transient temperatures after the first 14 days of composting. The glass transient temperature was increased by about 11 °C and melt temperature by about 5 °C. The elimination of cold crystallization, primary melting enthalpies, and enhancement of enthalpy balance (ΔH) was further evaluated. The next 14 days of composting did not evoke any significant changes. This could mean that the main process of migration and changes in structural order were already finished. No enthalpy changes could also mean low degradation of the amorphous part. Therefore, it was not possible to evaluate the glass transient temperature for the aged PLA/PEG. However, minimal changes in melting temperature within 28 days of composting were evaluated. Therefore, compared to as-produced PLA/PEG films, a lower release of PEG into the compost environment could be assumed. The reason for this could be the high migration tendency of PEG plasticizer evoked by the process of ageing. The aged PLA/PEG films consequently contained a lower content of plasticizer that could be released during composting. On the contrary, for as-produced PLA/PEG films, a slight increase in enthalpy balance (ΔH) was observed. The supposed low-level of releasing of plasticizer and high structure order level (minimal structural changes) could mean more intensive degradation. Also, for the dicarboxylic acid-based plasticizers, it was not possible to evaluate the glass transition temperatures for composted films. Nevertheless, from the shifting of melting temperature and evaluated enthalpies (ΔH), some differences are obvious. Similar to PLA/ATBC and PLA/PEG films, the increase in melt temperature of as-produced PLA/MC 2178 films were evaluated after the first 14 days of composting. Consequently, some release of plasticizer into the compost environment could be assumed. Further, the elimination of cold crystallization and enhancement of enthalpy balance (ΔH) was evaluated. The following 14 days of composting evoked another decrease in melt temperature and enhancement of enthalpy (ΔH). Regarding the previous summary, a more intensive degradation (chain scission) of amorphous parts than for PLA/PEG films could be assumed. The composted, aged PLA/MC 2178 films did not show any significant changes in melt temperature within the first 14 days. This result could also be ascribed (as well as for PLA/PEG films) to some migration tendency of MC2178 of plasticizer, evoked by the ageing process. However, after the next 14 days of composting, the PLA/ MC 2178 films showed a decrease in melt temperature and further enhancement of enthalpy (ΔH). Therefore, for aged PLA/MC2178, also, the degradation of the amorphous phase could be assumed to be the main aspect of this event. The as-produced and aged composted PLA films that contained plasticizer MC 2192 showed, after the first 14 days of composting, the highest decrease in melting temperature and enthalpy (ΔH) enhancement. Regarding this, and low migration/releasing tendency of MC2192 plasticizer, the highest degradation rate could be assumed for these films.

### *3.4. Thermogravimetric Analysis (TGA)*

The results of TGA are shown in Figure 5 and summarized in Table 5. Several studies reported significant decreases in thermal stability using ATBC and PEG plasticizers in PLA [10,11,32,43,44]. However, only a low decrease in thermal stability (initial decomposition temperature T5%) was observed for as-produced PLA/ATBC films. The reason could be different types of PLA (molecular weight, L and D isomer content, etc.) and their interaction with ATBC. Furthermore, applied technology (compression molding, extrusion, casting, etc.) as well as processing conditions, could also be contributing factors. Also, the addition of dicarboxylic acid-based plasticizer MC2192 to PLA films evoked only a low decrease in thermal stability. The incorporation of MC2178 plasticizer with the lower molecular weight prompted a higher decrease in thermal stability. The initial decomposition temperature (T5%) decreased by about 16 °C when compared to neat PLA. The highest change in thermal stability was observed for PLA films plasticized with PEG. The initial decomposition temperature (T5%) decreased by about 53 °C. The evaluated result of thermal stability did not show any considerable changes between as-produced and aged neat PLA films. Furthermore, minimal differences were evaluated between as-produced and aged PLA films plasticized with MC2192. The aged PLA/ATBC films showed approximately a 9 °C decrease in initial decomposition temperature (T5%). This result could be ascribed to an event of chain scission that evokes a decrease in intramolecular forces. On the contrary, an increase in initial decomposition temperature (T5%) was found for aged PLA/PEG and PLA/MC2178 films. The initial decomposition temperature (T5%) increased by about 10 °C for PLA plasticized with MC 2178 and about 37 °C for PLA plasticized with PEG. The increase in thermal stability is, as well as a shift in transient temperatures and decrease in viscosity (MVR), evoked by the migration of plasticizer. Consequently, the previously-stated presumption (Sections 3.2 and 3.3) about the tendency of migration for these two plasticizers could be confirmed.

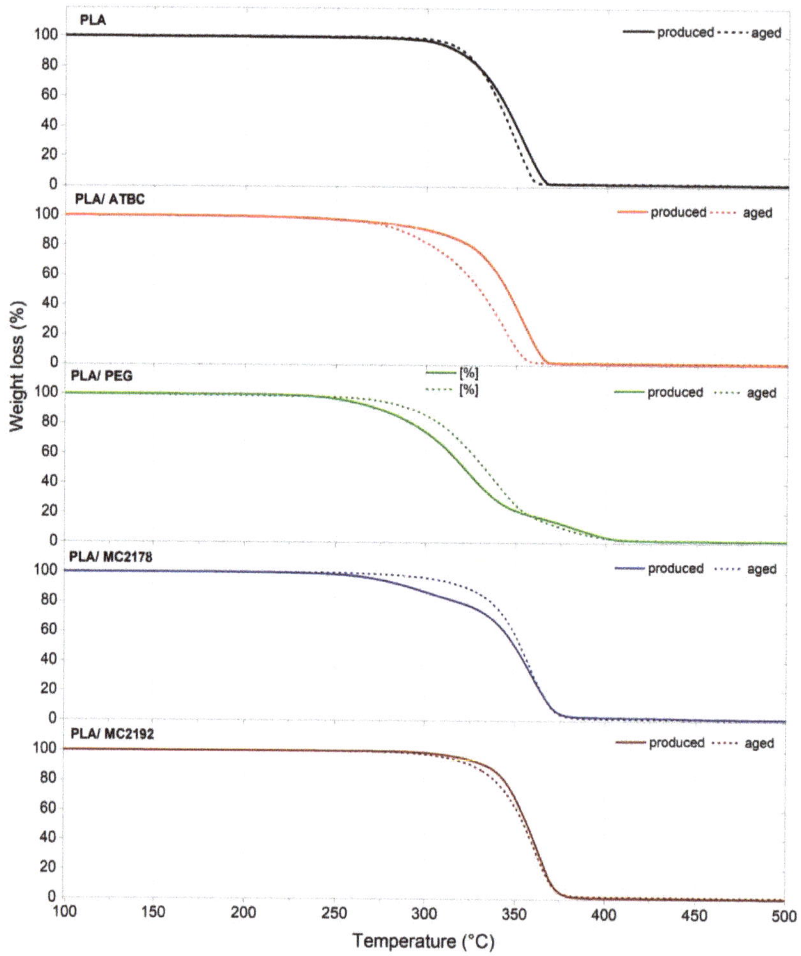


**Figure 5.** TGA curves of neat and plasticized PLA films after production and accelerated ageing.

**Table 5.** TGA data of neat and plasticized PLA films after production, accelerated ageing, and composting process.

| Sample Designation | Exposition Time of Composting | | | | | |
|---|---|---|---|---|---|---|
| | Initial State | | 14 Days | | 28 Days | |
| | $T_5$ (%) | $T_{50}$ (%) | $T_5$% (%) | $T_{50}$ (%) | $T_5$ (%) | $T_{50}$ (%) |
| Produced PLA | 313.4 | 349.9 | 310.9 | 342.5 | 293.9 | 330.6 |
| Aged PLA | 315.8 | 352.5 | 306.9 | 335.0 | 288.8 | 351.7 |
| Produced PLA/ATBC | 310.5 | 349.0 | 293.8 | 344.8 | 276.9 | 335.7 |
| Aged PLA/ATBC | 301.6 | 356.7 | 292.0 | 341.2 | 273.3 | 329.3 |
| Produced PLA/PEG 400 | 259.4 | 321.7 | 250.4 | 301.6 | 244.1 | 282.9 |
| Aged PLA/PEG 400 | 296.8 | 348.5 | 280.9 | 332.3 | 250.6 | 300.7 |
| Produced PLA/MC2178 | 297.9 | 355.8 | 295.1 | 340.4 | 283.7 | 340.1 |
| Aged PLA/MC2178 | 308.5 | 353.0 | 293.11 | 338.4 | 272.9 | 331.3 |
| Produced PLA/MC2192 | 309.6 | 357.0 | 299.1 | 350.1 | 254.3 | 305.7 |
| Aged PLA/MC2192 | 311.8 | 356.1 | 297.3 | 350.1 | 257.6 | 305.3 |

The thermophilic composting process of as-produced and aged neat PLA films showed major changes in thermal stability, similar to changes in transient temperatures and enthalpies (Section 3.3), after 28 days of composting (Table 5, Figures S3 and S4). About a 20 °C decrease in initial decomposition temperature (T5%) was found for as-produced neat PLA films, and about 28 °C decrease for aged ones. Consequently, considerable degradation (hydrolytic chain scission) could be assumed. On the other hand, as-produced and aged PLA/ATBC films showed considerable changes in thermal stability even after 14 days of composting. The initial decomposition temperature (T5%) of as-produced PLA/ATBC films decreased by about 17 °C and, for aged PLA/ATBC films, about 10 °C. Also, the further 14 days of composting evoked another considerable decrease (about 18 °C) in initial decomposition temperature (T5%). A relatively low decrease in initial decomposition temperature (about 10 °C) was observed for as-produced PLA/PEG films after the first 14 days of composting. Another 14 days of composition did not evoke any significant changes in thermal stability. Therefore, a lower level of degradation (chain scission) than for neat PLA and PLA/ATBC films could be assumed. Nevertheless, it is important to mention that the results of thermal stability are also be influenced by the migration of plasticizer. The aged PLA/PEG films showed a higher decrease in thermal stability. The initial decomposition temperature (T5%) decreased by about 16 °C after 14 days of composting and by about 46 °C after 28 days of composting. The low changes in thermal stability within 28 days of composting were evaluated for as-produced PLA films plasticized with dicarboxylic acid-based plasticizer MC2178. The ageing process of PLA/MC2178 films evoked obviously higher changes. The initial decomposition temperature (T5%) decreased by about 16 °C at the first 14 days of composting and about 36 °C after 28 days. Consequently, a higher degradation (chain scission) of aged PLA films plasticized with MC2178 and PEG than for as-produced ones could be assumed. The highest decrease in initial decomposition temperature (highest degradation) was observed for composted PLA films plasticized with MC2192. The 14 days of thermophilic composting evoked about a 10 °C decrease in initial decomposition temperatures (T5%) and, after 28 days of composting, decreased by about 56 °C. No significant differences were observed between as-produced and aged films.

### 3.5. *Scanning Electron Microscopy (SEM) Analysis*

The SEM images of surfaces of neat, as-produced PLA films and films after thermophilic composting (14 and 28 days) are shown in Figure 6. The PLA samples after ageing and subsequent thermophilic composting are shown in Figure 7. The neat PLA films showed smooth surfaces after the production and accelerated ageing. The exposition of as-produced films to thermophilic composting evoked only a slight increase in the roughness of their surface after the first 14 days. A significant change in thermal properties after 28 days of composting was observed. However, a very low enhancement of surface erosion could be seen from SEM images. This could mean that mainly bulk degradation occurred. Arrieta et al. [45] reported similar results. Minimal surface erosion was observed within 28 days of lab thermophilic composting of PLA films. No significant erosion of the surface was observed for aged PLA films after 14 or 28 days of composting. The SEM images of surfaces of as-produced and aged PLA/ATBC films and PLA/ATBC films after thermophilic composting (14 and 28 days) are shown in Figures 8 and 9.

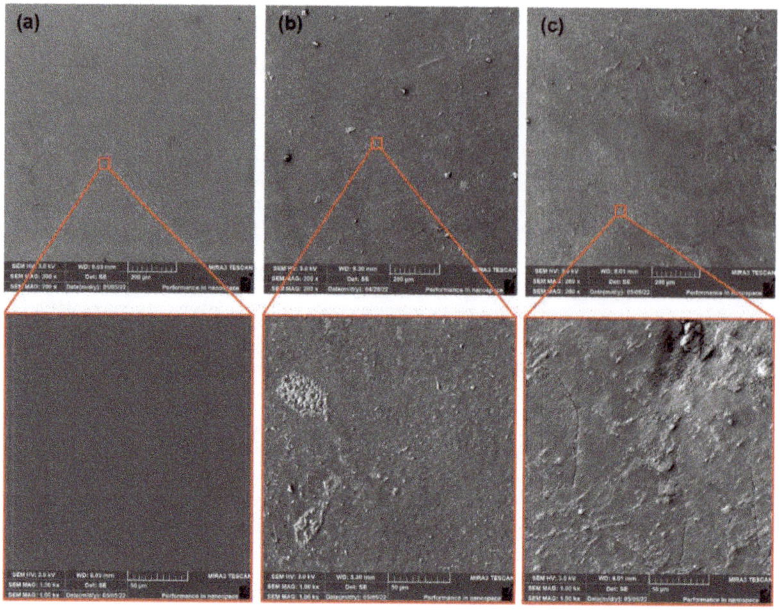

**Figure 6.** SEM images of as-produced neat PLA films (**a**) at initial state and after (**b**) 14 days of thermophilic composting, and (**c**) 28 days of thermophilic composting.

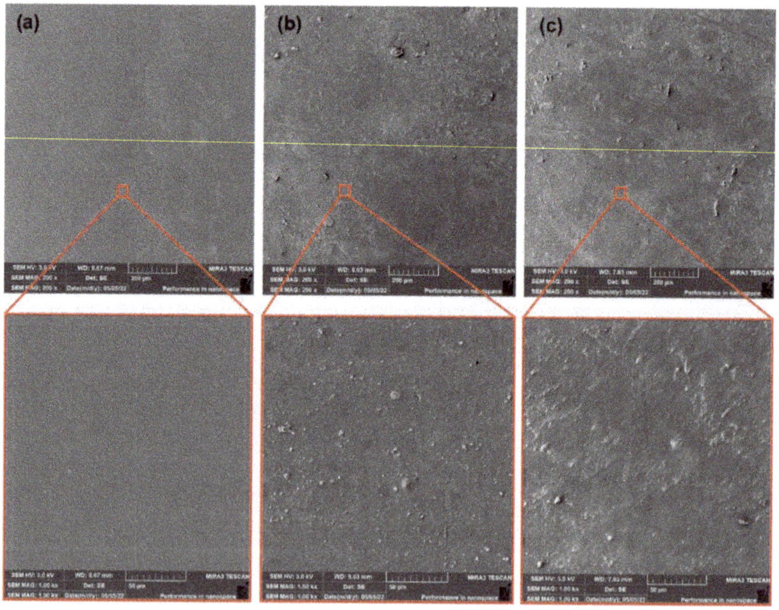

**Figure 7.** SEM images of aged neat PLA films (**a**) at initial state and after (**b**) 14 days of thermophilic composting, and (**c**) 28 days of thermophilic composting.

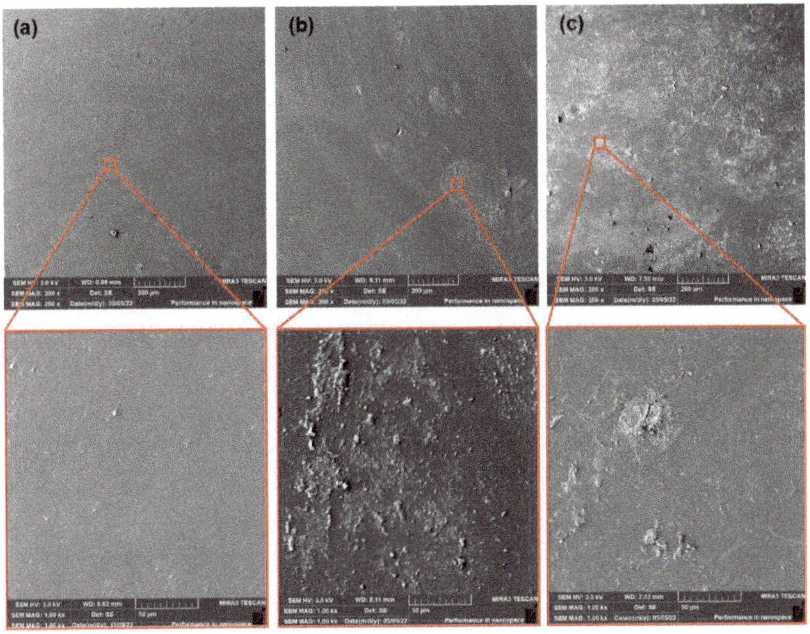


**Figure 8.** SEM images of as-produced PLA/ATBC films (**a**) at initial state and after (**b**) 14 days of thermophilic composting, and (**c**) 28 days of thermophilic composting.

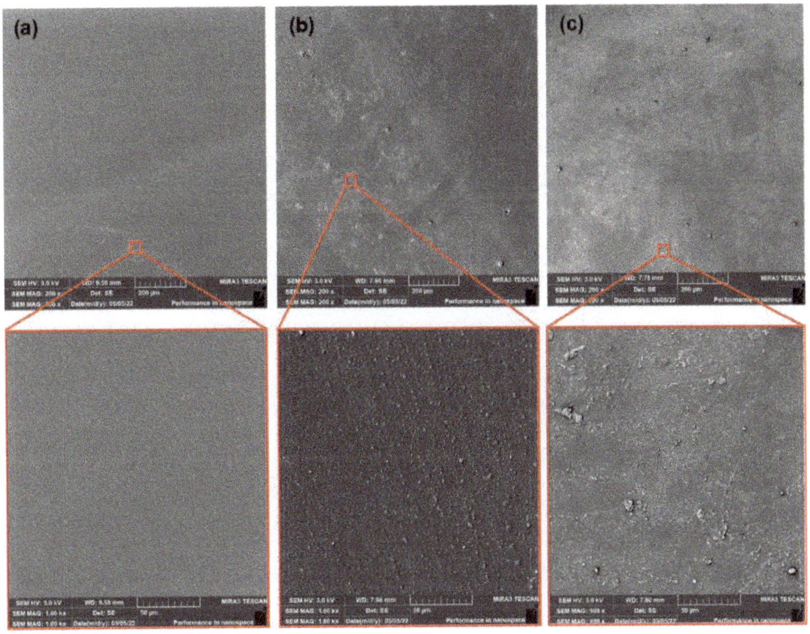


**Figure 9.** SEM images of aged PLA/ATBC films (**a**) at initial state and after (**b**) 14 days of thermophilic composting, and (**c**) 28 days of thermophilic composting.

The incorporation of PEG plasticizer into PLA films evoked more evident changes. The surfaces of as-produced PLA and PLA/ATBC films were considerably smoother than as-produced PLA/PEG films (Figure 10). The appearance of the "lunar surface" morphology was characteristic of these films. Low viscosity and poor miscibility of PEG with PLA at higher concentrations [14] could be a reason for this event. The PEG plasticizer, during extrusion through the head, is released on the surface and limits the process of calandering (entrapping of plasticizer between rolls and films). The first 14 days of composting evoked enormous changes in surface roughness. The surface was very rugged with a large number of "craters." The intensive release of water-soluble PEG plasticizer and the degradation of amorphous parts could be ascribed as the main phenomena of these changes. Another 14 days of composting did not evoke such dynamic changes. The level of roughness was similar to the first evaluated period. Regarding previous results (change in mechanical, rheological, and thermal properties), the ageing process evoked a significant migration of PEG plasticizer from PLA films. Therefore, an increase in brittleness could be expected. This assumption could be confirmed by the SEM surface images of aged PLA/PEG films (Figure 11). The appearance of cracks could be seen on the surface of aged films. Furthermore, due to the lower content of PEG plasticizer in PLA films, a lower level of release into the compost environment could be expected. Because of the lower level of roughness, a lower numbers of "craters" compared to as-produced films could be seen. This presumption could be confirmed by analysis of the surface images of aged PLA/PEG films after 14 days of composting. Despite being rough, a large amount of craterless surface could be seen after 28 days of composting. Thus, again, a minimal migration of PEG could be further expected.

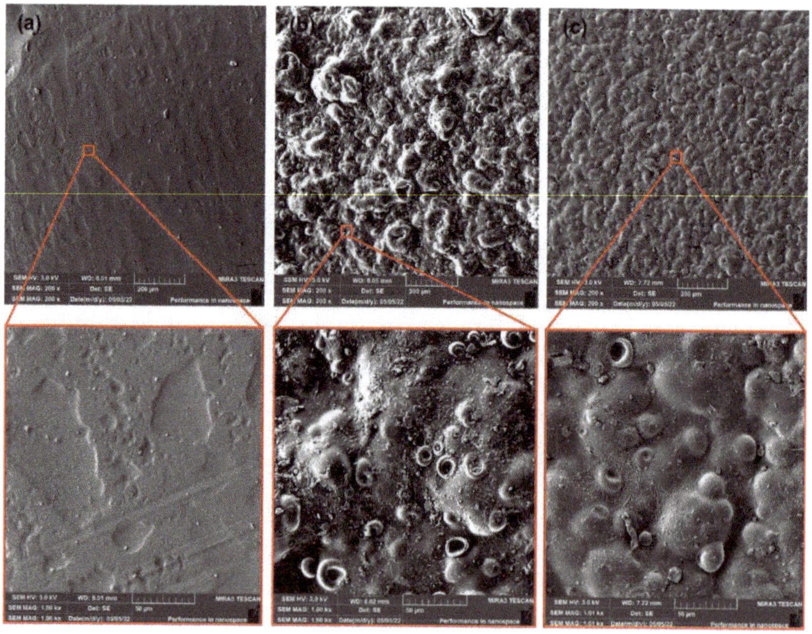


**Figure 10.** SEM images of as produced PLA/PEG films (**a**) at initial state and after (**b**) 14 days of thermophilic composting, and (**c**) 28 days of thermophilic composting.

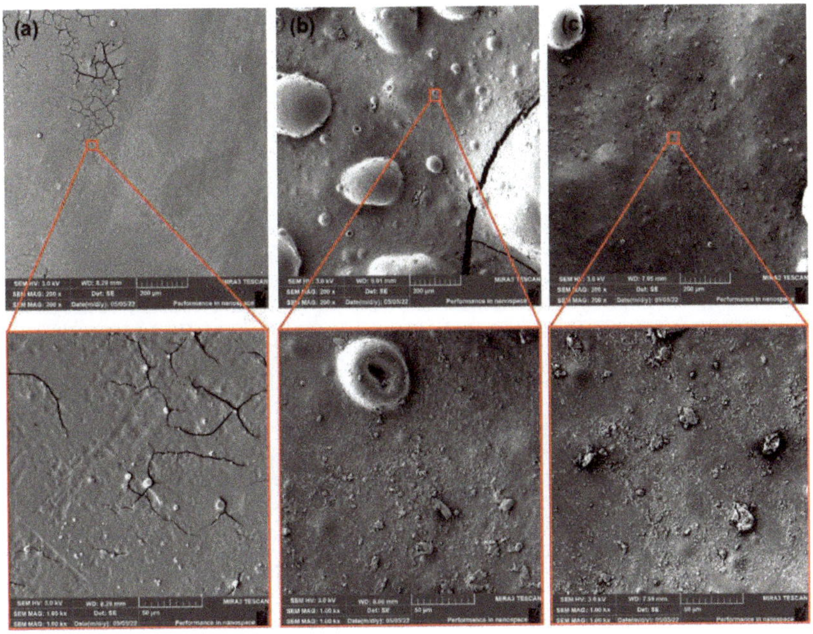

**Figure 11.** SEM images of aged PLA/PEG films (**a**) at initial state and after (**b**) 14 days of thermophilic composting, and (**c**) 28 days of thermophilic composting.

The PLA films that contained dicarboxylic acid-based plasticizer MC2178 showed (Figures 12 and 13) smooth surfaces after production and ageing process. The first 14 days of composting did not evoke any significant changes in surface roughness. However, the initiation of algae fibres growth was observed. If the as-produced and aged PLA/MC2178 films are compared, a slightly higher algae fibres ratio is observed for the aged films. The differences after 28 days of composting are more obvious. The higher activity of algae fibres could evoke faster disturbance of the surface, which could increase the kinetics of disintegration, which further, would lead to biotic attack and biodegradation of films. An even higher ratio of algae fibres was observed for PLA films plasticized with MC2192 after the first 14 days of composting (Figures 14 and 15). It is well known that the formation of lactic acid oligomers during chain scission of PLA increases the concentration of carboxylic acid end groups in the degradation medium. The catalytic action of these groups at their increasing content further results in a self-catalyzed and self-maintaining process [46]. Based on the increased content of carboxylic acid groups in both PLA/MC2178 and PLA/MC2192 when compared to other plasticizers, it could be assumed that they, during degradation, stimulate biotic attack. Furthermore, Ren et al. [22] reported that water molecules easily diffuse into amorphous regions, and these regions are also easily assimilated by microorganisms. Consequently, the easy assimilation of algae fibres could also be assumed. Another exposition (14 days) of PLA/MC2192 films to composting further enhanced the algae fibres activity. The hydrolytic degradation and microbial/enzymatic activity caused the flaking of fragments. Differences between as-produced and aged films were observed within the 28 days of composting.

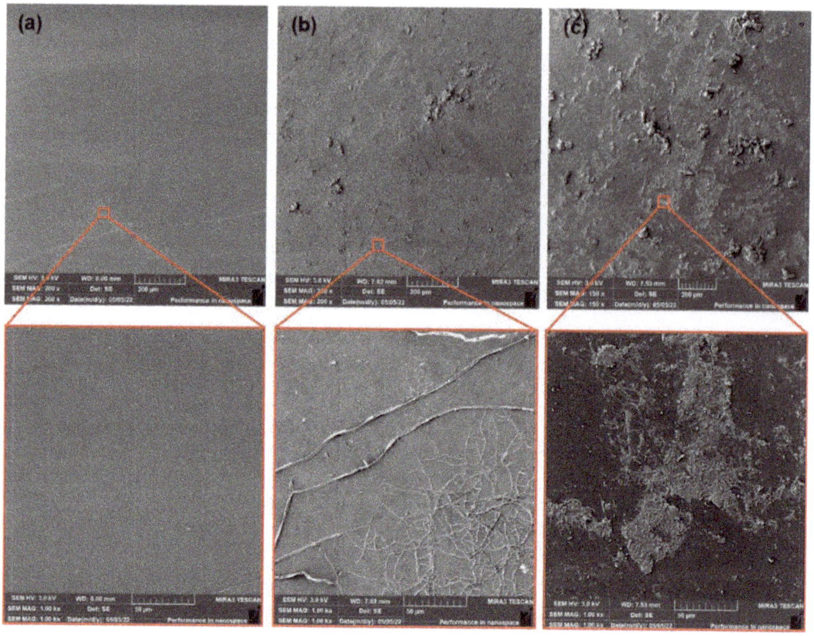


**Figure 12.** SEM images of as-produced PLA/MC 2178 films (**a**) at initial state and after (**b**) 14 days of thermophilic composting, and (**c**) 28 days of thermophilic composting.

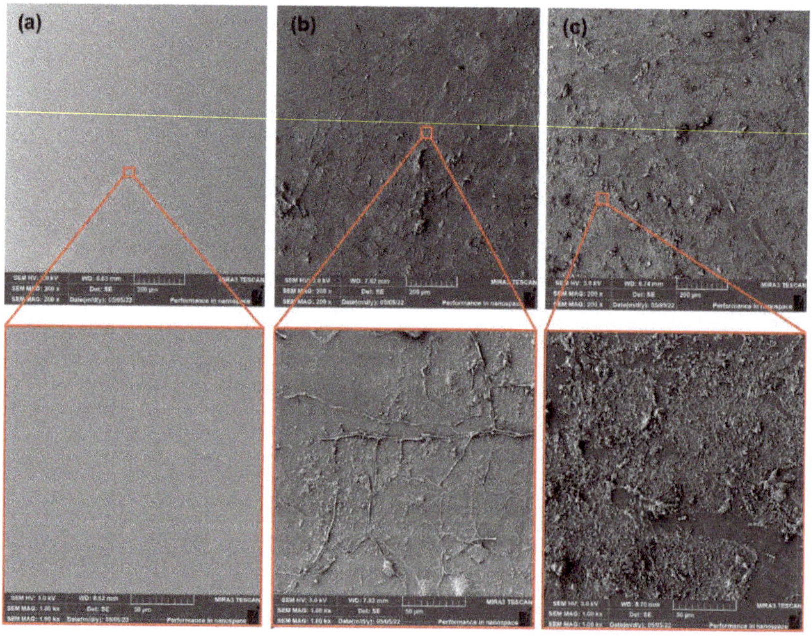


**Figure 13.** SEM images of aged PLA/MC2178 films (**a**) at initial state and after (**b**) 14 days of thermophilic composting, and (**c**) 28 days of thermophilic composting.

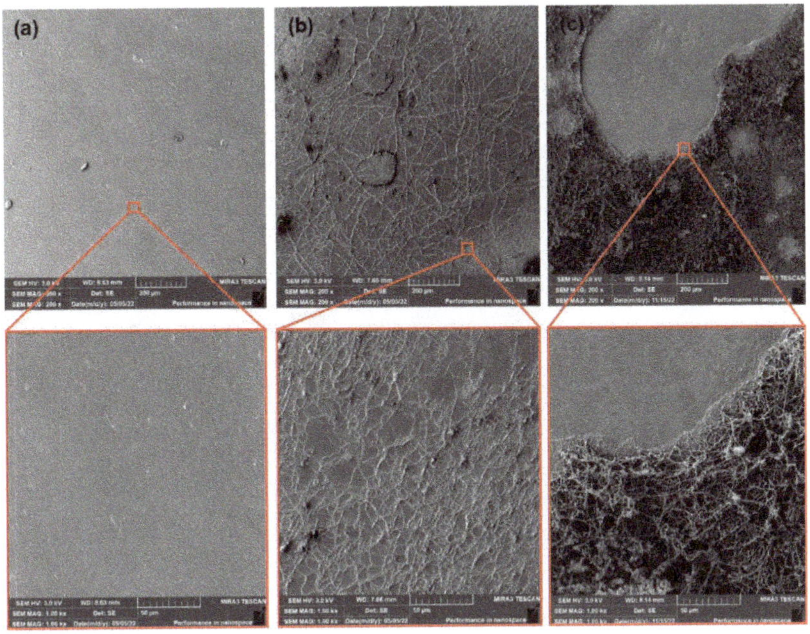

**Figure 14.** SEM images of as-produced PLA/MC2192 films (**a**) at initial state and after (**b**) 14 days of thermophilic composting, and (**c**) 28 days of thermophilic composting.

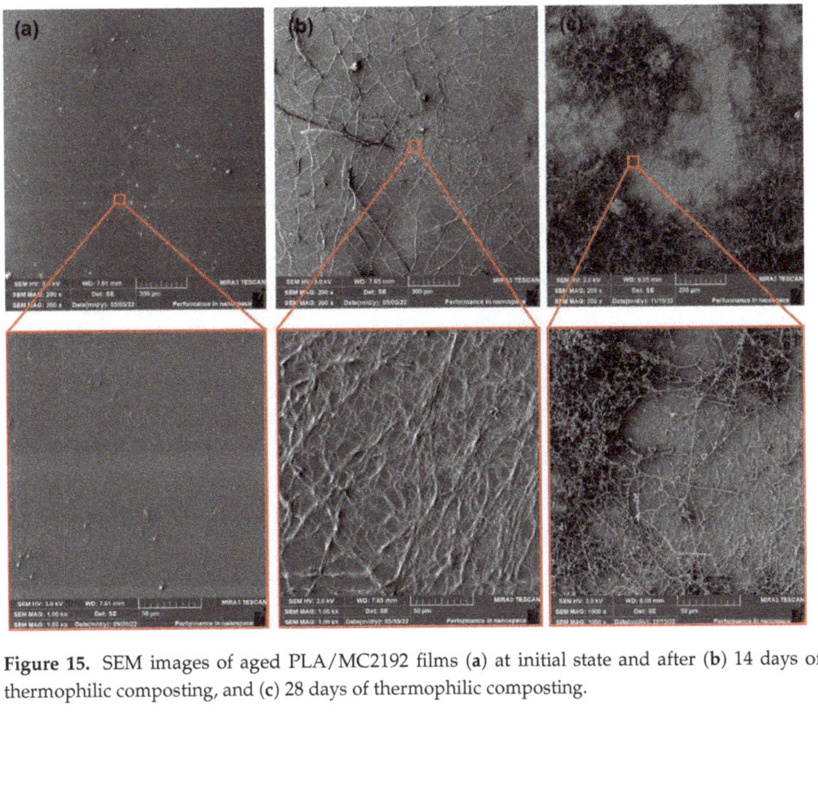

**Figure 15.** SEM images of aged PLA/MC2192 films (**a**) at initial state and after (**b**) 14 days of thermophilic composting, and (**c**) 28 days of thermophilic composting.

### 3.6. Fourier Transform Infrared Spectroscopy (FT-IR)

According to the evaluated results, no significant differences in the ATR-FT-IR spectra between as-produced and aged PLA and PLA-plasticized films were observed. Consequently, assessed and discussed are only the as-produced and composted ones (Figure 16). The ATR-FT-IR spectra of aged films are added in supplementary materials (Figures S5 and S6). Typical absorption bands for PLA, corresponding to the C=O stretching of ester groups at 1747 $cm^{-1}$, with asymmetric and symmetric $CH_3$ stretching at 2995 $cm^{-1}$ and 2945 $cm^{-1}$, the C-O stretching bands of –CH–O– at 1180 $cm^{-1}$, and –O–C=O groups at 1127 $cm^{-1}$, 1080 $cm^{-1}$, and 1043 $cm^{-1}$, respectively, were observed [15,47–49]. Moreover, similar to the work of Zaidi et al. [50], bending frequencies for $CH_3$ were identified at 1452 $cm^{-1}$, 1382 $cm^{-1}$, and 1359 $cm^{-1}$. As well, bands related to the C=O double-bound around 700 $cm^{-1}$ were further observed. When comparing the spectra of as-produced neat PLA films and PLA films plasticised with ATBC at the initial state and within 14 and 28 days of composting, no significant differences were noted. On the contrary, the spectra of as-produced PLA films plasticised with PEG and dicarboxylic acid-based plasticizers (MC2178 and MC2192) showed an obvious decrease in absorption intensity after 14 days of composting. According to results of Vasile et al. [51], the decrease in peak ratio declares chemical changes that could be evoked by the process of hydrolytic degradation (chain scission). Another 14 days of composting did not evoke any considerable changes in ATR-FT-IR spectra for PLA/PEG films. However, the dicarboxylic acid-based plasticizers in PLA showed another significant decrease in peak intensity. Especially, composted PLA films with MC2192 plasticizer showed a very low peak intensity after 28 days of composting. Consequently, extensive degradation could be, for these films, assumed.

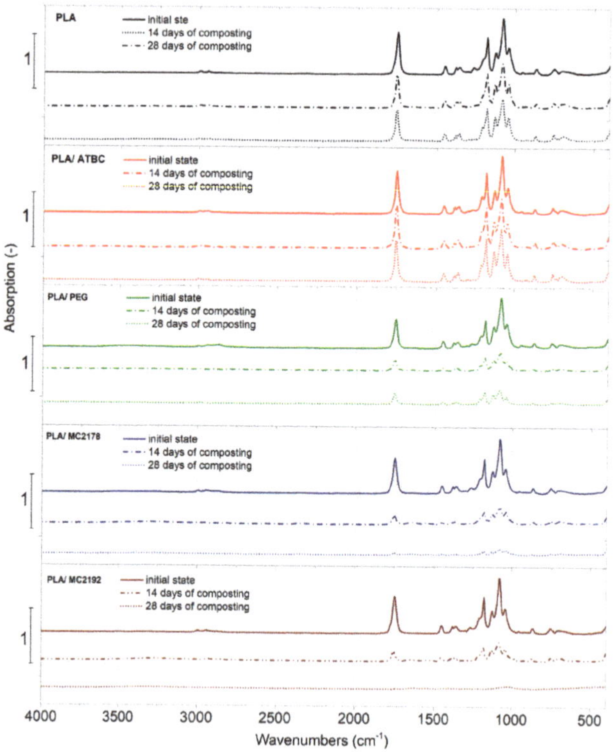


**Figure 16.** FTIR spectra of as-produced PLA films at initial state and after 14 days and 28 days of thermophilic composting.

Oliveira et al. [52] and Kammoun et al. [53] reported that, during PLA degradation, the appearance of hydroxyl bands around 3400 cm$^{-1}$ indicates the occurrence of hydrolysis degradation. Consequently, the comparison of changes in peak intensity at this area has been separately evaluated in Figure 17. Only minimal changes in hydroxyl bands were observed for neat PLA films and PLA films plasticized with ATBC within 28 days of composting. PEG is a water-soluble plasticizer with polar (-OH) groups. Therefore, the increase in hydroxyl bands was evaluated by the incorporation of this plasticizer into PLA films. Also, Rafie et al. [39] and Darie-Nita et al. [12] observed the enhancement of hydroxyl bands around 3400 cm$^{-1}$ if the PEG plasticizer was incorporated into PLA. According to the decreasing peak intensity of hydroxyl bands evaluated by the process of composting, the stated migration and release of plasticizer into the environment could be confirmed. Within the composting process, the significant increase in peak intensity of hydroxyl bands was for both dicarboxylic acid-based plasticizers, which were further observed. The rapid hydrolysis could consequently evoke faster disintegration, assimilation by microorganisms/enzymes, and faster biodegradation.

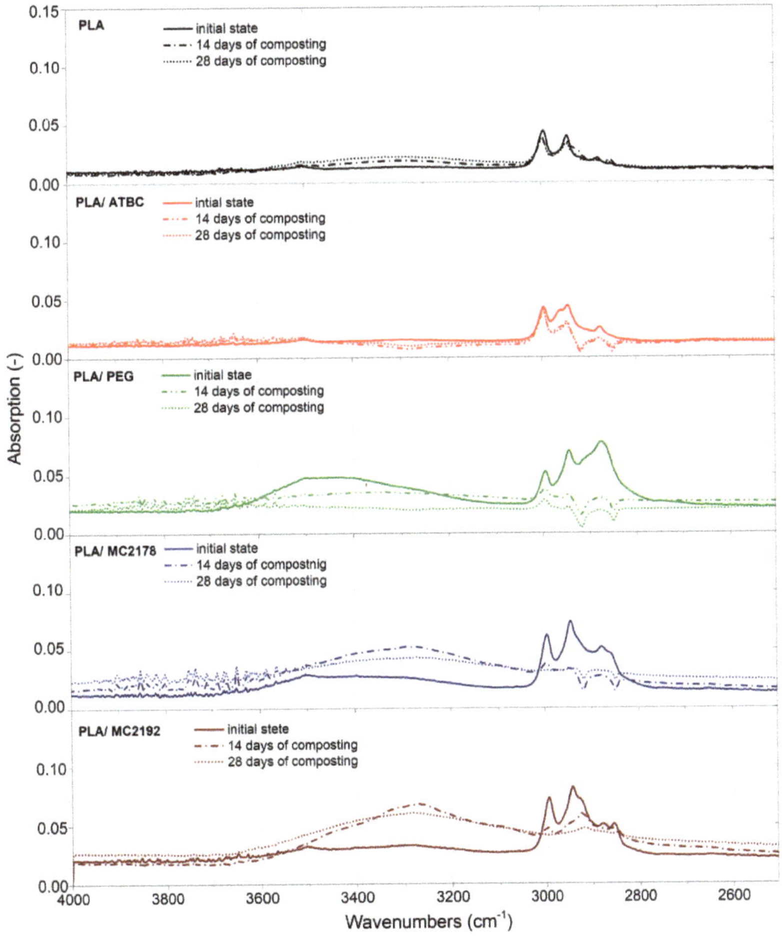


**Figure 17.** FTIR spectra of as-produced PLA films at initial state and after 14 days and 28 days of thermophilic composting that indicate hydroxyl bands.

### 3.7. Degree of Disintegration of Composted PLA Films

The evaluated results of weight loss of as-produced and aged PLA films after 14 and 28 days of thermophilic composting are shown in Figures 18 and 19. With respect to standard deviation, the as-produced and aged neat PLA films did not show any changes in weight after the first 14 days of thermophilic composting. Further exposition of as-produced and aged neat PLA films to the compost environment evoked considerable changes in thermal properties (DSC, TGA). However, any evident changes were obvious from SEM surface images. No considerable weight loss was found for these films after 28 days of composting. On the other hand, the as-produced PLA/PEG films showed about 20% weight loss after 14 days of composting. Regarding the previous conclusions, this weight loss could be ascribed mainly to the intensive release of plasticizer into the compost environment. Because the following exposition to the compost environment for another 14 days evoked only minimal changes in weight, the low tendency of as-produced PLA/PEG to disintegrate could be further assumed. Also, aged PLA/PEG films showed a high decrease in weight (about 21%) after the first 14 days of composting. Further composting of the aged PLA/PEG films evoked, in comparison to as-produced films, a higher weight decrease of about 28%. As it was introduced, during the aging process, some migration of PEG and minimal structure changes were found. Therefore, a lower ratio of releasing of plasticizer could be expected (SEM surface images, Figure 11). Consequently, the slightly higher decrease in the weight of aged PLA/PEG films was probably caused by higher disintegration rate. Significant differences in weight loss were evaluated for PLA films plasticised with dicarboxylic acid-based plasticizers within 28 days of composting. The higher molecular weight of MC2192 plasticizer evoked lower enhancement of crystalline structure of PLA films than for lower molecular ones (MC2178). PLA is polyester, where hydrolysis (bulk degradation) is one of the most critical factors for disintegration [54]. The water molecules easily diffuse through amorphous regions. Consequently, the PLA films plasticised with MC2192 should achieve a faster disintegration rate. The evaluated results for weight loss confirmed this presumption. The difference between the weight loss of as-produced films was, after 14 days of composting, low, being 3% weight loss for PLA/MC2178 films and 5% for PLA/MC2192 films. The only 4% enhancement of weight loss was after another 14 days of composting, evaluated for as-produced PLA/MC2178 films. However, about 47% weight loss was found for PLA films plasticised with MC2192. With respect to the evaluated standard deviation, we cannot state any relevant differences between as-produced and aged PLA/MC2192 films within 28 days of composting. Nevertheless, slightly higher increase in weight loss could be seen for aged PLA/MC2178 films than for as-produced ones. This could be caused (regarding the evaluated results of thermal analyses) by the releasing of plasticizer to the compost environment. Another reason could be higher microbial/enzymatic attack (the level of algae fibres observed on SEM surface images, Figures 12 and 13) that evoke a faster process of disintegration. The as-produced PLA/ATBC films was characterized by low enhancement of degree of crystallinity (similar level as for PLA/MC2192 films) and low tendency to migration. Therefore, easier hydrolysis and faster disintegration were expected. However, the evaluated weight loss was only about 3% after 14 days of composting and about 5% after 28 days of composting. Consequently, the chemical composition of plasticizers must be stated as another crucial aspect that influences the disintegration rate. The importance of chemical composition of plasticizer could be confirmed from results of weight loss of aged PLA/ATBC films. During the ageing process of PLA/ATBC films, a significant increase in structure order was concluded. Nevertheless, the weight reduction in aged PLA/ATBC films was achieved within 28 days of composting at a similar level to the as-produced ones.

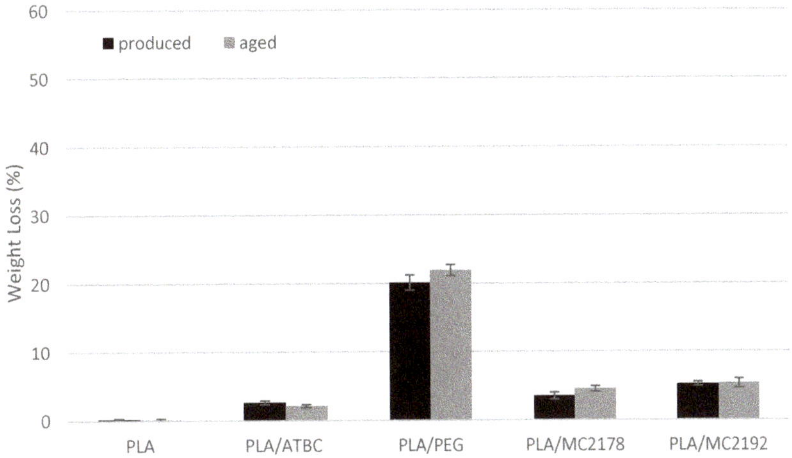


**Figure 18.** The weight loss of as-produced and aged PLA films after 14 days of composting.

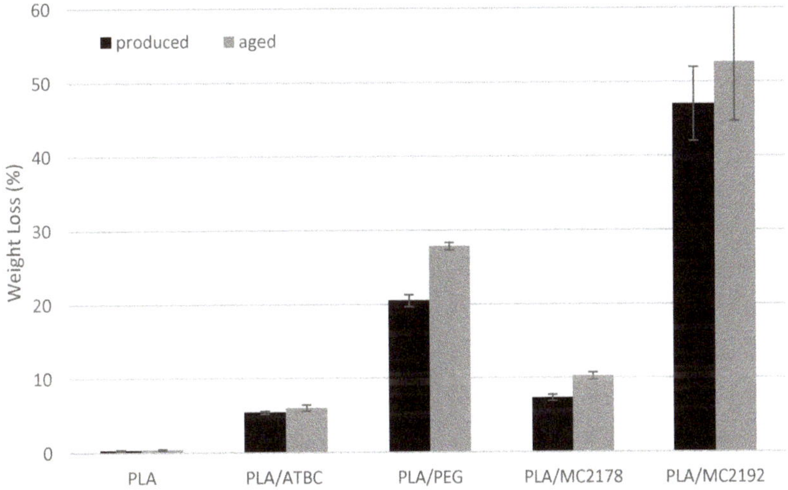


**Figure 19.** The weight loss of as-produced and aged PLA films after 28 days of composting.

### 3.8. Biodegradability under Thermophilic Composting

The evaluated biodegradation curves are shown in Figure 20. During the first 14 days of composting, similar courses and levels of biodegradation were evaluated for neat PLA films and PLA films plasticized with ATBC or dicarboxylic acid-based plasticizer MC 2178. Nevertheless, the PLA films plasticized with PEG and dicarboxylic acid-based plasticizer of higher molecular weight (MC2192) showed different biodegradation rates. The incorporation of PEG plasticizer caused a decrease in the biodegradation rate of PLA films. Polymer materials are microbially degradable in a two-step process. The first step consists of a reduction in the polymer chain into low molecular weight oligomers, dimers, and monomers that are short enough to be assimilated by microorganisms in the second step [54,55]. Consequently, the disintegration rate, which depends on the intensity of enzymatic and hydrolytic attack, has a crucial factor in biodegradation. As it was mentioned above, PEG is water-soluble hydrophilic plasticizer with polar groups (–OH). The hydrophilic character of plasticizer could increase the permeability of water and oxygen into the PLA films, as was presented in the study of Courgneau et al. [14]. Therefore, the incorporation of PEG plasticizer could increase the disintegration and biodegradation rate of PLA. However, Cao et al. [56] and Laboulfie et al. [57] reported that the water permeability depends on the molecular weight of the PEG plasticizer. The significantly lower water permeability was evaluated for films plasticized with a low molecular weight of PEG (PEG 300) compared to high molecular ones (PEG 4000). The reason for this could be hydrogen bonding between polar groups (-OH) of plasticizes and PLA. The high molecular weight PEGs might not be able to position themselves to create sufficient hydrogen bonds with polymers [58]. Other aspects that must be taken in consideration are the influence of plasticizer on microbial/enzymatic activity as well as the influence on the structure order (degree of crystallinity) of films. Kammoun et al. [53] evaluated the enhancement of the antibacterial activity of chitosan films if the PEG plasticizer was incorporated. However, it likely that the main reason for lower microbial activity (lower biodegradation rate) is the enhancement of crystallinity. As was introduced in Section 3.3, the low viscosity of PEG plasticizer evoked the highest enhancement of the degree of crystallinity from used plasticizers. The diffusion of water as well as enzymatic degradation take place primarily in the amorphous part. Consequently, slower biodegradation could be assumed.

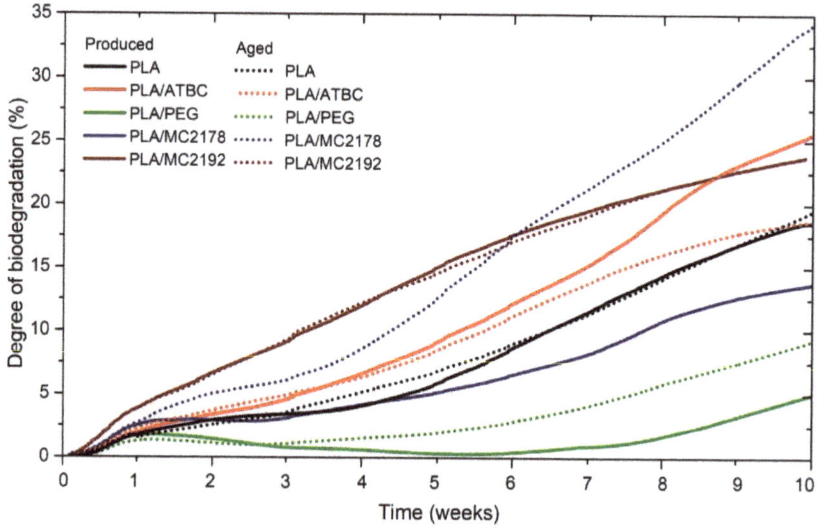


**Figure 20.** Biodegradation curves of as produced and aged PLA films under controlled thermophilic composting ISO 18455-1.

The high molecular weight of dicarboxylic acid-based plasticizer MC2192 caused the lowest enhancement of crystallinity degree of PLA films among used plasticizers. Thus, the higher content of amorphous parts, the higher molecular weight of plasticizer (lower interaction bonds with PLA), and the self-catalyzed action caused by increasing content of carboxylic acid end groups could be the main reason for the higher disintegration (weight loss) and biodegradation level. Comparing the biodegradation of films after accelerated ageing, no significant difference was observed at the first 14 days of composting. The exception was only PLA/MC2178, where a slight increase in biodegradation level has been found for aged films. This result corresponds to SEM surface images (Figures 12 and 13) where higher algae fibres activity was observed for aged films than for as-produced one. The biodegradation degree at 28 days of composing showed higher differences among the as-produced films. A degree of about 12% has been found for as-produced PLA/MC2192. The neat PLA films achieved about 4% biodegradation. The same results have been found for as-produced PLA/MC2178 films. Compared to neat PLA, a slight increase in the biodegradation degree was evaluated for as-produced PLA/ATBC films (biodegradation about 6%). ATBC is a water-insoluble plasticizer with better barrier properties against oxygen and water permeability than PEG [14,44]. Therefore, slower disintegration and biodegradation could be assumed. Nevertheless, the diffusion of water and oxygen in PLA film strongly depends on temperature. Courgenau et al. [14] evaluated extensive differences in water and oxygen diffusion of PLA/PEG and PLA/ATBC films at 25 °C. However, relatively low differences were evaluated at 38 °C. The increasing temperature has a similar effect as the plasticizing of PLA. The chain mobility is increased, and intermolecular forces decrease. Therefore easier diffusion of water and oxygen is achieved [14]. Similar to PLA/MC2192 films, the degree of crystallinity could be assumed to be the major factor for the higher biodegradation rate of PLA/ATBC. Compared to PLA/PEG films, a very small enhancement of the degree of crystallinity was observed for as-produced PLA/ATBC films. Consequently, easier diffusion of water molecules, easier assimilation of microorganisms, and faster disintegration and biodegradation were achieved. As-produced PLA/PEG films showed minimal disintegration and biodegradation degree. The differences in biodegradation between as-produced and aged films (expect PLA/MC2178) are still too small to make any valuable statement. However, at the end of the experiment (70 days), there are obvious differences between aged PLA/PEG and PLA/ATBC. Due to the intensive increase in degree of crystallinity for aged PLA/ATBC films, a considerably lower biodegradation rate than that for as-produced ones was achieved. The aged PLA/ATBC films achieved 19% biodegradation and as-produced ones reached up to 26% of biodegradation degree. According to the result of DSC analysis, no significant changes in crystallinity degree were observed for aged PLA/PEG films. Despite these findings, a slightly faster biodegradation rate is evident from the biodegradation curse, where about 5% enhancement of biodegradation level has been found for aged films. The reason for this could be a high tendency of PEG plasticizer to migration. Both Courgenau et al. [14] and Mariana et al. [44] reported that the migration of plasticizer evokes a decrease in hydrogen bonding between PLA/PEG and enhanced diffusion of water or oxygen. The highest level of biodegradation (34%) has been evaluated for aged PLA films plasticized with MC2178. On the contrary, only 14% biodegradation has been found for as-produced ones. This phenomenon could be ascribed to the high tendency to migration of MC2178 plasticizer, relatively small change in crystallinity degree during the aging process, and the chain scission of macromolecules which results in an easier water and oxygen diffusion process that evokes higher microbial/enzymatic attack. Another aspect could be the chemical composition and bonding between PLA and plasticizers. According to the SEM surface images, the dicarboxylic acid-based plasticizers that contain two carboxyl groups (–COOH) evoked evidently higher biotic activity. Consequently, faster disintegration and microbial degradation was achieved. The minimal differences in biodegradation were evaluated for as-produced and aged neat PLA films. Both films achieved about 18% biodegradation after 70 days of thermophilic composting. A slightly higher biodegradation rate (biodegradation of about 31% biodegradation within the same time period) was evaluated for as-produced neat PLA films in our previous study [40]. The faster biodegradation rate could be caused by the used PLLA. Luminy L130, used in this study, has a higher content of L-isomer (99% L-isomer stereochemical purity) than Ingeo 3001D (95 wt. % of L- lactide), which was used in the previous experiment. The stereochemical purity influences the structure order (Section 3.3) as well as biodegradation rate. Cadar et al. [59] reported some correlation between the composting and the level of biodegradation of PLA-based copolymers. The biodegradation of copolymers containing higher amounts of lactic acid was found to be faster than the biodegradation of copolymers containing smaller amounts.

## 4. Conclusions

According to the evaluated results, the molecular weight and chemical composition of plasticizers are the main aspects that predetermine the future mechanical, rheological, and thermal properties of PLA films. In addition to the effectiveness of plasticizer, the stability of properties is an extremely important element for their applicability. Therefore, the influence of ageing must not be neglected. Consequently, the influence of accelerated ageing on mechanical, rheological, and thermal properties of PLA films plasticized with acetyl tributyl citrate (ATBC), polyethylene glycol (PEG), and nature-based dicarboxylic acid-based plasticizers (MC2178 and MC2192) were evaluated. The simulation of solar radiation at ground level that is defined in standard IEC 60068-2-5 evoked minimal changes in mechanical properties of PLA films plasticized with 15 wt. % of dicarboxylic acid-based plasticizer MC2192 and relatively low changes in rheological properties. Also, from the thermal analysis, no considerable changes were reported. In contrast, for PLA films plasticized with a dicarboxylic acid-based plasticizer of lower molecular weight (MC2178), some mechanical, rheological, and structural changes and migration tendencies were observed. Even greater changes in thermal and mechanical properties were achieved for PLA/ATBC. However, the highest changes were evaluated if the PEG plasticizer (15 wt. %) was incorporated into PLA film. The migration of PEG plasticizer could be ascribed as the major reason behind these changes.

Furthermore, the influence of plasticizer and accelerated ageing on biodegradation during the thermophilic composting (ISO 14855-1) of PLA films was evaluated. From the estimated results, the chemical composition and structure order (degree of crystallinity) could be highlighted as the most important factors. PEG plasticizer is a water-soluble hydrophilic plasticizer with polar groups (–OH). Consequently, intensive disintegration and biodegradation could be assumed. However, a low biodegradation rate was observed during the thermophilic composting. The used PEG 400 is characterized by a low molecular weight that evoked high structure order changes (high degree of crystallinity) in PLA films. Because the diffusion of water, as well as enzymatic degradation, takes place primarily in the amorphous phase, the low disintegration and microbial activity were evaluated. The aged PLA/PEG films evoked, when compared to as-produced films, a slightly higher biodegradation rate. The reason for this could be a high tendency of PEG plasticizer to migrate. The migration of plasticizer spurs a decrease in hydrogen bonding between PLA/PEG and ensures easy diffusion of water or oxygen. In contrast, the high molecular weight of dicarboxylic acid-based plasticizer MC2192 caused very low enhancement of the crystallinity degree of PLA films. Consequently, the higher content of amorphous parts and high molecular weight (lower interaction bonds with PLA) caused faster disintegration and biodegradation. The plasticizer MC2178, due to its lower molecular weight compared to MC2192, evoked higher structure order enhancement of as-produced PLA films. Therefore, the lower biodegradation rate of films was evaluated. The accelerated ageing of PLA/MC2178 films evoked some migration/releasing tendency of plasticizers that caused easier hydrolysis and increases in microbial activity (biodegradation rate). The influence of structure order could be further confirmed by the results of the biodegradation of as-produced PLA/ATBC films. The water-insoluble ATBC plasticizer evoked only low crystallinity degree enhancement of PLA films. Consequently, a faster biodegradation rate than for PLA films, with a higher structure order (as produced PLA/PEG, PLA/MC2178), was observed. After the accelerated ageing process, a significant increase in structure order and a minimal tendency to migration was observed. Therefore, a decrease of biodegradation rate was observed.

**Supplementary Materials:** The following supporting information can be downloaded at: https://www.mdpi.com/article/10.3390/polym15010140/s1, Figure S1: FTIR spectra of aged PLA films at initial state and after 14 days and 28 days of thermophilic composting; Figure S2: FTIR spectra of aged PLA films at initial state and after 14 days and 28 days of thermophilic composting that indicate hydroxyl bands; Figure S3: DSC curves of as produced neat and plasticized PLA films within 28 days of thermophilic composting; Figure S4: DSC curves of aged neat and plasticized PLA films within 28 days of thermophilic composting; Figure S5: TGA curves of as produced neat and plasticized PLA films within 28 days of thermophilic composting; Figure S6: TGA curves of aged neat and plasticized PLA films within 28 days of thermophilic composting.

**Author Contributions:** Conceptualisation, P.B.; methodology, P.B., M.B. and J.N.; validation, P.B., M.B. and J.N.; formal analysis, P.B. and M.B.; investigation, P.B., M.B. and J.N.; data curation, P.B., M.B. and J.N.; writing—original draft preparation, P.B. and M.B.; writing—review and editing, P.B. and M.B.; supervision, P.L. and L.B.; project administration, P.B.; funding acquisition, P.B., J.N., M.B., L.B. and P.L. All authors have read and agreed to the published version of the manuscript.

**Funding:** This research was funded by Investment Funds in the frames of Operational Program Research, Development and Education—project Hybrid Materials for Hierarchical Structures [HyHi, Reg. No. CZ.02.1.01/0.0/0.0/16_019/0000843].

**Institutional Review Board Statement:** Not applicable.

**Informed Consent Statement:** Not applicable.

**Data Availability Statement:** The data presented in this study are available on request from the corresponding author.

**Conflicts of Interest:** The authors declare no conflict of interest.

## References

1. Plastic-the Facts 2021. [online]. B.m.: Plastic Europe Association of Plastic Manufactures. 2021. Available online: http://plasticseurope.org/knowledge-hub/plastics-the-facts-2021 (accessed on 1 September 2022).
2. Hao, Y.; Shafer, G. Models for Predicting Global Plastic Waste. *Aresty RURJ* **2021**, *1*, 57–61. [CrossRef]
3. Becker, N.; Siebert-Raths, A. *Biopolymers Facts and Statistics 2021*; Hochschule Hannover: Hannover, Germany, 2021.
4. Rahman, M.; Brazel, C.S. The Plasticizer Market: An Assessment of Traditional Plasticizers and Research Trends to Meet New Challenges. *Prog. Polym. Sci.* **2004**, *29*, 1223–1248. [CrossRef]
5. Vieira, M.G.A.; da Silva, M.A.; dos Santos, L.O.; Beppu, M.M. Natural-Based Plasticizers and Biopolymer Films: A Review. *Eur. Polym. J.* **2011**, *47*, 254–263. [CrossRef]
6. Jacobsen, S.; Fritz, H.-G. Plasticizing Polylactide—the Effect of Different Plasticizers on the Mechanical Properties. *Polym. Eng. Sci.* **1999**, *39*, 1303–1310. [CrossRef]
7. Al-Mulla, E.A.J.; Yunus, W.M.; Wan, Z.; Ibrahim, N.A.B.; Rahman, M.Z.A. Properties of Epoxidized Palm Oil Plasticized Polytlactic Acid. *J. Mater. Sci.* **2010**, *45*, 1942–1946. [CrossRef]
8. Carbonell-Verdu, A.; Samper, M.D.; Garcia-Garcia, D.; Sanchez-Nacher, L.; Balart, R. Plasticization Effect of Epoxidized Cottonseed Oil (ECSO) on Poly (Lactic Acid). *Ind. Crop. Prod.* **2017**, *104*, 278–286. [CrossRef]
9. Ljungberg, N.; Wesslen, B. Tributyl Citrate Oligomers as Plasticizers for Poly (Lactic Acid): Thermo-Mechanical Film Properties and Aging. *Polym. J.* **2003**, *44*, 7679–7688. [CrossRef]
10. Maiza, M.; Benaniba, M.T.; Quintard, G.; Massardier-Nageotte, V. Biobased Additive Plasticizing Polylactic Acid (PLA). *Polimeros* **2015**, *25*, 581–590. [CrossRef]
11. Hassouna, F.; Raquez, J.-M.; Addiego, F.; Dubois, P.; Toniazzo, V.; Ruch, D. New Approach on the Development of Plasticized Polylactide (PLA): Grafting of Poly (Ethylene Glycol)(PEG) via Reactive Extrusion. *Eur. Polym. J.* **2011**, *47*, 2134–2144. [CrossRef]
12. Darie-Niţă, R.N.; Vasile, C.; Irimia, A.; Lipşa, R.; Râpă, M. Evaluation of Some Eco-friendly Plasticizers for PLA Films Processing. *J. Appl. Polym. Sci.* **2016**, *133*, 43223. [CrossRef]
13. Chieng, B.W.; Azowa, I.N.; Yunus, W.; Wan, M.Z.; Hussein, M.Z. Effects of Graphene Nanopletelets on Poly (Lactic Acid)/Poly (Ethylene Glycol) Polymer Nanocomposites. In Proceedings of the Advanced Materials Research; Trans Tech Publications: Zürich, Switzerland, 2014; Volume 1024, pp. 136–139.
14. Courgneau, C.; Domenek, S.; Guinault, A.; Avérous, L.; Ducruet, V. Analysis of the Structure-Properties Relationships of Different Multiphase Systems Based on Plasticized Poly (Lactic Acid). *J. Polym. Environ.* **2011**, *19*, 362–371. [CrossRef]
15. Sessini, V.; Palenzuela, M.; Damián, J.; Mosquera, M.E. Bio-Based Polyether from Limonene Oxide Catalytic ROP as Green Polymeric Plasticizer for PLA. *Polym. J.* **2020**, *210*, 123003. [CrossRef]
16. Kodal, M.; Sirin, H.; Ozkoc, G. Long-and Short-Term Stability of Plasticized Poly (Lactic Acid): Effects of Plasticizers Type on Thermal, Mechanical and Morphological Properties. *Polym. Bull.* **2019**, *76*, 423–445. [CrossRef]
17. Tsou, C.-H.; Suen, M.-C.; Yao, W.-H.; Yeh, J.-T.; Wu, C.-S.; Tsou, C.-Y.; Chiu, S.-H.; Chen, J.-C.; Wang, R.Y.; Lin, S.-M. Preparation and Characterization of Bioplastic-Based Green Renewable Composites from Tapioca with Acetyl Tributyl Citrate as a Plasticizer. *Materials* **2014**, *7*, 5617–5632. [CrossRef]
18. Burgos, N.; Martino, V.P.; Jiménez, A. Characterization and Ageing Study of Poly (Lactic Acid) Films Plasticized with Oligomeric Lactic Acid. *Polym. Degrad. Stab.* **2013**, *98*, 651–658. [CrossRef]
19. Itävaara, M.; Karjomaa, S.; Selin, J.-F. Biodegradation of Polylactide in Aerobic and Anaerobic Thermophilic Conditions. *Chemosphere* **2002**, *46*, 879–885. [CrossRef]
20. Kale, G.; Kijchavengkul, T.; Auras, R.; Rubino, M.; Selke, S.E.; Singh, S.P. Compostability of Bioplastic Packaging Materials: An Overview. *Macromol. Biosci.* **2007**, *7*, 255–277. [CrossRef]

21. Kolstad, J.J.; Vink, E.T.; De Wilde, B.; Debeer, L. Assessment of Anaerobic Degradation of Ingeo$^{TM}$ Polylactides under Accelerated Landfill Conditions. *Polym. Degrad. Stab.* **2012**, *97*, 1131–1141. [CrossRef]
22. Ren, Y.; Hu, J.; Yang, M.; Weng, Y. Biodegradation Behavior of Poly (Lactic Acid)(PLA), Poly (Butylene Adipate-Co-Terephthalate)(PBAT), and Their Blends under Digested Sludge Conditions. *J. Polym. Environ.* **2019**, *27*, 2784–2792. [CrossRef]
23. Poly(ethylene glycol) Product Specification (Product Number 202398). [online]. B.m.: Merck KGaA, Darmstadt, Germany. 2022. Available online: http://www.sigmaaldrich.com (accessed on 1 March 2022).
24. Sequeira, M.C.; Pereira, M.F.; Avelino, H.M.; Caetano, F.J.; Fareleira, J.M. Viscosity Measurements of Poly (Ethyleneglycol) 400 [PEG 400] at Temperatures from 293 K to 348 K and at Pressures up to 50 MPa Using the Vibrating Wire Technique. *Fluid Phase Equilib.* **2019**, *496*, 7–16. [CrossRef]
25. Wypych, A. *Databook of Plasticizers*; Elsevier: Amsterdam, The Netherlands, 2017; ISBN 1-927885-15-9.
26. Citroflex A-4 Data Sheet. [online]. B.m.: Special Chem. 2022. Available online: http://polymer-additives.specialchem.com/product/a-vertellus-specialties-citroflex-a4 (accessed on 10 January 2022).
27. Gibbons, W.S.; Kusy, R.P. Influence of Plasticizer Configurational Changes on the Dielectric Characteristics of Highly Plasticized Poly (Vinyl Chloride). *Polym. J.* **1998**, *39*, 3167–3178. [CrossRef]
28. MC 2178 Technical Data Sheet. [online]. B.m.: Emery Oleochemicals. 2022. Available online: http://greenpolymeradditives.emeryoleo.com (accessed on 15 January 2022).
29. MC 2192 Technical Data Sheet. [online]. B.m.: Emery Oleochemicals. 2022. Available online: http://greenpolymeradditives.emeryoleo.com (accessed on 15 January 2022).
30. Average Values of Air [online]. Available online: http://www.qpro.cz/Prumerny-stav-vzduchu-dle-vyberu-hodin (accessed on 1 October 2022).
31. Râpă, M.; Miteluţ, A.C.; Tănase, E.E.; Grosu, E.; Popescu, P.; Popa, M.E.; Rosnes, J.T.; Sivertsvik, M.; Darie-Niţă, R.N.; Vasile, C. Influence of Chitosan on Mechanical, Thermal, Barrier and Antimicrobial Properties of PLA-Biocomposites for Food Packaging. *Compos. Part B Eng.* **2016**, *102*, 112–121. [CrossRef]
32. Gálvez, J.; Correa Aguirre, J.P.; Hidalgo Salazar, M.A.; Vera Mondragón, B.; Wagner, E.; Caicedo, C. Effect of Extrusion Screw Speed and Plasticizer Proportions on the Rheological, Thermal, Mechanical, Morphological and Superficial Properties of PLA. *Polymers* **2020**, *12*, 2111. [CrossRef] [PubMed]
33. Greco, A.; Ferrari, F. Thermal Behavior of PLA Plasticized by Commercial and Cardanol-Derived Plasticizers and the Effect on the Mechanical Properties. *J. Therm. Anal. Calorim.* **2021**, *146*, 131–141. [CrossRef]
34. Aliotta, L.; Vannozzi, A.; Panariello, L.; Gigante, V.; Coltelli, M.-B.; Lazzeri, A. Sustainable Micro and Nano Additives for Controlling the Migration of a Biobased Plasticizer from PLA-Based Flexible Films. *Polymers* **2020**, *12*, 1366. [CrossRef]
35. Rapa, M.; Darie-Nita, R.N.; Irimia, A.M.; Sivertsvik, M.; Rosnes, J.T.; Trifoi, A.R.; Vasile, C.; Tanase, E.E.; Gherman, T.; Popa, M.E. Comparative Analysis of Two Bioplasticizers Used to Modulate the Properties of PLA Biocomposites. *Mater. Plast.* **2017**, *54*, 610–615. [CrossRef]
36. Hu, Y.; Rogunova, M.; Topolkaraev, V.; Hiltner, A.; Baer, E. Aging of Poly (Lactide)/Poly (Ethylene Glycol) Blends. Part 1. Poly (Lactide) with Low Stereoregularity. *Polym. J.* **2003**, *44*, 5701–5710. [CrossRef]
37. Hu, Y.; Hu, Y.S.; Topolkaraev, V.; Hiltner, A.; Baer, E. Aging of Poly (Lactide)/Poly (Ethylene Glycol) Blends. Part 2. Poly (Lactide) with High Stereoregularity. *Polym. J.* **2003**, *44*, 5711–5720. [CrossRef]
38. Farah, S.; Anderson, D.G.; Langer, R. Physical and Mechanical Properties of PLA, and Their Functions in Widespread Applications—A Comprehensive Review. *Adv. Drug Deliv. Rev.* **2016**, *107*, 367–392. [CrossRef]
39. Rafie, M.A.F.; Marsilla, K.K.; Hamid, Z.A.A.; Rusli, A.; Abdullah, M.K. Enhanced Mechanical Properties of Plasticized Polylactic Acid Filament for Fused Deposition Modelling: Effect of in Situ Heat Treatment. *Prog. Rubber Plast. Rcycl. Technol.* **2020**, *36*, 131–142. [CrossRef]
40. Brdlík, P.; Borůvka, M.; Běhálek, L.; Lenfeld, P. Biodegradation of Poly (Lactic Acid) Biocomposites under Controlled Composting Conditions and Freshwater Biotope. *Polymers* **2021**, *13*, 594. [CrossRef] [PubMed]
41. Wu, H.; Nagarajan, S.; Zhou, L.; Duan, Y.; Zhang, J. Synthesis and Characterization of Cellulose Nanocrystal-Graft-Poly (D-Lactide) and Its Nanocomposite with Poly (L-Lactide). *Polym. J.* **2016**, *103*, 365–375. [CrossRef]
42. Jiménez, A.; Peltzer, M.; Ruseckaite, R. *Poly (Lactic Acid) Science and Technology: Processing, Properties, Additives and Applications*; Royal Society of Chemistry: London, UK, 2014; ISBN 1-84973-879-3.
43. Erceg, M.; KovaČiĆ, T.; KlariĆ, I. Thermal Degradation of Poly (3-Hydroxybutyrate) Plasticized with Acetyl Tributyl Citrate. *Polym. Degrad. Stab.* **2005**, *90*, 313–318. [CrossRef]
44. Arrieta, M.P.; Samper, M.D.; López, J.; Jiménez, A. Combined Effect of Poly (Hydroxybutyrate) and Plasticizers on Polylactic Acid Properties for Film Intended for Food Packaging. *J. Polym. Environ.* **2014**, *22*, 460–470. [CrossRef]
45. Arrieta, M.P.; López, J.; Rayón, E.; Jiménez, A. Disintegrability under Composting Conditions of Plasticized PLA–PHB Blends. *Polym. Degrad. Stab.* **2014**, *108*, 307–318. [CrossRef]
46. Gorrasi, G.; Pantani, R. Hydrolysis and Biodegradation of Poly (Lactic Acid). In *Synthesis, Structure and Properties of Poly (Lactic Acid)*; Springer: Cham, Switzerland, 2017; pp. 119–151.
47. Amorin, N.S.; Rosa, G.; Alves, J.F.; Gonçalves, S.P.; Franchetti, S.M.; Fechine, G.J. Study of Thermodegradation and Thermostabilization of Poly (Lactide Acid) Using Subsequent Extrusion Cycles. *J. App. Polym. Sci.* **2014**, *131*, 40023. [CrossRef]

48. Weng, Y.-X.; Wang, L.; Zhang, M.; Wang, X.-L.; Wang, Y.-Z. Biodegradation Behavior of P (3HB, 4HB)/PLA Blends in Real Soil Environments. *Polym. Test.* **2013**, *32*, 60–70. [CrossRef]
49. Lee, J.C.; Moon, J.H.; Jeong, J.-H.; Kim, M.Y.; Kim, B.M.; Choi, M.-C.; Kim, J.R.; Ha, C.-S. Biodegradability of Poly (Lactic Acid)(PLA)/Lactic Acid (LA) Blends Using Anaerobic Digester Sludge. *Macromol. Res.* **2016**, *24*, 741–747. [CrossRef]
50. Zaidi, L.; Kaci, M.; Bruzaud, S.; Bourmaud, A.; Grohens, Y. Effect of Natural Weather on the Structure and Properties of Polylactide/Cloisite 30B Nanocomposites. *Polym. Degrad. Stab.* **2010**, *95*, 1751–1758. [CrossRef]
51. Vasile, C.; Pamfil, D.; Râpă, M.; Darie-Niţă, R.N.; Mitelut, A.C.; Popa, E.E.; Popescu, P.A.; Draghici, M.C.; Popa, M.E. Study of the Soil Burial Degradation of Some PLA/CS Biocomposites. *Compos. Part B Eng.* **2018**, *142*, 251–262. [CrossRef]
52. Oliveira, M.; Santos, E.; Araújo, A.; Fechine, G.J.; Machado, A.V.; Botelho, G. The Role of Shear and Stabilizer on PLA Degradation. *Polym. Test.* **2016**, *51*, 109–116. [CrossRef]
53. Kammoun, M.; Haddar, M.; Kallel, T.K.; Dammak, M.; Sayari, A. Biological Properties and Biodegradation Studies of Chitosan Biofilms Plasticized with PEG and Glycerol. *Int. J. Biol. Macromol.* **2013**, *62*, 433–438. [CrossRef] [PubMed]
54. Aitor, L.; Erlantz, L. A Review on the Thermomechanical Properties and Biodegradation Behaviour of Polyester. *Eur. Polym. J.* **2019**, *121*, 109296.
55. Salomez, M.; George, M.; Fabre, P.; Touchaleaume, F.; Cesar, G.; Lajarrige, A.; Gastaldi, E. A Comparative Study of Degradation Mechanisms of PHBV and PBSA under Laboratory-Scale Composting Conditions. *Polym. Degrad. Stab.* **2019**, *167*, 102–113. [CrossRef]
56. Cao, N.; Yang, X.; Fu, Y. Effects of Various Plasticizers on Mechanical and Water Vapor Barrier Properties of Gelatin Films. *Food Hydrocoll.* **2009**, *23*, 729–735. [CrossRef]
57. Laboulfie, F.; Hémati, M.; Lamure, A.; Diguet, S. Effect of the Plasticizer on Permeability, Mechanical Resistance and Thermal Behaviour of Composite Coating Films. *Powder Technol.* **2013**, *238*, 14–19. [CrossRef]
58. Turhan, K.N.; Sahbaz, F.; Güner, A. A Spectrophotometric Study of Hydrogen Bonding in Methylcellulose-based Edible Films Plasticized by Polyethylene Glycol. *J. Food Sci.* **2001**, *66*, 59–62. [CrossRef]
59. Cadar, O.; Paul, M.; Roman, C.; Miclean, M.; Majdik, C. Biodegradation Behaviour of Poly (Lactic Acid) and (Lactic Acid-Ethylene Glycol-Malonic or Succinic Acid) Copolymers under Controlled Composting Conditions in a Laboratory Test System. *Polym. Degrad. Stab.* **2012**, *97*, 354–357. [CrossRef]

*polymers*

MDPI

*Article*

# Ketorolac-Loaded PLGA-/PLA-Based Microparticles Stabilized by Hyaluronic Acid: Effects of Formulation Composition and Emulsification Technique on Particle Characteristics and Drug Release Behaviors

**Amaraporn Wongrakpanich [1], Nichakan Khunkitchai [2], Yanisa Achayawat [2] and Jiraphong Suksiriworapong [1,*]**

[1] Department of Pharmacy, Faculty of Pharmacy, Mahidol University, Bangkok 10400, Thailand
[2] Doctor of Pharmacy Program, Faculty of Pharmacy, Mahidol University, Bangkok 10400, Thailand
* Correspondence: jiraphong.suk@mahidol.edu

**Abstract:** This study aimed to develop ketorolac microparticles stabilized by hyaluronic acid based on poly(lactide-co-glycolide) (PLGA), poly(lactide) (PLA), and their blend for further application in osteoarthritis. The polymer blend may provide tailored drug release and improved physicochemical characteristics. The microparticles were prepared by water-in-oil-in-water (w/o/w) double emulsion solvent evaporation using two emulsification techniques, probe sonication (PS) and high-speed stirring (HSS), to obtain the microparticles in different size ranges. The results revealed that the polymer composition and emulsification technique influenced the ketorolac microparticle characteristics. The PS technique provided significantly at least 20 times smaller average size (1.3–2.2 μm) and broader size distribution (1.5–8.5) than HSS (45.5–67.4 μm and 1.0–1.4, respectively). The encapsulation efficiency was influenced by the polymer composition and the emulsification technique, especially in the PLA microparticles. The DSC and XRD results suggested that the drug was compatible with and molecularly dissolved in the polymer matrix. Furthermore, most of the drug molecules existed in an amorphous form, and some in any crystalline form. All of the microparticles had biphasic drug release composed of the burst release within the first 2 h and the sustained release over 35 days. The obtained microparticles showed promise for further use in the treatment of osteoarthritis.

**Keywords:** microparticle; ketorolac; PLGA; PLA; drug release; osteoarthritis

**Citation:** Wongrakpanich, A.; Khunkitchai, N.; Achayawat, Y.; Suksiriworapong, J. Ketorolac-Loaded PLGA-/PLA-Based Microparticles Stabilized by Hyaluronic Acid: Effects of Formulation Composition and Emulsification Technique on Particle Characteristics and Drug Release Behaviors. *Polymers* **2023**, *15*, 266. https://doi.org/10.3390/polym15020266

Academic Editor: Vsevolod Aleksandrovich Zhuikov

Received: 21 November 2022
Revised: 29 December 2022
Accepted: 30 December 2022
Published: 4 January 2023



## 1. Introduction

Microparticles have generally been developed and extensively applied to treat various diseases. It is of interest to fabricate biodegradable polymeric microparticles since numerous compounds, including peptides, pharmaceutical proteins, and hydrophobic medicines, can be encapsulated in the microparticles and administered into the body. Customizing the physicochemical properties of polymers enables control of the release characteristics of such delivery systems, which can be varied over the desirable period from days to months. Poly(lactide-co-glycolide) (PLGA) and poly(lactide) (PLA) are the most frequently utilized for the fabrication of delivery systems, including microparticles, thanks to their biodegradability, biocompatibility, various unique characteristics, and human use approval by USFDA [1]. PLGA or PLA microparticles are typically created using various methods, for example, emulsification solvent extraction/evaporation, spray drying, and microfluidic [2]. Emulsification solvent evaporation has been extensively employed to create microparticles after the solidification of particle core by solvent evaporation. Numerous techniques, such as high-speed stirring, and ultrasonication, can be used to emulsify and generate tiny oil droplets. Depending on the formulation and process setting, microparticles emulsified by high-speed homogenization or ultrasonication often have a wide range of particle sizes and microstructural morphology [3]. Diffusion, particle erosion, or a combination of these processes play a significant role in the release of drugs from biodegradable polymeric microparticles. Different size microparticles have illustrated unique characteristics of drug encapsulation and release [4].

Osteoarthritis (OA) is a chronic degenerative disorder characterized by low-grade systemic inflammation and the degeneration of joint-related tissues such as articular cartilage [5]. OA produces joint pain, which is often exacerbated by weight-bearing and activity, as well as stiffness following inactivity [6,7]. Weight control, medication, and supportive therapies are the most common treatments for OA. In some cases, intra-articular injection therapies or surgery may be required. Ketorolac tromethamine is one of the non-steroidal anti-inflammatory drugs (NSAIDs) used to treat OA. The drug solution can be administered via intraarticular injection (IA), which produces equipotent pain relief and functional improvement to IA corticosteroids [8,9]. Unlike corticosteroids, injecting ketorolac intraarticularly shows no sign of cartilage damage. The major issue of ketorolac is a short half-life in the body; hence more frequent administration is required leading to patient non-compliance. In addition, hyaluronic acid can also be co-administered with ketorolac, which enables a more rapid analgesic onset with no serious complications [10,11]. Moreover, injecting hyaluronic acid along with ketorolac can enhance joint space narrowing and bone marrow density [12].

Ketorolac microparticles have been widely investigated using different types of polymeric materials and preparation methods. Polymethacrylate (Eudragit®) microparticles for oral administration were fabricated by oil-in-oil solvent evaporation [13]. This method yielded high encapsulation efficiency with a particle size range of 75–225 μm. Parenteral ketorolac-loaded albumin microspheres were also produced by the emulsion cross-linking method. A wide range of encapsulation efficiency from 21–59% was obtained [14]. Other ketorolac microparticles have also been studied based on ethyl cellulose [15], Carbopol, polycarbophil, chitosan [16], PLGA, PLA, poly(ε-caprolactone) (PCL) [17–19] as well as the blend of two polymers such as chitosan/gelatin [20]. To regulate the physicochemical properties of polymers, a blend of at least two polymers has been extensively utilized. Despite the fact that polymer blends are rarely miscible, they may produce new materials or combinations with enhanced performance. On the contrary, immiscible blends can provide materials with phase separation in the microstructure. Eventually, the physicochemical characteristics of the materials rely on the degree of compatibility between the blend components. A few reports have produced the ketorolac microparticles from PLGA, PLA, and their blends with PCL [17–19]. The encapsulation efficiency of ketorolac was reported to be dependent on blending content, the inherent viscosity of the polymer, the lactide ratio in PLGA polymer, and particle size. Higher encapsulation efficiency was obtained when the formulation was composed of pure PLGA or PLA, a higher ratio of lactide, more inherent viscosity, and larger particle size [17,18]. Moreover, the blends of PCL with PLGA or PLA provided sustained release characters depending on the PLGA or PLA contents. Increasing PLGA or PLA ratios retarded the drug release from a few days to a few months [17,18]. Nonetheless, the PLGA/PLA blend microparticles for encapsulating ketorolac have not been reported. PLGA and PLA blends may provide better physicochemical behaviors of materials and lead to a customized degradation rate, modifiable drug release, and better mechanical characteristics.

Therefore, this study aimed to develop ketorolac microparticles based on PLGA, PLA, and their blend to deliver ketorolac in the presence of hyaluronic acid. As previously described, hyaluronic acid can be co-administered with ketorolac, and the developed formulations are intended to be administered intraarticularly in alleviating osteoarthritis. A key determinant in the quality of microparticles prepared by the emulsification evaporation method is a surfactant or stabilizer. In this study, hyaluronic acid was used as a stabilizer in water phase 1 and concurrently used with polyvinyl alcohol (PVA), a typical stabilizer for the double emulsion method [21], in a water phase 2 to ensure the formation of double emulsion droplets. Two techniques, probe sonication and high-speed stirring, were employed for the water-in-oil-in-water (w/o/w) emulsification to render two size ranges of the microparticles. Different compositions and emulsification techniques may provide unique properties of the microparticles. To the best of our knowledge, comparisons of different emulsification techniques and microparticle size ranges have rarely been reported. Furthermore, no published study has reported ketorolac-loaded PLGA/PLA microparticles and PLGA- and PLA-based microparticles formulating with hyaluronic acid and PVA as stabilizers. Hence, the characteristics of the prepared microparticles were evaluated in terms of size, size distribution, morphology, encapsulation efficiency, yield, thermal behaviors, crystallinity nature, and release behaviors.

## 2. Materials and Methods

### 2.1. Materials

Poly(DL-lactide-co-glycolide) (PLGA, 50:50, inherent viscosity 0.59 dL/g, MW 53.4 kDa) and poly(L-lactide) (PLA, inherent viscosity 1.16 dL/g, MW 158.0 kDa) were purchased from Durect Corporation, Birmingham, UK. Ketorolac (MW 255.27 g/mol) in the form of ketorolac tromethamine (water solubility 200 mg/mL, log P 2.1 [22]) was obtained from Cayman Chemical Company, Ann Arbor, MI, USA. Hyaluronic acid (300–500 kDa, Nanjing Gemsen International Co. ltd, Nanjing, China), polyvinyl alcohol (PVA, Mowiol® 8-88, MW 67 kDa, Sigma-Aldrich Pte. Ltd., Singapore), acetone (Honeywell Burdick & Jackson, Morris Plains, NJ, USA), dichloromethane (DCM, Honeywell Burdick & Jackson, Morris Plains, NJ, USA), methanol (high-performance liquid chromatography (HPLC) grade, Honeywell Burdick & Jackson, Muskegon, MI, USA), and sterile water for irrigation (SWI, General Hospital Products. Public Co., Ltd., Pathum Thani, Thailand) were used as received.

### 2.2. Microparticle Preparation

The microparticles were prepared using the w/o/w double emulsification solvent evaporation method using two emulsification techniques, namely probe sonication and high-speed stirrer, to obtain the different sizes of the microparticles. Three different formulations were prepared by both techniques, resulting in 6 individual formulations.

#### 2.2.1. Probe Sonication (PS) Technique

A water phase 1, 0.8 mL, containing 1% *w/v* hyaluronic acid with or without 50 mg of ketorolac was emulsified in an oil phase containing 200 mg of polymer dissolved in 2 mL of DCM by probe sonicator (Vibra-Cell Processors VCX 130, Sonics & Materials, Inc., Newtown, UK) at 40% amplitude for 60 s. The primary emulsion was further emulsified in 8 mL of water phase 2 containing 0.375% *w/v* PVA and 1% *w/v* hyaluronic acid (pH 3.0) by a probe sonicator at 40% amplitude, 60 s. The w/o/w emulsion was then poured into 7 mL of 1% *w/v* hyaluronic acid (pH 3.0) under continuous stirring, and then the solvent was evaporated under reduced pressure. The microparticles were collected and subjected to lyophilization (Crist Alpha 1–4 freeze dryer, SciQuip, Newtown, UK) for 72 h. The lyophilized microparticles were kept at −20 °C overnight for further study.

#### 2.2.2. High-Speed Stirring (HSS) Technique

The primary emulsion was prepared using a similar protocol as the PS technique. After obtaining the primary emulsion, the w/o/w emulsion was prepared by mixing the primary emulsion with 8 mL of water phase 2 using Ultra-Turrax® homogenizer (T 25 digital, IKA®, Staufen im Breisgau, Germany) at a speed of 4000 rpm for 1 min. Then, it was mixed with 7 mL of 1% *w/v* hyaluronic acid (pH 3.0) by a magnetic stirrer in a fume hood for the particles to solidify. The microparticles were obtained after lyophilization for 72 h. The lyophilized microparticles were kept at −20 °C overnight for further study.

### 2.3. *Characterization of Particles*

#### 2.3.1. Particle Size and Size Distribution

The particle size (d (0.5) and d (0.9)) and size distribution (span) were measured using Mastersizer 2000E (Malvern Instrument Ltd., Malvern, UK). The sample was prepared by dispersing the dry microparticles in SWI at a concentration of 20 mg/mL. During the measurement, the sample was diluted in reverse osmosis water. The measurement was performed in triplicate.

#### 2.3.2. Zeta Potential

The zeta potential of the particles was determined by Zetasizer NanoZS ((Malvern Instrument Ltd., Malvern, UK). The sample was prepared using the same method for particle size measurement. The measurement was performed in triplicate.

#### 2.3.3. Drug Loading Content

The drug loading (%DL) and entrapment efficiency (%EE) of ketorolac-loaded microparticles was measured as follows. A known amount of lyophilized microparticles was degraded in 0.3 N sodium hydroxide and sonicated for 15 min. The sample's pH was adjusted to 7.0 with diluted hydrochloric acid, and the volume of the sample was adjusted to 5 mL with 50% $v/v$ methanol in SWI. The sample was filtered and diluted before HPLC analysis. The %DL and %EE were computed according to Equations (1) and (2), respectively.

$$\%\text{DL} = \frac{\text{Analyzed amount of drug}}{\text{Weight of sample}} \times 100 \quad (1)$$

$$\%\text{EE} = \frac{\text{Analyzed amount of drug}}{\text{Initial weight of drug}} \times 100 \quad (2)$$

The validated HPLC analysis was performed using an isocratic mode, Agilent 1200 series HPLC instrument (Agilent Technologies Inc., Santa Clara, CA, USA) according to the United State Pharmacopoeia with some modifications [23]. The drug was eluted through ACE® C18-1104 reverse phase column (150 × 4.6 mm, 5 µm, Advanced Chromatography Technologies Ltd., Scotland, UK) with a guard column. The mixture of acetic acid in water (1:44 $v/v$) and methanol (40:60% $v/v$) at a flow rate of 1 mL/min was used as a mobile phase. The drug was detected by a diode array detector at a wavelength of 265 nm.

#### 2.3.4. Yield

After the lyophilization of the microparticles, the processing yield of the microparticles was determined compared to the initial amount of solid content used in each formulation.

#### 2.3.5. Morphology Observation

The surface morphology of the microparticles was visualized using a field-emission scanning electron microscope (FESEM, Schottky FESEM JSM-7610F, JEOL Ltd., Tokyo, Japan). The sample was fixed on a glass slide mounted on the SEM stub and coated with platinum. The SEM images were captured using an emission current of 2.0 kV at magnifications of 10,000×–20,000×.

### 2.4. *Differential Scanning Calorimetry*

The thermal behaviors of the formulations were evaluated using a differential scanning calorimeter (DSC 3+, Mettler Toledo Limited, Greifensee, Switzerland). The sample was placed in an aluminum pan with a lid. The measurement was performed under nitrogen flow. The sample was heated from room temperature to 200 °C (1st heating), then cooled down to 20 °C and re-heated to 200 °C (2nd heating) at a temperature rate of 20 °C/min. The diffractogram of the 2nd heating was analyzed for enthalpy of heating (ΔH), the onset of endothermic peak ($T_m$), and the glass transition temperature ($T_g$).

### 2.5. X-ray Diffraction

The crystallinity of the drug in the microparticle formulations was analyzed by an X-ray diffraction instrument (Miniflex, Rigaku Americas Holding Company, Inc., Wilmington, MA, USA). The dry sample was placed on a glass slide. The measurement was conducted at two-theta ($2\theta$) in the range of 0°–60° and a rate of 0.06 degree/s.

### 2.6. In Vitro Release Study

The release of ketorolac from the formulations was investigated using the dialysis method in a simulated physiological fluid pH 7.4. Briefly, the freshly lyophilized microparticles were reconstituted in SWI. Two milliliters of the microparticles were put into a dialysis bag (MWCO 6–8 kDa, CelluSep®T2 Membrane Filtration Products, Inc., Seguin, TX, USA) and immersed in 20 mL of phosphate-buffered saline pH 7.4 containing 0.02% sodium azide as a preservative. The study was conducted at 37 °C and 100 rpm for 5 weeks. At each time point, 1 mL of the release fluid was taken, and an equal volume of fresh warm medium was replaced immediately. The sample was analyzed by HPLC as previously described.

### 2.7. Statistical Analysis

The data are expressed as mean±SD from at least three measurements. One-way ANOVA or Student's *t*-test was used to compare the means of multiple groups or two groups, respectively. The difference was considered to be significant if the $p$-value < 0.05. The release profiles were compared using repeated measures two-way ANOVA to compare the percent cumulative drug release at any time and the release profiles of different formulations [24].

## 3. Results

### 3.1. Physicochemical Properties of Microparticles Prepared by Different Emulsification Techniques

In this study, we aimed to fabricate two different size series of microparticles by the w/o/w emulsification–solvent evaporation method using different emulsification techniques, namely PS and HSS. Each series was composed of three formulations using different compositions of PLGA and PLA, namely, 100% PLGA, 50% PLGA:50% PLA, and 100% PLA. PLA exhibits higher hydrophobicity than PLGA. The different polymer compositions can affect the physicochemical properties and release behaviors of the drug-loaded microparticles. The critical step in microparticle preparation is the emulsification of w/o primary emulsion, which may affect the stability of the emulsion and the capacity of drug encapsulation. To avoid any issue in this critical step, we prepared w/o primary emulsion by probe sonication for both techniques. Then, the w/o/w secondary emulsion was prepared by different emulsification techniques. Hyaluronic acid (1%) was used as a stabilizer in water phases 1 and 2; however, a small amount of PVA at a final concentration of 0.2% $w/v$ was also added in water phase 2 as an auxiliary stabilizer. In our preliminary study (data not shown), water phase 2 without pH adjustment produced very low drug encapsulation efficiency. It has been reported that the low encapsulation efficiency of ketorolac is attributed to the high water solubility of ketorolac tromethamine. This leads to the drug partition into continuous water phase 2 during emulsification [18,25]. Since ketorolac is a weak acid (pKa 3.5) [26], the solubility can be reduced in an acid-dispersing phase. The acidified external phase increases the encapsulation efficiency of ketorolac [17]. Therefore, in this study, water phase 2 was adjusted to pH 3.0 before the emulsification process, while water phase 1 was used without pH adjustment.

All of the blank microparticles prepared using the PS technique had a median size (d (0.5)) of 3.12–7.12 µm while the HSS technique yielded a significantly larger microparticle size with the range of 56.32–97.44 µm (Table S1, $p$-value < 0.05). All of the blank microparticles had negative zeta potential (ZP) ranging from −3.8 mV to −9.8 mV. The PS technique yielded a broader size distribution than the HSS technique ($p$-value < 0.05). Therefore, these two techniques could produce microparticles with different desired size ranges.

Three formulations (PLGA, PLGA/PLA, and PLA) of ketorolac-loaded microparticles were prepared using both PS and HSS techniques. The results are summarized in Table 1. After encapsulation, all of the formulations had significantly smaller particle sizes (d (0.5) and d (0.9)) ($p$-value < 0.05) when compared to the blank microparticles. At the same time, their span value was insignificantly different ($p$-value > 0.05), suggesting no effect of the drug on the size distribution of the microparticles. The zeta potential of most drug-loaded formulations became less negative than the blank ones, attributable to the presence of the drug on the microparticle surface.

**Table 1.** Characteristics of ketorolac-loaded microparticles.

| Formulation | Particle Size (μm) | | Span | Zeta Potential (mV) | %Yield | %Drug Loading | %Entrapment Efficiency |
|---|---|---|---|---|---|---|---|
| | d (0.5) * | d (0.9) * | | | | | |
| PLGA (PS) | 2.16 ± 0.00 | 5.89 ± 0.01 | 2.20 ± 0.01 | −3.9 ± 0.1 | 49.98 ± 8.59 | 8.83 ± 0.65 | 76.32 ± 5.58 |
| PLGA/PLA (PS) | 2.12 ± 0.11 | 19.20 ± 6.70 | 8.50 ± 2.83 | −3.5 ± 0.4 | 55.56 ± 4.13 | 9.24 ± 1.32 | 79.83 ± 11.38 |
| PLA (PS) | 1.36 ± 0.01 | 3.08 ± 0.19 | 1.47 ± 0.13 | −3.5 ± 0.1 | 55.83 ± 7.08 | 9.45 ± 0.34 | 81.63 ± 2.94 |
| PLGA (HSS) | 45.50 ± 0.06 | 64.76 ± 0.15 | 1.04 ± 0.00 | −4.4 ± 0.3 | 46.79 ± 6.81 | 8.93 ± 2.15 | 77.15 ± 18.61 |
| PLGA/PLA (HSS) | 67.43 ± 0.40 | 101.32 ± 1.32 | 0.97 ± 0.01 | −3.8 ± 0.2 | 58.18 ± 5.47 | 8.60 ± 0.37 | 74.28 ± 3.20 |
| PLA (HSS) | 60.88 ± 0.54 | 103.51 ± 1.67 | 1.44 ± 0.01 | −3.6 ± 0.2 | 50.29 ± 3.16 | 7.82 ± 0.56 | 67.52 ± 4.84 |

* d (0.5) and d (0.9) mean the size below which 50% and 90% of the sample are contained, respectively.

To investigate the effect of the polymer composition on the microparticle characteristics, three different polymer compositions of drug-loaded microparticles were prepared. PLA exerts higher hydrophobicity than PLGA. The blending of PLGA and PLA at a 1:1 mass ratio would modify the hydrophilic–hydrophobic balance of the individual polymers. It was found that PLA microparticles had the smallest d (0.5), d (0.9), and span values among all formulations prepared by the PS technique. Meanwhile, the blending of PLGA and PLA resulted in the largest d (0.9) and span values. Interestingly, when preparing by the HSS technique, the PLGA/PLA and PLA microparticles possessed comparable d (0.5) and d (0.9), and the smallest particles were obtained from PLGA microparticles.

All of the microparticles prepared by the PS technique had at least 20 times smaller sizes than those by HSS. The PS technique yielded smaller d (0.5) and d (0.9) but wider size distribution than the HSS technique ($p$-value < 0.05). These two techniques basically have different principles. Probe sonication generates a physical vibration and a high shear force by ultrasonic sound waves from the tip of the probe [27]. Differences in the sound intensity (compression–rarefaction cycles) or the sound wave will affect the efficiency of emulsion droplet size reduction and particle deagglomeration [28]. The shear force decreased with the distance far from the probe; thus, the emulsification may not be uniformly distributed throughout the emulsion mixture resulting in wide size distribution. Nevertheless, HSS facilitates emulsification by consistent shear force. The emulsion mixture is forced through a shaft space and consistently shear the emulsion into small droplets. This technique generally provides uniform shear force throughout the sample and thus yields the uniform size of the particles. The bigger microparticles did not result from the technique but from the designed low-speed homogenizer in order to prepare the microparticles in different sizes. Increasing the Ultra-Turrax® homogenizer speed or increasing the homogenization time, which exerts higher energy density and shear stress into the system, could reduce the particle size.

In addition to the particle emulsification technique, other factors can affect the particle size and size distribution, such as the viscosity of the internal phase. The higher the viscosity of the internal phase, the more difficult it was to break down into small droplets. PLA used in this study had higher molecular weight and inherent viscosity than PLGA. Thus, it was anticipated that the PLA microparticles would be larger than the PLGA microparticles. This phenomenon occurred in large microparticles. The microparticles made of PLA (HSS) and PLGA/PLA (HSS) showed larger particle sizes than the microparticles made of PLGA (HSS). However, all of the particles fabricated using the PS technique had small sizes in the same range (1–2 μm). The PS technique is potentially more powerful than the HSS and generates enough energy to break down the emulsion into very small droplets.

The formulation compositions had an insignificant effect on %yield, %DL and %EE ($p$-value > 0.05). This occurred in both particles prepared by PS and HSS techniques. However, there was a correlation between %EE and the polymer composition. For the PS technique, %EE tended to increase with the PLA component. On the other hand, the drug encapsulation efficiency by HSS was lower in the pure PLA formulation. One of the factors determining the quality of the microparticles manufactured by solvent evaporation is the rate of solvent evaporation [29]. A higher evaporation rate limits the rate of drug diffusion and reduces drug losses. Additionally, it has been established that the viscosity of the oil phase and the size of emulsified oil droplets affect the solvent evaporation rate and, thus, the microparticle encapsulation efficiency. Despite PLA having the highest molecular weight and inherent viscosity among other formulations, the PLA (PS) microparticles exhibited the highest %encapsulation efficiency among all of the PS formulations. This was possibly attributed to the more predominant effect of particle size than the viscosity of the oil phase. The very small size of the PLA particles had the largest surface area and, thus, the highest solvent evaporation rate preventing drug leakage during preparation [18]. On the contrary, the PLA microparticles produced by HSS had the lowest %EE compared to other HSS formulations due to the lowest surface area of the particles and the highest viscosity of PLA, slowing down the evaporation of the solvent and thus allowing drug loss during solvent evaporation. This result is consistent with the previous reports [18,29]. Different emulsification methods showed no effect on %yield, %DL, and %EE, except for the PLA microparticles. The %EE of the PLA (PS) formulation was significantly greater than that of the PLA (HSS) microparticles. The PS technique generated smaller droplets with a larger surface area, allowing for a higher solvent evaporation rate. Finally, it limited drug leakage and improved encapsulation efficiency, as previously described.

In conclusion, the emulsification technique clearly affected the particle size and size distribution. The encapsulation efficiency of ketorolac was influenced by the polymer composition and the emulsification technique. The molecular weight and the intrinsic viscosity of the polymers, as well as the solvent evaporation rate, play important roles in the entrapment efficiency of the big microparticles. On the other hand, the solvent evaporation rate insignificantly impacts the entrapment efficiency of the small microparticles with a comparable size range.

The morphology of the ketorolac-loaded microparticles was visualized by FESEM, as shown in Figure 1. Clearly, the microparticles fabricated using the PS technique were smaller than HSS-fabricating microparticles. Regardless of the emulsification techniques, all particles had an almost spherical shape. Overall, the PLA formulations had smoother surfaces than the PLGA formulations, although some PLA particles had pores, possibly due to the sample preparation.

### 3.2. *Differential Scanning Calorimetry (DSC)*

The thermal behaviors of the microparticles were analyzed to study the possible interaction and polymorphism of the components. The DSC thermograms are illustrated in Figure 2, and the thermal parameters are summarized in Table 2. It has been reported that PLGA and PLA have unique characteristics depending on their molecular weight and composition [1]. Our results revealed that PLGA had a $T_g$ at 42.8 °C without any endothermic peak, suggesting its amorphous nature. This is in agreement with the literature that PLGAs containing less than 70% glycolide are amorphous [30]. Meanwhile, PLA had a $T_g$ at 53.3 °C, a double cold crystallization exotherm ($T_{cc}$) at 110 °C and 121 °C and a recrystallization melting endothermic peak at 173 °C (156–180 °C) and an enthalpy of 45.5 J/g, suggesting the semicrystalline nature of the polymer. The cold crystallization of PLA is attributed to the nucleation of the melt state when heating from the glassy state [31,32].

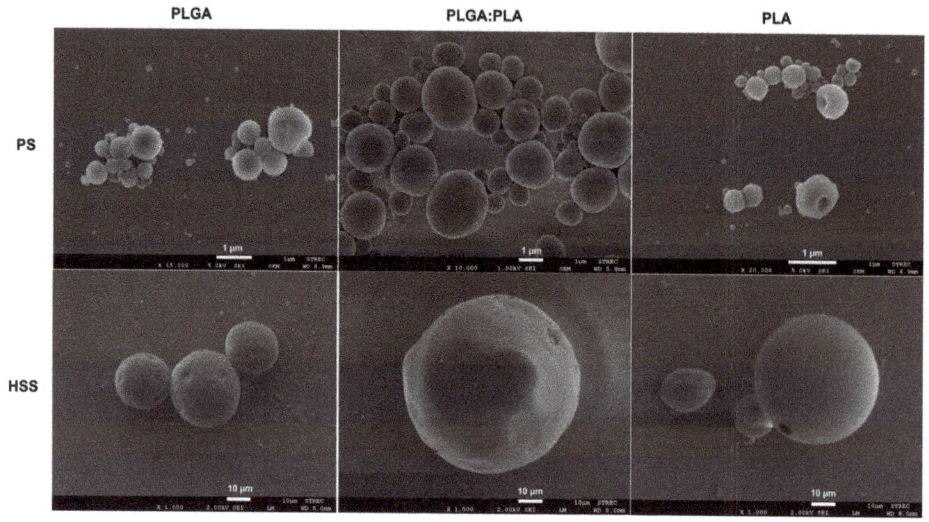


**Figure 1.** FE-SEM images of ketorolac-loaded microparticles. Scale bars in PS and HSS techniques represent 1 μm and 10 μm, respectively.

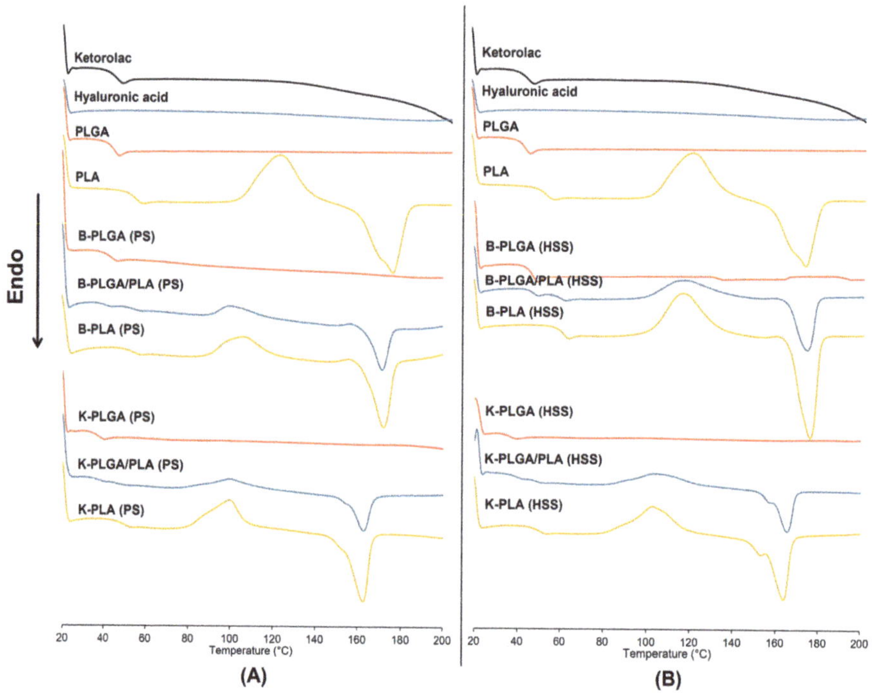


**Figure 2.** Differential scanning calorimetry thermograms of blank (B-PLGA, B-PLGA/PLA, and B-PLA) and ketorolac-loaded microparticles (K-PLGA, K-PLGA/PLA, and K-PLA) prepared by PS (**A**) and HSS (**B**) techniques.

**Table 2.** Temperatures of glass transition ($T_g$), cold crystallization ($T_{cc}$), and melting ($T_m$) and enthalpies of cold crystallization ($\Delta H_{cc}$) and melting ($\Delta H_m$) of individual components, blank and drug-loaded microparticles recorded from the second heating.

| Formulations | $T_g$ (°C) | $T_{cc}$ (°C) [a] | $\Delta H_{cc}$ (J/g) | $T_m$ (°C) [b] | $\Delta H_m$ (J/g) |
|---|---|---|---|---|---|
| Ketorolac | 43.72 | | | | |
| PLGA | 42.79 | | | | |
| PLA | 53.29 | (110.67) 121.00 | 45.93 | (167.67) 173.00 | 45.48 |
| **Blank microparticles** | | | | | |
| PLGA (PS) | 41.80 | | | | |
| PLGA/PLA (PS) | 40.82, 53.95 | 100.67 | 5.48 | 171.33 | 8.28 |
| PLA (PS) | 51.90 | (99.67) 106.67 | 11.49 | 172.67 | 15.00 |
| PLGA (HSS) | 45.70 | | | | |
| PLGA/PLA (HSS) | 46.29, 58.72 | 117.33 | 20.82 | 174.00 | 20.06 |
| PLA (HSS) | 59.80 | 116.33 | 40.56 | 175.33 | 39.86 |
| **Drug-loaded microparticles** | | | | | |
| PLGA (PS) | 35.87 | | | | |
| PLGA/PLA (PS) | 35.85, 48.01 | (84.93) 100.00 | 8.26 | (155.67) 163.00 | 12.11 |
| PLA (PS) | 48.73 | (88.33) 99.67 | 19.52 | (152.33) 162.67 | 21.82 |
| PLGA (HSS) | 34.94 | | | | |
| PLGA/PLA (HSS) | 38.65, 48.95 | (90.67) 104.33 | 12.70 | (157.00) 165.67 | 12.58 |
| PLA (HSS) | 48.79 | (89.67) 102.67 | 20.56 | (153.67) 164.00 | 168.26 |

[a,b] A number in parenthesis represents a small peak of the lower temperature of cold crystallization and melting, respectively.

The $T_g$s of PLA and PLGA of the blank microparticles slightly changed from the polymers (Table 2). HSS increased the $T_g$s of PLGA and PLA, whereas PS had a slight impact on the glass transition of the polymers. The blended polymers of the PLA/PLGA microparticles had a minimal change of glass transition, suggesting the partial miscibility of PLGA and PLA in the blend. All of the PLA-containing microparticles had lower $T_{cc}$, $\Delta H_{cc}$, and $\Delta H_m$ compared to those of the pure PLA polymer, while their $T_m$ of PLA was retained. A single cold crystallization peak was detected in the PLA and PLGA/PLA formulations except in PLA (PS) formulation, suggesting that the cold crystallization of PLA-containing microparticles occurred through heterogeneous nucleation [31]. The $T_{cc}$ of PLGA/PLA microparticles further decreased compared to the PLA microparticles because of the nucleating effect by the PLGA phase and PLA crystalline domains. The microparticles prepared by HSS had higher $T_{cc}$, $\Delta H_{cc}$, $T_m$, and $\Delta H_m$ than the PS technique. Thus, both emulsification techniques affected the thermal behaviors due to the energy applied to break down the droplets during the preparation. PS generated higher energy, while HSS employed lower mechanical energy to break up the droplets prior to the solidification step. Moreover, the smaller droplet size by PS had a much higher surface area than those by HSS for solvent removal, leading to more rapid solidification of the particle core and a shorter time for polymer chain rearrangement. Thus, the PS technique had a greater impact on the thermal behaviors of the polymers than HSS.

In the case of drug-loaded microparticles, the presence of the drug dramatically reduced the $T_g$s of PLGA and PLA by 4–7 °C and 3–4 °C, respectively (Table 2). A decrease in the $T_g$ of all of the drug-loaded microparticles compared to the blank formulations and the pure materials may suggest the compatibility of the drug and the polymer. The drug was molecularly dissolved in the polymer matrix due to the plasticizing effect [4,33]. Compared to the blank microparticles prepared by the PS technique, the drug-loaded microparticles had almost unchanged $T_{cc}$, increased $\Delta H_{cc}$ and $\Delta H_m$, and decreased $T_m$. By HSS, the drug-loaded microparticles had lower $T_{cc}$, $\Delta H_{cc}$, $T_m$, and $\Delta H_m$ than the blank microparticles. Interestingly, all of the ketorolac-loaded PLA-containing microparticles had double cold crystallization and melting events, suggesting that the encapsulation of the drug interfered with the cold crystallization of the polymer, leading to heterogeneous and homogeneous nucleation [34,35]. The physical mixture of the blank PLA microparticles and ketorolac shows similar crystallization patterns and melting events in the thermograms to the blank PLA microparticles, except that the double melting peak was observed. Ketorolac showed only $T_g$ without an endothermic peak in the second heating; however, it had two melting peaks at 161 and 168 °C in the first heating cycle (Figure S1), suggesting the crystalline nature of the drug, which turned to amorphous after rapid cooling. The double melting event of the drug-loaded PLA microparticles was possibly attributed to either the presence of the crystalline state of the drug or the thicker heterogeneous lamellar layers of the crystals. Therefore, we further investigated the crystallinity of the drug in the microparticle samples by XRD.

### 3.3. *X-ray Diffractometry (XRD)*

The crystallinity of the drug-loaded microparticles was investigated by powder XRD. As shown in Figure 3, major characteristic peaks of ketorolac tromethamine appeared at 8.76°, 13.98°, 18.06°, 18.66°, 19.32°, and 20.52° confirming the crystalline nature of the drug. The diffractograms of all ketorolac-loaded microparticles showed that the intensity of these peaks evidently reduced. This result suggested that the majority of the drug was in an amorphous form and coexisted with some crystalline form, possibly due to an unencapsulated drug. The peaks at 31.68° and 45.48° belonged to hyaluronic acid and appeared in all diffractograms of all microparticles.

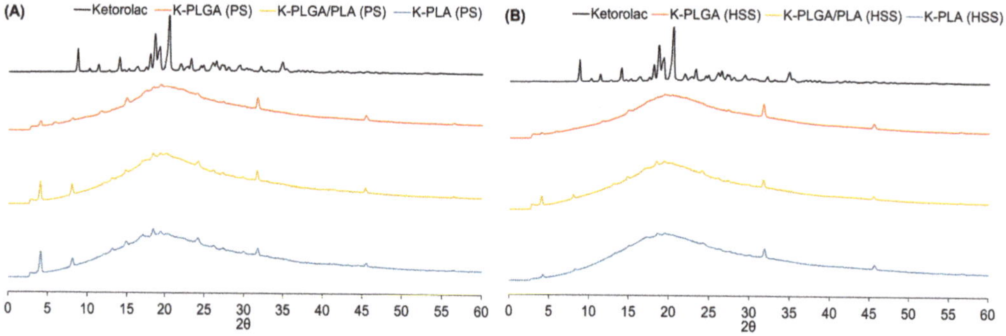


**Figure 3.** X-ray diffractograms of ketorolac-loaded microparticles (K-PLGA, K-PLGA/PLA, and K-PLA) prepared by PS (**A**) and HSS (**B**) methods.

### 3.4. *In Vitro Drug Release Study*

The release of ketorolac from different formulations was studied in phosphate-buffered saline (PBS) pH 7.4 using the dialysis method. In the previous reports [8,9,36], ketorolac was intraarticularly administered once a week for 5 weeks. Therefore, our release study was conducted for 35 days. The release profiles are illustrated in Figure 4. All microparticles exhibited biphasic release consisting of the initial rapid release phase followed by the sustained release of the drug over 35 days. The initial drug release from all PS microparticles by 64–77% was extremely fast within the first 2 h (Figure 4A). Meanwhile, only 39–55% of the drug was released from the HSS microparticles, followed by the slow release of the drug with different release rates, depending on the formulations, until the end of the study (Figure 4B). The difference in the initial release rate of PS and HSS microparticles may primarily be due to their different size and surface area. It has been stated that the drug release of PLGA microparticles is known to be principally influenced by a number of variables, including particle size, porosity, and polymer molecular weight [37]. Many reports have demonstrated that particle size is a primary determinant of drug release rate [4,38,39]. The release of drugs from the different sizes of PLGA microparticles was affected by the rates of swelling and water penetration of the particles, mainly related to the surface area [40]. The smaller PLGA microparticles (4–40 μm) had a burst release of 20% within the first day owing to their faster swelling rate than the larger particles (40–125 μm). Another study demonstrated that the small size range microparticles (<20 and 20–50 μm) had rapid and complete drug release within the first week, while the larger size microparticles (50–100 and >100 μm) showed slow release within the first week [4]. The initial fast release of the ketorolac-loaded PS microparticles may be attributed to the rapid rates of swelling and water penetration entailing matrix porosity and permeability of the microparticles. In addition to the water penetration rate, the drug adsorbing on the surface of the microparticles and the lyophilization process can affect the initial release. It is believed that the initial rapid release is associated with the release of drug molecules trapped close to the microparticle surface and a great initial drug concentration gradient between the particles and the aqueous medium [41]. During lyophilization, the drug migration may cause a heterogeneous drug distribution in the polymer matrix and result in rapid or burst release [42]. The changes in pore size, geometry, and pore interconnectivity during the freezing and lyophilization process may also contribute to this issue [43]. In the case of the HSS microparticles, the initial burst release contributed to the heterogeneous size distribution in the formulations. The PLGA, PLGA/PLA, and PLA (HSS) microparticles contained d(0.1) of 17.5, 36.1, and 15.9 μm, meaning that 10% of the samples had a size below the mentioned value. The concomitant presentation of small particles released the drug more quickly and increased the burst effect [37]. Our results are in agreement with the previous report [4]. The unfractionated microparticles exhibited the release pattern combining those of small and large microparticles. The burst release mainly originated from the smaller microparticles, while the sustained release phase arose from the large microparticles.

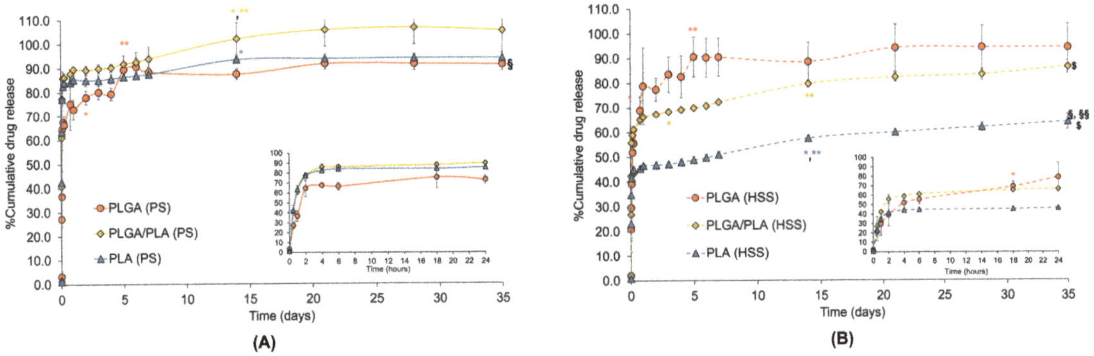


**Figure 4.** The cumulative release profiles of ketorolac from ketorolac-loaded microparticles prepared by PS (**A**) and HSS (**B**) methods in PBS, pH 7.4, over 35 days (mean ± SD, n = 3). An inset represents the release profiles within 24 h. Solid lines and dotted lines represent the drug release from the microparticles prepared by PS and HSS techniques, respectively. *, ** Significantly different when comparing %cumulative drug release at the time forward with 4 and 24 h, respectively; § significantly different when comparing the release profile with that of PLGA/PLA microparticles prepared by the same technique; §§ significantly different when comparing the release profile with that of PLGA microparticles prepared by the same technique; $ significantly different when comparing the release profiles of the similar component microparticles prepared by PS and HSS techniques.

Comparing the different emulsification techniques when using the same formulations, the release profiles of the ketorolac-loaded PLGA microparticles prepared by both techniques were comparable over 35 days ($p$-value > 0.05). Nevertheless, both formulations had significantly different initial drug releases within the first 4 h ($p$-value < 0.05). The PLGA (PS) microparticles had a faster drug release of 67.5 ± 1.5% than the PLGA (HSS) microparticles (52.0 ± 9.1%) due to the smaller size and faster rate of water penetration as previously described. After 4 h of drug release, both formulations exhibited indifferent release profiles ($p$-value > 0.05). Both formulations had comparable $T_g$ values below 37 °C, which was the studied temperature. It is possible that the glass transition of the particle core had a greater impact on the drug release from the PLGA microparticles than their particle size. Considering the microparticles made from the PLGA/PLA and PLA polymers, the smaller PS microparticles enabled significantly faster drug releases than the larger HSS microparticles ($p$-value < 0.05) due to a higher surface area and a greater rate of water penetration, as aforementioned.

Comparing the different formulations prepared by the same method, the drug release from the PLGA/PLA microparticles in the PS series was tentatively faster than in the PLA microparticles. The PLGA microparticles yielded the slowest drug release. In this study, PLGA has higher hydrophilicity and smaller molecular weight than PLA. Based on the properties of the polymers at a similar size of microparticles, the blended PLGA/PLA should have the drug release slower than pure PLGA but quicker than pure PLA. However, the result was inconsistent with the hypothesis, possibly due to polydisperse size distribution, thermal behavior, and the partial compatibility of drug and polymer. The PLGA/PLA microparticles had various sizes (1–19 μm) with a high span value of 8.50 ± 2.83. The small size fraction may be responsible for the fast release of the drug in the heterogeneous size distribution sample. Furthermore, faster drug release from PLGA/PLA microparticles was also possibly attributed to their lower $T_g$ than the studied temperature (from PLGA blended, 35.85 °C) and the partial compatibility between polymer and drug (from PLA blended). At the studied temperature (37 °C), higher than $T_g$, the particle core transformed to a rubbery state, leading to faster drug diffusion and release. The $\Delta C_p$ (specific heat change) in the glass transition region of PLGA/PLA microparticles (0.151 and 0.101 J/g for PLGA and PLA regions, respectively) was lower than that of PLGA microparticles (0.211 J/g), suggesting a smaller free volume when transitioning to a rubbery state and thus a minimal volume for the drug [33]. In addition, blending PLGA with PLA reduced the compatibility between drug and polymers, as evidenced by two $T_g$s of the PLGA/PLA blend in the thermograms. Thus, these circumstances contributed to a more rapid release of the drug from the PLGA/PLA microparticles than the PLGA and the PLA formulations. Despite the fact that the PLGA microparticles had a larger size than the PLA microparticles, the PLGA microparticles showed the slowest drug release. The compatibility of the PLGA with ketorolac, the water-soluble drug, was more critical than $T_g$ and particle size. The higher drug release of the PLA microparticles than the PLGA microparticles contributed to the incompatibility of ketorolac with PLA and the smallest size of the particles.

In the HSS series, the PLA microparticles released the drug slower than the PLGA and PLGA/PLA microparticles ($p$-value < 0.05). The particle size and hydrophilic/hydrophobic nature of the polymers play an important role in the drug release profile of large microparticles. The PLGA microparticles had a smaller size and higher hydrophilicity than the others, resulting in a higher rate of water uptake into the particles. In addition, they also had a $T_g$ lower than 37 °C. Thus, these factors made the polymer matrix more permeable and facilitated the drug release from the microparticles. On the other hand, the PLA microparticles showed the slowest and lowest drug release ($p$-value < 0.05). The slow release of the drug began after the initial 4 h at a constant rate. At the end of the 5-week study, only 64.1 ± 3.2% of the drug was released from the PLA microparticles. We further studied the drug release of this formulation for additional 3 weeks. The release of ketorolac gradually reached 81.6 ± 6.2% by the end of week 8 (data not shown). Their sustained release pattern was possibly attributable to the higher crystalline nature, larger molecular weight, and more hydrophobicity of PLA compared to the PLGA/PLA blend and the pure PLGA. Therefore, these factors hindered water penetration, polymer matrix permeability, drug diffusion, and drug release from the particle core [37].

In addition to the effects of the particle size and the hydrophilic/hydrophobic nature of the polymers on the drug release profiles of the HSS microparticles, other possible factors affecting the release profiles of ketorolac are pore closing, polymer degradation, and particle erosion [37,44–46]. The pore closing of the PLGA microparticles occurred during the incubation of the microparticles in a PBS medium at 37 °C [44]. This effect was more predominant with increasing temperature, while it was unlikely to be observed at a lower temperature. The pore closing effect was attributed to the flexibility of the polymers at the studied temperature. If the polymer has a $T_g$ nearby or lower than the studied temperature, the polymer chains possess the flexibility for pore closing. Consequently, the drug release was suddenly changed to a slower rate. In our study, the microparticles of PLGA, with a comparable MW with the previous report [44], exhibited a $T_g$ value lower than 37 °C, at which the pore closing may occur upon incubating the particles in the aqueous release medium. However, the blending of PLGA and PLA in the PLGA/PLA microparticles led to the partial miscibility of the matrix core and may limit the chain flexibility of PLGA. Thus, the polymer chains of PLGA in the blend may become less flexible than the PLGA microparticles while having more flexibility than the PLA microparticles for pore closing. On the other hand, the pore closing effect was less pronounced in the case of the PLA microparticles since they possessed a $T_g$ value much higher than 37 °C, which limits the flexibility of the polymer chain during incubation. Regarding polymer degradation and erosion, it is well-known that the drug release from PLGA or PLA microparticles is governed by drug diffusion, polymer degradation, and polymer erosion [37,40,46,47]. Both polymers undergo mainly hydrolysis of the ester linkages [46], but their degradation and erosion rates are considerably different depending on polymer crystallinity, lactide-to-glycolide mole ratio, polymer molecular weight, water absorption, and $T_g$ [37,45,46]. It is believed that water hydration in the matrix of an amorphous PLGA is greater than that of semicrystalline PLA, and thus, PLGA is more accessible to water and susceptible to hydrolysis than the homopolymer PLA [46]. The polymer degradation process normally begins much earlier than the polymer erosion. In the previous report, PLGA degradation occurred in a few days, while its erosion took a few weeks [40,46,47]. Our formulations contained different compositions of the polymers. So, this factor may contribute to the different release profiles of ketorolac from the HSS microparticles since they were much larger than the PS microparticles. The drug release of PLGA microparticles was the highest, followed by the PLGA/PLA and PLA microparticles. Thus, it was postulated that the matrix of PLGA formulations might be degraded faster than the PLGA/PLA and PLA formulations, respectively. Although our study has not investigated the degradation and erosion of the particles, it was hypothesized that the microparticles might be degraded or eroded during the release study. Nonetheless, further studies are required to prove these hypotheses.

According to the previously reported clinical trials [8,9,36], ketorolac and hyaluronic acid were concomitantly given once a week by intraarticular injection for 5 weeks. It is known that ketorolac has a short half-life of 4–6 h, and the use of the drug is needed frequent administration. Based on the study design of ketorolac and hyaluronic acid in OA patients, the developed microparticles could be employed in the patients once a month since the release of ketorolac could be retarded for 35 days. The sustained release of ketorolac could maintain the drug in articular fluid for a desirable period. From our results mentioned above, the fast release PS microparticles can be combined with the sustained release HSS formulations to achieve the therapeutic level as fast as the first day of injection and for the whole treatment period. The PLA (PS) microparticles, having a fast release of ketorolac within 24 h and narrow size distribution, can be a choice for combining with any HSS formulations, depending on the desired extent and rate of drug release.

## 4. Conclusions

This study demonstrated the effects of polymer composition and emulsification technique on the ketorolac-loaded microparticles. The ketorolac-loaded PLGA-/PLA-based microparticles with different size ranges were successfully prepared by PS and HSS techniques. Our study was designed based on the intraarticular injection of ketorolac and hyaluronic acid once a week continuously for 5 weeks in OA patients. The PS microparticles exhibited higher drug release within 24 h, while the HSS microparticles demonstrated the sustained release of ketorolac over 35 days. The combination of fast release PS microparticles and sustained release HSS formulations can be used once a month as an alternative regimen in OA patients, which may enhance patient compliance and minimize drug usage and administration costs. The obtained microparticles demonstrated potentiality for the treatment of OA. However, an efficacy investigation of these combinations in animals and patients is required before further application.

**Supplementary Materials:** The following supporting information can be downloaded at: https://www.mdpi.com/article/10.3390/polym15020266/s1, Figure S1: DSC thermogram of ketorolac after the first and second heating; Table S1: Characteristics of blank microparticles prepared by PS and HSS techniques.

**Author Contributions:** Conceptualization, A.W. and J.S.; Formal analysis, A.W., N.K., Y.A. and J.S.; Funding acquisition, J.S.; Investigation, N.K. and Y.A.; Methodology, A.W. and J.S.; Writing—original draft, A.W. and J.S.; Writing—review and editing, A.W. and J.S. All authors have read and agreed to the published version of the manuscript.

**Funding:** This research was funded by the Faculty of Pharmacy, Mahidol University.

**Institutional Review Board Statement:** Not applicable.

**Informed Consent Statement:** Not applicable.

**Data Availability Statement:** The data presented in this study are available on request from the corresponding author.

**Acknowledgments:** The partial financial support from the Faculty of Pharmacy, Mahidol University is acknowledged. We would like to thank Mettler-Toledo (Thailand) Limited for the DSC analysis.

**Conflicts of Interest:** The authors declare no conflict of interest.

## References

1. Jain, R.A. The manufacturing techniques of various drug loaded biodegradable poly(lactide-co-glycolide) (PLGA) devices. *Biomaterials* **2000**, *21*, 2475–2490. [CrossRef] [PubMed]
2. Chew, S.A.; Hinojosa, V.A.; Arriaga, M.A. 11—bioresorbable polymer microparticles in the medical and pharmaceutical fields. In *Bioresorbable Polymers for Biomedical Applications*; Perale, G., Hilborn, J., Eds.; Woodhead Publishing: Sawston, UK, 2017; pp. 229–264.
3. Rosca, I.D.; Watari, F.; Uo, M. Microparticle formation and its mechanism in single and double emulsion solvent evaporation. *J. Control. Release* **2004**, *99*, 271–280. [CrossRef]
4. Chen, W.; Palazzo, A.; Hennink, W.E.; Kok, R.J. Effect of particle size on drug loading and release kinetics of gefitinib-loaded PLGA microspheres. *Mol. Pharm.* **2017**, *14*, 459–467. [CrossRef] [PubMed]
5. Yin, B.; Ni, J.; Witherel, C.E.; Yang, M.; Burdick, J.A.; Wen, C.; Wong, S.H.D. Harnessing tissue-derived extracellular vesicles for osteoarthritis theranostics. *Theranostics* **2022**, *12*, 207–231. [CrossRef]
6. Sen, R.; Hurley, J.A. Osteoarthritis. In *Statpearls*; StatPearls Publishing LLC: Treasure Island, FL, USA, 2022.
7. Mahajan, A.; Verma, S.; Tandon, V. Osteoarthritis. *J. Assoc. Physicians India* **2005**, *53*, 634–641. [PubMed]
8. Xu, J.; Qu, Y.; Li, H.; Zhu, A.; Jiang, T.; Chong, Z.; Wang, B.; Shen, P.; Xie, Z. Effect of intra-articular ketorolac versus corticosteroid injection for knee osteoarthritis: A retrospective comparative study. *Orthop. J. Sports Med.* **2020**, *8*, 2325967120911126. [CrossRef]
9. Park, K.D.; Kim, T.K.; Bae, B.W.; Ahn, J.; Lee, W.Y.; Park, Y. Ultrasound guided intra-articular ketorolac versus corticosteroid injection in osteoarthritis of the hip: A retrospective comparative study. *Skelet. Radiol.* **2015**, *44*, 1333–1340. [CrossRef]
10. Lee, S.C.; Rha, D.W.; Chang, W.H. Rapid analgesic onset of intra-articular hyaluronic acid with ketorolac in osteoarthritis of the knee. *J. Back Musculoskelet. Rehabil.* **2011**, *24*, 31–38. [CrossRef]
11. Koh, S.H.; Lee, S.C.; Lee, W.Y.; Kim, J.; Park, Y. Ultrasound-guided intra-articular injection of hyaluronic acid and ketorolac for osteoarthritis of the carpometacarpal joint of the thumb: A retrospective comparative study. *Medicine* **2019**, *98*, e15506. [CrossRef]
12. Badawi, A.A.; El-Laithy, H.M.; Nesseem, D.I.; El-Husseney, S.S. Pharmaceutical and medical aspects of hyaluronic acid–ketorolac combination therapy in osteoarthritis treatment: Radiographic imaging and bone mineral density. *J. Drug Target.* **2013**, *21*, 551–563. [CrossRef]
13. Saraf, S.; Verma, A.; Tripathi, A.; Saraf, S. Fabrication and evaluation of sustained release microspheres of ketorolac tromethamine. *Int. J. Pharm. Pharm. Sci.* **2010**, *2*, 44–48.
14. Mathew, S.T.; Devi, S.G.; Kv, S. Formulation and evaluation of ketorolac tromethamine-loaded albumin microspheres for potential intramuscular administration. *AAPS PharmSciTech* **2007**, *8*, 14. [CrossRef]
15. Wagh, P.; Mujumdar, A.; Naik, J.B. Preparation and characterization of ketorolac tromethamine-loaded ethyl cellulose micro-/nanospheres using different techniques. *Part. Sci. Technol.* **2019**, *37*, 347–357. [CrossRef]

16. Nagda, C.D.; Chotai, N.P.; Nagda, D.C.; Patel, S.B.; Patel, U.L. Preparation and characterization of spray-dried mucoadhesive microspheres of ketorolac for nasal administration. *Curr. Drug Del.* **2012**, *9*, 205–218. [CrossRef] [PubMed]
17. Sinha, V.R.; Trehan, A. Development, characterization, and evaluation of ketorolac tromethamine-loaded biodegradable microspheres as a depot system for parenteral delivery. *Drug Deliv.* **2008**, *15*, 365–372. [CrossRef]
18. Sinha, V.R.; Trehan, A. Formulation, characterization, and evaluation of ketorolac tromethamine-loaded biodegradable microspheres. *Drug Deliv.* **2005**, *12*, 133–139. [CrossRef]
19. Bhaskaran, S. Poly (lactic acid) microspheres of ketorolac tromethamine for patenteral controlled drug delivery system. *Indian J. Pharm. Sci.* **2001**, *63*, 538–540.
20. Basu, S.K.; Kavitha, K.; Rupeshkumar, M. Evaluation of ketorolac tromethamine microspheres by chitosan/gelatin B complex coacervation. *Sci. Pharm.* **2010**, *78*, 79–92. [CrossRef]
21. Giri, T.; Choudhary, C.; Ajaz, A.; Alexander, A.; Badwaik, H.; Tripathi, D. Prospects of pharmaceuticals and biopharmaceuticals loaded microparticles prepared by double emulsion technique for controlled delivery. *Saudi Pharm. J.* **2013**, *21*, 125–141. [CrossRef]
22. Pubchem compound summary for cid 3826, ketorolac. In *Pubchem [Internet]*; National Library of Medicine (US), National Center for Biotechnology Information: Bethesda, MD, USA, 2004; Volume 2022.
23. Zaghloul, N.; Mahmoud, A.A.; Elkasabgy, N.A.; El Hoffy, N.M. PLGA-modified syloid®-based microparticles for the ocular delivery of terconazole: In-vitro and in-vivo investigations. *Drug Deliv.* **2022**, *29*, 2117–2129. [CrossRef]
24. Chantaburanan, T.; Teeranachaideekul, V.; Chantasart, D.; Jintapattanakit, A.; Junyaprasert, V.B. Effect of binary solid lipid matrix of wax and triglyceride on lipid crystallinity, drug-lipid interaction and drug release of ibuprofen-loaded solid lipid nanoparticles (SLN) for dermal delivery. *J. Colloid Interface Sci.* **2017**, *504*, 247–256. [CrossRef] [PubMed]
25. Selek, H.; Sahin, S.; Ercan, M.T.; Sargon, M.; Hincal, A.A.; Kas, H.S. Formulation and in vitro/in vivo evaluation of terbutaline sulphate incorporated in PLGA (25/75) and L-PLA microspheres. *J. Microencaps.* **2003**, *20*, 261–271.
26. Bianco, A.W.; Constable, P.D.; Cooper, B.R.; Taylor, S.D. Pharmacokinetics of ketorolac tromethamine in horses after intravenous, intramuscular, and oral single-dose administration. *J. Vet. Pharmacol. Ther.* **2016**, *39*, 167–175. [CrossRef] [PubMed]
27. Ghauri, A.; Ghauri, I.; Elhissi, A.M.A.; Ahmed, W. Chapter 14—Characterization of cochleate nanoparticles for delivery of the anti-asthma drug beclomethasone dipropionate. In *Advances in Medical and Surgical Engineering*; Ahmed, W., Phoenix, D.A., Jackson, M.J., Charalambous, C.P., Eds.; Academic Press: Cambridge, MA, USA, 2020; pp. 267–277.
28. Kohli, R. Chapter 2—applications of strippable coatings for removal of surface contaminants. In *Developments in Surface Contamination and Cleaning: Applications of Cleaning Techniques*; Kohli, R., Mittal, K.L., Eds.; Elsevier: Amsterdam, The Netherlands, 2019; pp. 49–96.
29. Goedemoed, J.H.; Mense, E.H.G.; de Groot, K.; Claessen, A.M.E.; Scheper, R.J. Development of injectable antitumor microspheres based on polyphosphazene. *J. Control. Release* **1991**, *17*, 245–257. [CrossRef]
30. Dinarvand, R.; Sepehri, N.; Manoochehri, S.; Rouhani, H.; Atyabi, F. Polylactide-co-glycolide nanoparticles for controlled delivery of anticancer agents. *Int. J. Nanomed.* **2011**, *6*, 877–895. [CrossRef]
31. Carmagnola, I.; Nardo, T.; Gentile, P.; Tonda-Turo, C.; Mattu, C.; Cabodi, S.; Defilippi, P.; Chiono, V. Poly(lactic acid)-based blends with tailored physicochemical properties for tissue engineering applications: A case study. *Int. J. Polymer. Mater.* **2015**, *64*, 90–98. [CrossRef]
32. Szuman, K.; Krucińska, I.; Boguń, M.; Draczyński, Z. PLA/PHA-biodegradable blends for pneumothermic fabrication of nonwovens. *Autex Res. J.* **2016**, *16*, 119–127. [CrossRef]
33. Park, P.I.P.; Jonnalagadda, S. Predictors of glass transition in the biodegradable poly-lactide and poly-lactide-co-glycolide polymers. *J. Appl. Polym. Sci.* **2006**, *100*, 1983–1987. [CrossRef]
34. Wang, Y.; Mano, J.F. Role of thermal history on the thermal behavior of poly(L-lactic acid) studied by DSC and optical microscopy. *J. Therm. Anal. Calorim.* **2005**, *80*, 171–175. [CrossRef]
35. Zhou, Z.H.; Liu, X.P.; Liu, Q.Q.; Liu, L.H. Influence of temperature and time on isothermal crystallization of poly-L-lactide. *Int. J. Polym. Mater. Polym. Biomater.* **2008**, *57*, 878–890. [CrossRef]
36. Jurgensmeier, K.; Jurgensmeier, D.M.D.; Kunz, D.E.; Fuerst, P.G.; Warth, L.C.; Daines, S.B. Intra-articular injections of the hip and knee with triamcinolone vs ketorolac: A randomized controlled trial. *J. Arthroplast.* **2021**, *36*, 416–422. [CrossRef] [PubMed]
37. Yoo, J.; Won, Y.-Y. Phenomenology of the initial burst release of drugs from PLGA microparticles. *ACS Biomater. Sci. Eng.* **2020**, *6*, 6053–6062. [CrossRef]
38. 5—Mathematical models of drug release. In *Strategies to Modify the Drug Release from Pharmaceutical Systems*; Bruschi, M.L. (Ed.) Woodhead Publishing: Sawston, UK, 2015; pp. 63–86.
39. Busatto, C.; Pesoa, J.; Helbling, I.; Luna, J.; Estenoz, D. Effect of particle size, polydispersity and polymer degradation on progesterone release from PLGA microparticles: Experimental and mathematical modeling. *Int. J. Pharm.* **2018**, *536*, 360–369. [CrossRef] [PubMed]
40. Sansdrap, P.; Moës, A.J. In vitro evaluation of the hydrolytic degradation of dispersed and aggregated poly(DL-lactide-co-glycolide) microspheres. *J. Control. Release* **1997**, *43*, 47–58. [CrossRef]
41. Gasmi, H.; Siepmann, F.; Hamoudi, M.C.; Danede, F.; Verin, J.; Willart, J.F.; Siepmann, J. Towards a better understanding of the different release phases from PLGA microparticles: Dexamethasone-loaded systems. *Int. J. Pharm.* **2016**, *514*, 189–199. [CrossRef] [PubMed]

42. Huang, X.; Brazel, C.S. On the importance and mechanisms of burst release in matrix-controlled drug delivery systems. *J. Control. Release* **2001**, *73*, 121–136. [CrossRef]
43. Kim, T.H.; Park, T.G. Critical effect of freezing/freeze-drying on sustained release of FITC-dextran encapsulated within PLGA microspheres. *Int. J. Pharm.* **2004**, *271*, 207–214. [CrossRef]
44. Kang, J.; Schwendeman, S.P. Pore closing and opening in biodegradable polymers and their effect on the controlled release of proteins. *Mol. Pharm.* **2007**, *4*, 104–118. [CrossRef]
45. Rapier, C.E.; Shea, K.J.; Lee, A.P. Investigating PLGA microparticle swelling behavior reveals an interplay of expansive intermolecular forces. *Sci. Rep.* **2021**, *11*, 14512. [CrossRef]
46. Park, T.G. Degradation of poly(lactic-co-glycolic acid) microspheres: Effect of copolymer composition. *Biomaterials* **1995**, *16*, 1123–1130. [CrossRef]
47. Kenley, R.A.; Lee, M.O.; Mahoney, T.R., II; Sanders, L.M. Poly(lactide-co-glycolide) decomposition kinetics in vivo and in vitro. *Macromolecules* **1987**, *20*, 2398–2403. [CrossRef]

polymers

MDPI

Article

# The Role of Dissolution Time on the Properties of All-Cellulose Composites Obtained from Oil Palm Empty Fruit Bunch

Mohd Zaim Jaafar [1], Farah Fazlina Mohd Ridzuan [1], Mohamad Haafiz Mohamad Kassim [1,2,*] and Falah Abu [3,4,*]

1 Bioresource Technology Division, School of Industrial Technology, Universiti Sains Malaysia, Penang 11800, Penang, Malaysia
2 Green Biopolymer, Coatings & Packaging Cluster, School of Industrial Technology, Universiti Sains Malaysia, Penang 11800, Penang, Malaysia
3 Department of Ecotechnology, School of Industrial Technology, Faculty of Applied Sciences, Universiti Teknologi MARA (UiTM) Shah Alam, Shah Alam 40450, Selangor, Malaysia
4 Smart Manufacturing Research Institute (SMRI), Universiti Teknologi MARA (UiTM) Shah Alam, Shah Alam 40450, Selangor, Malaysia
* Correspondence: mhaafiz@usm.my (M.H.M.K.); falah@uitm.edu.my (F.A.)

**Abstract:** All-cellulose composite (ACC) films from oil palm empty fruit bunches (OPEFBs) were successfully fabricated through the surface selective dissolution of cellulose fibers in 8 wt% LiCl/DMAc via the solution casting method. The effect of dissolution time on the properties of the ACC films was assessed in the range of 5–45 min. The results showed that under the best conditions, there were sufficiently dissolved fiber surfaces that improved the interfacial adhesion while maintaining a sizable fraction of the fiber cores, acting as reinforcements for the material. The ACC films have the highest tensile strength and modulus of elasticity of up to 35.78 MPa and 2.63 GPa after 15 min of dissolution. Meanwhile, an X-ray diffraction analysis proved that cellulose I and II coexisted, which suggests that the crystallite size and degree of crystallinity of the ACC films had significantly declined. This is due to a change in the cellulose structure, which results in fewer voids and enhanced stress distribution in the matrix. Scanning electron microscopy revealed that the interfacial adhesion improved between the reinforcing fibers and matrices as the failure behavior of the film composite changed from fiber pullout to fiber breakage and matrix cracking. On the other hand, the thermal stability of the ACC film showed a declining trend as the dissolution time increased. Therefore, the best dissolution time to formulate the ACC film was 15 min, and the obtained ACC film is a promising material to replace synthetic polymers as a green composite.

**Keywords:** all-cellulose composite; surface selective dissolution; interfaceless; self-reinforced composite; green composite

**Citation:** Jaafar, M.Z.; Mohd Ridzuan, F.F.; Mohamad Kassim, M.H.; Abu, F. The Role of Dissolution Time on the Properties of All-Cellulose Composites Obtained from Oil Palm Empty Fruit Bunch. *Polymers* **2023**, *15*, 691. https://doi.org/10.3390/polym15030691

Academic Editors: Vsevolod Aleksandrovich Zhuikov and Rosane Michele Duarte Soares

Received: 27 December 2022
Revised: 17 January 2023
Accepted: 25 January 2023
Published: 30 January 2023



## 1. Introduction

Global attention has been drawn to the urgent appeal for new eco-friendly composite materials to replace petroleum-based ones. According to the 2020 JEC Group report, a non-profit association dedicated to the promotion of composite materials, the global production of composites and plastics currently amounts to 10 and 350 million tons a year, respectively [1]. A challenge now arises in finding technically and economically viable recovery routes for these materials, which were developed on a large scale in the last century and have now become unavoidable. As asserted by the Organization for Economic Cooperation and Development (OECD) in its 2022 report, global plastic waste generation has more than doubled from 2000 to 2019 to 353 million tons [2]. These statistics demonstrate the urgent need to produce novel biocomposite materials with biodegradable properties, such as cellulose, to substitute materials derived from petrochemicals and fossil fuels.

Cellulose is one of the most abundant materials on earth and has eco-friendly features that make it useful as a reinforcement agent for biocomposite materials [3]. However, in most instances, the composite matrix is still frequently based on polymers generated from petroleum [4]. Research into using this bio-based material to create biocomposites has faced obstacles since hydrophilic fibers and hydrophobic matrices cannot coexist due to interfacial adhesion [5–9]. As a result, numerous methods have been used with varying degrees of success to overcome it. These methods include physical and chemical modifications of the fiber and matrix to improve interfacial bonding [10], including silane and alkali treatments, acetylation, chemical grafting, corona discharge, as well as using sub-micron and nano-sized forms of cellulose [11,12]. Although these methods can enhance the mechanical qualities of green composites, they also increase the expense and complexity of their manufacturing [13].

The newly developed green composite called all-cellulose composite (ACC) looks promising for formulating green composites that aim to eliminate the chemical incompatibilities between the reinforcement agent and matrix phases by utilizing cellulose for both components [14,15]. Inspired by the classic mono-component polymer composites developed by Capiati and Porter (1975) [16], Nishino et al. (2004) [17] posed the concept of all-cellulose composites. ACCs can be considered bio-derived mono-component composites, which is the same source of cellulosic materials that would need to be used for the reinforcing and matrix phases. In other words, it is a new class of materials that, unlike other biocomposites, are made entirely from cellulosic materials, leading to chemically identical matrix and reinforcing phases [13]. The following two processing methods are suggested for the manufacture of an ACC: (i) complete cellulose dissolution, followed by mixing with extra reinforcing cellulosic materials, and (ii) partial cellulose dissolution to generate a matrix phase in situ around the residual fiber core [17–21]. Additionally, cellulose-dissolving utilizing solvent is a distinguishing feature of ACC engineering in contrast to other traditional composites due to cellulose's inherent property, i.e., it does not melt. Lithium chloride/N,N-dimethylacetamide (LiCl/DMAc), NaOH, and ionic liquids (ILs) are some of the most popular solvents used by researchers [13,14]. Despite being environmentally acceptable solvents, it is a challenge to find industrial applications for ionic liquids (ILs) due to their complex and expensive manufacture. On the other hand, toxicity concerns restrict the usage of LiCl/DMAc in academic studies [13].

Although Nishino introduced the first ACC in 2004 by utilizing Ramie fiber, it was only recently that it was given serious consideration. ACCs are mono-material composites made from cellulose. Therefore, they show excellent interfacial compatibility, are fully recyclable, and are environmentally friendly [15,22,23]. As a result, ACCs may represent the next stage in the creation of composite materials with greater sustainability. The world's primary source of cellulose, forests, would be in peril if cellulose became the only component of this type of composite. For forests to be sustainable, the reliance on plant cellulose, particularly wood, must be significantly reduced through the use of biomass or agricultural waste.

Numerous research has focused on the development of ACCs using non-wood or biomass, including food crops, cotton [23,24], canola straw [25], flax [4], bagasse [26], hemp [14], pineapple leaf [27], Napier grass [28], and many others [24,28,29]. The substantial amount of biomass produced by the palm oil sector is one of the components of this category that has tremendous relevance to the Malaysian situation. In Malaysia, palm oil is a well-known and important commodity [30]. According to the Malaysian Palm Oil Board (MPOB), 400 hectares were cultivated in 1920, leading to five million plantations today in Malaysia [31]. Enormous amounts of lignocellulose residues from the oil palm industry, such as oil palm trunks (OPTs), oil palm fronds (OPFs), and oil palm empty fruit bunches (OPEFBs), are generated at an estimated 15 million tons annually, becoming a major economic pillar for the country [32]. This biomass is readily available at a minimal cost. Attempts have been made to turn these wastes into value-added products; one example is converting the lignocellulosic residue into a paper-making pulp. Hence, other utilization alternatives are sought in an attempt to optimize the use of this biomass. Therefore, the development of ACCs from this biomass would be the best approach.

During the past decade, creating ACC films from OPEFBs has been the subject of only a small number of studies [33–39]. Several earlier studies had utilized MCCs and nanoscale. However, to date, OPEFBs in the form of bleached pulps, utilizing facile preparation, which is a straightforward pulping and bleaching method, have not been addressed yet. In addition, this form of pure cellulose and simple method can reduce energy and costs and avoid the complexity of ACC preparations. Therefore, the objective of this research is to determine the best dissolution time in an attempt to enhance the performance of ACC films made from OPEFB-bleached pulps. This article reports the effect of dissolution time in the range of 5–45 min on the tensile properties of ACC films. It is expected that the produced ACC films would have the potential to be used for green composite productions and applications.

## 2. Materials and Methods

### *2.1. Materials*

Oil palm empty fruit bunches (OPEFBs) were obtained from United Oil Palm Sdn Bhd, Nibong Tebal, Penang, Malaysia. All of the reagent-grade chemicals, including N-N dimethyl acetamide (DMAc), lithium chloride (LiCl), sodium hydroxide (NaOH), methanol, and acetone, were derived from Nacalai Tesque Co., Kyoto, Japan, and used as received.

### *2.2. Isolation and Extraction of OPEFB Cellulose*

In this work, non-cellulosic components were removed from cellulose samples made from OPEFB biomass sources. By using soda pulping, extended delignification by bleaching, and hemicellulose removal, cellulose was obtained from the OPEFBs. Based on the processes disclosed by Wanrosli et al. [40], the pulping and bleaching sequence of the OPEFB pulp was carried out. To isolate the cellulose, the dried fiber was first delignified according to ASTM D 1104-56 to produce holocellulose, followed by the removal of the hemicellulose fraction according to ASTM D 1103-60.

#### 2.2.1. Soda Pulping Process

The pulping process was achieved through the use of a digestor (model IBSUTEK ZAT 92) from RB Supply Enterprise, Penang, Malaysia. The OPEFB was cooked at 26% NaOH with a ratio of 1:8 (OPEFB: liquor) at 170 °C for 2 h. A hydropulper was then used to refine the pulp, which was then washed with water, screened, dried, and noted as OPEFB-P.

#### 2.2.2. Holocellulose Production

A total of 5 g of pulp from the OPEFB-P was transferred into a 1000-mL conical flask, and 160 mL of distilled water, 1.5 g of sodium chlorite ($NaClO_2$), and 10 drops of acetic acid (10% wt/v) were added. The sample was kept inside a water bath at 70 °C for 3 h. Next, 1.5 g of $NaClO_2$, 10 drops of acetic acid, and 160 mL of distilled water were added alternately for each hour. After 3 h, cooled distilled water was added to stop the chemical reaction in the flask. The precipitate was then filtered with a glass crucible (2G2). The remaining pulp in the glass crucible was washed using cold distilled water, followed by acetone, and it was left to dry in the oven for 24 h.

#### 2.2.3. Bleaching Process

The 2.0 g of the holocellulose (hemicellulose + cellulose) produced was treated with 50 mL of 17.5% sodium hydroxide at 20 °C for 2 h, and then 70 mL of distilled water was added. This was to separate the hemicellulose from the holocellulose, leaving the cellulose. The insoluble cellulose was filtered and washed with 50 mL of 8.3% sodium hydroxide.

### 2.3. Preparation of All-Cellulose Composite Film

In another article, the surface selective dissolution procedure used to create all-cellulose composites was described [19,41,42]. The OPEFB pulp was immersed in distilled water, acetone, and DMAc, respectively, for 1 h each at room temperature to activate the cellulose fibers. The activated EFB pulp was immersed in the LiCl/DMAc solution at room temperature for four different dissolution times (5, 15, 30, and 45 min) with a 1% initial cellulose concentration (wt. vol-1). The solution was then cast on a glass plate with a blade and allowed to dry for 24 h in the open air. After that, the obtained ACC was immersed in distilled water for 24 h and further dried in the open air overnight.

### 2.4. X-ray Diffraction

The Bruker D2 Phaser X-ray diffractometer (XRD), Bruker, Germany, equipped with the LYNXEYE 1D solid-state ultra-fast detector with Cu-K$\alpha$ radiation at room temperature was used to analyze the crystallinity of the ACC films. The samples were analyzed using a step size of 0.02 and a dwell time of 0.1 s per step. The following equation was used to calculate the crystallinity index (CrI) for all ACC composite films [35]:

$$\text{Crystallinity Index (CrI)} = ( I - I' / I ) / ( I ) \times 100$$

where I = the height of the peak assigned to (200) planes, measured in the range $2\theta$ = 20–23°, and I′ = the height of the peak assigned to (100) planes, located at $2\theta$ = 12–16°.

### 2.5. Fourier Transform Infrared (FTIR) Analysis

Fourier transform infrared (FTIR) spectroscopy was performed using a Nicolet Avatar Model 360 spectrometer (Thermo Nicolet Corporation, Madison, WI, USA). The FTIR analysis was measured in the transmittance mode from 500 nm to 4000 nm.

### 2.6. Scanning Electron Microscopy (SEM)

The tensile fracture surfaces of the composite samples were examined by scanning electron microscopy (Quanta FEG 650, FEI, Brno, Czech Republic). The samples were placed on the SEM holder using double-sided electrically conducting carbon adhesive tape and gold-coated with a Polaron SEM coating unit. The SEM micrographs were obtained under conventional secondary imaging conditions with an acceleration voltage of 10 kV.

### 2.7. Tensile Testing

Tensile tests were conducted according to ASTM D 882 using a texture analyzer (TA.XT plus, Stable Micro Systems, Surrey, UK) with sample dimensions of 150 mm × 600 mm (width × length). The samples were equilibrated in a desiccator at room temperature with an RH range of 30–50% prior to the tensile testing. The tensile strength (TS), Young's modulus (E), and elongation at break (EAB) were recorded. The filmstrip was clamped between the tensile grips with an initial grip separation of 100 mm, and the test speed was set at 0.8 mm $s^{-1}$ using a load cell of 30 kg. An average of five replicates were recorded.

### 2.8. Differential Scanning Calorimetry (DSC)

A differential scanning calorimetry analysis was performed using a Perkin Elmer Pyris 7 thermal analyzer (Radeberg, Germany). About 6 mg of the sample were put into the sample pan, after which they were sealed and heated at 10 °C/min in a nitrogen flux from 30 to 400 °C. The data obtained from the samples were recorded continuously along with the temperature and time intervals.

### 2.9. Thermogravimetric Analysis (TGA)

About 10 mg of the sample was heated at 10 °C/min in a nitrogen flux (20 mL/min) from room temperature to 800 °C in a nitrogen flux atmosphere using a Mettler Toledo

TGA/DSC 1 (STAR e system) (Schwerzenbach, Switzerland). The percentage of sample weight loss versus temperature was recorded using a thermogram.

## 3. Results and Discussion

Following this point, the samples were identified by their initial cellulose concentrations (preceded by "ACC") and their dissolution times (preceded by "5–45"). The initial cellulose concentration for each of the other ACCs is set to 1% (wt.vol$^{-1}$) to ensure a single variable. For example, sample ACC5 designates a composite prepared with a dissolution time of 5 min in an 8% LiCl/DMAc solution. Table 1 summarizes the formulation of the all-cellulose composite films in the present work.

**Table 1.** Formulation of the all-cellulose composite films in the present work under different conditions.

| Samples | Dissolution Time, td (min) | Initial Cellulose Concentration, C (%) |
|---|---|---|
| ACC5 | 5 | 1 |
| ACC15 | 15 | 1 |
| ACC30 | 30 | 1 |
| ACC45 | 45 | 1 |

### *3.1. Tensile Properties*

Figure 1 shows the effect of dissolution time on the tensile strength (TS) of all-cellulose composite films. The mean values of the tensile strength, elongation at break (EAB), and modulus of elasticity are shown in Table 2. Overall, it can be observed that the longer the dissolution time, the higher the TS of the ACC films. The results showed that the TS of the initial ACC5 increased from 11.96 MPa to 35.78 MPa at its highest when the dissolution time reached 15 min, increasing by 66.58% in comparison to the initial film with a 5-min dissolution period. This finding is in line with the results reported for ACCs from other cellulose sources, such as flax, cotton fiber, canola straw, pineapple leaf, bacterial cellulose, and bagasse, in response to the impact of longer dissolution times that lead to significant improvements in the mechanical behavior of ACC films [4,24,25,27,42]. The tensile properties of the ACCs generated from those cellulose sources were considerably altered with the longer dissolution times, up to an ideal period.

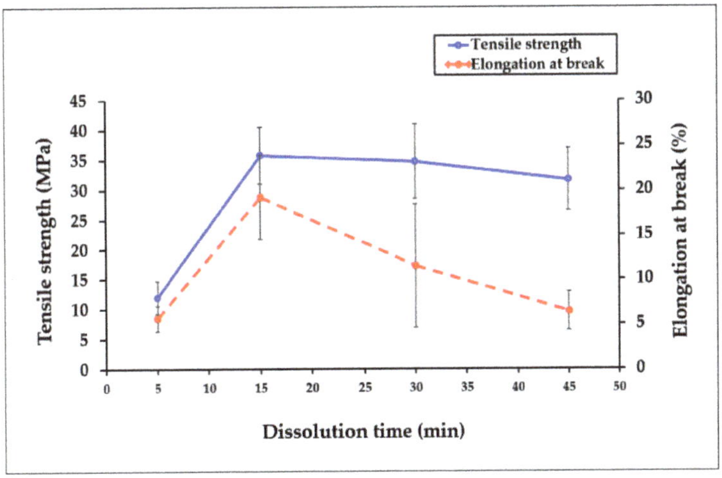


**Figure 1.** Mechanical properties of the ACC film composites made from OPEFB at dissolution times ranging from 5 to 45 min.

**Table 2.** Mechanical properties of the all-cellulose composite films prepared at different dissolution times.

| ACC Films | Tensile Strength (MPa) | Young's Modulus (GPa) | Elongation at Break (%) |
|---|---|---|---|
| ACC5 | 11.96 ± 2.7 | 0.73 ± 2.1 | 5.66 ± 1.4 |
| ACC15 | 35.78 ± 4.7 | 2.63 ± 2.2 | 19.22 ± 4.7 |
| ACC30 | 34.72 ± 6.3 | 2.94 ± 4.1 | 11.53 ± 6.9 |
| ACC45 | 31.72 ± 5.2 | 2.75 ± 1.2 | 6.46 ± 2.2 |

The properties of the ACC films depend on the matrix, reinforcing phase, and interface [22]. The improvement in the ACC films also filled the voids between the fibers and cracks, which is probably because of the dissolved matrix [4,15,43]. This suggests that the reinforcing phase and the polymer matrix interact strongly, preventing the mobility of polymer chains. Furthermore, because of the improved contact between the fibers, the strain at break of the all-cellulose composites was observed to significantly rise with the longer dissolution periods. As a result, there was an increase in the process of stress transfer from the matrix to the reinforcing material, hence, fostering the TS traits [27].

In contrast, the TS of the films showed a slight drop from 35.78 MPa to 31.72 MPa when the dissolving time was prolonged from 15 to 45 min, although the TS value remained superior to the initial film with the 5-min dissolution time. The excessive dissolution of cellulose reduces the fraction of the reinforcing phase and, hence, weakens the mechanical properties [22]. As the dissolving time increases, the TS of the ACC film decreases due to its high wettability [41,42]. Similar to the findings of Yousefi et al. (2011) [25], the TS decreases when the dissolution time is prolonged due to the aggregation of the cellulose particles. Cellulose cannot perform its fundamental function when the particles are gathered. Thus, the TS typically decreases.

Due to the various features that affect the EAB and TS values, the EAB trend typically opposes the TS trend for many types of composites. EAB measures a material's ductility, while TS provides information about a material's strength. The addition of more fillers would typically raise the composite's rigidity in a conventional composite system, reducing the material's elongation at break (flexibility). In the ACC system, higher dissolution times result in increased matrix production and a lower reinforcing phase volume. As a result, according to typical composite results, the EAB result should decrease as the dissolution time increases. In this study, however, it is intriguing to note that EAB exhibits a directly proportionate tendency to TS, which is somewhat raised when the dissolution time is prolonged from 5 to 15 min before gradually decreasing as time passes.

This might be explained by the fact that the ACC was created using only one material, EFB cellulose, which served as both the matrix and the reinforcement. This resulted in an uncontrollable amount of dissolved (for the matrix) and undissolved (for the reinforcement) cellulose in the structure of the ACC films, also known as the fiber volume fraction. Estimating the fiber volume fraction of ACCs made using partial dissolving techniques presents various difficulties. In contrast to conventional fiber-reinforced composites, the amount of fiber needed to make ACCs will not be the same after processing because some of the fiber will have dissolved to become the matrix [26].

In this study, the fiber volume fraction is inferred rather than measured by varying the dissolving time; the higher the fiber volume percent produced, the longer the dissolution time. Consequently, the fiber volume fraction produced the matrix and reinforcement ratio that affected the EAB outcome. Despite the dissolving time being prolonged to 45 min, there was not enough reinforcing phase inside the matrix domain to permit appreciable changes in the elongation at break of the ACC films. As a result, the fraction that contributes to elasticity may also change since the "actual composition" of the reinforcing phase in this situation may not be proportional to the EFB content.

In addition to the variable elongation at break value, environmental factors including moisture absorption, film absorption rate, and varying void contents due to the random dispersion of the EFB particles in the ACC film structures could also be factors. These elements have an impact on the biopolymer chains' flexibility and conformation when they are deformed by tensile forces. These explanations could account for the surprising elongation at break results of these ACC films [39].

Figure 2 shows the effect of dissolution time on the modulus of elasticity of the all-cellulose composite films. The modulus of elasticity value increases for up to 30 min of the dissolution time and then drops afterwards. The reason for the decrement could probably be due to the increase in the OPEFB contents, which can cause aggregation of the particles. Additionally, the particles that were not dispersed well in the solvent system were reported by Zailuddin et al. [35].

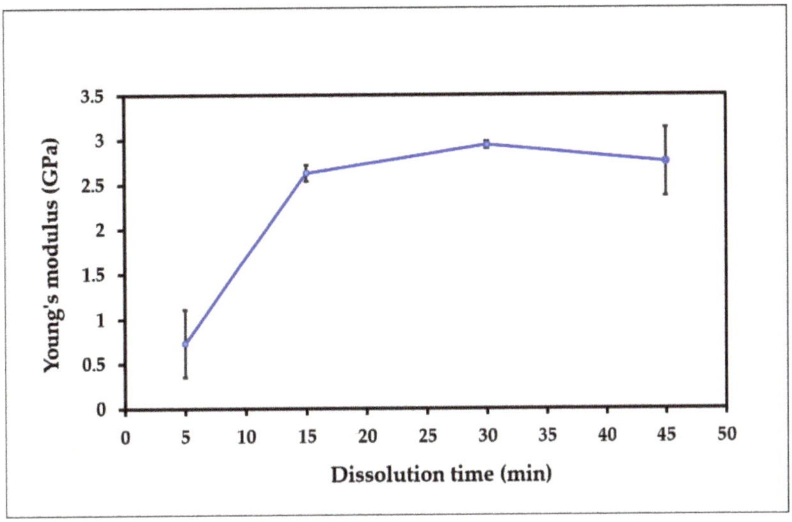


**Figure 2.** Modulus of elasticity of the ACC films at various dissolution times (5–45 min).

### 3.2. FTIR Analysis

FTIR spectroscopy was used to distinguish between the cellulose allomorphs and the celluloses of different crystallinities. In Figure 3, the FTIR spectra of the understudied samples plotted in absorbance form within the range of 4000–500 cm$^{-1}$ are shown. The effect of the dissolution time on the structural changes of the ACC films was analyzed using FTIR, which reflects the changes in the functional groups. It can be observed that all four spectra had common bands, as both the matrix and the reinforcement agent in the ACCs are cellulosic. All samples displayed two main absorbance regions and are found at high (3000–3600 cm$^{-1}$) and low (500–1800 cm$^{-1}$) wavenumbers, respectively.

Table 3 shows the list of characteristic FTIR peaks and the corresponding motions of the organic bonds in this study. Based on the spectra, it is evident that the O–H groups, which make up the majority of the OPEFB ACC films, reside in a broad absorption band in the 3700–3100 cm$^{-1}$ area. Besides that, the broad peak in the wavenumber range of 3000–3600 cm$^{-1}$ is believed to be generated by the O(3)H–O(5) intramolecular hydrogen bond [44], and the peak at about 2850 cm$^{-1}$ is attributed to the stretching vibration of C–H bonds in the methyl and methylene groups [45]. The peak located at about 1420 cm$^{-1}$ is assigned to methyl group deformation and the lignin aromatic ring vibrations, whereas the peak at about 1255 cm$^{-1}$ is associated with C–O stretching in the guaiacyl ring [27]. The peaks at 1310 cm$^{-1}$ and 1370 cm$^{-1}$ are respectively assigned to the $CH_2$ wagging motion and the –OH in-plane bending of the crystalline form of cellulose [44]. The peaks centered at about 1160 cm$^{-1}$, 1025 cm$^{-1}$, and 895 cm$^{-1}$ are attributed to C–O–C asymmetric bridge stretching, C–O–C pyranose ring skeletal vibration, and β-glucosidic linkage, respectively [27]. These peaks are attributed to cellulose II [27], and the increase in their intensities results in more formation of the structure. It was claimed that the band of the O–H region (3000–3600 cm$^{-1}$) contrasts strongly with that of the materials produced by partial dissolution [44]. Accordingly, the monotonic shift of the band to lower wavenumbers by increasing the dissolution time indicates the generation of more new inter- and intramolecular hydrogen bonds, meaning a higher transformation from cellulose I to cellulose II [46]. It is well known that cellulose II is thermodynamically more stable than cellulose I and exists in antiparallel strains with intersheet hydrogen bonding [47].

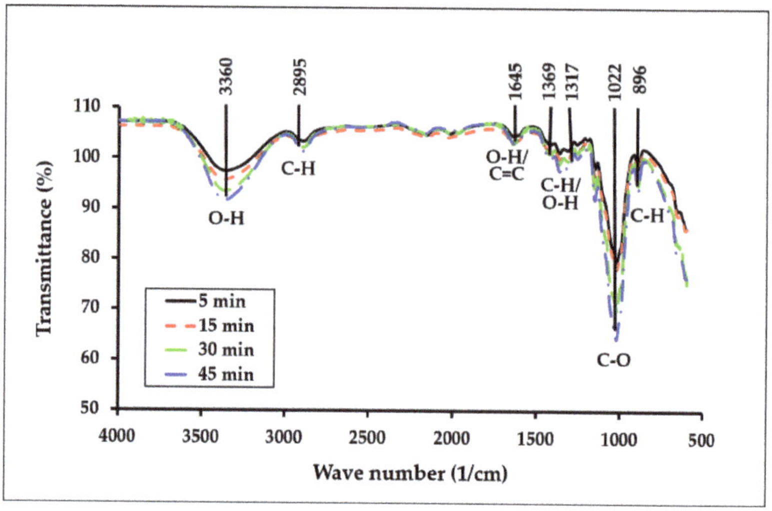


**Figure 3.** FTIR spectra of the ACCs film composites at different dissolution time.

**Table 3.** Characteristic FTIR peaks and corresponding motions of organic bonds.

| Wave Number (cm$^{-1}$) | Bond and Motion |
|---|---|
| 3600–3000 | O(3)H–O(5) intramolecular hydrogen bond |
| 2850 | stretching vibration of C–H bonds in methyl and methylene groups |
| 1420 | methyl group deformation and the lignin aromatic ring vibrations |
| 1370 | –OH in-plane bending of crystalline form of cellulose |
| 1310 | $CH_2$ wagging motion |
| 1255 | C–O stretching in guaiacyl ring |
| 1160 | C–O–C asymmetric bridge stretching |
| 1025 | C–O–C pyranose ring skeletal vibration |
| 895 | β-glucosidic linkage |

The calculation of the total crystallinity index (TCI) was performed from the ratio of the absorption peaks at 1370/2850 cm$^{-1}$ [48]. These two peaks were chosen because of their low susceptibility to water. The variations of the indices for the understudied samples are shown in Figure 4.

For instance, the understudied samples have TCI values in the range of 0.95–0.98. Nelson and O'Connor [49] reported a value of 0.84 for hydrolyzed cotton cellulose and a value of 0.90 for purified cotton yarn. The same authors reported values for partly mercerized cotton in the 0.54–0.58 range [50]. Carillo et al. [48] reported values in the 0.64–0.87 range for regenerated fibers of different origins, whilst Duchemin et al. [44] reported values in the 0.66–0.95 range for filter paper (FP) and micro-fibrillated cellulose (MFC). From the TCI value variations, it can be seen that the parameter has a decreasing trend versus the dissolution time. The sample underwent a monotonic TCI loss, starting from the highest value of about 0.98 initially and decreasing to a value of about 0.96 after 45 min. It is in good agreement with the TGA data and shows that the sample with a lower initial concentration (1%) is sensitive to the dissolution time. Hence, the FTIR results confirmed that the structural change of cellulose from cellulose I to cellulose II can be detected in the composites by increasing the dissolution time in the samples.

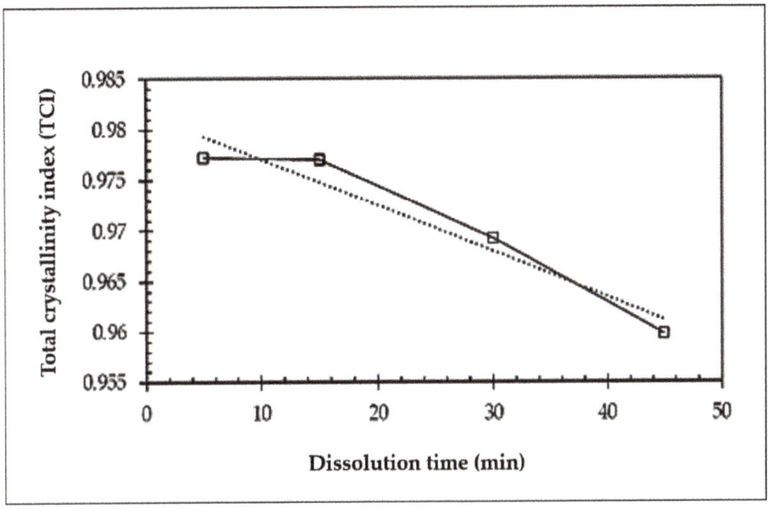


**Figure 4.** TCI variations obtained from the ratio of the absorption peaks at 1370/2850 $cm^{-1}$ for the understudied samples.

In order to get more information about the structure of the materials and the crystal structure of the ACCs, the ACC films prepared at various dissolution times were examined by X-ray diffraction.

### 3.3. *X-ray Diffraction (XRD)*

The XRD curve of the all-cellulose composite films with different cellulose contents and at dissolution times is presented in Figure 5. The X-ray diffraction (XRD) analysis suggests that the degree of crystallinity of the ACC films significantly declined as the structure of the dissolved cellulose changed from cellulose I to cellulose II.

All ACC samples showed the characteristic peaks at $2\theta \approx 12.0°$ for the (1 0 1) plane, $2\theta \approx 13°$ for the (1 0 1) plane, and $2\theta = 23.0°$ for the (0 0 2) plane [14]. For qualitative analysis of the XRD patterns, crystallinity index (CrI) evaluations were calculated by the following equation (Equation (1)) [51]:

$$\text{Crystallinity Index (CrI)} = (I - I') \ / \ (I) \times 100 \quad (1)$$

where I is the diffraction intensity assigned to (2 0 0) plane of cellulose $I_\beta$ ($2\theta \approx 23°$). $I'$ is the intensity measured at $2\theta = 18°$, where the maximum happens in a diffractogram for the non-crystalline cellulose. Crystallite size (width) of the cellulose was assessed by Scherrer's equation (Equation (2)) [52]:

$$D = K\lambda \ / \ \beta\cos(\theta) \quad (2)$$

where D is the crystal size, K is Scherrer's constant (0.9), $\lambda$ is the X-ray wavelength (0.15418 nm), $\theta$ is the Bragg angle for the (2 0 0) reflection and $\beta$ is the corrected integral width.

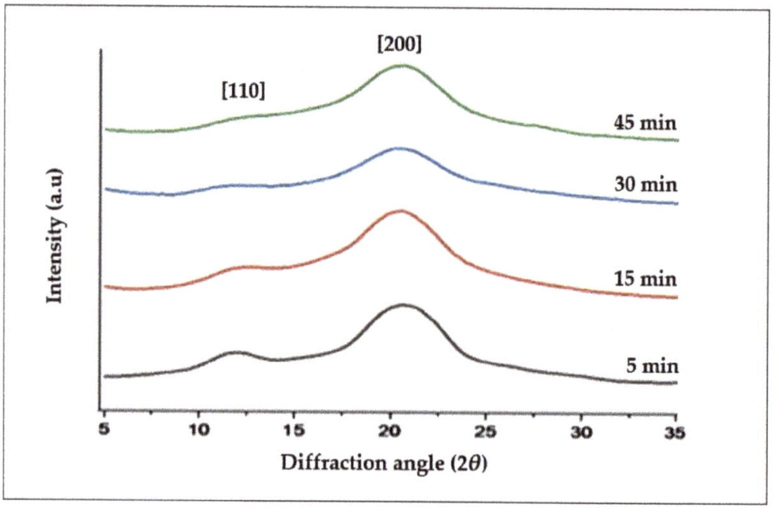


**Figure 5.** XRD patterns of the ACC film composites at different dissolution time.

A comparison of the X-ray diffraction patterns of the all-cellulose composites prepared with 5, 15, 30, and 45 min of dissolution time shows that the crystalline peak intensities of the composites mostly decreased with the dissolution time. The intensity decrement trend with increasing dissolution time is clearer for the crystalline peaks of $(1\,\bar{1}\,0)$ and (1 1 0) plans in the 2θ range of 10–15°. In fact, at longer dissolution times, larger fractions of the fibers are dissolved to form an amorphous matrix phase (non-crystalline domains). This confirms the findings of previous studies [6,24,53,54] and indicates that the increase in dissolution time would result in a transformation of cellulose I to cellulose II, as demonstrated by the FTIR analysis. This phenomenon directly results in a reduction in the overall crystallinity index and crystallite size of the composites, which were calculated from the XRD data, as shown in Figure 6 and tabulated in Table 4.

According to Figure 6a, the highest crystalline index values in almost all dissolution times belong to the sample with the initial dissolution time, which was 5 min. In addition, from this figure, it can be seen that the rate of the CrI loss for the sample was significant, indicating a higher sensitivity to the dissolution time, as seen in the TGA and FTIR results. The decreasing trend of the CrI parameter with increasing the dissolution time can be attributed to the effect of more solvents penetrating the spaces between the crystallites and dissolved fibers, as well as the outer chains of the crystallites [25].

The processing of the X-ray diffraction data using the Scherrer's equation (Figure 6b) also reveals that the lateral crystallite size normal to the (2 0 0) plane in cellulose composites is reduced with the immersion time. In other words, for all the ACC samples prepared with dissolution times of 5, 15, 30, and 45 min, more crystalline cellulose is dissolved and, hence, smaller cellulose crystallites remain with larger non-crystalline domains being formed. For the ACC30 crystallite size, which is the smallest size, there is an interesting consequence. It is possible that the fiber inside sample ACC30 has a smaller diameter or fewer entanglements than those in the other samples. As a result, the unusually deep penetration of the solvent caused it to become amorphous more quickly and lose more crystallite size. This was also illustrated by the FTIR spectra of the composites (Figure 3), which indicate that the absorption peak intensity of cellulose II increased with the dissolution time.

**Table 4.** Degree of crystallinity and crystallite size of the ACCs of the ACC films.

| Sample | Degree of Crystallinity (%) | Crystallite Size (nm) |
|---|---|---|
| ACC5 | 42.62 | 3.7 |
| ACC15 | 41.21 | 3.9 |
| ACC30 | 32.08 | 2 |
| ACC45 | 28.8 | 3.9 |

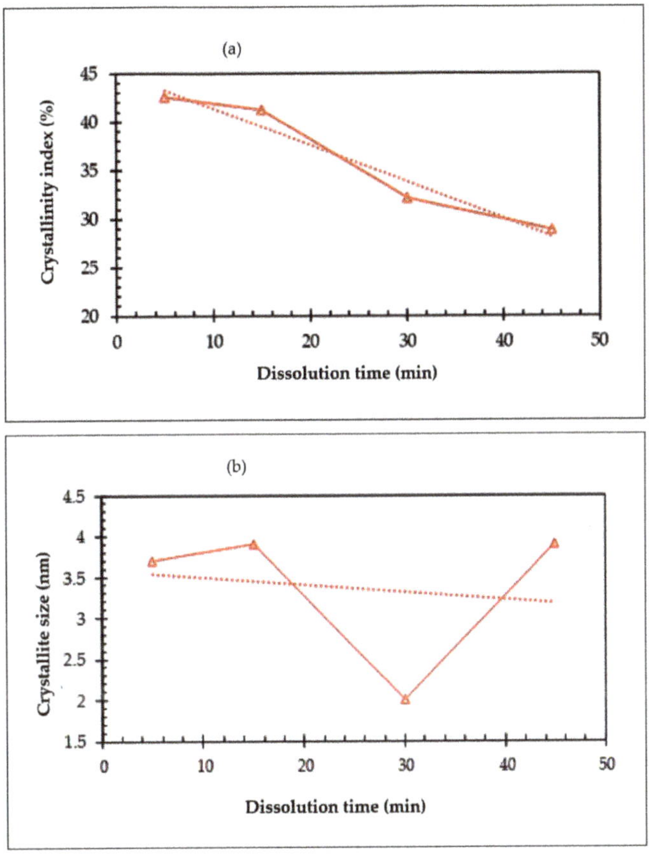


**Figure 6.** The variations in crystallinity index and crystallite size. (**a**) Crystallinity index; (**b**) Crystallite size.

*3.4. Scanning Electron Microscopy (SEM)*

Figure 7a–d presents the SEM images of the surface of the ACC films as a function of the dissolution time. The matrix was created from the fibers' outer portions and joined nearby fibers together after a 5-min dissolving period, as it can be clearly seen. Due to insufficient wetting of the fibers at this early stage, the TS value is the lowest compared to others, as predicted, and it is still possible to see microfibers that have not yet dissolved.

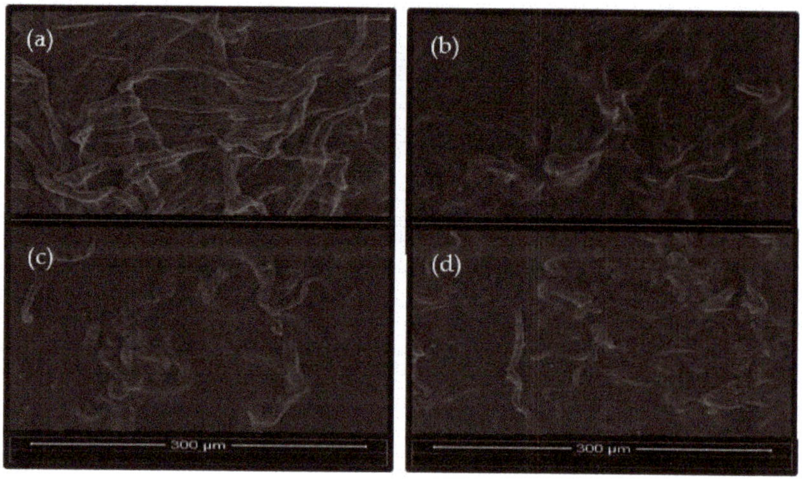


**Figure 7.** SEM micrographs of the surface morphology of the all-cellulose composites at different dissolution times: (**a**) ACC5, (**b**) ACC15, (**c**) ACC30, and (**d**) ACC45. (magnification ×500).

Theoretically, with the increase in the dissolution time, larger amounts of the matrix were formed, as reported [23,55]. At 15 min of immersion, there was a skyrocketing increase in the TS value. This can be explained by the formation of enough matrix from the disintegration of cellulose to cover the spaces between the fibers. In the SEM images, there is a noticeable reduction in the outlines and sinking of the fibers' "vein". However, the TS value gradually decreased as the dissolution time increased. This is due to the films becoming more brittle as a result of the excessive amount of matrix created. Hence, the key to understanding the failure behavior lies in the optimum 15-min dissolution time. This SEM analysis supports the TS results, whereby the dissolution time in the range of 15–45 min has a higher tensile strength compared to the 5-min initial dissolution time.

Morphology studies using SEM confirmed the XRD and TS results, i.e., that the coexistence of cellulose I and II enhanced the interfacial adhesion between the reinforcing fibers and matrices as the failure behavior of the film composite changed from fiber pullout to fiber breakage. In addition, a better surface consists of fewer voids.

The study of the cross-sectional characteristics of ACC composite films can provide a fair indication of the degree of interfacial adhesion between the fibers in a composite. In this case, the composites prepared with longer dissolution times exhibit better TS values. The fracture surfaces of the ACC films after tensile deformation are shown in Figure 8. Following the five minutes of dissolution, the microvoids and fiber pullout deformation are visible, as shown in Figure 8a. At the dissolution period of five minutes, the matrix (dissolved cellulose) was unable to fill all the gaps between the fibers and all adjacent fibers. Thus, the mechanical characteristics of the composites started to lower due to the inefficient stress transfer from the matrix to the reinforcement. This failure mechanism has been reported for cellulose mats with high porosity and weak interfiber bonding [27]. The ACC films showed a smoother and more homogenous surface at the dissolution time of 15 min, which is the greatest TS value of any other film. Therefore, stress can be efficiently transferred from the matrix to the reinforcement (undissolved cellulose microfibers). However, the failure mode changed to fiber breakage after the 15-min dissolution time, showing higher mechanical properties resulting from the better interaction between the microfibers and the cellulose matrix.

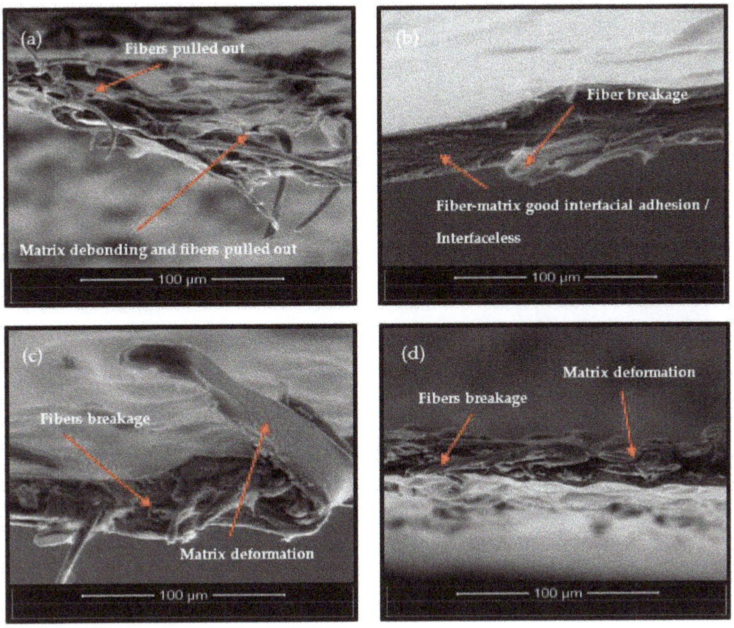


**Figure 8.** Cross-sectional SEM images of all-cellulose composite films at different dissolution times: (**a**) ACC5, (**b**) ACC15, (**c**) ACC30, and (**d**) ACC45. (magnification ×500).

### *3.5. Thermogravimetric TGA*

In order to examine the influence of the dissolution times on the thermal stability of the film composites, a thermogravimetric analysis of the OPEFB cellulose at different dissolution times in the temperature range of 30–800 °C was performed. Figure 9 presents the TGA/DTG curves of the all-cellulose composites at different dissolution times.

The weight loss in the early temperature stage (<100 °C) can be attributed to the evaporation of moisture physically adsorbed to the ACC structure [27,56]. The weight loss for all types of ACC films remained nearly unchanged at temperatures between 100 and 240 °C. This could be due to the removal of the remaining hemicellulose and extractives during the pulping and bleaching processes and the fiber isolation and extraction. Meanwhile, the sharp drop in weight in the 250–350 °C temperature range corresponds to the loss of crystal water and the thermal degradation of the main cellulose materials [24]. The variations in the amount of residual char of the ACC samples at 800 °C ($R_{800}$), the 50% weight loss temperatures ($T_{w50}$) of the samples, and the temperature of the minimum point of the DTG peak ($DTG_{min}$) are shown in Figure 10; the value is also included in Table 5.

**Table 5.** Thermal properties of the ACC films.

| Sample | Degradation Temperature, °C $T_{w50}$ | $DTG_{min}$, °C | Residual Char at 800 °C, % $R_{800^\circ C}$ |
|---|---|---|---|
| ACC5 | 335.5 | 327.66 | 15.95 |
| ACC15 | 319.83 | 315.34 | 15.49 |
| ACC30 | 312 | 304.17 | 18.86 |
| ACC45 | 327.66 | 319.83 | 14.66 |

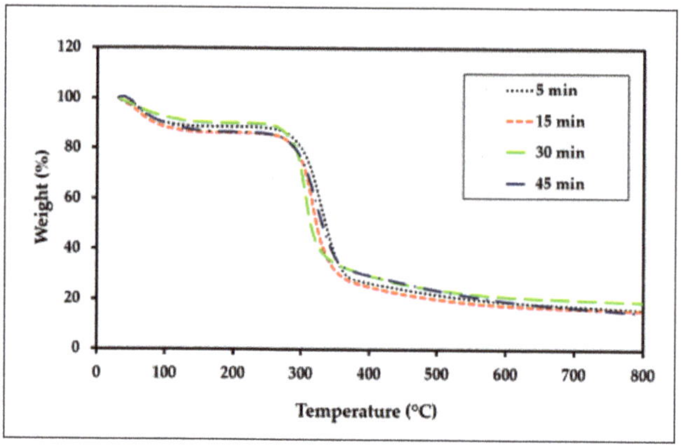


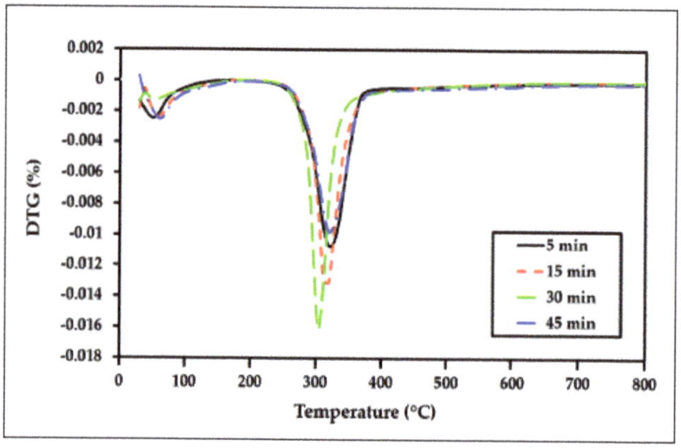


**Figure 9.** TGA−DTG curves of all-cellulose composites after 5, 15, 30, and 45 min of dissolution.

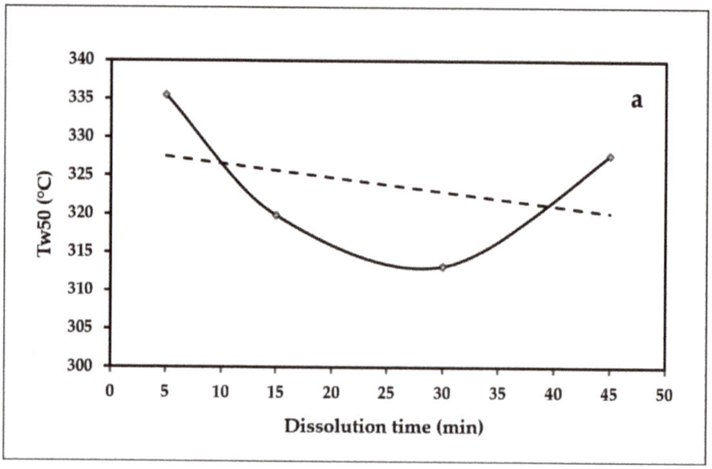


**Figure 10.** *Cont.*

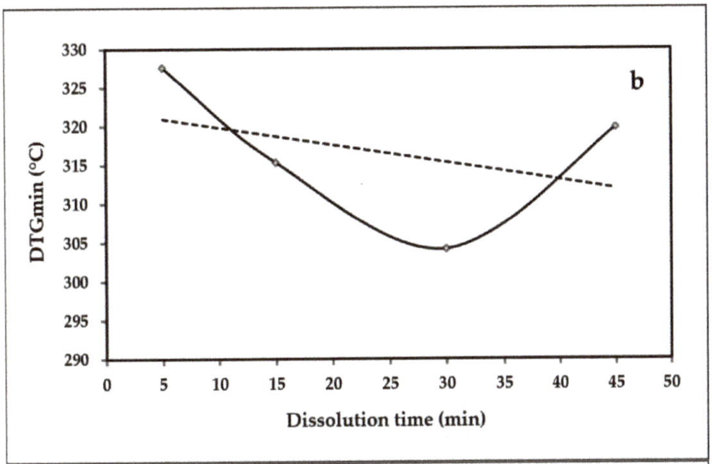


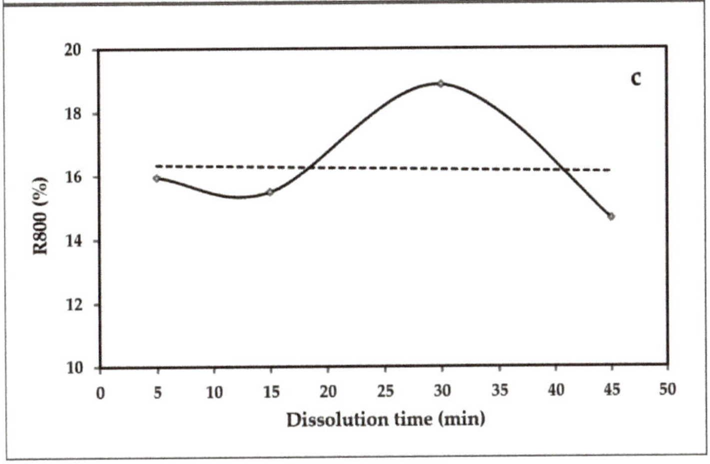


**Figure 10.** The variations in (**a**) the 50% weight loss temperatures ($T_{w50}$), (**b**) the temperature of the minimum point of the DTG peak ($DTG_{min}$), and (**c**) the amount of residual char at 800 °C ($R_{800}$).

It is clear from Figure 10a,b that the general trends of the $T_{w50}$ and $DTG_{min}$ are decreasing, and longer dissolution times lead to a decrease in the temperatures, as previously indicated by other researchers [24,26,27]. This is because the long dissolution time results in an almost complete dissolution of the cellulose fibers. On the other hand, the general variation trends, shown in Figure 10c, indicate that the amount of residual char at 800 °C increased with increasing the dissolution time. The increased char residue due to the increased dissolution time can be related to the fact that the surface of the cellulose structure exhibits flame retardant properties, which acted as a catalyst for the dehydration and thermal degradation of the ACCs, resulting in the increased char residue in the presence of more cellulose-based structures (lower dissolution times) [57].

### 3.6. *Differential Scanning Calorimetry (DSC)*

Figure 11 shows the DSC thermograms of the understudied samples from 30 °C to 400 °C. The endothermic peak in the temperature range of 250–370 °C was attributed to cellulose melting, which corresponded to the breakage of glycosidic bonds and the depolymerization of the cellulose [58]. This indicated that the molecular structure had changed, as observed in the TGA curves, and was also related to thermal decomposition.

Moreover, the thermal transitions in this temperature range are associated with the onset of thermal degradation of the cellulose, which consists of the rearrangement of molecular chains, followed by molecular chain cleavage of glucosidic bonds and intermolecular cross-linking [59]. This is in agreement with the work of Miranda et al. [60], which used simultaneous differential thermal analysis (simultaneous DSC-TGA).

Figure 12 presents the variations of the maximum point temperature of each peak and the obtained ΔH from the DSC diagrams versus the dissolution time for the samples. The values are tabulated in Table 6. According to Figure 12a, the general trend of $T_{max}$ decreases with increasing the dissolution time of the samples. It can be explained by the breakage of more intramolecular and intermolecular hydrogen bonding in the molecular chain during the dissolution process, resulting in a decrease in crystallinity and an endothermic transition. Besides that, Liu et al. [61] also identified that regenerated cellulose films have a lower endothermic peak than the original cotton pulp due to the breakage of hydrogen bonds. From Figure 12b, it is clear that the dissolution time has no significant effect on the ΔH values.

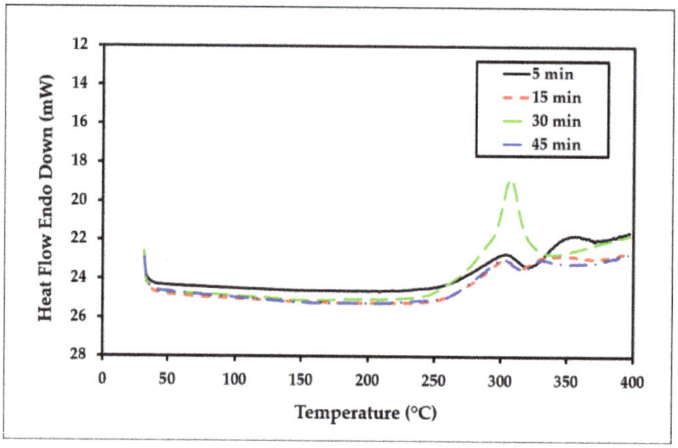


**Figure 11.** DSC thermograms of the all-cellulose composites after 5, 15, 30, and 45 min of dissolution time.

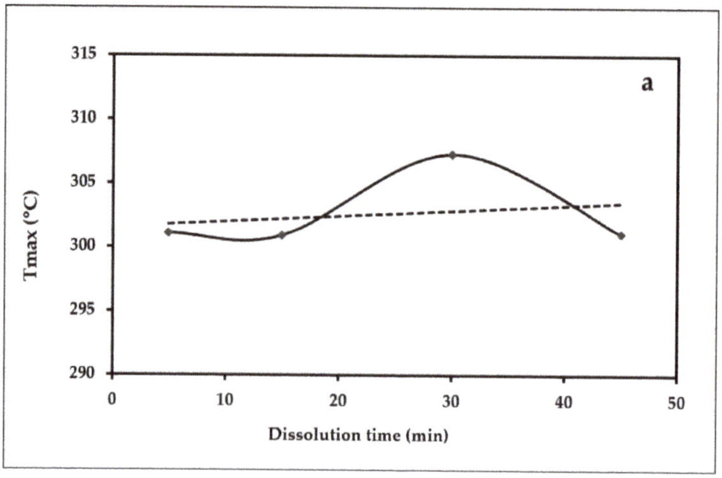


**Figure 12.** *Cont.*

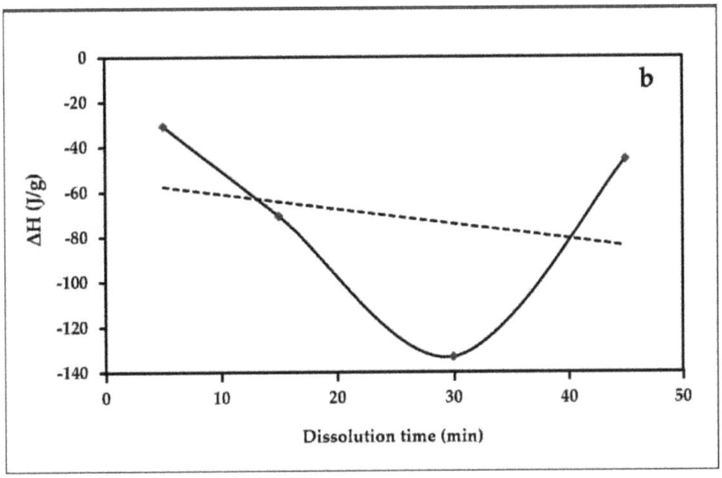


**Figure 12.** The variations in (**a**) $T_{max}$ and (**b**) ΔH versus dissolution time.

**Table 6.** The variations in (a) $T_{max}$ and (b) ΔH versus dissolution time.

| Sample | Maximum Point Temperature, °C $T_{max}$ | Enthalpy, (J/g) ΔH |
|---|---|---|
| ACC5 | 301.09 | −30.79 |
| ACC15 | 300.91 | −70.53 |
| ACC30 | 307.32 | −123.24 |
| ACC45 | 301.07 | −45.64 |

## 4. Conclusions

The relationships between the dissolution time and the properties of the all-cellulose composites obtained by the partial dissolution of OPEFBs in LiCl/DMAc were successfully investigated. The following are the main findings that can be summarized from the performed analyses:

- The amount of dissolved fiber surfaces is adequate to provide sufficient interfacial adhesion to the composite, while a considerable fraction of the fiber cores remain, reinforcing the material.
- The best dissolution time was discovered to be 15 min, which has the highest tensile strength.
- A decrease in the crystallite size and degree of crystallinity was observed with an increase in the dissolution time in the ACC films. The initial crystallinity of the OPEFB-bleached pulp affected the processing and the properties of the all-cellulose composites.
- The thermal stability of the ACC films shows a declining trend as the dissolution time is increased.

The best composite produced in this study has comparable tensile qualities to other ACC films described in past studies, making it a good candidate to be used in the production of green composites.

**Author Contributions:** Writing—original draft preparation, M.Z.J.; writing—review and editing, M.Z.J., M.H.M.K. and F.A.; conceptualization, visualization, methodology, formal analysis, and validation, M.Z.J. and M.H.M.K.; software and data curation, M.Z.J., F.F.M.R. and M.H.M.K.; project administration, supervision, and funding acquisition, F.A. and M.H.M.K. All authors have read and agreed to the published version of the manuscript.

**Funding:** This work was supported by Fundamental Research Grant Scheme FRGS/1/2022/STG05/USM02/10 from the Ministry of Higher Education, Malaysia, and the Research University Grant Scheme (RU) 1001/PTEKIND/8011108 allocated by Universiti Sains Malaysia.

**Acknowledgments:** The authors are grateful to the Bioresource Technology Division of the School of Industrial Technology for the facilities provided during the fabrication of the nanocomposite films.

**Conflicts of Interest:** This manuscript, or its contents in some other form, has not been published previously by any of the authors and/or is not under consideration for publication in another journal at the time of submission. All authors declare that there is no conflict of interest regarding the publication of this paper.

## References

1. OECD. Only Nine Percent of Plastic Recycled Worldwide. Available online: https://phys.org/news/2022-02-percent-plastic-recycled-worldwide-oecd.html (accessed on 16 December 2022).
2. JEC Group. New Trends in Composites and Plastics Recycling. Available online: https://www.jeccomposites.com/news/new-trends-in-composites-and-plastics-recycling/ (accessed on 16 December 2022).
3. Osman, A.F.; Ashafee, A.; Moh. T., L.; Adnan, S.A.; Alakrach, A. Influence of Hybrid Cellulose/Bentonite Fillers on Structure, Ambient, and Low Temperature Tensile Properties of Thermoplastic Starch Composites. *Polym. Eng. Sci.* **2020**, *60*, 810–822. [CrossRef]
4. Chen, F.; Sawada, D.; Hummel, M.; Sixta, H.; Budtova, T. Unidirectional All-Cellulose Composites from Flax via Controlled Impregnation with Ionic Liquid. *Polymers* **2020**, *12*, 1010. [CrossRef]
5. Spörl, J.M.; Batti, F.; Vocht, M.-P.; Raab, R.; Müller, A.; Hermanutz, F.; Buchmeiser, M.R. Ionic Liquid Approach toward Manufacture and Full Recycling of All-Cellulose Composites. *Macromol. Mater. Eng.* **2017**, *303*, 1700335. [CrossRef]
6. Gupta, H.; Kanaujia, K.K.; Abbas, R.S.M.; Shukla, R. A Review on the Mechanical Properties of Natural Fibre Reinforced Polypropylene Composites. *Int. Res. J. Eng. Technol.* **2019**, *6*, 337–342.
7. Nazrin, A.; Sapuan, S.M.; Zuhri, M.Y.M.; Ilyas, R.A.; Syafiq, R.; Sherwani, S.F.K. Nanocellulose Reinforced Thermoplastic Starch (TPS), Polylactic Acid (PLA), and Polybutylene Succinate (PBS) for Food Packaging Applications. *Front. Chem.* **2020**, *8*, 213. [CrossRef] [PubMed]
8. Jumaidin, R.; Khiruddin, M.A.A.; Asyul Sutan Saidi, Z.; Salit, M.S.; Ilyas, R.A. Effect of Cogon Grass Fibre on the Thermal, Mechanical and Biodegradation Properties of Thermoplastic Cassava Starch Biocomposite. *Int. J. Biol. Macromol.* **2020**, *146*, 746–755. [CrossRef]
9. Sari, N.H.; Pruncu, C.I.; Sapuan, S.M.; Ilyas, R.A.; Catur, A.D.; Suteja, S.; Sutaryono, Y.S.; Pullen, G. The Effect of Water Immersion and Fibre Content on Properties of Corn Husk Fibres Reinforced Thermoset Polyester Composite. *Polym. Test.* **2020**, *91*, 106751. [CrossRef]
10. Amiandamhen, S.O.; Meincken, M.; Tyhoda, L. Natural Fibre Modification and Its Influence on Fibre-Matrix Interfacial Properties in Biocomposite Materials. *Fibers Polym.* **2020**, *21*, 677–689. [CrossRef]
11. Azizi Samir, M.A.S.; Alloin, F.; Dufresne, A. Review of Recent Research into Cellulosic Whiskers, Their Properties and Their Application in Nanocomposite Field. *Biomacromolecules* **2005**, *6*, 612–626. [CrossRef]
12. Bondeson, D.; Syre, P.; Niska, K.O. All Cellulose Nanocomposites Produced by Extrusion. *J. Biobased Mater. Bioenergy* **2007**, *1*, 367–371. [CrossRef]
13. Duchemin, B.J.C.; Newman, R.H.; Staiger, M.P. Structure–Property Relationship of All-Cellulose Composites. *Compos. Sci. Technol.* **2009**, *69*, 1225–1230. [CrossRef]
14. Huber, T.; Müssig, J.; Curnow, O.; Pang, S.; Bickerton, S.; Staiger, M.P. A Critical Review of All-Cellulose Composites. *J. Mater. Sci.* **2011**, *47*, 1171–1186. [CrossRef]
15. Baghaei, B.; Skrifvars, M. All-Cellulose Composites: A Review of Recent Studies on Structure, Properties and Applications. *Molecules* **2020**, *25*, 2836. [CrossRef]
16. Capiati, N.J.; Porter, R.S. The Concept of One Polymer Composites Modelled with High Density Polyethylene. *J. Mater. Sci.* **1975**, *10*, 1671–1677. [CrossRef]
17. Nishino, T.; Matsuda, I.; Hirao, K. All-Cellulose Composite. *Macromolecules* **2004**, *37*, 7683–7687. [CrossRef]
18. Nishino, T.; Arimoto, N. All-Cellulose Composite Prepared by Selective Dissolving of Fiber Surface. *Biomacromolecules* **2007**, *8*, 2712–2716. [CrossRef]
19. Gindl, W.; Keckes, J. All-Cellulose Nanocomposite. *Polymer* **2005**, *46*, 10221–10225. [CrossRef]
20. Gindl, W.; Martinschitz, K.J.; Boesecke, P.; Keckes, J. Structural Changes during Tensile Testing of an All-Cellulose Composite by in Situ Synchrotron X-Ray Diffraction. *Compos. Sci. Technol.* **2006**, *66*, 2639–2647. [CrossRef]
21. Gindl, W.; Schöberl, T.; Keckes, J. Structure and Properties of a Pulp Fibre-Reinforced Composite with Regenerated Cellulose Matrix. *Appl. Phys. A* **2006**, *83*, 19–22. [CrossRef]
22. Li, J.; Nawaz, H.; Wu, J.; Zhang, J.; Wan, J.; Mi, Q.; Yu, J.; Zhang, J. All-Cellulose Composites Based on the Self-Reinforced Effect. *Compos. Commun.* **2018**, *9*, 42–53. [CrossRef]

23. Cheng, G.; Zhu, P.; Li, J.; Cheng, F.; Lin, Y.; Zhou, M. All-Cellulose Films with Excellent Strength and Toughness via a Facile Approach of Dissolution-Regeneration. *J. Appl. Polym. Sci.* **2018**, *136*, 46925. [CrossRef]
24. Arévalo, R.; Picot, O.T.; Wilson, R.M.; Soykeabkaew, N.; Peijs, T. All-Cellulose Composites by Partial Dissolution of Cotton Fibres. *J. Biobased Mater. Bioenergy* **2010**, *4*, 129–138. [CrossRef]
25. Yousefi, H.; Faezipour, M.; Nishino, T.; Shakeri, A.; Ebrahimi, G. All-Cellulose Composite and Nanocomposite Made from Partially Dissolved Micro-and Nanofibers of Canola Straw. *Polym. J.* **2011**, *43*, 559–564. [CrossRef]
26. Ghaderi, M.; Mousavi, M.; Yousefi, H.; Labbafi, M. All-Cellulose Nanocomposite Film Made from Bagasse Cellulose Nanofibers for Food Packaging Application. *Carbohydr. Polym.* **2014**, *104*, 59–65. [CrossRef]
27. Tanpichai, S.; Witayakran, S. All-Cellulose Composites from Pineapple Leaf Microfibers: Structural, Thermal, and Mechanical Properties. *Polym. Compos.* **2016**, *39*, 895–903. [CrossRef]
28. Senthil Muthu Kumar, T.; Rajini, N.; Obi Reddy, K.; Varada Rajulu, A.; Siengchin, S.; Ayrilmis, N. All-Cellulose Composite Films with Cellulose Matrix and Napier Grass Cellulose Fibril Fillers. *Int. J. Biol. Macromol.* **2018**, *112*, 1310–1315. [CrossRef]
29. Gindl-Altmutter, W.; Keckes, J.; Plackner, J.; Liebner, F.; Englund, K.; Laborie, M.-P. All-Cellulose Composites Prepared from Flax and Lyocell Fibres Compared to Epoxy–Matrix Composites. *Compos. Sci. Technol.* **2012**, *72*, 1304–1309. [CrossRef]
30. Norrrahim, M.N.F.; Ariffin, H.; Hassan, M.A.; Ibrahim, N.A.; Yunus, W.M.Z.W.; Nishida, H. Utilisation of Superheated Steam in Oil Palm Biomass Pretreatment Process for Reduced Chemical Use and Enhanced Cellulose Nanofibre Production. *Int. J. Nanotechnol.* **2019**, *16*, 668. [CrossRef]
31. Suriani, M.J.; Radzi, F.S.M.; Ilyas, R.A.; Petrů, M.; Sapuan, S.M.; Ruzaidi, C.M. Flammability, Tensile, and Morphological Properties of Oil Palm Empty Fruit Bunches Fiber/Pet Yarn-Reinforced Epoxy Fire Retardant Hybrid Polymer Composites. *Polymers* **2021**, *13*, 1282. [CrossRef]
32. Abdul Khalil, H.P.S.; Marliana, M.M.; Issam, A.M.; Bakare, I.O. Exploring Isolated Lignin Material from Oil Palm Biomass Waste in Green Composites. *Mater. Des.* **2011**, *32*, 2604–2610. [CrossRef]
33. Isroi; Cifriadi, A.; Panji, T.; Wibowo, N.A.; Syamsu, K. Bioplastic Production from Cellulose of Oil Palm Empty Fruit Bunch. *IOP Conf. Ser. Earth Environ. Sci.* **2017**, *65*, 012011. [CrossRef]
34. Zailuddin, N.L.I.; Osman, A.F.; Husseinsyah, S.; Ariffin, Z.; Badrun, F.H. Mechanical Properties and X-Ray Diffraction of Oil Palm Empty Fruit Bunch All-Nanocellulose Composite Films. In *Proceedings of the Second International Conference on the Future of ASEAN (ICoFA) 2017—Volume 2*; Springer: Singapore, 2018; pp. 505–514. [CrossRef]
35. Zailuddin, N.L.I.; Osman, A.F.; Rahman, R. Morphology, Mechanical Properties, and Biodegradability of All-Cellulose Composite Films from Oil Palm Empty Fruit Bunch. *SPE Polym.* **2020**, *1*, 4–14. [CrossRef]
36. Zailuddin, N.L.I.; Osman, A.F.; Rahman, R. Effects of Formic Acid Treatment on Properties of Oil Palm Empty Fruit Bunch (OPEFB)-Based All Cellulose Composite (ACC) Films. *J. Eng. Sci.* **2020**, *16*, 75–95. [CrossRef]
37. Gea, S.; Andita, D.; Rahayu, S.; Nasution, D.Y.; Rahayu, S.U.; Piliang, A.F. Preliminary Study on the Fabrication of Cellulose Nanocomposite Film from Oil Palm Empty Fruit Bunches Partially Solved into Licl/Dmac with the Variation of Dissolution Time. *J. Phys. Conf. Ser.* **2018**, *1116*, 042012. [CrossRef]
38. Gea, S.; Panindia, N.; Piliang, A.F.; Sembiring, A.; Hutapea, Y.A. All-Cellulose Composite Isolated from Oil Palm Empty Fruit Bunch. *J. Phys. Conf. Ser.* **2018**, *1116*, 042013. [CrossRef]
39. Husseinsyah, S.; Zailuddin, N.L.I.; Osman, A.F.; Li Li, C.; Alrashdi, A.A.; Alakrach, A. Methyl Methacrylate (MMA) Treatment of Empty Fruit Bunch (EFB) to Improve the Properties of Regenerated Cellulose Biocomposite Films. *Polymers* **2020**, *12*, 2618. [CrossRef] [PubMed]
40. Rosli, W.D.W.; Leh, C.P.; Zainuddin, Z.; Tanaka, R. Optimisation of Soda Pulping Variables for Preparation of Dissolving Pulps from Oil Palm Fibre. *Holzforschung* **2003**, *57*, 106–113. [CrossRef]
41. Soykeabkaew, N.; Arimoto, N.; Nishino, T.; Pejis, T. All-Cellulose Composites by Surface Selective Dissolution of Aligned Ligno-Cellulosic Fibres. *Compos. Sci. Technol.* **2008**, *68*, 2201–2207. [CrossRef]
42. Soykeabkaew, N.; Sian, C.; Gea, S.; Nishino, T.; Peijs, T. All-Cellulose Nanocomposites by Surface Selective Dissolution of Bacterial Cellulose. *Cellulose* **2009**, *16*, 435–444. [CrossRef]
43. Korhonen, O.; Sawada, D.; Budtova, T. All-Cellulose Composites via Short-Fiber Dispersion Approach Using NaOH–Water Solvent. *Cellulose* **2019**, *26*, 4881–4893. [CrossRef]
44. Duchemin, B.; Le Corre, D.; Leray, N.; Dufresne, A.; Staiger, M.P. All-Cellulose Composites Based on Microfibrillated Cellulose and Filter Paper via a NaOH-Urea Solvent System. *Cellulose* **2015**, *23*, 593–609. [CrossRef]
45. Shahmoradi, A.R.; Talebibahmanbigloo, N.; Javidparvar, A.A.; Bahlakeh, G.; Ramezanzadeh, B. Studying the Adsorption/Inhibition Impact of the Cellulose and Lignin Compounds Extracted from Agricultural Waste on the Mild Steel Corrosion in HCl Solution. *J. Mol. Liq.* **2020**, *304*, 112751. [CrossRef]
46. Wei, Q.Y.; Lin, H.; Yang, B.; Li, L.; Zhang, L.Q.; Huang, H.D.; Zhong, G.J.; Xu, L.; Li, Z.M. Structure and Properties of All-Cellulose Composites Prepared by Controlling the Dissolution Temperature of a NaOH/Urea Solvent. *Ind. Eng. Chem. Res.* **2020**, *59*, 10428–10435. [CrossRef]
47. Perez, S.; Samain, D. Structure and Engineering of Celluloses. *Adv. Carbohydr. Chem. Biochem.* **2010**, *1*, 25–116.
48. Carrillo, F.; Colom, X.; Suñol, J.J.; Saurina, J. Structural FTIR Analysis and Thermal Characterisation of Lyocell and Viscose-Type Fibres. *Eur. Polym. J.* **2004**, *40*, 2229–2234. [CrossRef]

49. Nelson, M.L.; O'Connor, R.T. Relation of Certain Infrared Bands to Cellulose Crystallinity and Crystal Latticed Type. Part, I. Spectra of Lattice Types I, II, III and of Amorphous Cellulose. *J. Appl. Polym. Sci.* **1964**, *8*, 1311–1324. [CrossRef]
50. Nelson, M.L.; O'Connor, R.T. Relation of Certain Infrared Bands to Cellulose Crystallinity and Crystal Lattice Type. Part II. A New Infrared Ratio for Estimation of Crystallinity in Celluloses I and II. *J. Appl. Polym. Sci.* **1964**, *8*, 1325–1341. [CrossRef]
51. Segal, L.; Creely, J.J.; Martin, A.E.; Conrad, C.M. An Empirical Method for Estimating the Degree of Crystallinity of Native Cellulose Using the X-Ray Diffractometer. *Text. Res. J.* **1959**, *29*, 786–794. [CrossRef]
52. Patterson, A.L. The Scherrer Formula for X-Ray Particle Size Determination. *Phys. Rev.* **1939**, *56*, 978–982. [CrossRef]
53. Sun, L.; Chen, J.Y.; Jiang, W.; Lynch, V. Crystalline Characteristics of Cellulose Fiber and Film Regenerated from Ionic Liquid Solution. *Carbohydr. Polym.* **2015**, *118*, 150–155. [CrossRef]
54. Pang, J.; Wu, M.; Zhang, Q.; Tan, X.; Xu, F.; Zhang, X.; Sun, R. Comparison of Physical Properties of Regenerated Cellulose Films Fabricated with Different Cellulose Feedstocks in Ionic Liquid. *Carbohydr. Polym.* **2015**, *121*, 71–78. [CrossRef] [PubMed]
55. Tang, X.; Liu, G.; Zhang, H.; Gao, X.; Li, M.; Zhang, S. Facile Preparation of All-Cellulose Composites from Softwood, Hardwood, and Agricultural Straw Cellulose by a Simple Route of Partial Dissolution. *Carbohydr. Polym.* **2021**, *256*, 117591. [CrossRef] [PubMed]
56. Roshanghias, A.; Sodeifian, G.; Javidparvar, A.A.; Tarashi, S. Construction of a Novel Polytetrafluoroethylene-Based Sealant Paste: The Effect of Polyvinyl Butyral (PVB) and Nano-Alumina on the Sealing Performance and Construction Formulations. *Results Eng.* **2022**, *14*, 100460. [CrossRef]
57. Jiang, Z.; Tang, L.; Gao, X.; Zhang, W.; Ma, J.; Zhang, L. Solvent Regulation Approach for Preparing Cellulose-Nanocrystal-Reinforced Regenerated Cellulose Fibers and Their Properties. *ACS Omega* **2019**, *4*, 2001–2008. [CrossRef] [PubMed]
58. Ciolacu, D.; Ciolacu, F.; Popa, V.I. Amorphous Cellulose—Structure and Characterization. *Cellul. Chem. Technol.* **2011**, *45*, 13.
59. Yeng, L.C.; Wahit, M.U.; Othman, N. Thermal and Flexural Properties of Regenerated Cellulose (RC)/Poly(3-Hydroxybutyrate) (PHB) Biocomposites. *J. Teknol.* **2015**, *75*, 107–112. [CrossRef]
60. Miranda, M.I.G.; Bica, C.I.D.; Nachtigall, S.M.B.; Rehman, N.; Rosa, S.M.L. Kinetical Thermal Degradation Study of Maize Straw and Soybean Hull Celluloses by Simultaneous DSC–TGA and MDSC Techniques. *Thermochim. Acta* **2013**, *565*, 65–71. [CrossRef]
61. Liu, Z.; Wang, H.; Li, Z.; Lu, X.; Zhang, X.; Zhang, S.; Zhou, K. Characterization of the Regenerated Cellulose Films in Ionic Liquids and Rheological Properties of the Solutions. *Mater. Chem. Phys.* **2011**, *128*, 220–227. [CrossRef]

polymers

MDPI

*Article*

# Nitrocellulose Based Film-Forming Gels with Cinnamon Essential Oil for Covering Surface Wounds

**Lauryna Pudžiuvelytė [1,2], Evelina Drulytė [2] and Jurga Bernatonienė [1,2,*]**

[1] Institute of Pharmaceutical Technologies, Medical Academy, Lithuanian University of Health Sciences, Sukileliu pr. 13, LT-50161 Kaunas, Lithuania
[2] Department of Drug Technology and Social Pharmacy, Medical Academy, Lithuanian University of Health Sciences, Sukileliu pr. 13, LT-50161 Kaunas, Lithuania
* Correspondence: jurga.bernatoniene@lsmuni.lt

**Abstract:** Acute and chronic wounds caused by assorted reasons impact patient's quality of life. Films are one of the main types of moisture retentive dressings for wounds. To improve the healing of the wound, films must ensure there is no microorganism contamination, protect from negative environmental effects, and support optimal moisture content. The aim of this study was to formulate optimal film-forming gel compositions that would have good physico-chemical properties and be suitable for wound treatment. Nitrocellulose, castor oil, ethanol (96%), ethyl acetate, and cinnamon leaf essential oil were used to create formulations. During the study, the drying rate, adhesion, flexibility, tensile strength, cohesiveness, swelling, water vapor penetration, pH value, and morphology properties of films were examined. Results showed that optimal concentrations of nitrocellulose for film-forming gel production were 13.4% and 15%. The concentrations of nitrocellulose and cinnamon leaf essential oil impacted the films' physicochemical properties (drying rate, swelling, adhesion, flexibility, etc.). The swelling test showed that films of formulations could absorb significant amounts of simulant wound exudate. Film-forming gels and films showed no microbial contamination and were stable three months after production.

**Keywords:** nitrocellulose; film-forming gel; film; cinnamon leaf essential oil; wounds

**Citation:** Pudžiuvelytė, L.; Drulytė, E.; Bernatonienė, J. Nitrocellulose Based Film-Forming Gels with Cinnamon Essential Oil for Covering Surface Wounds. *Polymers* **2023**, *15*, 1057. https://doi.org/10.3390/polym15041057

Academic Editors: Vsevolod Aleksandrovich Zhuikov and Rosane Michele Duarte Soares

Received: 29 December 2022
Revised: 17 February 2023
Accepted: 17 February 2023
Published: 20 February 2023



## 1. Introduction

A large part of society suffers from wounds caused by assorted reasons. They can be acute, formed by mechanical damage to the skin, injuries, and chronic, which are caused by certain diseases (diabetes, varicose, cancer, and others) and classified by etiology into four categories, each with its own typical location, depth, and appearance: arterial, diabetic, pressure, and venous [1–3]. Both types of wounds cause discomfort to patients: swelling, bleeding, exudation, purulence, non-healing, and infections; some chronic wounds can take decades to heal, thus contributing to secondary conditions such as depression, and can lead to isolation and family distress [2–4]. Semi-solid dosage forms such as creams, ointments, gels, and patches have traditionally been used to heal the wounds and deliver drug or natural active ingredients transdermally. However, the main disadvantage of these preparations is the limited contact with skin due to the rub-off by clothes during daily activities [5]. In this field, there is a need to obtain transdermal dosage forms which would avoid skin irritation and contact the skin closely for a prolonged time to deliver active compounds [5,6]. In this regard, the film forming system (FFS) is a novel approach which can be used as an alternative to conventional topical and transdermal formulations due to its strong adhesion to the skin, sustained drug delivery to the affected site, and the reduced need for frequent applications [6,7]. Film-forming gels are semi-solid formulations of both solid and liquid materials; usually transparent and colourless, they are cosmetically appealing [8]. A semi-solid form is obtained by treating the liquid phase with gelling agents, but the polymers must also have film-forming properties [9,10]. Film-forming gels form a thin, transparent film in situ by solvent evaporation [11]. Addition of certain materials into the composition of such gels allows to select various desired properties for the films, for example: easy application, high adherence to skin, resistance to water, air permeability, flexibility, the ability to absorb wound exudate and control its moisture, softness, and easy, painless removal of the films at the patient's request [8,11]. It is also important that the resulting protective layer traps external pathogenic microorganisms, thus preventing them from penetrating the injured area and causing an infection [1]. The production of a film-forming gel with some or all of the above-mentioned properties would result in a dosage form optimally suited for protecting any wound from environmental factors and infection [8,11]. Natural substances used against infection could be essential oils [12]. Cinnamon (*Cinnamomum*, family *Lauraceae*) leaf or bark essential oils have various biological activities including antibacterial, antioxidant, antifungal, and cytotoxic [13,14]. Eugenol, benzyl benzoate, linalool, borneol cinnamaldehyde, trans-cinnamyl acetate, benzene dicarboxylic acid, α-pinene, and coumaric acid are the main chemical compounds determined in the composition of cinnamon essential oil [13,15].

The production of gels that form films, as found in the literature, is usually using the mass method. This production method is quite simple and does not require high technical costs. The liquid phase of film-forming gels is treated with gelling polymers capable of forming films, and other substances of significant importance to the effect of the drug are added after allowing the gel to form.

The aim of this work is to design nitrocellulose films containing cinnamon essential oil, to perform their optimization and to evaluate the physico-chemical, mechanical, and biopharmaceutical properties of the films.

## 2. Materials and Methods

### 2.1. Materials

The main ingredients used during experiments were 96% alcohol (AB "Vilniaus degtinė", Vilnius, Lithuania), castor oil, ethyl acetate, sodium chloride, calcium chloride, buffer solution (pH = 7.5) (Sigma-Aldrich, Darmstadt, Germany), cinnamon leaf essential oil (UAB "Naujoji Barmunė", Vilnius, Lithuania), jojoba oil (UAB "Kvapai", Vilnius, Lithuania), purified water (Ph. Eur. 01/2008:0008, LSMU laboratory), gelatin (UAB "Klingai", Vilnius, Lithuania).

### 2.2. Preparation of Gels Forming Films

The base of a film-forming gel was prepared from ethyl acetate, ethanol, and castor oil. Essential oil was mixed with ethanol, then ethyl acetate was added and mixed well again. The solution was mixed with castor oil. After stirring the mixture was clear and homogeneous. Then it was poured into a glass with a gelling agent—nitrocellulose and mixed again until a homogeneous consistency. The obtained product was poured into a container and sealed [1,6]. When the gel is spread on the coating surface and the films are formed, no cracks or wrinkles should be visible. The film must be colorless and transparent to be less noticeable and for the condition of the wound to be visible [16].

### 2.3. Evaluation of Film Drying Rate

A small amount of the gel was placed on a Parafilm at 33–36°C and the time needed for the gel to form a film was recorded. If the film formed too quickly or too slowly, the formulation was adjusted [9]. After a drying time of 2 min, a glass plate was placed on the film and observed. If the film was dry there should be no liquid/gel residue on the plate. If the film was not dry, the test was repeated and the time for drying extended. A modified method was used to perform the test [6,16].

### 2.4. Assessment of Film Surface and Homogeneity Properties

Microscopy was used to evaluate the surface morphology of the formed films [17]. The films were cut into $3 \times 5$ mm$^2$ pieces and placed on a microscope slide. Images were obtained using a microscope with a working distance of 15 mm and a magnification of 40×. No lumps or uneven surface should be visible during the test.

### 2.5. Assessment of Surface Adhesiveness and Wound Adhesion

The surface adhesion of films was determined when the obtained film was completely dry. A piece of cotton wool gently touched to its surface, should not stick to the film. Adhesion was considered high if a lot of cotton fibers remained, medium if only a few fibers remained, and the film was non-sticky if no fibers remained at all [6,9]. In addition to the above-mentioned method, there was another, more specific method for evaluating the adhesiveness of films: a texture analyzer (Stable Micro Systems Texture Analyzer TA. XT plus) was used. A completely dry piece of film measuring 2 × 2 $cm^2$ was placed on a special fixed plastic table and pressed with a transparent plastic plate with a 1 cm diameter cavity in the middle, leaving an open part of the film. A stick with a flat end slightly less than 1 cm in diameter was attached to the metal support on top of the texture analyzer. The parameters were determined—the applied force during contact was 5 g, the contact time between the stick and the film—5 s, the speed of the stick movement during the test—1 mm/s. The texture analyzer lowered the stick until it pressed against the surface of the film, raised it up after 5 s and measured the force required to tear the end of the stick from the surface of the film.

The films prepared for adhesion test were cut into square strips (2 × 2 $cm^2$) and attached to a probe with a diameter of 35 mm. Before the test, a simulation of wound was prepared from 20 g of 6.67% (V/V) gelatin in a 90 mm Petri dish and kept at 4 °C overnight. Gelatin surface was covered with 500 μL of an aqueous phosphate solution (pH = 7.5). The films were kept in contact with the gelatin for 1 min. The probe was set before and during the test for 0.5 mm/s speed, after the test speed was 1 mm/s with a force of 1 N. The peak adhesion force represents the maximum force required to remove the film from the simulated wound surface, the area under the curve (AUC) represents the overall adhesion, and the cohesion represents the distance (mm) to tear the film away from the wound [17].

### 2.6. Evaluation of Film Flexibility

The flexibility of the films was evaluated by resistance and tensile tests. A texture analyzer (Stable Micro Systems Texture Analyzer TA. XT plus) was used for both tests. Resistance testing was performed using a ball with a metal needle attached to a metal support. Film in 3 × 3 $cm^2$ shape (n = 6 for each composition) was transferred to the middle of a special device stand with a round cavity of 1 cm diameter in the way that no wrinkles and unstressed areas were visible. The film was pressed so that the ball would not move and distort the results when hit it. The test parameters were determined—the ball's downward pressure distance—40 mm. The test showed the force required to pierce the film, thus showing its flexibility. The tensile strength was calculated based on the formula [17]:

$$\text{Tensile strength}\left(\text{N/mm}^2\right) = \frac{\text{applied force (N)}}{\text{initial cross} - \text{sec tional area } (\text{mm}^2)} \quad (1)$$

A tensile test was performed, where the distance and force required to break the film was measured [1]. The tensile test of the films was performed with a texture analyzer by attaching films measuring 65 × 30 $mm^2$ from both ends to clamps on opposite sides, approximately 3 cm apart (n = 6 for each composition). The clamps were held on a metal base. One of the clamps was locked, the other could move. During the measurement, the moving clamp pulls the film upwards. The stretching speed was 6 mm/s, the lifting distance of the moving clamp—150 mm. Tensile strength was measured when the film was torn [16,17].

### 2.7. Evaluation of Water Vapor Permeation and Swelling

The evaluation of water vapor permeation test was based on a modified Mu et al. method [1,18]. A round shape piece of film (n = 3 for each composition) was placed on a plastic tube of 30 mm diameter with 8 mL of distilled water. The tube was placed in an incubator, kept at a temperature of 38 ± 2 °C with a humidity of 60 ± 5% for 24 h. Water vapor transmission rate was calculated by the formula [19]:

$$\mathrm{WVTR} = \frac{\mathrm{W/t}}{\mathrm{A}} \tag{2}$$

WVTR—water vapor transmission rate; it is the amount of water vapor penetrated in grams per square centimeter in 1 h interval (g $cm^{-2}$ $d^{-1}$)
W—change in weight (g),
t—time (d),
A—sample area ($cm^2$).

Swelling was determined by weighing three 2 × 2 $cm^2$ strips of film and immersing them in 10 mL of simulated wound fluid. Simulated wound fluid was prepared from 0.184 g of calcium chloride, 4.149 g of sodium chloride, and 1000 mL of purified water [20]. After films immersion in simulated wound fluid, changes in weight were observed every 15 min for 2 h. The hydrated films were gently blotted with filter paper to remove excess simulated wound fluid from the surface and then weighed again. The swelling index was calculated by the formula [17]:

$$\text{Swelling index (\%)} = \frac{\mathrm{W_s} - \mathrm{W_d}}{\mathrm{W_d}} \times 100 \tag{3}$$

$W_s$—the weight of the film after hydration,
$W_d$—the weight of the film before hydration.

### 2.8. Microbiological Evaluation of Gel Quality

Microbiological evaluation of the gels was performed on 5, 5K, 6, and 6K samples. The physiological solution was dispensed into 5 mL individual tubes and used for preparation of suspension of the following bacteria: *Staphylococcus aureus* and *Pseudomonas aeruginosa*. All bacteria were isolated from clinical material. The broth liquid medium was dispensed into test tubes to give a final volume of 10 mL (with a sample of film-forming gel). The medium was sterilized. For each bacterial culture Mueller Hinton broth was used. The tubes were inoculated with 10 µL of bacterial suspension with the film-forming gel. After 48 h of incubation, each tube was inoculated with 10 µL of suspension on soy-tryptone agar (Thermo Fisher, Hampshire, UK) [21]. After this bacterial growth (bacterial colonies growing (+)/no growing (−)) was counted.

### 2.9. Determination of Stability and pH Value of Film-Forming Gels

The stability of gels forming films was determined after 1 month under accelerated conditions (equivalent to 3 months under normal conditions). The manufactured and packaged gels of the formulations 5 and 6 were stored at 38 ± 2 °C for 28 days. The appearance of films was recorded on the day of production, then on days 7, 14, 21, and 28 [22].

Meanwhile, the pH of the samples was determined immediately after gels preparation. Five grams of gel and 95 mL of purified water were put together, mixed for 10 min, filtered and the pH value measured [22].

### 2.10. Qualitative and Quantitative Evaluation of Essential Oil Composition of Cinnamon Leaves

For the qualitative and quantitative analysis of the essential oil of cinnamon leaves a Shimadzu GC-MS-QP2010 gas chromatograph-mass spectrometer system (Shimadzu, Tokyo, Japan) equipped with a Shimadzu autoinjector AOG-5000 (Shimadzu, Tokyo, Japan) was used. A capillary column RXI-5MS (30 m × 0.25 mm × 0.25 µm film thickness) was used. Sample of 0.2 µL was injected into the split/splitless injector with a split ratio 1:50. The operating conditions were as follows: the chamber temperature was maintained at 40 °C for 3 min and raised at 5 °C/min rate to 120 °C. After 3 min the temperature was reduced at 2 °C/min rate to 180 °C, maintained for 3 min. After 3 min the temperature was raised for the last time to 230 °C at 5 °C/min rate and again maintained for 3 min. Temperature of injector, detector and ion stream was 250 °C. Helium was used as the carrier gas. The identification of the components was based on the comparison of their mass spectrum with the spectra specified in the databases. Quantitative analysis of components was carried out by peak area normalization measurements [23].

### *2.11. Statistical Analysis*

The results obtained in the tests were evaluated by statistical analysis. Data processing program IBM SPSS Statistics (version 27.0.1) (IBM Corporation, New York, NY, USA) was used to determine arithmetic means, standard deviations, and statistical significance. Significance was assessed by Kruskal-Wallis, one-way ANOVA (k samples) tests (significant when $p < 0.05$).

## 3. Results and Discussion

### *3.1. Effect of ingredients on Film-Forming Gel Formation and Appearance of Films*

At the beginning of the study, empirical tests were conducted to find the optimal control composition. Table 1 shows that during the test, seven formulations and two controls with different amounts of used substances were produced, but only two met the visual and physical requirements. For further studies only two formulations were chosen.

**Table 1.** Composition of film-forming gels.

| Ingredients | Formulations | | | | | | | | |
|---|---|---|---|---|---|---|---|---|---|
| | 5 K | 6 K | 1 | 2 | 3 | 4 | 5 | 6 | 7 |
| Nitrocellulose | 13.4% | 15% | 11% | 15.4% | 15.4% | 16% | 13.4% | 15% | 15% |
| Ethanol (96%) | 26.1% | 24.7% | 25.7% | 24.6% | 24.6% | 24.7% | 26.1% | 24.7% | 24.7% |
| Ethyl acetate | 52.2% | 51.9% | 51.3% | 54.8% | 51.7% | 51% | 52.2% | 51.9% | 51.9% |
| Castor oil | 6.3% | 6.4% | 10% | 4% | 6.3% | 6.3% | 6.3% | 6.4% | - |
| Jojoba oil | - | - | - | - | - | - | - | - | 6.4% |
| CLEO | - | - | - | - | - | - | 2% | 2% | - |

CLEO—cinnamon leaf essential oil (active ingredient), K—control.

In the literature, concentration of cellulose derivatives in film-forming gels varied from 2% to 30% [24,25]. However, after making the product according to the first composition, which contained 11% nitrocellulose and 10% castor oil, the gel turned out liquid. The film formed, but the drying process was long, and the surface felt very sticky when dry. It also did not peel nicely from the Parafilm. Based on these observations, other formulations with higher amounts of nitrocellulose and lower amounts of castor oil were tested. By increasing the nitrocellulose content in the second composition to 15.4% and after reducing castor oil to 4% the gel was thicker, but the film was of white color. After it dried, large cracks appeared (Figure 1).

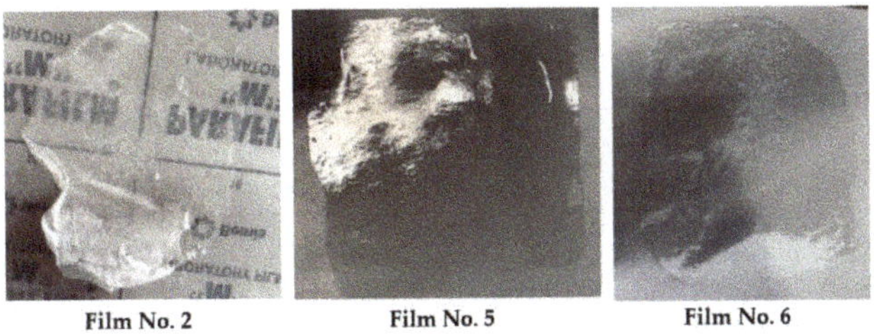


**Figure 1.** Appearance of films.

The visual investigation concluded that the film does not bend at all. This could be due to the smaller amount of castor oil. Increasing the castor oil content to 6.3% in formulation 3 produced a strong film that dried quickly and peeled off the surface, but the color remained off-white. The reason for that could be that the amount of nitrocellulose was too high. The same film appearance was obtained with formulation 4: 16% of nitrocellulose produced a thick gel, but the film was white, uneven, sticky, and cracked. Formulation 5 consisted of 13.4% nitrocellulose, the gel was significantly more liquid than in formulation 4, and it also formed transparent, elastic, fast-drying, and solid film (Figure 1).

However, stickiness was still noticeable, which can be annoying when in contact with clothes, so the amount of nitrocellulose in the composition was increased to 15%. Film 6 showed the best results—a colorless, fast-drying, a little sticky, solid, strong, but at the same time elastic film was formed (Figure 1).

During the research, castor oil was replaced with jojoba oil, trying to obtain high-quality, fast-drying films. The composition was 15% nitrocellulose, 24.7% ethanol, 51.9% ethyl acetate, and 6.4% jojoba oil. Visually, the gel containing jojoba oil was similar to gel 6, but the differences became visible after pouring it in a thin layer and observing the film formation process. After a few minutes, the film started to wrinkle, shrink at the center, and uncured gel remained at the edges. After removal from the surface, and trying to stretch the film, it was inelastic and started to turn white, so for further experiments jojoba oil was not used. The condition of the films is shown in Figure 2.

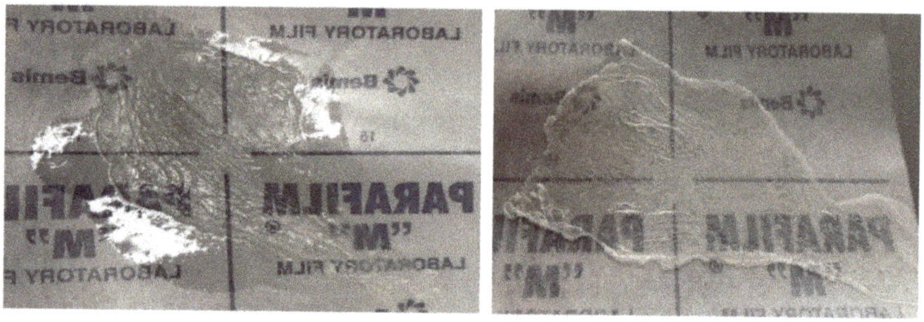

**Figure 2.** Film No. 7 with jojoba oil. On the left side—appearance of the film after film-forming gel preparation and application on the surface, on the right side—appearance of the film after 1 h.

According to the results, the gels of formulations 5 and 6 formed films with optimal properties. They were selected for further studies after adding 2% of cinnamon leaf essential oil as an active antibacterial agent. Due to the rapid formation of films and excellent appearance, these products would be suitable for use by patients, and the transparency of films will allow for monitoring of the wound's condition.

### 3.2. Results of Drying Time Test of Films

After selecting the most suitable composition, the drying time test of the films formed by the gels at 33–36 °C temperature was carried out. Drying time is a crucial factor in evaluating the quality of films. During the study, three (1–3) tests were performed by pouring the gels in a 0.8 mm layer and three (4–6) in a 1.5 mm layer. The drying of the formed films was assessed visually at the beginning—it was observed when the surface became no longer gel-like, and drying time was recorded with a timer. After that, a microscope slide was placed on top of it to check if the film was completely formed. Drying time results presented in Table 2 show that the film formation time for the gel layers of 0.8 mm was almost double for the fully prepared formulations compared to the controls: in both cases, control compositions dried faster ($p < 0.05$). A statistically significant difference ($p < 0.05$) was observed when comparing samples with the active substance—the films formed by the gels of formulation 6 dried faster than of formulation 5. On the contrary, three tests when casting 1.5 mm gel layer (tests marked with an * in Table 2) showed that both compositions needed more time to form films and the difference between the obtained data was not statistically significant ($p > 0.05$). Due to the data on the drying time of 0.8 mm thickness films, it could be assumed that the addition of essential oil to the composition slowed down the drying of the films, but comparing the drying of 1.5 mm thickness films in the same way, it was found that there is no statistically significant difference between drying times ($p > 0.05$). Researchers have found that films containing PVP/PVA materials dried about 3 min [26]. According to Velaga et al. [27], films prepared from hydroxypropyl methylcellulose and polyvinyl alcohol had various drying times depending on drying temperature: hydroxypropyl methylcellulose was 6.39, 4.65, and 3.62 min at 40, 60, and 80 °C, respectively, and the polyvinyl alcohol was 6.98, 4.18, and 3.03 min at 40, 60, and 80 °C, respectively. Different results were obtained by Tapia-Blacido et al. [28], who prepared the film from *Amaranthus cruentus* seed flour. The drying time of 80 ± 5 µm thickness films depended on drying conditions (temperature and relative humidity)—0 °C, 40% RH; 30 °C, 70% RH; 50 °C, 40% RH; 50 °C, 70% RH; 25.9 °C, 55% RH; 54.1 °C, 55% RH; 40 °C, 33.8% RH; 40 °C, 76.2% RH; and 40 °C, 55% RH. The fastest drying of film was obtained using 50 °C, 40% RH—4.2 h, and the slowest drying was at temperature 30 °C, RH 70%–14.6 h [28]. Longer drying time depends on lower solution evaporation because of lower temperature and higher humidity.

**Table 2.** Results of the drying rate of films with different compositions.

| Tests | Formulations | | | |
|---|---|---|---|---|
| | 5 K | 6 K | 5 | 6 |
| 1 test * | 1 min 04 s | 0 min 56 s | 2 min 36 s | 1 min 27 s |
| 2 test * | 1 min 08 s | 0 min 44 s | 2 min 14 s | 1 min 25 s |
| 3 test * | 1 min 25 s | 0 min 52 s | 1 min 58 s | 1 min 15 s |
| 4 test ** | 2 min 51 s | 2 min 20 s | 2 min 47 s | 2 min 55 s |
| 5 test ** | 2 min 43 s | 2 min 12 s | 3 min 02 s | 2 min 20 s |
| 6 test ** | 2 min 36 s | 2 min 10 s | 3 min 10 s | 2 min 30 s |

*—gel poured 0.8 mm layer; **—gel poured 1.5 mm layer; K—the control formulation, which does not contain cinnamon leaf essential oil.

### 3.3. Results of Evaluation of the Surface and Homogeneity Properties of the Films

The morphological properties of the films were evaluated by microscopic analysis at 40 times magnification. Figure 3B shows the surface of the 5K film. The smallest particles were 3.17 µm wide, the largest—10.3 µm. Figure 3A shows the surface of composition 5. The smallest diameter of the particles was 8.8 µm, and biggest—23.93 µm. The average diameter of analyzed objects was 14.69 µm. To determine the effect of essential oil on the morphology of the films, control films were also evaluated. The average of particles is two times bigger than in composition 5 (6.14 µm).

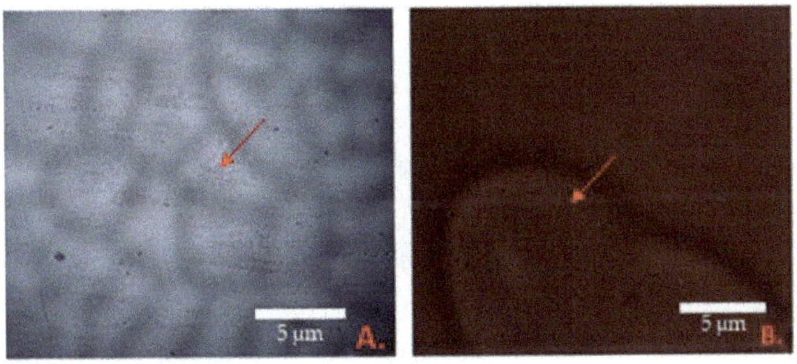


**Figure 3.** The surface of the film 40× magnification: (**A**) film No. 5; (**B**) film No. 5K (control compositions); arrows show the dimples of the film surface.

The diameter of the particles from formulation 6 are shown in Figure 4. Control formulation 6 is shown in B: the smallest particles of this sample were 2.20 ± 0.85 µm and the largest—12.44 ± 0.71 μm. The smallest particles of composition 6 (Figure 4A) were 8.21 ± 0.69 μm, the largest—20.2 ± 76 μm, and the average—14.12 ± 0.88 μm. The dimples of the formulation 6K films were significantly smaller than of formulation 6—the average diameter was 7.26 μm (Figure 4B).

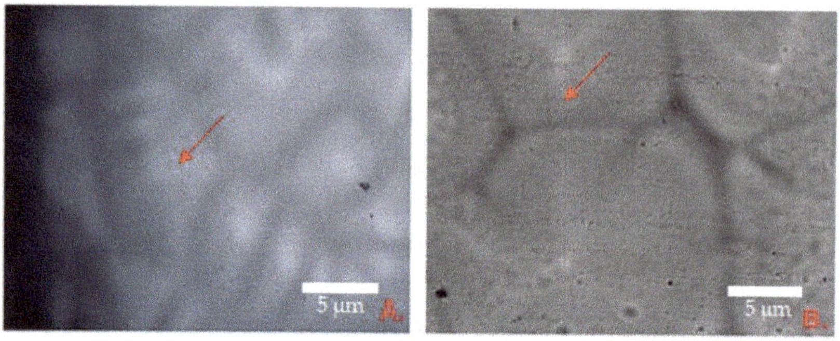


**Figure 4.** The surface of the film 40× magnification: (**A**) film No. 6; (**B**) film No. 6K (control compositions); arrows show the dimples of the film surface.

According to the results, the dimples of films are very small. Comparing the averages of surface indentations of the films of the fifth and sixth compositions, the similarity of the diameters is noticeable, so it was not possible to single out the superior composition in this case. Adding the essential oil to the composition caused the formation of pits twice as large. Small bends are natural, indicating that volatile components were removed during the formation of the films. The films were formed qualitatively because none of the gels formed clumps when drying.

*3.4. The Results of the Tests on the Adhesiveness and Adhesion of the Films to the Surface*

The film surface adhesion test was based on the amount of force required to peel the surface of the texture analyzer stick from the film surface (Figure 5).

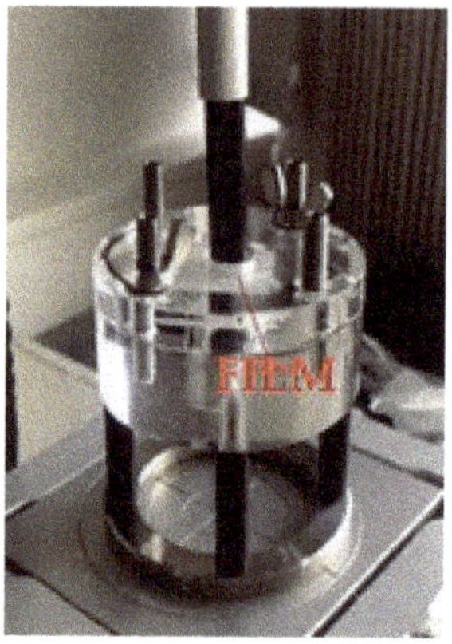


**Figure 5.** An apparatus to perform adhesion test for films.

The tests showed a statistically significant difference between the averages of the force required to peel the film from the surface between the 5 and 5K composition films (0.15 N and 0.07 N, respectively) ($p < 0.05$). A statistically significant difference ($p > 0.05$) was found between the composition 6 and 6K. Figure 6 shows the film adhesion comparison between all formulations. Between compositions 5 and 6, adhesiveness remains very similar, so the difference is not significant ($p > 0.05$).

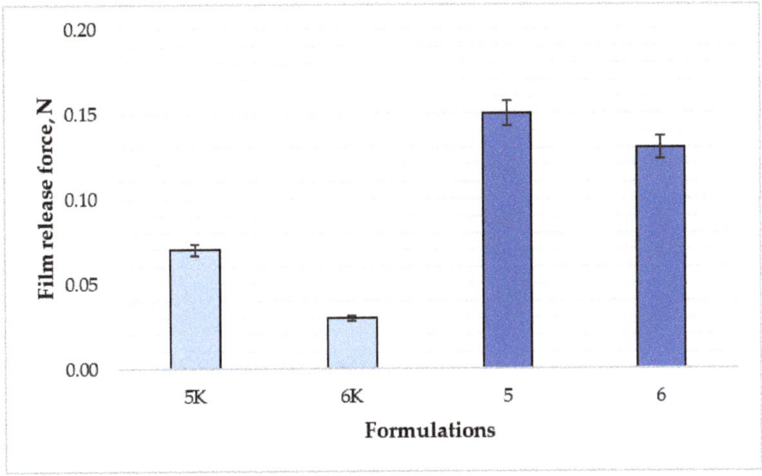


**Figure 6.** Comparison of film adhesion between formulations (n = 6). The composition values are given in Table 1. $p < 0.05$ between 5 and 5K; 6 and 6K compositions.

Unlike surface adhesion, the inner surface of films must be sufficiently adhesive to adhere to the wound. To evaluate this property, a wound adhesion test was used, which shows three parameters—average film removal force peak, cohesiveness, and overall adhesion. The results of formulations 5 and 5K, in Figures 7 and 8, revealed that there were no differences between maximum removal force and total surface adhesion results ($p > 0.05$), while the cohesiveness (indicated in red line Figure 8) was 1.45 mm for formulation 5, and 1.3 mm for formulation 5K.

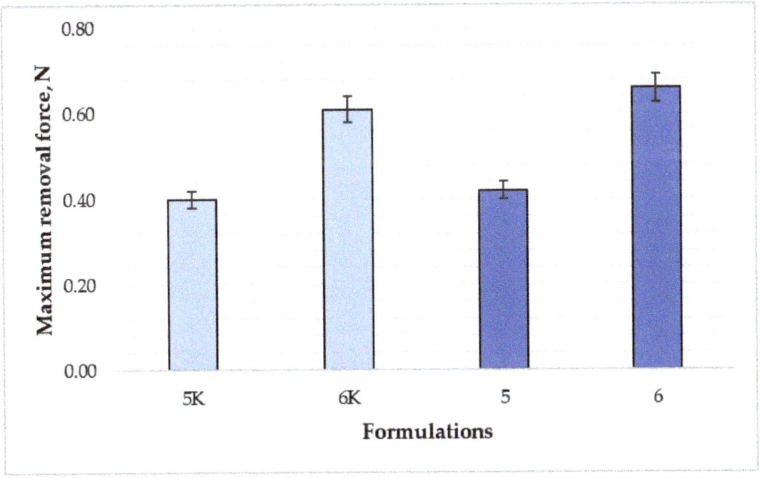


**Figure 7.** Comparison of film removal force by composition (n = 8). The composition values are given in Table 1. $p < 0.05$ between 6 and 6K; 5 and 6 compositions; $p > 0.05$ between 5 and 5K compositions.

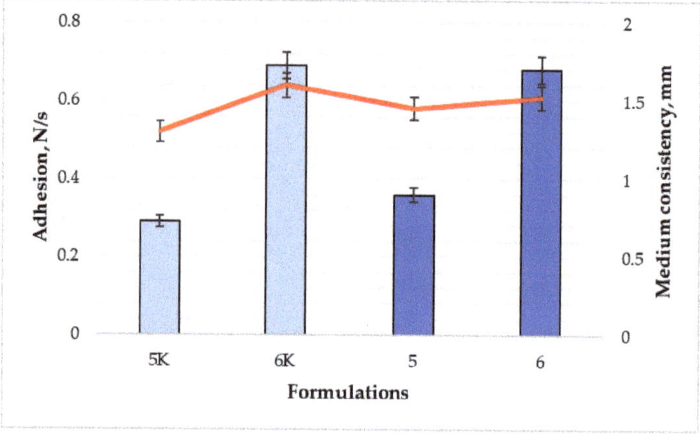


**Figure 8.** Comparison of adhesion and cohesiveness between formulations (n = 8). The composition values are given in Table 1. $p < 0.05$ between 5 and 5K compositions; $p > 0.05$ between 6 and 6K; 5 and 6 compositions.

For formulations 6 and 6K (Figure 7), the results showed a statistically significant difference only for maximum film removal force (0.66 N and 0.61 N, respectively) ($p < 0.05$). The data of cohesiveness and adhesion were similar, and the difference was not significant ($p > 0.05$). According to the results, essential oil could affect the adhesion to the surface of the films. The adhesion test results of nitrocellulose films are quite like sodium alginate films. Fiume et al. evaluated the effect of protein content on the adhesion properties of films: removal force peak was 0.49 N, adhesion—0.15 N/s, and cohesiveness—1.6 mm [29]. This data corresponded to the 5K composition film results.

According to the obtained results, the addition of cinnamon leaf essential oil increased the surface adhesion of the films but did not affect the wound adhesion parameter. The adhesion of formulation 5 increased about two times, and formulation 6—more than four times (Figure 8). The internal adhesion test showed that the gels of the sixth formulation formed films with a total adhesion and maximum average removal force almost twice that of the 5th formulation. Coherence was not significantly different. These results allow us to confidently assert that the sixth gel composition, which contains more nitrocellulose, would adhere more strongly to the damaged area, thus forming a reliable barrier. Additionally, it would be more convenient to apply on the surface of the wound, because it would not stick to clothes, bedding, or other surfaces.

### 3.5. Results of Film Flexibility Tests

Two tests were performed to determine the flexibility of the films. The differences in the flexibility of the films of compositions 5 and 6 and control compositions were tested. The purpose of the compression test was to find out the maximum force that can be pressed down on the film before it tears (Figure 9).

Flexibility properties of the films were compared (Figure 10). The difference between control compositions, 5 and 6 compositions were statistically significant ($p < 0.05$). Tensile strength of 6K was the highest of all the samples—0.5735 N/mm$^2$, sample 5K—0.3846 N/mm$^2$. Data show that the tensile strength of film 6 was 2.7 times higher than film 5 (0.1948 and 0.0728 N/mm$^2$, respectively). The addition of essential oil of cinnamon leaves and higher concentration of nitrocellulose strongly reduced the elasticity and strength of the films. Momoh et al. [17] claimed that the optimal properties of the films are achieved when there is a balance between strength (brittleness) and elasticity (flexibility)—tensile strength of alginate film was 6.12 ± 0.11 N/mm$^2$. According to results from Marangoni Junior et al. [30], a higher concentration of film-forming gel additives, such as propolis extract and silica, ensured higher tensile strength of sodium alginate films: only sodium alginate film—12.9 MPa, sodium alginate and 3% propolis extract—16.5 MPa, sodium alginate, 3% propolis extract, and 5% silica—18.2 MPa and sodium alginate, 3% propolis extract, and 10% silica—19.6 MPa. Nitrocellulose films have lower tensile strength because of different polymer (nitrocellulose vs. sodium alginate) structure and different compounds used in composition (castor oil, CLEO vs. propolis extract, silica). According to other studies, tensile strength could increase after adding higher concentrations of polymers, plasticizers, and other bonding compounds.

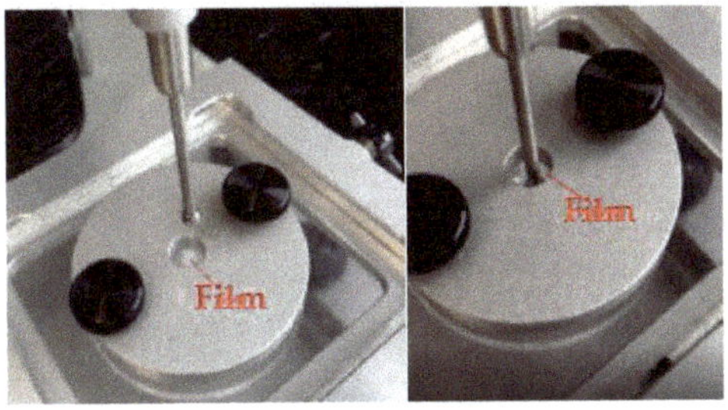


**Figure 9.** Pressure resistance test for films.

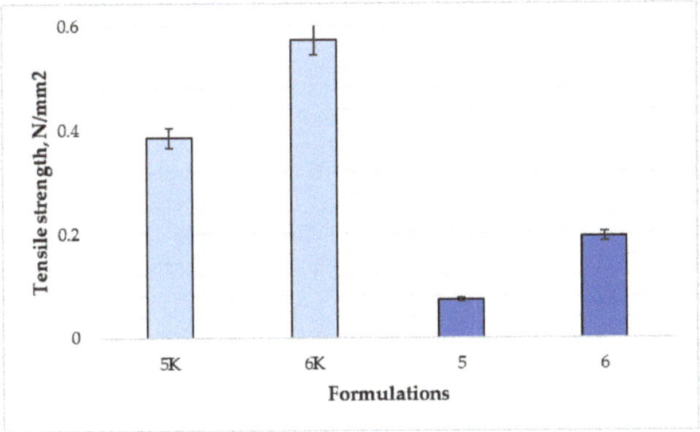


**Figure 10.** Comparison of tensile strength by composition (n = 6). The composition values are given in Table 1. $p < 0.05$ between 5 and 5K; 6 and 6K; 5 and 6 compositions.

The second test for assessing flexibility is tensile. In this test, the film was attached to the clamps from two sides and the upper clamp was pulled up until it teared (Figure 11).

**Figure 11.** Film before (left) and after (right) tensile test.

The results showed (Figure 12) statistically significant differences in tensile values between all the tested films ($p < 0.05$). The tensile value of formulations 5 and 5K was 3.24 $N/mm^2$ and 3.33 $N/mm^2$, respectively, for formulations 6 and 6K—3.34 $N/mm^2$ and 3.23 $N/mm^2$, respectively.

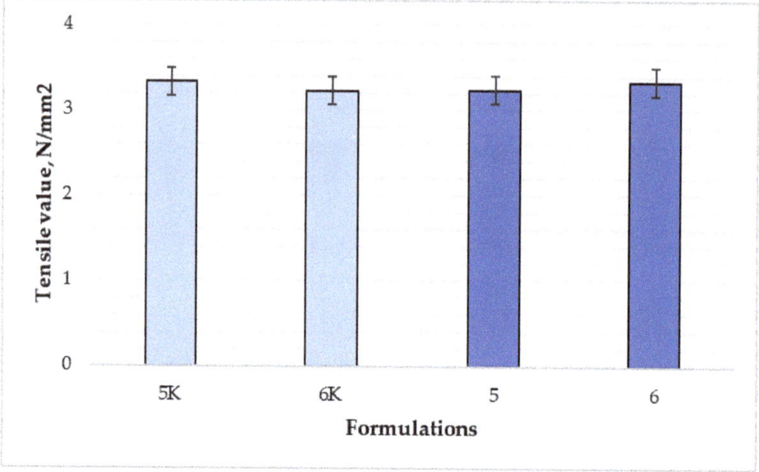


**Figure 12.** Comparison of tensile value of films by composition (n = 6). The composition values are given in Table 1. $p < 0.05$ between 5 and 5K; 6 and 6K; 5 and 6 compositions.

The results of the films' flexibility showed that the resistance and tensile tests do not correlate and the effect of the essential oil the elasticity of the films could not be proven. Furthermore, due to the conflicting results of the two tests, it would not be correct to claim that the flexibility of composition 6 is better than composition 5.

### 3.6. Water Vapor Penetration Test Results

The water vapor permeability test is an important indicator that shows how the film is able to retain moisture in the wound. During the test, samples were placed in an incubator and kept for 24 h. After 24 h, the samples were analyzed and weighed. The subjects were slightly pale after the test. This was most likely due to the ability of nitrocellulose to interact with water molecules and turn from colorless to white (Figure 13).

**Figure 13.** Water vapor transmission rate (WVTR) test. Films before (left) and after (right).

Figure 14 shows the results of WVTR for compositions 5, 5K, 6, and 6K. Film formulation 5K was able to transmit 0.0196 g of water vapor per 1 $cm^2$ per day, and formulation 5K—0.0206 g ($p < 0.05$). Formulations 6 and 6K transmitted less water vapor than formulations 5 and 5K. There was no difference in WVTR between 6 and 6K films ($p > 0.05$). Results from Maragoni Junior et al. [30] show that, the addition of propolis extract and silica provided a decrease in WVTR: the control film (708.7 ± 60.9 g $m^{-2}$ $day^{-1}$) and the film containing the highest concentration of silica (10%) (595.2 ± 3.5 g $m^{-2}$ $day^{-1}$) (reduction about 16% in this variable). Furthermore, the authors found that the reduction of WVTR may have been influenced by the tendency of increasing film thickness, particularly at the highest filler load [30]. Tapia-Blacido et al. [28] found that WVTR of films prepared with sorbitol is lower than that of glycerol-containing films: the better water vapor barrier properties of films containing sorbitol as plasticizer compared with those of the films containing glycerol might be because sorbitol is less hygroscopic.

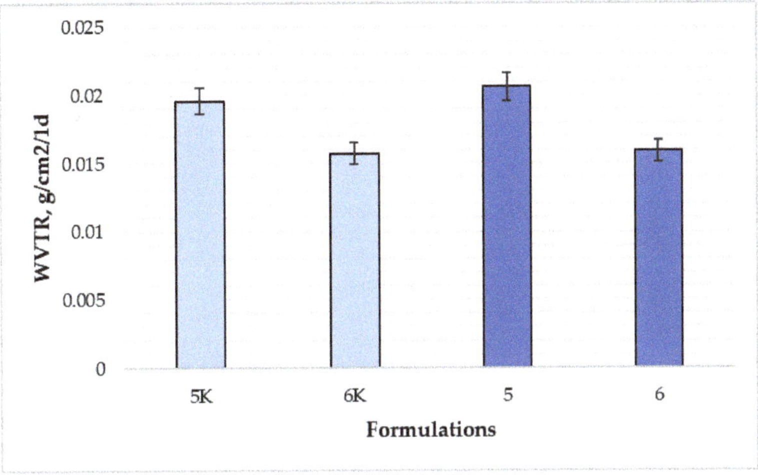


**Figure 14.** Comparison of average water vapor permeation through the film by composition (n = 3). The composition values are given in Table 1. $p < 0.05$ between compositions 5 and 6; $p > 0.05$ between 5 and 5K; 6 and 6K compositions.

According to obtained data, all film formulations are very poorly permeable to moisture, which may cause discomfort to patients. The moisture that accumulates in the wound will not be removed, causing the tissues to swell, stretch, and not heal. The amount of nitrocellulose in the composition affected the transmission of water vapor, while the addition of essential oil had no impact on it.

### *3.7. Swelling Test Results of the Films*

A test that analyzes film ability to absorb wound exudate was carried out for 2 h by weighing the films, which had previously been dried with filter paper, every 15 min. Figure 15 shows that formulation 6 maintained the lowest swelling index (4.55%) during the first 15 min of the test. At 45 min of the test, the percentage of wound fluid absorption increased to 30.3%. Then the swelling index stopped for a while and only between the 75 and 90 min of the study it started to increase again. The highest swelling index was after 2 h for composition 5K. Comparing compositions 5 and 5K, it was observed that the increase in the swelling index of both took place evenly, but composition 5 was less swollen. Only a small difference in the level of absorption of wound fluid was observed between compositions 6 and 6K. The composition of 6K initially had only a 1.25% higher swelling index than composition 6 and the difference increased to just 4.32% over time. Thus, the 6 formulation showed the lowest swelling ability after the 2 h test. Fan et al. [31] data matches our results because the swelling rate decreases with increasing the concentration of nitrocellulose, which decreases from 19.86% at 2% solid content to 3.13% at 16% solid content.

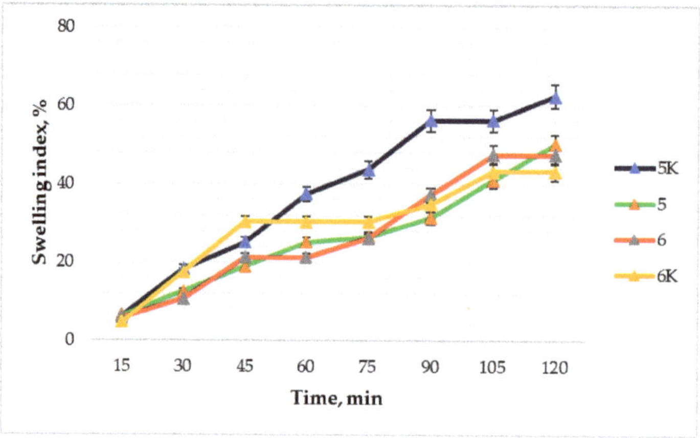


**Figure 15.** Comparison of swelling index, % variation with time (n = 3). The composition values are given in Table 1.

Man et al. [12] studied the properties of sodium alginate films and found that for effective wound healing, the films must be able to absorb a large amount of fluid secreted by the wound: the swelling index of the film with sodium alginate was 85% after 2 h.

According to the results, the control formulations showed slightly better wound fluid absorption capacity, suggesting that cinnamon leaf essential oil slightly reduced the swelling properties of the films. In the compositions with the active substance, the swelling index was 43.18 and 50.2% (compositions 6 and 5, respectively). The lower swelling of the formulation 6 could be explained by the higher amount of nitrocellulose leading to an inferior interaction with water.

*3.8. Microbiological Test Results*

Six samples were tested in this test for the microbiological contamination—gels of formulations 5 and 6 stored for a month at 38 ± 2 °C, freshly prepared gels of formulations 5 and 6, and freshly prepared control samples of formulations 5 and 6 without essential oil. *S. aureus* and *P. aeruginosa* bacteria were searched for in the samples mentioned. The obtained results confirmed the expectations—no bacteria were detected in any of the samples (Table 3, Figure 16). The essential oil may not have been the most important factor in the absence of bacteria. Control formulations without the active substance also inhibited the growth of microorganisms in the gels. The absence of bacteria in the gels proves that the product is suitable for wound treatment, and it does not cause soft tissue infections, such as purulent/non-suppurative Impetigo, ecthyma gangrene, staphylococcal scalded skin syndrome (SSSS), rosacea, etc. [32]. Mu et al. [1] found that nanopores formed by nitrocellulose inhibit the penetration of bacteria through the film, but this microbiological study could indicate that the gelling substance also has microorganism-killing properties, and the essential oil of cinnamon leaves was not the most important factor determining the proper microbiological quality of the gel.

**Table 3.** Results of film-forming gels microbiological analysis.

| Bacteria Strains | Formulations | | | | | |
|---|---|---|---|---|---|---|
| | **5M** | **6M** | **5FP** | **6FP** | **5KFP** | **6KFP** |
| *S. aureus* | - | - | - | - | - | - |
| *P. aeruginosa* | - | - | - | - | - | - |

M—film-forming gels, stored for a month at 38 ± 2 °C; FP—freshly prepared film-forming gels; K—control formulations; —no bacteria was detected.

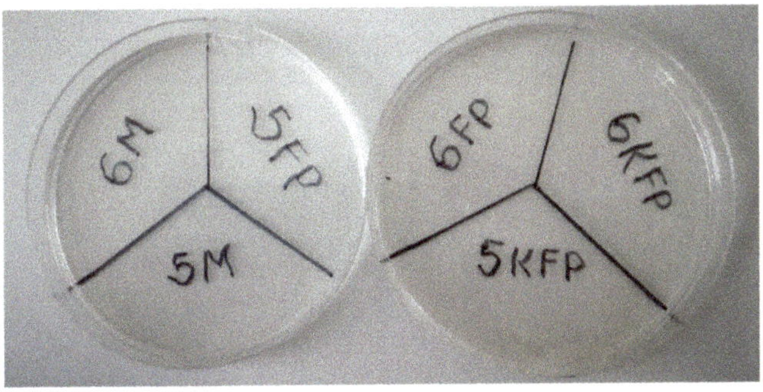


**Figure 16.** No bacteria growing was detected on film-forming gels. M—film-forming gels, stored for a month at 38 ± 2 °C; FP—freshly prepared film-forming gels; K—control formulations.

### *3.9. Stability Test Results*

The stability of the film-forming gels and films of formulations 5 and 6 was tested for 28 days at 38 ± 2 °C. Monitoring changes in stability took place on the day of production, and on days 7, 14, 21, and 28. The color, integrity, and surface of the films did not change during stability test (Table 4). The films remained intact, there were no visible cracks or thinning, no change of color. Film-forming gels after 28 days had no color, odor, viscosity, texture changes (Table 5). Film-forming gels based on various grades hydroxypropyl methylcellulose (HPMC) remained stable for 3 months at 40 °C and 75% RH for accelerated stability and at 4 °C in the fridge [33] and hydroxypropyl cellulose-based film-forming gels—6 months in the same conditions [16]. Bharti et al. [34] used HPMC as the base for film-forming gels to prepare films with buspirone hydrochloride nanoparticles. After 90 days at 25 °C and 60% relative humidity there was no difference seen in the physical appearance, surface pH, and disintegration time of films [34].

**Table 4.** The stability test for films No. 5 and No. 6.

| Days | Film Formulation 5 | Film Formulation 6 |
|---|---|---|
| Production day (0) |  | 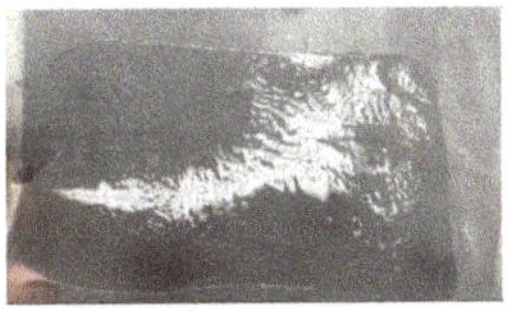 |
| 7 | 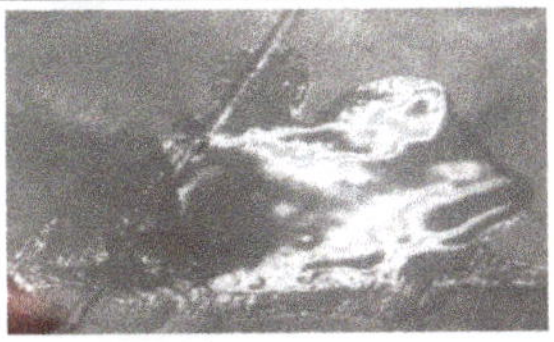 | 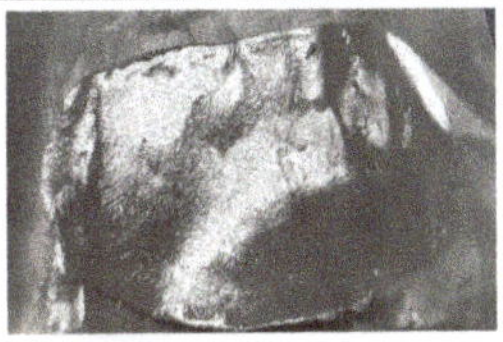 |
| 14 | 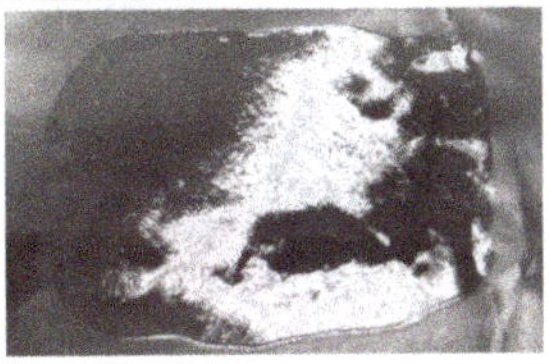 | 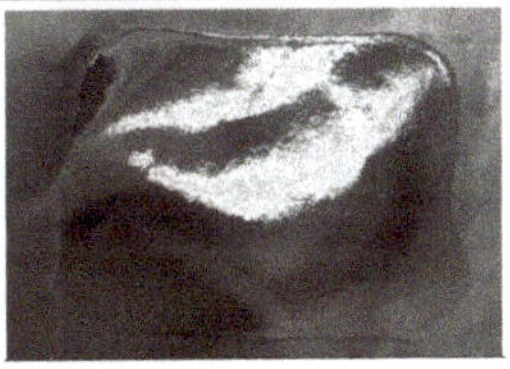 |

**Table 4.** *Cont.*

| Days | Film Formulation 5 | Film Formulation 6 |
|---|---|---|
| 21 | 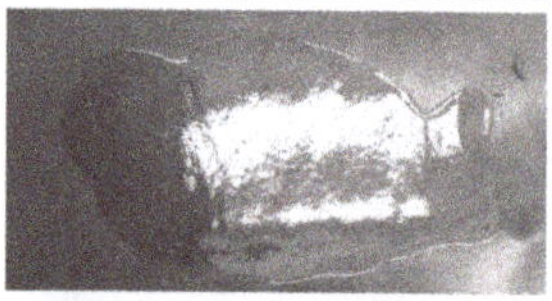 |  |
| 28 | 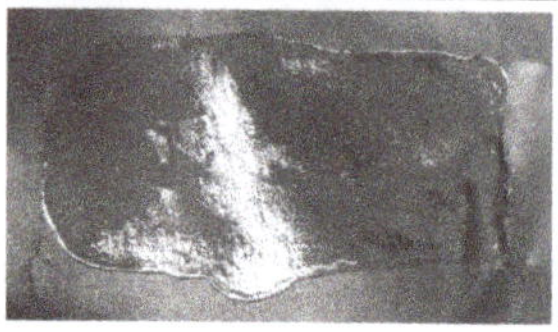 | 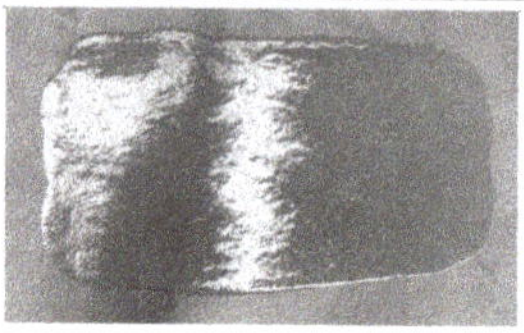 |

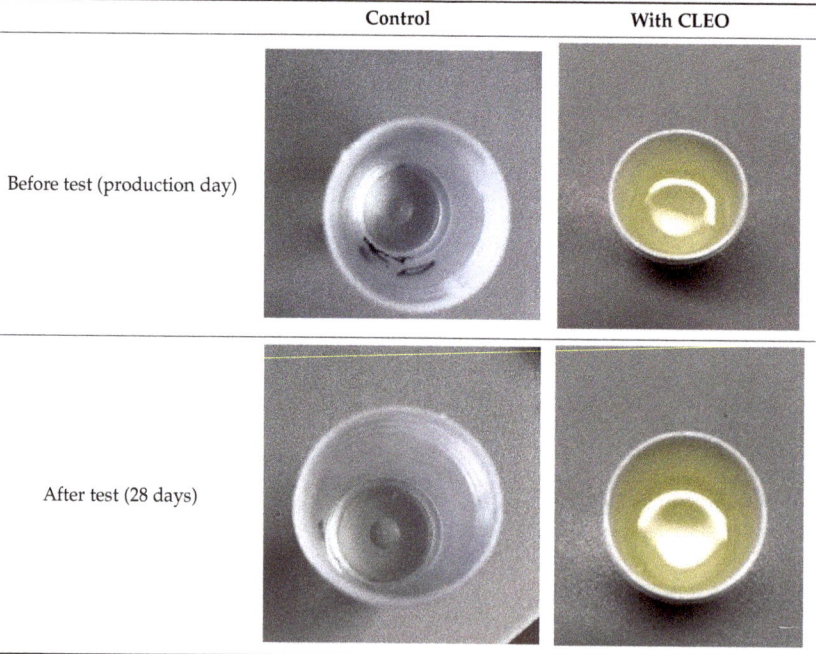

**Table 5.** Status of the control and non-control gels formulation before and after the stability study.

| | Control | With CLEO |
|---|---|---|
| Before test (production day) |  |  |
| After test (28 days) |  |  |

CLEO—cinnamon leaf essential oil (active ingredient).

### *3.10. Evaluation of the pH Value of Film-Forming Gels*

The normal pH value of human skin is slightly acidic. It is an important factor for the maintenance of normal skin condition and microflora [35]. In the case of chronic wounds, the pH indicator becomes more alkaline (between 7.15 and 8.1). It has been shown that if pH value is higher, the wound healing is slower [36]. To make sure that the product will not irritate the skin and at the same time promote the healing of skin, pH value of film-forming gels was determined. After making aqueous solutions of film-forming gels, six pH meter measurements were performed for each formulation. Formulation 5 showed pH value of 4.28, while the pH value of formulation 5K was 4.29. Formulation 6 had a pH value of 4.45 and formulation 6K had a pH of 4.49. These slight differences between the control and non-control formulations indicated that the essential oil of cinnamon leaves does not affect the pH value of the formulation. Comparing the differences in the pH values of formulations 5 and 6 (Figure 17), it shows that the pH value of formulation 6 is slightly more alkaline and the difference between formulations 5 and 6 is statistically significant ($p < 0.05$). Both products have a pH value below 5 and are therefore considered suitable for use on the skin.

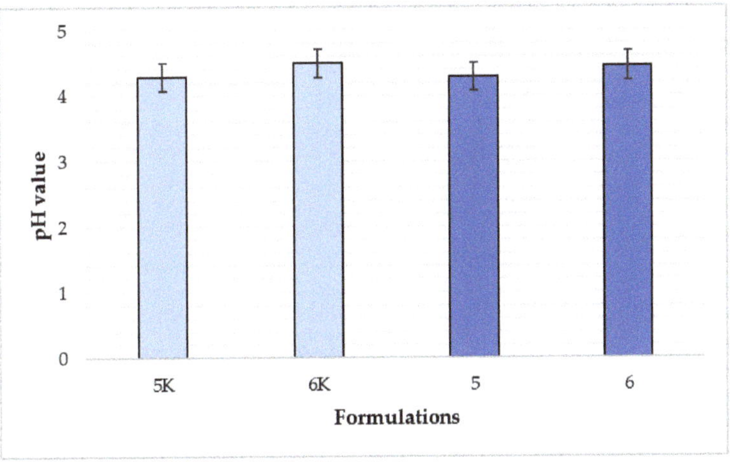


**Figure 17.** Comparison of pH values by composition (n = 6). The composition values are given in Table 1. $p < 0.05$ between compositions 5 and 6.

### 3.11. *Results of Qualitative and Quantitative Analysis of Cinnamon Leaf Essential Oil*

Qualitative and quantitative analysis of cinnamon leaf essential oil (*Cinnamonum zeilanicum*) by GC-MS showed that the product contains 4 components. As shown in the chromatogram (Figure 18), the highest peak was 1, 3, 4—eugenol (94.14%).

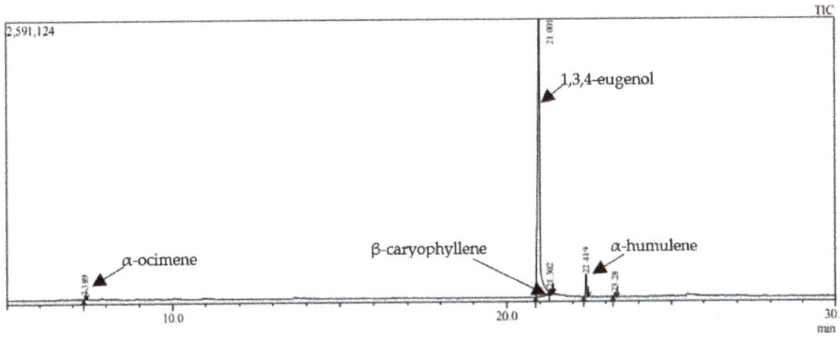


**Figure 18.** A GC-MS chromatogram of the cinnamon leaf essential oil.

Other components' presence was significantly less—3.96% α-humulene, 0.57% α-ocimene, and 0.55% β-caryophyllene. Eugenol has anesthetic, antioxidant, anti-inflammatory, antimicrobial activities [37]. Eugenol may give to the film-forming gel excellent antibacterial and anti-inflammatory properties, and applying the product to the wound would improve the healing processes.

## 4. Conclusions

Physico-chemical and mechanical analyses of film-forming gels and films is important for the creation of high quality and suitable-for-use dermatological films. A range of factors could impact the formation of films, appearance, mechanical properties, stability, and release of active ingredients. According to the findings of previous research, films containing nitrocellulose have proven to be suitable for use as a delivery form due to its high physico-chemical, mechanical, biopharmaceutical properties and cosmetic attractiveness. Nitrocellulose based films showed acceptable drying rate, adhesion, flexibility, tensile strength, cohesiveness, swelling, water vapor penetration, and pH value. The results of these parameters confirm the ability to use these films on skin. This study has shown that the concentration of nitrocellulose, castor oil, and essential oil could impact the drying ratio, mechanical properties, and other parameters. The results of the study have provided fundamental insights into the preparation of film-forming gels and films that are highly relevant for diverse industries such as pharmaceutical, food, cosmetics, and agriculture. Future work will focus on extending the applicability of hydrophilic and hydrophobic polymers cast from aqueous and organic solvents, investigating the effects of the thickness of the film, different drying conditions, active substance release in vitro, anti-inflammatory, and antibacterial effects on the wounds.

**Author Contributions:** Writing—original draft preparation, writing—review and editing, visualization, supervision, L.P. Formal analysis, investigation, E.D. Conceptualization, methodology, J.B. All authors have read and agreed to the published version of the manuscript.

**Funding:** This research received no external funding.

**Institutional Review Board Statement:** Not applicable.

**Informed Consent Statement:** Not applicable.

**Data Availability Statement:** Not applicable.

**Conflicts of Interest:** Nothing to declare by any of the authors.

## References

1. Mu, X.; Yu, H.; Zhang, C.; Chen, X.; Cheng, Z.; Bai, R.; Wu, X.; Yu, Q.; Wu, C.; Diao, Y. Nano-porous nitrocellulose liquid bandage modulates cell and cytokine response and accelerates cutaneous wound healing in a mouse model. *Carbohydr. Polym.* **2016**, *136*, 618–629. [CrossRef] [PubMed]
2. Bowers, S.; Franco, E. Chronic Wounds: Evaluation and Management. *Am. Fam. Physician* **2020**, *101*, 8.
3. Powers, J.G.; Higham, C.; Broussard, K.; Phillips, T.J. Wound healing and treating wounds. *J. Am. Acad. Dermatol.* **2016**, *74*, 607–625. [CrossRef] [PubMed]
4. Gardner, S.E.; Frantz, R.A.; Doebbeling, B.N. The validity of the clinical signs and symptoms used to identify localized chronic wound infection. *Wound Repair Regen.* **2001**, *9*, 178–186. [CrossRef] [PubMed]
5. Parhi, R.; Goli, V.V.N. Design and optimization of film-forming gel of etoricoxib using research surface methodology. *Drug Deliv. Transl. Res.* **2020**, *10*, 498–514. [CrossRef]
6. Kathe, K.; Kathpalia, H. Film forming systems for topical and transdermal drug delivery. *Asian J. Pharm. Sci.* **2017**, *12*, 487–497. [CrossRef]
7. Van Bocxlaer, K.; McArthur, K.-N.; Harris, A.; Alavijeh, M.; Braillard, S.; Mowbray, C.E. Film-Forming Systems for the Delivery of DNDI-0690 to Treat Cutaneous Leishmaniasis. *Pharmaceutics* **2021**, *13*, 516. [CrossRef]
8. Pünnel, L.C.; Lunter, D.J. Film-Forming Systems for Dermal Drug Delivery. *Pharmaceutics* **2021**, *13*, 932. [CrossRef]
9. Tran, T.T.D.; Tran, P.H.L. Controlled Release Film Forming Systems in Drug Delivery: The Potential for Efficient Drug Delivery. *Pharmaceutics* **2019**, *11*, 290. [CrossRef]
10. Bornare, S.S.; Aher, S.S.; Saudagar, R.B. A review: Film forming gel novel drug delivery system. *Int. J. Curr. Pharm. Sci.* **2018**, *10*, 25. [CrossRef]
11. Kale, S.B.; Bachhav, R.S. A Review on Film Forming Gel. *Int. J. Trend Sci. Res. Dev.* **2021**, *10*, 966–974.
12. Man, A.; Santacroce, L.; Iacob, R.; Mare, A.; Man, L. Antimicrobial Activity of Six Essential Oils Against a Group of Human Pathogens: A Comparative Study. *Pathogens* **2019**, *8*, 15. [CrossRef]
13. Farias, A.P.P.; dos Monteiro, O.S.; da Silva, J.K.R.; Figueiredo, P.L.B.; Rodrigues, A.A.C.; Monteiro, I.N. Chemical composition and biological activities of two chemotype-oils from Cinnamomum verum J. Presl growing in North Brazil. *J. Food Sci. Technol.* **2020**, *57*, 3176–3183. [CrossRef]
14. Shahina, Z.; El-Ganiny, A.M.; Minion, J.; Whiteway, M.; Sultana, T.; Dahms, T.E.S. Cinnamomum zeylanicum bark essential oil induces cell wall remodelling and spindle defects in *Candida Albicans*. *Fungal Biol. Biotechnol.* **2018**, *5*, 3. [CrossRef]
15. Kallel, I.; Hadrich, B.; Gargouri, B.; Chaabane, A.; Lassoued, S.; Gdoura, R.; Bayoudh, A.; Ben Messaoud, E. Optimization of Cinnamon (*Cinnamomum zeylanicum* Blume) Essential Oil Extraction: Evaluation of Antioxidant and Antiproliferative Effects. *Evid.-Based Complement. Altern. Med.* **2019**, *2019*, 6498347. [CrossRef]

16. Vij, N.N.; Saudagar, R.B. Formulation, development and evaluation of film-forming gel for prolonged dermal delivery of terbinafine hydrochloride. *Int. J. Pharma Sci. Res.* **2014**, *5*, 18.
17. Momoh, F.U.; Boateng, J.S.; Richardson, S.C.W.; Chowdhry, B.Z.; Mitchell, J.C. Development and functional characterization of alginate dressing as potential protein delivery system for wound healing. *Int. J. Biol. Macromol.* **2015**, *81*, 137–150. [CrossRef]
18. Guo, R.; Du, X.; Zhang, R.; Deng, L.; Dong, A.; Zhang, J. Bioadhesive film formed from a novel organic–inorganic hybrid gel for transdermal drug delivery system. *Eur. J. Pharm. Biopharm.* **2011**, *79*, 574–583. [CrossRef]
19. Basha, R.K.; Konno, K.; Kani, H.; Kimura, T. Water Vapor Transmission Rate of Biomass Based Film Materials. *Eng. Agric. Environ. Food* **2011**, *4*, 37–42. [CrossRef]
20. Pagano, C.; Puglia, D.; Luzi, F.; Michele, A.D.; Scuota, S.; Primavilla, S. Development and Characterization of Xanthan Gum and Alginate Based Bioadhesive Film for Pycnogenol Topical Use in Wound Treatment. *Pharmaceutics* **2021**, *13*, 324. [CrossRef]
21. Nemati, M.; Hamidi, A.; Maleki Dizaj, S.; Javaherzadeh, V.; Lotfipour, F. An Overview on Novel Microbial Determination Methods in Pharmaceutical and Food Quality Control. *Adv. Pharm. Bull.* **2016**, *6*, 301–308. [CrossRef] [PubMed]
22. Gupta, M.; Rout, P.K.; Misra, L.N.; Gupta, P.; Singh, N.; Darokar, M.P. Chemical composition and bioactivity of *Boswellia serrata* Roxb. essential oil in relation to geographical variation. *Plant Biosyst.-Int. J. Deal. All Asp. Plant Biol.* **2017**, *151*, 623–629. [CrossRef]
23. Wang, X.; Gong, L.; Jiang, H. Study on the Difference between Volatile Constituents of the Different Parts from *Elsholtzia ciliata* by SHS-GC-MS. *Am. J. Anal. Chem.* **2017**, *08*, 625–635. [CrossRef]
24. Geng, H.; Yuan, Z.; Fan, Q.; Dai, X.; Zhao, Y.; Wang, Z. Characterisation of cellulose films regenerated from acetone/water coagulants. *Carbohydr. Polym.* **2014**, *102*, 438–444. [CrossRef] [PubMed]
25. Pawar, V.M.; Nadkarni, V.S. Preparation of thin films of cellulose acetate-nitrocellulose blend for solid state nuclear track detection using spin coating technique. *J. Radioanal. Nucl. Chem.* **2020**, *323*, 1329–1338. [CrossRef]
26. Kim, D.W.; Kim, K.S.; Seo, Y.G.; Lee, B.-J.; Park, Y.J.; Youn, Y.S. Novel sodium fusidate-loaded film-forming hydrogel with easy application and excellent wound healing. *Int. J. Pharm.* **2015**, *495*, 67–74. [CrossRef]
27. Velaga, S.P.; Nikjoo, D.; Vuddanda, P.R. Experimental Studies and Modeling of the Drying Kinetics of Multicomponent Polymer Films. *AAPS PharmSciTech* **2018**, *19*, 425–435. [CrossRef]
28. Tapia-Blácido, D.R.; do Amaral Sobral, P.J.; Menegalli, F.C. Effect of drying conditions and plasticizer type on some physical and mechanical properties of amaranth flour films. *LWT-Food Sci. Technol.* **2013**, *50*, 392–400. [CrossRef]
29. Fiume, M.M.; Bergfeld, W.F.; Belsito, D.V.; Hill, R.A.; Klaassen, C.D.; Liebler, D.C. Safety Assessment of Nitrocellulose and Collodion as Used in Cosmetics. *Int. J. Toxicol.* **2016**, *35*, 50S–59S. [CrossRef]
30. Marangoni Júnior, L.; Jamróz, E.; Gonçalves S de, Á.; da Silva, R.G.; Alves, R.M.V.; Vieira, R.P. Preparation and characterization of sodium alginate films with propolis extract and nano-SiO2. *Food Hydrocoll. Health* **2022**, *2*, 100094. [CrossRef]
31. Fan, W.; Zhou, J.; Ding, Y.; Xiao, Z. Fabrication and mechanism study of the nitrocellulose aqueous dispersions by solvent displacement method. *J. Appl. Polym. Sci.* **2023**, *140*, e53290. [CrossRef]
32. Bagnoli, F.; Rappuoli, R.; Grandi, G. Staphylococcus aureus: Microbiology, Pathology, Immunology, Therapy and Prophylaxis. In *Current Topics in Microbiology and Immunology*; Springer: Berlin/Heidelberg, Germany, 2017; Volume 409. [CrossRef]
33. Ranade, S.; Bajaj, A.; Londhe, V.; Kao, D.; Babul, N. Fabrication of Polymeric Film Forming Topical Gels. *Int. J. Pharm. Sci. Rev. Res.* **2014**, *26*, 306–313.
34. Bharti, K.; Mittal, P.; Mishra, B. Formulation and characterization of fast dissolving oral films containing buspirone hydrochloride nanoparticles using design of experiment. *J. Drug Deliv. Sci. Technol.* **2019**, *49*, 420–432. [CrossRef]
35. Lambers, H.; Piessens, S.; Bloem, A.; Pronk, H.; Finkel, P. Natural skin surface pH is on average below 5, which is beneficial for its resident flora. *Int. J. Cosmet. Sci.* **2006**, *28*, 359–370. [CrossRef]
36. Percival, S.L.; McCarty, S.; Hunt, J.A.; Woods, E.J. The effects of pH on wound healing, biofilms, and antimicrobial efficacy: pH and wound repair. *Wound Repair Regen.* **2014**, *22*, 174–186. [CrossRef]
37. Khalil, A.A.; ur Rahman, U.; Khan, M.R.; Sahar, A.; Mehmood, T.; Khan, M. Essential oil eugenol: Sources, extraction techniques and nutraceutical perspectives. *RSC Adv.* **2017**, *7*, 32669–32681. [CrossRef]

*polymers*

MDPI

*Article*

# Protective Properties of Copper-Loaded Chitosan Nanoparticles against Soybean Pathogens *Pseudomonas savastanoi* pv. *glycinea* and *Curtobacterium flaccumfaciens* pv. *flaccumfaciens*

**Rashit Tarakanov [1,*], Balzhima Shagdarova [2], Tatiana Lyalina [2], Yuliya Zhuikova [2], Alla Il'ina [2], Fevzi Dzhalilov [1] and Valery Varlamov [2,*]**

1 Department of Plant Protection, Russian State Agrarian University—Moscow Timiryazev Agricultural Academy, 127434 Moscow, Russia
2 Research Center of Biotechnology, Russian Academy of Sciences, 119071 Moscow, Russia
* Correspondence: tarakanov.rashit@mail.ru (R.T.); varlamov@biengi.ac.ru (V.V.)

**Abstract:** Soybeans are a valuable food product, containing 40% protein and a large percentage of unsaturated fatty acids ranging from 17 to 23%. *Pseudomonas savastanoi* pv. *glycinea* (Psg) and *Curtobacterium flaccumfaciens* pv. *flaccumfaciens* (Cff) are harmful bacterial pathogens of soybean. The bacterial resistance of soybean pathogens to existing pesticides and environmental concerns requires new approaches to control bacterial diseases. Chitosan is a biodegradable, biocompatible and low-toxicity biopolymer with antimicrobial activity that is promising for use in agriculture. In this work, a chitosan hydrolysate and its nanoparticles with copper were obtained and characterized. The antimicrobial activity of the samples against Psg and Cff was studied using the agar diffusion method, and the minimum inhibitory concentration (MIC) and minimum bactericidal concentration (MBC) were determined. The samples of chitosan and copper-loaded chitosan nanoparticles ($Cu^{2+}$ChiNPs) significantly inhibited bacterial growth and were not phytotoxic at the concentrations of the MIC and MBC values. The protective properties of chitosan hydrolysate and copper-loaded chitosan nanoparticles against soybean bacterial diseases were tested on plants in an artificial infection. It was demonstrated that the $Cu^{2+}$ChiNPs were the most effective against Psg and Cff. Treatment of pre-infected leaves and seeds demonstrated that the biological efficiencies of ($Cu^{2+}$ChiNPs) were 71% and 51% for Psg and Cff, respectively. Copper-loaded chitosan nanoparticles are promising as an alternative treatment for bacterial blight and bacterial tan spot and wilt in soybean.

**Keywords:** copper-loaded chitosan nanoparticles; seed treatment; antibacterial properties; *Pseudomonas; Curtobacterium;* bacterial tan spot; bacterial blight

**Citation:** Tarakanov, R.; Shagdarova, B.; Lyalina, T.; Zhuikova, Y.; Il'ina, A.; Dzhalilov, F.; Varlamov, V. Protective Properties of Copper-Loaded Chitosan Nanoparticles against Soybean Pathogens *Pseudomonas savastanoi* pv. *glycinea* and *Curtobacterium flaccumfaciens* pv. *flaccumfaciens*. *Polymers* **2023**, *15*, 1100. https://doi.org/10.3390/polym15051100

Academic Editor: Luminita Marin

Received: 5 February 2023
Revised: 17 February 2023
Accepted: 20 February 2023
Published: 22 February 2023



## 1. Introduction

Soybean (*Glycine max* (L.) Merr.) is a legume crop in the *Fabaceae* family. The importance of this crop stems from the fact that it is a valuable source of high-quality protein for human and livestock nutrition. Soybeans are a complete source of protein with essential amino acids as well as unsaturated fatty acids, dietary fiber, isoflavones, anthocyanins and vitamins [1]. In 2020, 353.5 million tons of soybean were harvested from 126.9 million ha worldwide, with an average yield of 27.8 q/ha [2]. However, yield growth is limited by several factors, most notably crop infestation, pests and diseases. More than 45 species of fungi, 15 species of viruses and 6 species of phytopathogenic bacteria cause economically significant diseases in soybean [3,4].

Bacterial infestation reduces yields by up to 40% and is the most destructive disease [5]. The Gram-negative bacterium *Pseudomonas savastanoi* pv. *glycinea* syn-*Pseudomonas syringae* pv. *glycinea* (Psg) is the causative agent of soybean bacterial blight [6]. At the present time, the area of distribution of the disease includes all the climatic zones and 41 countries in which the disease has been detected [7]. Psg not only causes specific symptoms on the upper leaves and pods, but it can also infect all above-ground parts of the soybean. Symptoms of the disease include the appearance of oily necrotic spots surrounded by chlorotic halos that gradually coalesce to form zones of necrosis [8]. Infected seeds and, more rarely, plant residues are the reservoirs of infection. This disease can reduce the germination of infected seeds and the yield and the content of unsaturated fatty acids [9].

Another disease affecting soybean is bacterial spot and wilt caused by the Gram-positive bacterium *Curtobacterium flaccumfaciens* pv. *flaccumfaciens* (Cff). The bacterium is able to cause spotting on leaves, burns and the death of seedlings and adult plants as well as penetrate into the vascular system [10]. The main symptoms of infection are slow growth, dying off of shoots, burns and wilting of stems. This pathogen can affect a wide range of leguminous crops, including soybean [11], although the main host plant is the common bean (*Phaseolus vulgaris* L.). Cff leads to a decrease in the yield and quality of seeds [12,13]. Cff is listed by the European and Mediterranean Plant Protection Organization (EPPO) as a quarantine object of category A2 [14]. The main source of infection is infected seeds [15].

The control of bacterial diseases in soybean requires a complex approach. The primary source of inoculum for bacterial diseases is infected plant residues. Infested seeds are a secondary source of inoculum; therefore, their certification is necessary in order to prevent their entry into the field [9,16]. Other methods for controlling bacterial diseases include strict crop rotation, the use of resistant varieties, and the treatment of plants and seeds with biological and chemical agents [17–19]. As decided by the European Union, the synthetic pesticides with the highest toxicity should be replaced by substances with a lower environmental impact in order to phase out their use (Implementing Regulation (EU) 2015/408) [20]. Therefore, the use of chitosan and chitosan-based compounds against bacterial diseases is a promising approach [21].

Chitosan is a biodegradable, biocompatible and low-toxicity biopolymer characterized by antimicrobial, antiviral, antioxidant, sorption and chelating properties [22–25]. Chitosan is a polycation in acidic pH. The biopolymer is obtained by chemical deacetylation of chitin under alkaline conditions. Chitin is one of the most common polysaccharides found in crustacean shells, insect cuticles and the cell walls of fungi [26]. The protective effect of chitosan is demonstrated by a triple action: activation of host defenses, effect on microorganisms and film formation on the treated surface [27]. The enhancement of plant immune response under the action of chitosan is due to the fact that positively charged chitosan can interact with negatively charged pectin. Plant cells receive information about the destruction of the cell wall and the presence of pathogens by inducing a specific alarm signal arising from chitosan's effect on the supramolecular pectin structure. Plants react to chitosan–pectin dimeric complexes stronger than to individual components [28]. Chitosan can also directly inhibit the growth of many plant pathogens: phytopathogenic fungi [29], oomycetes [30] and bacteria [31]. Chitosan forms a protective film preventing the interaction of pathogens with the plant cell wall [32]. A number of commercial products based on chitosan, such as Armour-Zen (New Zealand), Chito Plant (Germany) and KaitoSol (United Kingdom) are used to inhibit the incidence of bacterial diseases in plants [33].

The application of chitosan nanoforms in the control of plant pathogens has become a trend in recent years [34]. Chitosan nanoparticles act as plant growth stimulators and antimicrobial agents against phytopathogenic microorganisms [35,36]. The mechanism of the antibacterial activity of chitosan nanoparticles is similar to that of chitosan and is primarily due to interaction with the cell wall and the bacterial cell membrane. In the case of chitosan nanoparticles, higher zeta potential values have a significant effect on bacterial growth inhibition when compared to the original forms of chitosan. In this vein, the smaller size and higher zeta potential of chitosan nanoparticles provide a higher level of antibacterial activity and attract increased interest from researchers as a means of combating bacteria [34,37]. Nanocomplexes of chitosan with metals, particularly with copper, are also actively studied. In the work [38] it was shown that copper-loaded chitosan-based nanoparticles with a size of 89 nm effectively inhibited the growth of *Xanthomonas axonopodis* pv. *punicae*, which causes the bacterial blight of pomegranate, at 1000 ppm and remained on par with standard streptocycline at 500 ppm.

Information on the antibacterial activity of copper-loaded chitosan nanoparticles and unmodified chitosan against Psg is scarce. Earlier, we described the antibacterial properties of chitosan hydrolysate against Psg in vitro at a concentration of 0.3% (*v/v*) [39]. In article [40], Cu-chitosan NPs were shown to have high antibacterial activity against Psg under in vitro conditions at concentrations of 400 ppm and 1000 ppm. There is no information on the efficacy of chitosan and chitosan nanoparticles with $Cu^{2+}$ against Cff.

The aim of this study was to determine the antibacterial activity of chitosan hydrolysate and chitosan nanoparticles with copper against *Pseudomonas savastanoi* pv. *glycinea* and *Curtobacterium flaccumfaciens* pv. *flaccumfaciens* and determine their effectiveness in the treatment of soybean seeds and plants artificially infected with bacteria.

## 2. Materials and Methods

### 2.1. Preparation of Chitosan Hydrolysate

Crab shell chitosan with a molecular weight (MW) of 1040 kDa and a deacetylation degree (DD) of 85% was purchased from Bioprogress (Shchelkovo, Russia). Chitosan hydrolysate (ChiH) was prepared by chemical depolymerization of crab shell chitosan using nitric acid as described previously [41], with some modifications. Briefly, 1 g of chitosan was dispersed in 20 mL of 6.5% nitric acid, incubated for 7 h at 70 °C with stirring, cooled to room temperature, and kept without stirring for 16 h at 23 °C. Then, the pH was adjusted to 5.0–5.2 with 25% ammonium hydroxide, and the mixture was diluted with distilled water to a final volume of 180 mL. The pH of obtained ChiH with concentration 5 mg/mL was 5.2.

To determine the MW and polydispersity index, DD ChiH was preliminarily dialyzed against $H_2O$. The MW of ChiH was determined via high-performance gel permeation chromatography in an S 2100 Sykam chromatograph (Sykam, Eresing, Germany) using a separation column (8 mm × 300 mm; PSS NOVEMA Max analytical 1000 A) and a pre-column (8.0 mm × 50 mm) [42]. Pullulans were used as calibration standards. The DD of Chi was determined using proton nuclear magnetic resonance ($^1$H-NMR). Samples were prepared in deuterated water, and proton spectra were recorded on a Bruker AMX 400 spectrometer (Bruker, Watertown, MA, USA); 4,4-dimethyl-4-silapentane-sulfonic acid was used as a standard.

### 2.2. Preparation and Characterization of Chitosan Nanoparticles and Copper-Loaded Nanoparticles

Chitosan nanoparticles (ChiNPs) were formed using the ionotropic gelation method as described previously [43] with some modifications. Chi (2 g) with MW = 39 kDa, DD = 90% and a polydispersity index of 2.4 was solved in 300 mL of 1% acetic acid. The chitosan solution was filtered through a glass filter to remove mechanical impurities. Tripolyphosphate (TPP) solution (Sigma-Aldrich, Munich, Germany) (5 mg/mL) was added dropwise under vigorous stirring until opalescence occurred (A = 0.100, λ = 590 nm), which was estimated using a Spekol 11 spectrophotometer (Carl Zeiss Jena, Germany). Copper-loaded chitosan nanoparticles ($Cu^{2+}$ChiNPs) were obtained by dropwise addition of 25 mg/mL $CuSO_4$ solution up to A = 0.144 (λ = 590 nm). NP preparations were adjusted with 1% acetic acid to a concentration of Chi = 5 mg/mL, $CuSO_4$ = 0.83 mg/mL, pH 4.0. For the biological experiments, the ChiNPs and $Cu^{2+}$ChiNP particles were not further purified and used as nanoparticles contained within suspensions. To characterize the particles, the suspension was preliminarily centrifuged for 10 min at 1000× *g* and then the supernatants were centrifuged at 14,000× *g* for 20 min to separate the NP fraction. The yield of ChiNPs was 8%, and that of $Cu^{2+}$ChiNPs was 10%.

The mean hydrodynamic diameter of the NPs was determined via dynamic light scattering (DLS) in reflected light (scattering angle 180°) using a NANO-flex II analyzer (Colloid Metrix, Meerbusch, Germany), sample temperature 25 °C. The zeta potential of the NPs was characterized via DLS using a Zetasizer Nano (Malvern Panalytical, Malvern,

UK); all measurements were performed at 25 °C, and the scattering angle was equal to 173°. NP suspensions were preliminarily centrifuged (Centrifuge 5418, Eppendorf, Hamburg, Germany) at 1000× *g* for 10 min, and then the supernatants were centrifuged at 14,000× *g* for 20 min at RT to separate the NP fraction.

Dimensional characteristics of the NPs were explored using an atomic force microscope, INTEGRA Prima (NT-MDT SI, Zelenograd, Russia). Scanning was performed with a resolution of 512 × 512 points in the semicontact mode in air. The scanning frequency was 1.3 Hz. Golden NSG01 silicon probes (TipsNano, Zelenograd, Russia) with a tip average resonance frequency of 150 kHz, an average force constant of about 5.1 N/m, and a cantilever curvature radius of 6 nm were used. The data obtained via AFM were visualized using software NOVA 1.0.26.860 (NT-MDT SI, Zelenograd, Russia), analyzed using Image Analysis 3.5.0.2069 (NT-MDT SI, Zelenograd, Russia) and processed with OriginPro B.9.2.196 (OriginLab Corporation, Northampton, MA, USA).

### *2.3. Bacterial Strains*

The strains on which the experiments were carried out (strains Psg CFBP 2214 and Cff CFBP 3418) were obtained from the CFBP collection (Beaucouzé, France) and isolated and characterized by us in previous articles (Psg: G2 and G17, Cff: F-125-1 and F-30-1) [44,45]. These strains were pathogenic in soybean plants cv. Kasatka. The characterization of the Psg strains was performed by PCR analysis of the *cfl* (coronafacate ligase) gene [8] and analysis of the relationship of nucleotide sequences for the *cts* (citrate synthase) gene [46] with pathogen strains available in Genbank. Isolates belonging to Cff were determined via PCR analysis with genus-specific [47] and species-specific [48] primers.

### *2.4. Determination of Antibacterial Activity of Chitosan Samples*

#### 2.4.1. Determination of Antibacterial Activity via Agar Diffusion Method

The agar diffusion method [49] was used for the primary determination of antibacterial activity using all six strains mentioned above. Briefly, 100 μL of bacterial suspension with a concentration of $1 \times 10^8$ CFU/mL was applied to King's B medium, distributed with a sterile loop, and wells 8 mm in diameter were pierced with a sterile cork borer. The bottoms of the wells were sealed by pouring a drop of molten King's B medium (1.5% agar) into them. Then, 100 μL of sample was added to each at concentrations of 1, 5, 10, 25, 50, and 100% of the initial solutions. Samples with different concentrations were obtained by diluting the initial (100% solutions) in sterile water according to Table 1. The dishes were left at 4 °C for the diffusion of the solutions into agar (2 h) and then incubated at 28 °C for 48 h; ChiH and $CuSO_4$ solutions were used as controls. The experiment was repeated three times.

#### 2.4.2. Determination of Minimum Inhibitory Concentration (MIC)

MIC was determined according to [50] with modifications for all strains used in the article. In a 96-well sterile microtiter plate (Corning, Glendale, CA, USA), serial twofold dilutions of analyzed samples in King's B liquid medium were prepared; the volume was 100 μL. After the dilutions, 70 μL of bacterial suspension at a concentration of $10^4$ CFU/mL was dissolved in King's B liquid medium, 30 μL of 0.02% resazurin was added to each cell, and the mixture was thoroughly mixed. After 24 h, the plates were visually evaluated. The growth of bacteria in the medium was indicated by a color change from purple to pink. The lowest concentration at which a color change was observed was recorded as MIC. The experiment was repeated three times.

**Table 1.** Concentrations of samples obtained by dilution of the initial (100% solutions) in distilled water.

| Samples | Relative Concentrations of Samples, % (v/v) | Concentration of Chitosan, mg/mL | Concentration of $CuSO_4$, mg/mL |
|---|---|---|---|
| ChiH | 100 | 5 | - |
| | 75 | 3.75 | - |
| | 50 | 2.5 | - |
| | 25 | 1.25 | - |
| | 10 | 0.5 | - |
| | 1 | 0.05 | - |
| $Cu^{2+}$ChiH | 100 | 5 | 0.83 |
| | 75 | 3.75 | 0.62 |
| | 50 | 2.5 | 0.42 |
| | 25 | 1.25 | 0.21 |
| | 10 | 0.5 | 0.083 |
| | 1 | 0.05 | 0.0083 |
| ChiNPs | 100 | 5 | - |
| | 75 | 3.75 | - |
| | 50 | 2.5 | - |
| | 25 | 1.25 | - |
| | 10 | 0.5 | - |
| | 1 | 0.05 | - |
| $Cu^{2+}$ChiNPs | 100 | 5 | 0.83 |
| | 75 | 3.75 | 0.62 |
| | 50 | 2.5 | 0.42 |
| | 25 | 1.25 | 0.21 |
| | 10 | 0.5 | 0.083 |
| | 1 | 0.05 | 0.0083 |
| $CuSO_4$ | 100 | - | 0.83 |
| | 75 | - | 0.62 |
| | 50 | - | 0.42 |
| | 25 | - | 0.21 |
| | 10 | - | 0.083 |
| | 1 | - | 0.0083 |

2.4.3. Determination of Minimum Bactericidal Concentration (MBC)

The bactericidal activity of the chitosan samples in relation to all Psg and Cff strains used in the article was evaluated in accordance with the method of microdilution of broth described in CLSI 2015 [51] with modifications. For this, serial twofold dilutions of chitosan samples in King's B liquid medium were prepared in a 96-well microtiter plate (Corning, Corning, NY, USA), and the volume was 100 µL. After the dilutions, 100 µL of bacterial suspension at a concentration of $10^4$ CFU/mL was added to each cell and the contents were thoroughly mixed. The plates were sealed with parafilm and incubated on a shaker-incubator, ES 20 (Biosan, Riga, Latvia), at 180 rpm and 28 °C. After 24 h of cultivation, 10 µL of bacterial suspension from each cell was tenfold diluted in sterile water and dispersed onto YD agarized medium (YDC without $CaCO_3$) for subsequent titer calculation after 48 h. The experiment was repeated four times. A statistical analysis based on the results of the determination of MIC and MBC was not carried out because there were no differences within the repetitions.

2.4.4. Determination of Time–Kill Curves

Time–kill curves were determined as described in [52] with some modifications. One colony of each bacterium (Psg CFBP 2214 and Cff CFBP 3418) was pre-cultured in 4 mL of King's B liquid medium for 12 h at 28 °C and incubated on a shaker ES 20 (Biosan, Riga, Latvia) at 200 rpm. The cells were precipitated via centrifugation and titrated to a concentration of $10^4$ CFU/mL with sterile water. Bacterial titer control was carried out spectrophotometrically according to the $OD_{600}$ index measured using a Nanodrop One (Thermo Fisher Scientific, Waltham, MA, USA). The cell suspension was then transferred to

1.5 mL sterile test tubes, and preparations were added to concentrations of 1xMBC. After that, the tubes were placed in a Thermomixer C (Eppendorf, Hamburg, Germany) and cultivated at 27 °C and 350 rpm. After 0, 2, 5 and 30 min and 1, 2 and 24 h, 10 μL of the mixture was taken, diluted in sterile SPS buffer, and dispersed on King's B agarized medium. Colonies were counted after 48 h of cultivation at 28 °C. A suspension without antimicrobial agents was used as a negative control. The experiment was repeated three times.

### *2.5. Phytotoxicity on Soybean Seeds and Plants*

The phytotoxicity of the chitosan samples on soybean seeds was assessed by a germination test using the standard "over paper" method described by the International Seed Testing Association [53]. Soybean (cv. Kasatka) seeds were soaked in aqueous solutions of the chitosan samples at various concentrations for 10 min and then completely dried on sterile filter paper at room temperature under sterile conditions. The samples of (1) water, (2) ChiH, (3) $Cu^{2+}$ChiNPs, (4) $Cu^{2+}$ChiH, (5) ChiNPs and (6) $CuSO_4$ were diluted to concentrations of 1, 5, 10, 25, 50 and 100% (Table 1) of the stock solutions using sterile water.

Seeds soaked in sterile water were used as a negative control. Next, the seeds were incubated at a temperature of 25 °C with constant humidity. On the 8th day after treatment (DAT), germination was assessed; if a sprout with a well-developed root grew from a seed, it was considered to have germinated. The average percentage of seed germination was determined for all repetitions. The length of the roots was measured with a caliper after counting the germination rate and separating the cotyledons. The experiment consisted of 3 repetitions of 50 seeds in each group.

To test the phytotoxicity of the chitosan samples on plants, soybeans were grown to phase R1 (beginning bloom) in a turf–perlite mixture (Vieltorf, Velikiye Luki, Russia) in plastic pots for plant cultivation (volume 1 L, AgrofloraPack, Vologda, Russia).

Plants were kept in a greenhouse at 28/22 °C (14 h day/10 h night) under natural light and watered as needed. The foliar treatment was carried out with tested samples using a sprayer (with a drop size of ~300 μm) at a consumption rate of a sample solution of ~5 mL/plant (until all leaves were completely wetted).

Phytotoxicity was assessed after 7 days of incubation under the same conditions, according to the phytotoxicity scale [54], where: 0—no symptoms; 1—very slight discoloration; 2—more severe, but short; 3—moderate and longer; 4—medium and long; 5—moderately severe; 6—heavy; 7—very heavy; 8—almost destroyed; 9—destroyed; 10—completely destroyed. The experiment was repeated three times, with two plants in each repetition. The phytotoxicity rating was considered as the average score for each variant (the sum of the scores of each leaf/the number of analyzed leaves).

### *2.6. Control Psg and Cff Artificial Infection by Chitosan Samples*

All experiments on the use of the chitosan samples in the artificial infection of soybean seeds and leaves with bacterial diseases were carried out from May to August 2022 under the conditions of an experimental greenhouse using the Kasatka soybean cultivar (harvest year 2021; weight of 1000 seeds = 122.8 g). In these experiments, the strains Psg CFBP 2214 and Cff CFBP 3418 were used.

#### 2.6.1. Control Psg on Seeds

Artificial Psg infection of seeds was carried out according to the method in [45]. Briefly, a 72 h culture of Psg CFBP 2214 was suspended in sterile 10 mM $MgCl_2$ at ~$10^4$ CFU/mL. Soybean seeds were sterilized in 75% ethanol for 2 min, washed with an aqueous 50% solution of commercial bleach (sodium hypochlorite)/0.002% Tween 20 (*v/v*) for 8–10 min and distilled $H_2O$ until the chlorine was removed, and left in a humid chamber for 2 h to make them swell. The swollen seeds were pierced with a sterile toothpick, transferred to a flask with a bacterial suspension, vacuum treated at $-10^5$ Pa for 10 min and dried to remove excess liquid.

The infected seeds were immersed for 10 min in 50% solutions of (1) water, (2) ChiH, (3) $Cu^{2+}$ChiNPs, (4) $Cu^{2+}$ChiH, (5) ChiNPs and (6) $CuSO_4$ (Table 1). After that, the seeds were dried on paper towels to get rid of excess moisture.

The treated seeds of the experiment were sown in a peat–perlite mixture (Veltorf, Velikie Luki, Russia) in 40-cell plastic seed trays (cell volume 0.12 L, AgrofloraPak, Vologda, Russia). The plants were watered as needed and grown in a greenhouse in natural sunlight at 28/22 °C (14 h day/10 h night). Treatments in each experiment were organized according to the scheme of complete randomization. Each treatment had 5 replications with 40 seeds (1 tray per replication).

#### 2.6.2. Control Psg on Leaves

Psg infection of soybean plants was carried out according to the method in [55], using suspension infiltration with a 1113 AirControl airbrush (JAS, Ningbo, China). The bacterial suspension was prepared in the same way as it was for seed inoculation, but with the addition of Silwet Gold surfactant (Chemtura, Philadelphia, PA, USA) at a concentration of 0.01% (*w/w*). Infection was carried out with an average dose of 5 mL of suspension with a concentration of $10^9$ CFU/mL per trifoliate leaf. Plants were cultivated according to Section 2.6.1. in 0.5 L pots. Each treatment had three replications with 10 plants per replication.

The design of the experiment included the use of (1) water, (2) ChiH, (3) $Cu^{2+}$ChiNPs, (4) $Cu^{2+}$ChiH, (5) ChiNPs (6) and $CuSO_4$ (Table 1).

The percentage of plants that exhibited leaf symptoms was recorded. The LeafDoctor app (https://www.quantitative-plant.org/software/leaf-doctor, accessed on 21 July 2022) installed on an iPhone SE 2 was used to assess the development of the disease by the degree of infection of adult plants. For this, all plants were photographed and analyzed by moving the threshold slider until only symptomatic tissues were converted to blue and the percentage of affected tissue was calculated according to the developer's recommendations [56]. The same calculations were made in the seed treatment experiment after reaching stage V3 (35 days after sowing).

#### 2.6.3. Control Cff on Seeds

Inoculation through hilum injury described in [16] with modifications was used for seed infection by Cff. For this purpose, the hilum of each seed was pierced with a sterile needle, soaked in a bacterial suspension, placed in a vacuum, and then dried on paper towels under sterile conditions.

Soybean seeds were treated via immersion for 10 min in an aqueous solution of (1) water, (2) ChiH, (3) $Cu^{2+}$ChiNPs, (4) $Cu^{2+}$ChiH, (5) ChiNPs and (6) $CuSO_4$, then dried on paper napkins to get rid of excess moisture. Further actions with plants and growing conditions were similar to Section 2.6.1.

Bacterial wilt was scored for each plant at 15, 18, 21, 24, 27, and 31 days post-seeding on a scale of 0 to 5, where 0 = no wilt symptoms; 1 = wilting of one of the primary leaves; 2 = wilting of both primary leaves but not the first trifoliate; 3 = withering of the first trifoliate leaf; 4 = death of the seedling after the development of primary leaves; and 5 = no germination or complete wilting and loss of turgor (in adult plants) of soybean scales adapted by us in a previous study described in [57]. Using this scale and methodology [58], the AUPDC (area under progress disease curve) was calculated using MS Excel 2007.

#### 2.6.4. Control Cff on Leaves

The Cff infection of soybean plants and the method for calculating plant disease were similar to Section 2.6.2. The design of the experiment included the use of: (1) water, (2) ChiH, (3) $Cu^{2+}$ChiNPs, (4) $Cu^{2+}$ChiH, (5) ChiNPs and (6) $CuSO_4$. The calculation of the incidence rate, replicate and plant growth conditions was similar to Section 2.6.2.

### 2.7. Statistical Analysis

For all experiments, data analysis was carried out using the analysis of variance method using Statistica 12.0 (StatSoft, TIBCO, Palo Alto, CA, USA), comparing the average values using Duncan's criterion. The percentage data were converted to arcsine before processing. Graphs were created using GraphPad Prism 9.2.0 (GraphPad Software Inc., Boston, MA, USA).

## 3. Results and Discussion

### 3.1. Preparation Samples Based on Chitosan

Chitosan hydrolysate with the main fraction (MW 39 kDa, DD 90%, polydispersity index 2.4) was prepared from high-molecular-weight chitosan (MW 1040 kDa, DD 85%) by acid hydrolysis using nitric acid. We assume that the chitosan hydrolysate preparation considered in this work can be applied in practice in agriculture. In this regard, we attempted to simplify the method of preparation by not isolating a separate fraction of low-molecular-weight chitosan.

Along with the properties typical for Chi, ChiNPs had the valuable advantages of nanoparticles, namely their large surface area and small size [59,60]. Muthukrishnan et al. described the ability of chitosan nanoparticles to inhibit the growth of *Pyricularia grisea, Alternaria solani* and *Fusarium oxysporum* [61]. In the same work, chickpea seed treatment had positive morphological effects, such as an increase in germination percentage, seed strength index and vegetative biomass of seedlings.

The versatility of ChiNP activity against plant pathogens, particularly tomato, of both fungal and bacterial etiology was presented in [62]. It was shown that chitosan nanoparticles possessed antimicrobial activity towards a complex of tomato pathogens, which include fungi *Colletotrichum gelosporidies, F. oxysporum, Gibberella fujikuori, Sclerotinia sclerotiorum* and *Phytophthora capsici* and bacterium *Pectobacterium carotovorum* subsp. *carotovorum X. campestris* pv. *vesicatoria*. Recently, chitosan–metal nanocomplexes with improved antimicrobial activity were synthesized ($Ag^{+}$-ChiNPs, $Cu^{2+}$-ChiNPs, $Zn^{2+}$-ChiNPs, $Mn^{2+}$-ChiNPs and $Fe^{2+}$-ChiNPs) [63–66]. The work [67] shows the activity of chitosan nanoparticles against a number of pathogens of bacterial plant diseases, including *Agrobacterium tumefaciens, Erwinia* sp. and *X. campestris* with MIC values of 100, 500 and 500 ppm, respectively. All these results indicate that chitosan nanoparticles can be used in the field to protect various crops from pathogens of different etiologies.

To obtain chitosan nanoparticles, the main fraction of hydrolysis (MW 39 kDa) was dialyzed and freeze-dried. ChiNPs were obtained by ionotropic gelation under acidic conditions (pH 4.0). The formation of ChiNPs occurred due to the interaction of positively charged chitosan amino groups with TPP, which has phosphate groups with a negative charge. In contrast to the original technique [43], low-molecular-weight chitosan was used; thus, it was possible to use a more concentrated solution of Chi, but the formed particles were larger. In our work, the ChiNPs and ChiH samples had the same chitosan concentration of 5 mg/mL.

To obtain $Cu^{2+}$ChiNPs, copper sulfate was added to a suspension of nanoparticles up to a final concentration of 0.83 mg/mL. The formation of complexes can occur through adsorption, ion exchange and chelation. The interaction type is defined by the solution formulation, the pH value and the type of metal ion [68]. Chitosan is able to form complexes with some metal ions, predominantly through interactions with amino groups and hydroxy groups (especially in the C3 position), that promote sorption [69]. The dimensional characteristics and charge of the nanoparticles measured using the DLS method are shown in Table 2. The measurements of nanoparticle size were carried out using the particle number distribution. The hydrodynamic diameter of ChiNPs was larger compared to $Cu^{2+}$ChiNPs. The polydispersity of nanoparticles purified from unbound polymer (ChiNPs cf) was lower than that of ChiNPs and $Cu^{2+}$ChiNPs.

**Table 2.** Characteristics of ChiNPs and $Cu^{2+}$ChiNPs.

| Samples | Size, nm | Polydispersity Index | Zeta-Potential, mV |
|---|---|---|---|
| ChiNPs | 254 ± 37 | 0.499 | 37.8 ± 1.6 |
| $Cu^{2+}$ChiNPs | 153 ± 30 | 0.421 | 22.7 ± 0.4 |
| ChiNPs cf * | 251 ± 32 | 0.367 | 48.5 ± 0.6 |
| $Cu^{2+}$ChiNPs cf * | 157 ± 42 | 0.540 | 27.2 ± 0.6 |

*: ChiNPs or $Cu^{2+}$ChiNPs were preliminarily centrifuged for 10 min at 1000× *g*, and then supernatants were centrifuged at 14,000× *g* for 20 min to separate NP fraction.

The AFM method was used to characterize the size and morphology of the nanoparticles (Figure 1). The AFM-images of ChiNPs and $Cu^{2+}$ChiNPs (Figure 1(A1,A2,B1,B2)) show that the suspension contained a large amount of unbound polymer forming aggregates smaller than 15 nm. To characterize the main fraction of nanoparticles, it was separated from the unbound polymer by centrifugation (ChiNPs or $Cu^{2+}$ChiNPs were preliminarily centrifuged for 10 min at 1000× *g* and then supernatants were centrifuged at 14,000× *g* for 20 min to separate the NP fraction). As a result, the fractions of nanoparticles ChiNPs cf and $Cu^{2+}$ ChiNPs cf were isolated (Figure 1(C1,C2,D1,D2)). There were no significant differences in the particle sizes of ChiNPs cf and $Cu^{2+}$ChiNPs cf, which were 30–60 nm (Figure 1(C3,D3)). ChiNPs cf had an amorphous structure in contrast to the more compact structure of $Cu^{2+}$ChiNPs cf. When comparing the morphology of the synthesized nanoparticles, it was found that ChiNPs cf had a greater tendency to aggregate.

We assumed that when scaling up the nanoparticle formation technology for agricultural use, it would not be advisable to isolate the nanoparticles fraction from the reaction mixture. Therefore, biological efficacy tests were carried out with crude nanoparticle preparations, which were a mixture of nanoparticles and an unbound polymer/$Cu^{2+}$, but for brevity, we continued to use the abbreviations ChiNPs or $Cu^{2+}$ChiNPs for nanoparticle-containing samples.

### 3.2. *Antibacterial In Vitro Activity*

The primary antibacterial activity of chitosan samples was tested using the agar diffusion method towards three *P. savastanoi* pv. *glycinea* strains and three *C. flaccumfaciens* pv. *flaccumfaciens* strains.

#### 3.2.1. Determination of Antibacterial In Vitro Activity via Agar Diffusion Method

Pathogens had different sensitivities to chitosan that depended on the strain, sample type and dose (Figure 2). It should be noted that the analyzed substances exhibited a stronger antibacterial effect against Psg strains, whereas Cff strains were more resistant. $Cu^{2+}$ChiNPs were the most effective at all analyzed concentrations; the diameter of the inhibition zone of 100% $Cu^{2+}$ChiNPs suspension (5 mg/mL of chitosan and 0.83 mg/mL of copper) was 27 mm for Psg and about 15 mm for Cff (Figure S1). Although ChiH and $CuSO_4$ exhibited no antibacterial activity (Figure 2), the average diameter of the inhibition zone for $Cu^{2+}$ChiH (the combination of ChiH and $CuSO_4$) on Psg strains was about 5 mm. The low efficiency of the $CuSO_4$ solution can be explained by the low concentration. We suggest that the low effect of ChiH is due to the difficulty of diffusion of the chitosan polymer molecules in the nutrient medium at neutral pH, similar to the data reported in article [70]. It is most likely that the addition of copper sulfate to ChiH resulted in the formation of more compact complexes of chitosan with copper, which increased diffusion into the agar. For $Cu^{2+}$ChiH, ChiNPs, and $Cu^{2+}$ChiNPs, there were dose-dependent dynamics of increasing the zone of bacterial growth inhibition. Chitosan-based copper nanoparticles, obtained using a chemical reduction method, effectively inhibited growth of *X. axonopodis* pv. *punicae* at a concentration of 1000 [38], which is in agreement with our data.

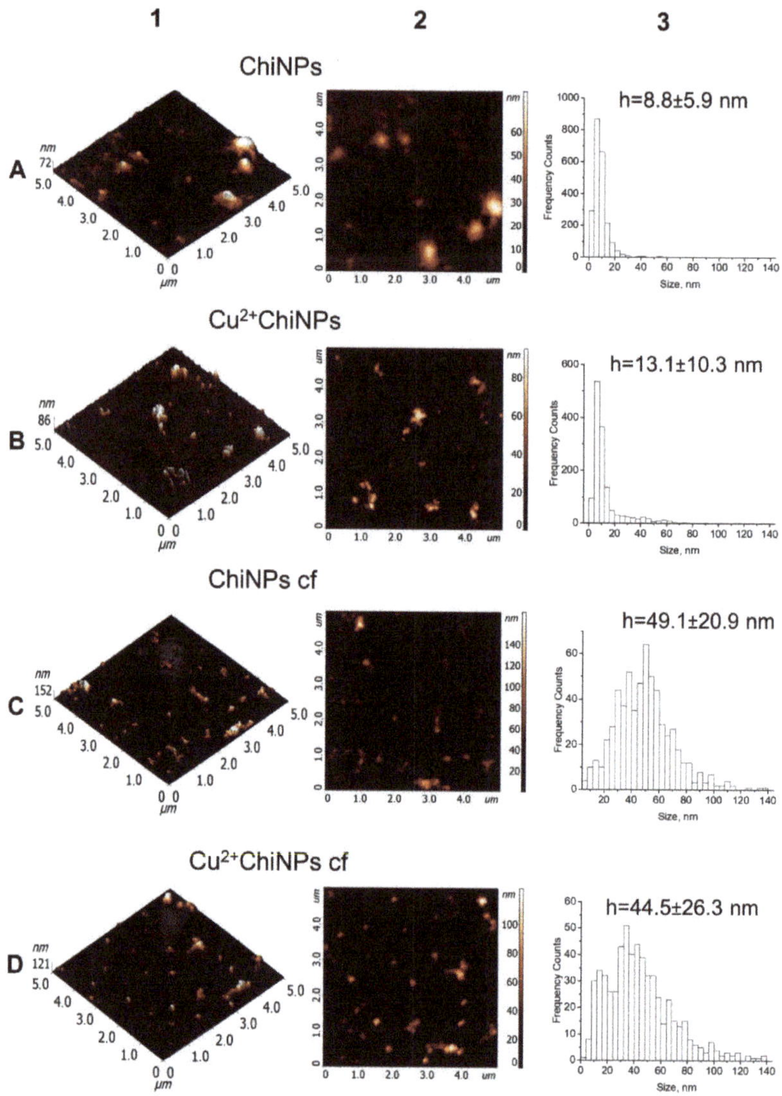


**Figure 1.** Dimensional characteristic of NPs. Column 1 (**A1–D1**)—3D AFM images of nanoparticles, column 2 (**A2–D2**)—2D AFM images of nanoparticles, column 3 (**A3–D3**)—histograms of the size distribution of nanoparticles and their average sizes according to AFM. Row **A**—ChiNPs, row **B**—$Cu^{2+}$ChiNPs, row **C**—ChiNPs cf, row **D**—$Cu^{2+}$ChiNPs cf.

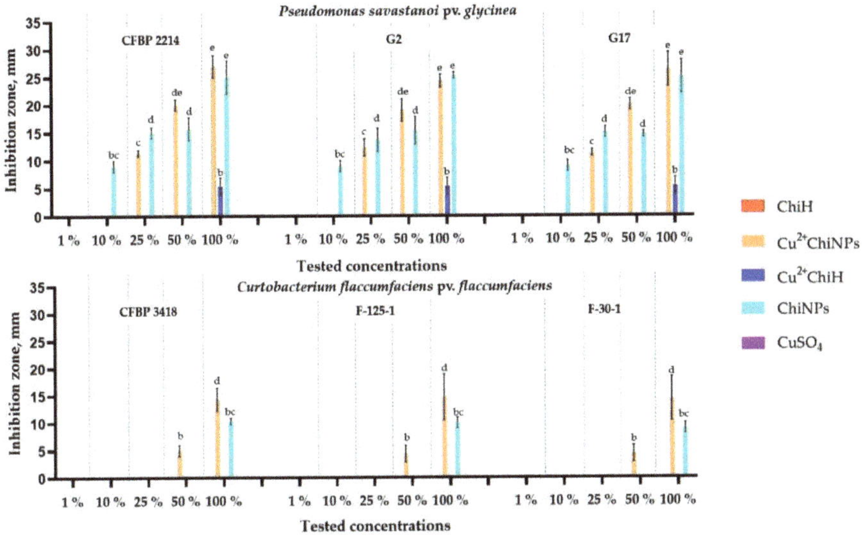


**Figure 2.** The diameters of the inhibition zones for each test substance, depending on the concentration of solutions against Psg and Cff strains (the average value for every strain of each bacterium) in an agar diffusion test. We added 100 μL of the sample to the well, and after 48 h of incubation at 28 °C, the zone of inhibition was measured. Different letters indicate a significant difference in values, according to Duncan's test, at $p = 0.05$. All tests were carried out three times. The standard deviation (SD) is shown for each bar.

### 3.2.2. Determination of Minimum Inhibitory and Bactericidal Concentrations

The MIC and the MBC of the chitosan samples are shown in Table 3. It was found that the inclusion of copper in the nanoparticles led to a decrease in MIC and MBC in relation both to copper sulfate and ChiNPs. The addition of copper ions to chitosan hydrolysate ($Cu^{2+}$ChiH) enhanced its antibacterial activity. However, the activity of $Cu^{2+}$ChiH was lower compared to $Cu^{2+}$ChiNPs.

**Table 3.** Inhibitory and bactericidal concentrations of chitosan samples and $CuSO_4$ against Psg and Cff strains.

| Samples | Minimal Inhibitory (MIC) and Bactericidal (MBC) Concentrations of Samples, μg/mL (Chitosan/Copper) | | | | | | | | | | | |
|---|---|---|---|---|---|---|---|---|---|---|---|---|
| | Psg Strains | | | | | | Cff Strains | | | | | |
| | CFBP 2214 | G2 | G17 | CFBP 2214 | G2 | G17 | CFBP 3418 | F-125-1 | F-30-1 | CFBP 3418 | F-125-1 | F-30-1 |
| | MIC | | | MBC | | | MIC | | | MBC | | |
| ChiH | 156/- | 156/- | 156/- | 625/- | 625/- | 625/- | 78/- | 78/- | 78/- | 312/- | 312/- | 312/- |
| $Cu^{2+}$ChiH | 78/13 | 78/13 | 78/13 | 78/13 | 78/13 | 39/6 | 19/3 | 19/3 | 19/3 | 312/52 | 312/52 | 0.321/52 |
| ChiNPs | 39/- | 39/- | 39/- | 156/- | 156/- | 156/- | 39/- | 39/- | 39/- | 156/- | 156/- | 156/- |
| $Cu^{2+}$ChiNPs | 19/3 | 19/3 | 19/3 | 78/13 | 78/13 | 78/13 | 19/3 | 19/3 | 19/3 | 78/13 | 78/13 | 78/13 |
| $CuSO_4$ | -/6 | -/13 | -/3 | -/13 | -/26 | -/13 | -/13 | -/13 | -/13 | -/52 | -/52 | -/52 |

Unfortunately, there are few works devoted to the study of the efficacy of chitosan nanoparticles loaded with copper against phytopathogenic bacteria. Therefore, we will also consider those works in which antibacterial activity was studied on human opportunistic bacteria. Du et al. investigated chitosan-based nanoparticles loaded with $Cu^{2+}$ ions obtained via ionotropic gelation. On the bacteria *E. coli*, *S. choleraesuis* and *S. aureus*, it was shown that the antibacterial activity of such nanoparticles was significantly higher compared to the activity of chitosan nanoparticles and $Cu^{2+}$ ions. In addition, Gram-negative bacteria were more sensitive than Gram-positive bacteria [71]. Antibacterial activity of CuO, $Cu_2O$ and $Cu^0$ nanoparticles obtained by using reducing agents has also been studied. For CuO nanoparticles, the bactericidal concentration against *Ralstonia solanacearum* causing bacterial wilt was 250 μg/mL [72]. The MBC values for CuO nanoparticles were 100 μg/mL for *S. aureus* (MRSA), 250 μg/mL for *E. coli* and 5000 μg/mL for *P. aeruginosa* in [73]. These data are consistent with our data.

It was found that ChiH was less active compared to ChiNPs. This was probably due to the fact that in King's B medium with a pH of 7.0–7.2, used in this test, the protonation of amino groups responsible for the manifestation of antibacterial activity decreases [35]. One of the mechanisms of chitosan action is considered to be its ability to form films around bacterial cells [74]. However, in our work, ChiH contained the main fraction with a low molecular weight, which decreases film-forming ability. Chitosan NPs exhibited higher antibacterial activity than chitosan, probably due to their higher surface-to-volume ratio and surface energy [35]. The higher activity of chitosan NPs compared to chitosan was previously reported by Qi et al. in [43].

From the $CuSO_4$ and $Cu^{2+}$ChiH test results, it is evident that the CFBP 2214 and G17 (Psg) strains had a greater sensitivity to copper than strain G2. This fact may be an indirect indicator of the diversity of strains, including sensitivity to bactericides in the country. MIC ChiH data show that bacteria of the Cff species were more sensitive to chitosan (78 mg/mL) compared to Psg (156 mg/mL). One of the possible reasons for these differences is the different structure of the bacterial cell wall. For example, in the paper [75], using four Gram-negative bacteria (*Escherichia coli*, *P. fluorescens*, *Salmonella typhimurium*, and *Vibrio parahaemolyticus*) and seven Gram-positive bacteria (*Listeria monocytogenes*, *Bacillus megaterium*, *B. cereus*, *Staphylococcus aureus*, *Lactobacillus plantarum*, *L. brevis*, and *L. bulgaricus*), it was shown that Gram-positive bacteria are more sensitive to chitosan.

### 3.2.3. Antibacterial In Vitro Activity by Determination of Time–Kill Curves

Another important parameter that determines the effectiveness of antibacterial agents is the rate of cell death, as described by time–kill curves. Figure 3 shows the time–kill curves for the Psg CFBP 2214 and Cff CFBP 3418 strains.

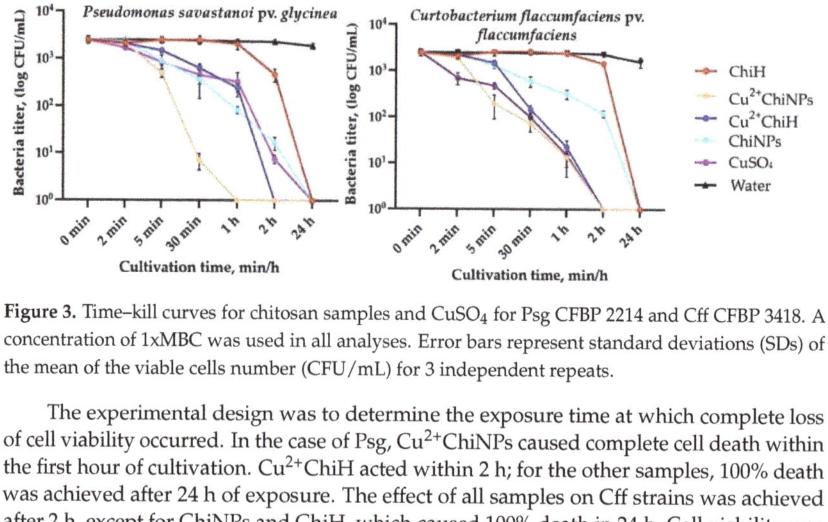


**Figure 3.** Time–kill curves for chitosan samples and $CuSO_4$ for Psg CFBP 2214 and Cff CFBP 3418. A concentration of 1xMBC was used in all analyses. Error bars represent standard deviations (SDs) of the mean of the viable cells number (CFU/mL) for 3 independent repeats.

The experimental design was to determine the exposure time at which complete loss of cell viability occurred. In the case of Psg, $Cu^{2+}$ChiNPs caused complete cell death within the first hour of cultivation. $Cu^{2+}$ChiH acted within 2 h; for the other samples, 100% death was achieved after 24 h of exposure. The effect of all samples on Cff strains was achieved after 2 h, except for ChiNPs and ChiH, which caused 100% death in 24 h. Cell viability was virtually unchanged in the presence of water. A similar kinetic of ChiNPs action was shown by Dash et al., where complete killing of *B. subtilis* and *S. aureus* was not achieved within 4 h [76]. At the same time, in nearly all variants, 50% cell death occurred within 30 min. Thus, $Cu^{2+}$ChiNPs exhibited the most rapid bactericidal effect, causing the complete death of bacteria in liquid nutrient medium within 1 h for Psg and 2 h for Cff. Christena et al. also found that CuNPs had a bactericidal effect on *S. aureus* at a concentration of 2xMIC and *P. aeruginosa* at a concentration of 1xMIC. Four hours after treatment with CuNPs, a five-fold logarithmic decrease in CFU was observed for *Staphylococcus*, and a three-fold logarithmic decrease in CFU was demonstrated for *Pseudomonas* [77]. Thus, the determination of time–kill curves shows that $Cu^{2+}$ChiNPs have a greater potential to fight bacteria due to their high kill rate compared to the initial forms of chitosan and copper.

### 3.3. Phytotoxicity on Seeds and Leaves

To determine the limiting concentration of the samples for the treatment of soybean plants, phytotoxicity tests were performed at concentrations of 0, 25, 50, 75 and 100 % of the stock solutions (according to Table 1).

The effect of the sample concentrations on seed germination and root length of soybean seedlings is shown in Figure 4A,B. The phytotoxicity of samples at various concentrations was determined by the average values of germination and root length. The obtained values were compared with a water-treated control. For all samples, the phytotoxic effect was observed at concentrations above 50% of the stock solutions, corresponding to 2.5 mg/mL of chitosan and 0.42 mg/mL of $CuSO_4$.

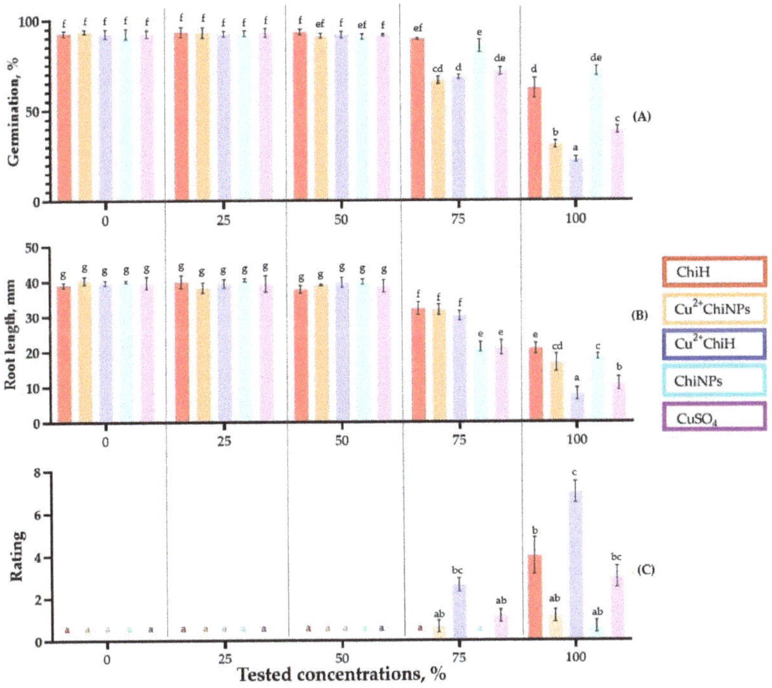


**Figure 4.** Phytotoxicity of chitosan samples on soybean leaves and seeds. Germination values (**A**) and root length (**B**) of soybean seeds after treatment with different concentrations of samples 8 d after treatment. The average score of the phytotoxicity integral value on soybean leaves for chitosan samples was measured at 72 h after treatment (**C**). Values represent the mean of three independent trials, error bars represent the standard deviation. Values marked by different letters have a significant difference, according to Duncan's criteria, at $p = 0.05$.

At 50% concentration of the samples (2.5 mg/mL of chitosan and 0.42 mg/mL of $CuSO_4$), an insignificant decrease in germination and root length was observed. When treated at initial concentrations (5 mg/mL of chitosan and 0.83 mg/mL of $CuSO_4$), $Cu^{2+}$ChiH had the strongest reduction in seed germination, and ChiNPs had the least phytotoxic effect. $Cu^{2+}$ChiH had the most toxic effect on root length, and ChiH was the least toxic. It is important that the inclusion of copper in the nanoparticles increased their antibacterial activity and reduced the phytotoxicity of copper.

Phytotoxicity on soybean leaves was tested by spraying samples at different concentrations. For all samples, a dose-dependent increase in phytotoxicity with increasing concentration was determined.

As in the case of seeds, safe non-phytotoxic concentrations for leaf treatment were 50% of the initial solutions (2.5 mg/mL of chitosan and 0.42 mg/mL of $CuSO_4$) for all analyzed samples (Figure 4C). $Cu^{2+}$ChiH had the highest phytotoxicity; when treating with a 100% solution (5 mg/mL of chitosan, 0.83 mg/mL of $CuSO_4$), phytotoxicity symptoms in the form of leaf blights were observed, with the average phytotoxicity score reaching 7.0, which corresponds to very heavy leaf damage (Figure 4C and Figure S2).

The high phytotoxicity of ChiH was probably due to the presence of salts in the form of ammonium nitrate and $NO_3^-$ as counter ions on the amino groups of chitosan. The phytotoxicity of copper in $Cu^{2+}$ChiNPs was much lower compared to $Cu^{2+}$ChiH and $CuSO_4$ solution. This is probably due to the slow release of copper from the nanoparticles compared to $CuSO_4$ solution, as confirmed in the study by Young et al. [78]. Sathiyabama et al. found no symptoms of phytotoxicity when finger millet (*Eleusine Coracana* (L.) Gaertn) was treated with copper–chitosan nanoparticle solution [79], which is consistent with our data. At the same time, metal particles without chitosan exhibited phytotoxic properties, such as in the work of Stampoulis et al., where treatment with copper nanoparticles ($Cu^0$) at a concentration of 1 mg/mL resulted in a 90% reduction in biomass of zucchini plants compared to untreated control plants [80]. In contrast, Shende et al. found that treatment of pigeon pea (*Cajanus cajan* L.) with CuNP solution at a concentration of 20 ppm resulted in an increase in height, root length, fresh and dry weight and plant productivity index [81]. This may be due to both the lower copper concentration and the green method of particle production using plant extracts.

Thus, to comply with the principle of a single difference, further studies on the control of soybean bacterial diseases using chitosan-containing samples were performed using 50% solutions (2.5 mg/mL of chitosan and 0.42 mg/mL of $CuSO_4$) that found no statistically significant indicators of phytotoxicity on soybean.

### 3.4. *The Efficiency of Chitosan Samples against Psg and Cff Infection on Leaves and Seeds*

The repeatability of the «Psg-soybean» and «Cff-soybean» pathosystem models has been described and explained in detail in our previous publications [45,57], and the experimental conditions were identical. Soybean leaves preliminarily infected with Psg and Cff suspensions were treated with chitosan samples. Disease spread on soybean leaves was measured 12 days after treatment using Leaf Doctor software.

Chitosan samples reduced the degree of leaf lesions from Psg by 15–71% compared with water-treated controls (Figure 5A,B). $Cu^{2+}$ChiNPs resulted in a 71% decrease in lesion area compared to controls, while $Cu^{2+}$ChiH contributed only up to 50%. Treatment with $CuSO_4$, ChiH and ChiNPs did not cause a significant reduction in leaf lesions (15–20%) compared to control.

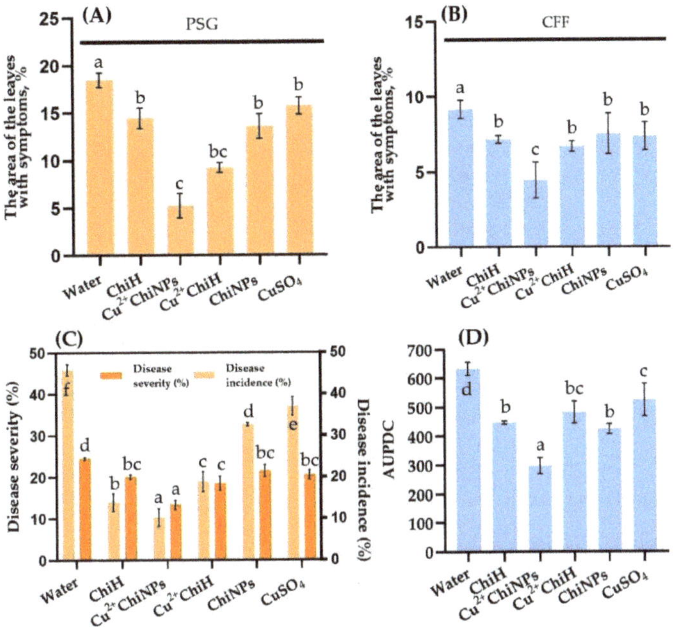


**Figure 5.** Bacterial blight (**A**,**C**) and bacterial tan spot and wilting (**B**,**D**) of soybean, caused by artificial inoculation of Psg and Cff after treatment with chitosan samples. (**A**,**B**): disease severity on inoculated green plants; (**C**): disease severity and incidence after inoculation of soybean seed by Psg; (**D**): values of AUPDC after inoculation of soybean seeds by Cff. Values are averages from three independent tests, error bars show standard deviation. Columns with a significant difference are marked with different letters, Duncan's test, $p = 0.05$.

The average leaf area with disease symptoms in the control group infected with Cff was inferior to the Psg infected group but remained at a high level (9.2% and 18.5%, respectively). The highest efficiency was observed for $Cu^{2+}$ChiNPs (51.3% reduction of lesion area), while the efficiency of other samples ranged from 17.8 to 26.9% compared with control (Figure S3).

Treatment of soybean seeds pre-infected with Psg using the chitosan samples exhibited a significant decrease in seedling infection frequency and disease development rate.

In the case of water-treated plants, rapid disease development was observed (Figure 5C). With daily overwatering of plants, a secondary infection was created, similar in severity to an outbreak of the disease in the field.

The biological effectiveness of $Cu^{2+}$ChiNPs treatment was 77% (disease incidence) or 45.3% (disease severity) compared to control. $Cu^{2+}$ChiH treatment reduced disease development ~1.3-fold and disease incidence more than 2-fold. $CuSO_4$ solution and ChiNPs treatments were the least effective. Their effectiveness on disease development was 19.3% and on disease incidence 16.3%. The Cff infected control group of seeds exhibited symptoms of wilting and yellowing of soybean leaves with an average AUPDC = 609 score (Figure 5D). In general, the treatment efficacy of all samples was lower for Cff than for Psg. Thus, the $Cu^{2+}$ChiNPs treatment was the best, with a biological efficiency of 53% compared to the control, while ChiNPs reduced AUPDC by only 33%. Treatment of seeds with $CuSO_4$ demonstrated a low biological effect; the efficiency was 17%.

The effectiveness of treatment of plants with chitosan copper-loaded nanoparticles strongly depends on many factors, one of which is the concentration of active substances and the type of pathogen. For example, in the work of Swati et al., the treatment of soybean plants with Cu-chitosan NP at a concentration of 0.02–0.12% reduced the severity of the bacterial pustule by 50.0–33.3% and 55.3–34.0% in the pot and in the field, respectively [82]. Kumar et al. studied the effectiveness of copper–chitosan-based nanoparticles in the treatment of banana plants against *F. oxysporum* f. sp. *cubense*. At a concentration of 0.20 mg/mL, high efficacy was shown, which amounted to a 73% reduction in symptoms compared to the untreated control [83].

Thus, our results demonstrate the protective effects of copper-loaded chitosan nanoparticles on soybean seed and leaf from bacterial blight and rust-brown bacterial spot and wilt. Further research is needed to improve the efficacy of soybean treatments by optimizing delivery technology, determining biosafety and developing the formulation for commercial use.

## 4. Conclusions

In this article, the synthesis of different chitosan samples (chitosan hydrolysate, chitosan hydrolysate with copper, chitosan nanoparticles and copper-loaded chitosan nanoparticles) and evaluation of their antibacterial action in vitro and in an artificial infection of soybean bacterial diseases were carried out.

The $Cu^{2+}$ChiNPs sample demonstrated the greatest antibacterial activity, with maximum inhibition zone diameters of 27 mm and 15 mm and the shortest total bacterial kill times of 1 h and 2 h for *Pseudomonas savastanoi* pv. *glycinea* and *Curtobacterium flaccumfaciens* pv. *flaccumfaciens*, respectively. Evaluation of all samples for their phytotoxicity by treatment of soybean leaves and seeds demonstrated that they are safe for soybean plants at the concentrations of 2.5 mg/mL of chitosan and 0.42 mg/mL of $CuSO_4$ or less.

In the process of studying the protective properties of samples against an artificial infection background of two major bacterial diseases of soybean, it was found that treatment with $Cu^{2+}$ChiNPs solution of seeds and leaves that had been previously infected by bacterial diseases is an effective tool to reduce pathogen damage in soybean.

These results are encouraging because the studied samples could potentially be used as an element of protection of soybean against the diseases of bacterial etiology mentioned in this study. However, potential side effects on non-target organisms should be evaluated and field trials should be conducted before using substances as pesticides to control phytopathogenic bacteria on an industrial scale.

**Supplementary Materials:** The following supporting information can be downloaded at: https://www.mdpi.com/article/10.3390/polym15051100/s1, Figure S1: Primary testing of the antibacterial properties of $Cu^{2+}$Chi-NPs against Psg and Cff strains by agar diffusion. We added 100 µL of solution to the wells, and the inhibition zone was measured after 48 h of incubation at 28 °C. A: growth inhibition of Pseudomonas savastanoi pv. glycinea CFBP 2214; B: growth inhibition of Curtobacterium flaccumfaciens pv. flaccumfaciens; Figure S2: Phytotoxicity of 100% solution $Cu^{2+}$ChiH (B) and water treatment (A) 72 h after treatment of soybean leaves. Characteristic leaves from the groups are presented; Figure S3: Psg and Cff symptoms on soybean leaves 12 d after inoculation with an airbrush. (A) Water treatment of infected leaves (positive control; Psg infection); (B) treatment with $Cu^{2+}$ChiNPs (Psg infection); (C) water treatment of infected leaves (positive control; Cff infection); (D) treatment with $Cu^{2+}$ChiNPs (Cff infection). Characteristic leaves from the groups are presented.

**Author Contributions:** Conceptualisation, R.T., B.S. and T.L.; methodology, F.D. and V.V.; phytopathological studies, R.T. and F.D.; nanoparticle obtaining and analysis, B.S., T.L. and Y.Z.; software, B.S. and A.I.; validation, formal analysis and data curation, R.T., B.S. and T.L.; visualization, R.T., Y.Z., B.S. and T.L.; writing—original draft preparation, R.T., B.S. and T.L.; writing—review and editing, A.I., F.D. and V.V.; supervision, B.S. and F.D.; project administration and funding acquisition, B.S. All authors have read and agreed to the published version of the manuscript.

**Funding:** This research was partially supported by the Russian Foundation for Basic Research, project no. 20-016-00205.

**Institutional Review Board Statement:** Not applicable.

**Informed Consent Statement:** Not applicable.

**Data Availability Statement:** The data that support the findings of this study are available on request from the corresponding author.

**Acknowledgments:** We are grateful to Irina Safenkova (Research Center of Biotechnology, Russian Academy of Sciences, Moscow, Russia) and Anna Nechaeva (Mendeleev University of Chemical Technology of Russia) for analysis of NP samples via dynamic light scattering.

**Conflicts of Interest:** The authors declare no conflict of interest.

## References

1. Chen, K.I.; Erh, M.H.; Su, N.W.; Liu, W.H.; Chou, C.C.; Cheng, K.C. Soyfoods and soybean products: From traditional use to modern applications. *Appl. Microbiol. Biotechnol.* **2012**, *96*, 9–22. [CrossRef]
2. FAO. *World Food and Agriculture—Statistical Yearbook 2021*; FAO: Roma, Italy, 2021; ISBN 978-92-5-134332-6.
3. Hartman, G.L. Diseases of Soybean (Glycine max [L.] Merr.). Available online: https://www.apsnet.org/edcenter/resources/commonnames/Pages/Soybean.aspx (accessed on 23 November 2022).
4. Bull, C.T.; De Boer, S.H.; Denny, T.P.; Firrao, G.; Fischer-Le Saux, M.; Saddler, G.S.; Scortichini, M.; Stead, D.E.; Takikawa, Y. Comprehensive list of names of plant pathogenic bacteria, 1980–2007. *J. Plant Pathol.* **2010**, *92*, 551–592.
5. Jagtap, G.P.; Dhopte, S.B.; Dey, U. Bio-efficacy of different antibacterial antibiotic, plant extracts and bioagents against bacterial blight of soybean caused by *Pseudomonas syringae* pv. *glycinea*. *Sci. J. Microbiol.* **2012**, *1*, 1–9.
6. Zhang, J.; Wang, X.; Lu, Y.; Bhusal, S.J.; Song, Q.; Cregan, P.B.; Yen, Y.; Brown, M.; Jiang, G.L. Genome-wide Scan for Seed Composition Provides Insights into Soybean Quality Improvement and the Impacts of Domestication and Breeding. *Mol. Plant* **2018**, *11*, 460–472. [CrossRef]
7. *Pseudomonas savastanoi* pv. *glycinea* (PSDMGL) [Overview] | EPPO Global Database. Available online: https://gd.eppo.int/taxon/PSDMGL (accessed on 24 November 2022).
8. Ignjatov, M.; Milošević, M.; Nikolić, Z.; Vujaković, M.; Petrović, D. Characterization of *Pseudomonas savastanoi* pv. *glycinea* isolates From Vojvodina. *Phytopathol. Pol.* **2007**, *45*, 43–54.
9. Shepherd, L.M.; Block, C.C. CHAPTER 13: Detection of *Pseudomonas savastanoi* pv. *glycinea* in Soybean Seeds. In *Detection of Plant-Pathogenic Bacteria in Seed and Other Planting Material*, 2nd ed.; The American Phytopathological Society: Saint Paul, MN, USA, 2017; pp. 85–88.
10. Huang, H.C.; Erickson, R.S.; Balasubramanian, P.M.; Hsieh, T.F.; Conner, R.L. Resurgence of bacterial wilt of common bean in North America. *Can. J. Plant Pathol.* **2009**, *31*, 290–300. [CrossRef]
11. Soares, R.M.; Fantinato, G.G.P.; Darben, L.M.; Marcelino-Guimarães, F.C.; Seixas, C.D.S.; de Souza Carneiro, G.E. First report of *Curtobacterium flaccumfaciens* pv. *flaccumfaciens* on soybean in Brazil. *Trop. Plant Pathol.* **2013**, *38*, 452–454. [CrossRef]
12. Huang, H.C.; Mondel, H.H.; Erickson, R.S.; Chelle, C.D.; Balasubramanian, P.M.; Kiehn, F.; Conner, R.L. Resistance of common bean (*Phaseolus vulgaris* L.) cultivars and germplasm lines to the purple variant of bacterial wilt (*Curtobacterium flaccumfaciens* pv. *flaccumfaciens*). *Plant Pathol. Bull.* **2007**, *16*, 91–95.
13. Camara, R.C.; Vigo, S.C.; Maringoni, A.C. Plant-to-seed transmission of *Curtobacterium flaccumfaciens* pv. *flaccumaciens* in a dry bean cultivar. *J. Plant Pathol.* **2009**, *91*, 549–554. [CrossRef]
14. EPPO A2 List. Available online: https://www.eppo.int/ACTIVITIES/plant_quarantine/A2_list (accessed on 28 November 2022).
15. Hsieh, T.-F.; Huang, H.C.; Erickson, R.S. Bacterial wilt of common bean: Effect of seedborne inoculum on disease incidence and seedling vigour. *Seed Sci. Technol.* **2006**, *34*, 57–67. [CrossRef]
16. Hsieh, T.-F.; Huang, H.-C.; Mündel, H.-H.; Erickson, R.S. A rapid indoor technique for screening common Bean (*Phaseolus vulgaris* L.) for resistance to bacterial wilt [*Curtobacterium flaccumfaciens* pv. *flaccumfaciens* (Hedges) Collins and Jones]. *Rev. Mex. Fitopatol.* **2003**, *21*, 364–369.
17. Monteil, C.L.; Yahara, K.; Studholme, D.J.; Mageiros, L.; Méric, G.; Swingle, B.; Morris, C.E.; Vinatzer, B.A.; Sheppard, S.K. Population-genomic insights into emergence, crop adaptation and dissemination of *Pseudomonas syringae* pathogens. *Microb. Genom.* **2016**, *2*, e000089. [CrossRef]
18. Silva Júnior, T.A.F.; Negrão, D.R.; Itako, A.T.; Maringoni, A.C. Pathogenicity of Curtobacterium flaccumfaciens pv. flaccumfaciens to several plant species. *J. Plant Pathol.* **2012**, *94*, 427–430.
19. Urrea, C.A.; Harveson, R.M. Identification of Sources of Bacterial Wilt Resistance in Common Bean (*Phaseolus vulgaris*). *Plant Dis.* **2014**, *98*, 973–976. [CrossRef]
20. Orzali, L.; Valente, M.T.; Scala, V.; Loreti, S.; Pucci, N. Antibacterial activity of essential oils and trametes versicolor extract against *Clavibacter michiganensis* subsp. *michiganensis* and *Ralstonia solanacearum* for seed treatment and development of a rapid in vivo assay. *Antibiotics* **2020**, *9*, 628. [CrossRef]
21. El Hadrami, A.; Adam, L.R.; El Hadrami, I.; Daayf, F. Chitosan in plant protection. *Mar. Drugs* **2010**, *8*, 968–987. [CrossRef]
22. Bernkop-Schnürch, A.; Dünnhaupt, S. Chitosan-based drug delivery systems. *Eur. J. Pharm. Biopharm.* **2012**, *81*, 463–469. [CrossRef]
23. Aranaz, I.; Harris, R.; Heras, A. Chitosan Amphiphilic Derivatives. Chemistry and Applications. *Curr. Org. Chem.* **2010**, *14*, 308–330. [CrossRef]

24. Khan, A.; Ali, N.; Bilal, M.; Malik, S.; Badshah, S.; Iqbal, H.M.N. Engineering Functionalized Chitosan-Based Sorbent Material: Characterization and Sorption of Toxic Elements. *Appl. Sci.* **2019**, *9*, 5138. [CrossRef]
25. Ali, N.; Khan, A.; Malik, S.; Badshah, S.; Bilal, M.; Iqbal, H.M.N. Chitosan-based green sorbent material for cations removal from an aqueous environment. *J. Environ. Chem. Eng.* **2020**, *8*, 104064. [CrossRef]
26. Dhillon, G.S.; Kaur, S.; Brar, S.K.; Verma, M. Green synthesis approach: Extraction of chitosan from fungus mycelia. *Crit. Rev. Biotechnol.* **2013**, *33*, 379–403. [CrossRef] [PubMed]
27. Romanazzi, G.; Feliziani, E.; Sivakumar, D. Chitosan, a biopolymer with triple action on postharvest decay of fruit and vegetables: Eliciting, antimicrobial and film-forming properties. *Front. Microbiol.* **2019**, *9*, 2745. [CrossRef] [PubMed]
28. Cabrera, J.C.; Boland, A.; Cambier, P.; Frettinger, P.; van Cutsem, P. Chitosan oligosaccharides modulate the supramolecular conformation and the biological activity of oligogalacturonides in Arabidopsis. *Glycobiology* **2010**, *20*, 775–786. [CrossRef] [PubMed]
29. Al-Hetar, M.Y.; Zainal Abidin, M.A.; Sariah, M.; Wong, M.Y. Antifungal activity of chitosan against *Fusarium oxysporum* f. sp. *cubense*. *J. Appl. Polym. Sci.* **2011**, *120*, 2434–2439. [CrossRef]
30. Han, C.; Shao, H.; Zhou, S.; Mei, Y.; Cheng, Z.; Huang, L.; Lv, G. Chemical composition and phytotoxicity of essential oil from invasive plant, *Ambrosia artemisiifolia* L. *Ecotoxicol. Environ. Saf.* **2021**, *211*, 111879. [CrossRef] [PubMed]
31. Cuong, H.N.; Tung, H.T.; Minh, N.C.; Van Hoa, N.; Phuong, P.T.D.; Trung, T.S. Antibacterial activity of chitosan from squid pens (*Loligo chenisis*) against *Erwinia carotovora* from soft rot postharvest tomato fruit. *J. Polym. Mater.* **2017**, *34*, 319–330.
32. Khalifa, I.; Barakat, H.; El-Mansy, H.A.; Soliman, S.A. Preserving apple (*Malus domestica* var. Anna) fruit bioactive substances using olive wastes extract-chitosan film coating. *Inf. Process. Agric.* **2017**, *4*, 90–99. [CrossRef]
33. Ramkissoon, A.; Francis, J.; Bowrin, V.; Ramjegathesh, R.; Ramsubhag, A.; Jayaraman, J. Bio-efficacy of a chitosan based elicitor on *Alternaria solani* and *Xanthomonas vesicatoria* infections in tomato under tropical conditions. *Ann. Appl. Biol.* **2016**, *169*, 274–283. [CrossRef]
34. Chandrasekaran, M.; Kim, K.D.; Chun, S.C. Antibacterial activity of chitosan nanoparticles: A review. *Processes* **2020**, *8*, 1173. [CrossRef]
35. Kong, M.; Chen, X.G.; Xing, K.; Park, H.J. Antimicrobial properties of chitosan and mode of action: A state of the art review. *Int. J. Food Microbiol.* **2010**, *144*, 51–63. [CrossRef]
36. Maluin, F.N.; Hussein, M.Z. Chitosan-Based Agronanochemicals as a Sustainable Alternative in Crop Protection. *Molecules* **2020**, *25*, 1611. [CrossRef]
37. Nguyen, T.V.; Nguyen, T.T.H.; Wang, S.L.; Vo, T.P.K.; Nguyen, A.D. Preparation of chitosan nanoparticles by TPP ionic gelation combined with spray drying, and the antibacterial activity of chitosan nanoparticles and a chitosan nanoparticle–amoxicillin complex. *Res. Chem. Intermed.* **2017**, *43*, 3527–3537. [CrossRef]
38. Chidanandappa; Nargund, V.B. Green synthesis of chitosan based copper nanoparticles and their bio-efficacy against bacterial blight of pomegranate. *Int. J. Curr. Microbiol. Appl. Sci.* **2020**, *9*, 1298–1305. [CrossRef]
39. Tarakanov, R.; Shagdarova, B.; Varlamov, V.; Dzhalilov, F. Biocidal and resistance-inducing effects of chitosan on phytopathogens. In Proceedings of the E3S Web of Conferences, Orel, Russian, 24–25 February 2021; Knyazev, S., Loretts, O., Kukhar, V., Panfilova, O., Tsoy, M., Eds.; EDP Sciences: Les Ulis, France, 2021; Volume 254, p. 05007.
40. Swati; Choudhary, M.K.; Joshi, A.; Saharan, V. Assessment of Cu-Chitosan Nanoparticles for its Antibacterial Activity against *Pseudomonas syringae* pv. *glycinea*. *Int. J. Curr. Microbiol. Appl. Sci.* **2017**, *6*, 1335–1350. [CrossRef]
41. Shagdarova, B.T.; Ilyina, A.V.; Lopatin, S.A.; Kartashov, M.I.; Arslanova, L.R.; Dzhavakhiya, V.G.; Varlamov, V.P. Study of the Protective Activity of Chitosan Hydrolyzate Against Septoria Leaf Blotch of Wheat and Brown Spot of Tobacco. *Appl. Biochem. Microbiol.* **2018**, *54*, 71–75. [CrossRef]
42. Lopatin, S.A.; Derbeneva, M.S.; Kulikov, S.N.; Varlamov, V.P.; Shpigun, O.A. Fractionation of chitosan by ultrafiltration. *J. Anal. Chem.* **2009**, *64*, 648–651. [CrossRef]
43. Qi, L.; Xu, Z.; Jiang, X.; Hu, C.; Zou, X. Preparation and antibacterial activity of chitosan nanoparticles. *Carbohydr. Res.* **2004**, *339*, 2693–2700. [CrossRef]
44. Tarakanov, R.I.; Lukianova, A.A.; Pilik, R.I.; Evseev, P.V.; Miroshnikov, K.A.; Dzhalilov, F.S.-U.; Tesic, S.; Ignatov, A. First report of *Curtobacterium flaccumfaciens* pv. *flaccumfaciens* causing a bacterial tan spot of soybean in Russia. *Plant Dis.* **2022**. *online ahead of print*. [CrossRef]
45. Tarakanov, R.I.; Lukianova, A.A.; Evseev, P.V.; Toshchakov, S.V.; Kulikov, E.E.; Ignatov, A.N.; Miroshnikov, K.A.; Dzhalilov, F.S.-U. Bacteriophage Control of *Pseudomonas savastanoi* pv. *glycinea* in Soybean. *Plants* **2022**, *11*, 938. [CrossRef]
46. Sarkar, S.F.; Guttman, D.S. Evolution of the Core Genome of *Pseudomonas syringae*, a Highly Clonal, Endemic Plant Pathogen. *Appl. Environ. Microbiol.* **2004**, *70*, 1999–2012. [CrossRef]
47. Evseev, P.; Lukianova, A.; Tarakanov, R.; Tokmakova, A.; Shneider, M.; Ignatov, A.; Miroshnikov, K. *Curtobacterium* spp. and *Curtobacterium flaccumfaciens*: Phylogeny, genomics-based taxonomy, pathogenicity, and diagnostics. *Curr. Issues Mol. Biol.* **2022**, *44*, 889–927. [CrossRef]
48. Tegli, S.; Sereni, A.; Surico, G. PCR-based assay for the detection of *Curtobacterium flaccumfaciens* pv. *flaccumfaciens* in bean seeds. *Lett. Appl. Microbiol.* **2002**, *35*, 331–337. [CrossRef]
49. Islam, M.; Masum, S.; Rayhan, K.; Haque, Z. Antibacterial activity of crab-chitosan against *Staphylococcus aureus* and *Escherichia coli*. *J. Advaced Sci. Res.* **2011**, *2*, 63–66.

50. Sowjanya, P.; Srinivasa, B.P.; Lakshmi, N.M. Phytochemical analysis and antibacterial efficacy of *Amaranthus tricolor* (L.) methanolic leaf extract against clinical isolates of urinary tract pathogens. *Afr. J. Microbiol. Res.* **2015**, *9*, 1381–1385. [CrossRef]
51. CLSI, C.L.S.I. Methods for dilution antimicrobial susceptibility tests for bacteria that grow aerobically; approved standard—ninth edition. CLSI document M07-A9. *Clin. Lab. Standars Inst.* **2015**, *32*, 18.
52. Foerster, S.; Unemo, M.; Hathaway, L.J.; Low, N.; Althaus, C.L. Time-kill curve analysis and pharmacodynamic modelling for in vitro evaluation of antimicrobials against Neisseria gonorrhoeae. *BMC Microbiol.* **2016**, *16*, 1–11. [CrossRef]
53. ISTA International Rules of Seed. Testing (Supplement rules). *Seed Sci. Technol.* **1999**, *27*, 178.
54. Nalini, S.; Parthasarathi, R. Optimization of rhamnolipid biosurfactant production from *Serratia rubidaea* SNAU02 under solid-state fermentation and its biocontrol efficacy against Fusarium wilt of eggplant. *Ann. Agrar. Sci.* **2018**, *16*, 108–115. [CrossRef]
55. Shine, M.; Fu, D.-Q.; Kachroo, A. Airbrush infiltration method for *Pseudomonas syringae* Infection Assays in Soybean. *Bio-Protocol* **2015**, *5*, e1427. [CrossRef]
56. Sibiya, M.; Sumbwanyambe, M. An algorithm for severity estimation of plant leaf diseases by the use of colour threshold image segmentation and fuzzy logic inference: A proposed algorithm to update a "Leaf Doctor" application. *AgriEngineering* **2019**, *1*, 205–219. [CrossRef]
57. Tarakanov, R.I.; Lukianova, A.A.; Evseev, P.V.; Pilik, R.I.; Tokmakova, A.D.; Kulikov, E.E.; Toshchakov, S.V.; Ignatov, A.N.; Dzhalilov, F.S.-U.; Miroshnikov, K.A. Ayka, a Novel Curtobacterium Bacteriophage, Provides Protection against Soybean Bacterial Wilt and Tan Spot. *Int. J. Mol. Sci.* **2022**, *23*, 913. [CrossRef]
58. Madden, L.V.; Hughes, G.; van den Bosch, F. *The Study of Plant Disease Epidemics*; The American Phytopathological Society: Saint Paul, MN, USA, 2017; ISBN 978-0-89054-505-8.
59. Ma, Z.; Garrido-Maestu, A.; Jeong, K.C. Application, mode of action, and in vivo activity of chitosan and its micro- and nanoparticles as antimicrobial agents: A review. *Carbohydr. Polym.* **2017**, *176*, 257–265. [CrossRef]
60. Rozman, N.A.S.; Tong, W.Y.; Leong, C.R.; Tan, W.N.; Hasanolbasori, M.A.; Abdullah, S.Z. Potential Antimicrobial Applications of Chitosan Nanoparticles (ChNP). *J. Microbiol. Biotechnol.* **2019**, *29*, 1009–1013. [CrossRef]
61. Sathiyabama, M.; Parthasarathy, R. Biological preparation of chitosan nanoparticles and its in vitro antifungal efficacy against some phytopathogenic fungi. *Carbohydr. Polym.* **2016**, *151*, 321–325. [CrossRef]
62. OH, J.-W.; Chun, S.C.; Chandrasekaran, M. Preparation and in vitro characterization of chitosan nanoparticles and their broad-spectrum antifungal action compared to antibacterial activities against phytopathogens of tomato. *Agronomy* **2019**, *9*, 21. [CrossRef]
63. Chen, Q.; Jiang, H.; Ye, H.; Li, J.; Huang, J. Preparation, antibacterial, and antioxidant activities of silver/chitosan composites. *J. Carbohydr. Chem.* **2014**, *33*, 298–312. [CrossRef]
64. Qian, J.; Pan, C.; Liang, C. Antimicrobial activity of Fe-loaded chitosan nanoparticles. *Eng. Life Sci.* **2017**, *17*, 629–634. [CrossRef]
65. Badawy, M.E.I.; Lotfy, T.M.R.; Shawir, S.M.S. Preparation and antibacterial activity of chitosan-silver nanoparticles for application in preservation of minced meat. *Bull. Natl. Res. Cent.* **2019**, *43*, 83. [CrossRef]
66. Katas, H.; Lim, C.S.; Nor Azlan, A.Y.H.; Buang, F.; Mh Busra, M.F. Antibacterial activity of biosynthesized gold nanoparticles using biomolecules from *Lignosus rhinocerotis* and chitosan. *Saudi Pharm. J.* **2019**, *27*, 283–292. [CrossRef]
67. Esyanti, R.R.; Farah, N.; Bajra, B.D.; Nofitasari, D.; Martien, R.; Sunardi, S.; Safitri, R. Comparative study of nano-chitosan and synthetic bactericide application on chili pepper (*Capsicum annuum* L.) infected by xanthomonas campestris. *Agrivita* **2020**, *42*, 13–23. [CrossRef]
68. Vold, I.M.N.; Vårum, K.M.; Guibal, E.; Smidsrød, O. Binding of ions to chitosan—Selectivity studies. *Carbohydr. Polym.* **2003**, *54*, 471–477. [CrossRef]
69. Mekahlia, S.; Bouzid, B. Chitosan-Copper (II) complex as antibacterial agent: Synthesis, characterization and coordinating bond-activity correlation study. *Phys. Procedia* **2009**, *2*, 1045–1053. [CrossRef]
70. Konovalova, M.; Shagdarova, B.; Zubov, V.; Svirshchevskaya, E. Express analysis of chitosan and its derivatives by gel electrophoresis. *Prog. Chem. Appl. Chitin Its Deriv.* **2019**, *24*, 84–95. [CrossRef]
71. Du, W.L.; Niu, S.S.; Xu, Y.L.; Xu, Z.R.; Fan, C.L. Antibacterial activity of chitosan tripolyphosphate nanoparticles loaded with various metal ions. *Carbohydr. Polym.* **2009**, *75*, 385–389. [CrossRef]
72. Chen, J.; Mao, S.; Xu, Z.; Ding, W. Various antibacterial mechanisms of biosynthesized copper oxide nanoparticles against soilborne *Ralstonia solanacearum*. *RSC Adv.* **2019**, *9*, 3788–3799. [CrossRef]
73. Ren, G.; Hu, D.; Cheng, E.W.C.; Vargas-Reus, M.A.; Reip, P.; Allaker, R.P. Characterisation of copper oxide nanoparticles for antimicrobial applications. *Int. J. Antimicrob. Agents* **2009**, *33*, 587–590. [CrossRef]
74. Fernandes, J.C.; Eaton, P.; Gomes, A.M.; Pintado, M.E.; Xavier Malcata, F. Study of the antibacterial effects of chitosans on *Bacillus cereus* (and its spores) by atomic force microscopy imaging and nanoindentation. *Ultramicroscopy* **2009**, *109*, 854–860. [CrossRef]
75. No, H.K.; Young Park, N.; Ho Lee, S.; Meyers, S.P. Antibacterial activity of chitosans and chitosan oligomers with different molecular weights. *Int. J. Food Microbiol.* **2002**, *74*, 65–72. [CrossRef]
76. Dash, S.; Kumar, M.; Pareek, N. Enhanced antibacterial potential of berberine via synergism with chitosan nanoparticles. *Mater. Today Proc.* **2019**, *31*, 640–645. [CrossRef]
77. Christena, L.R.; Mangalagowri, V.; Pradheeba, P.; Ahmed, K.B.A.; Shalini, B.I.S.; Vidyalakshmi, M.; Anbazhagan, V.; Subramanian, N.S. Copper nanoparticles as an efflux pump inhibitor to tackle drug resistant bacteria. *RSC Adv.* **2015**, *5*, 12899–12909. [CrossRef]

78. Young, M.; Santra, S. Copper (Cu)–Silica Nanocomposite Containing Valence-Engineered Cu: A New Strategy for Improving the Antimicrobial Efficacy of Cu Biocides. *J. Agric. Food Chem.* **2014**, *62*, 6043–6052. [CrossRef]
79. Sathiyabama, M.; Manikandan, A. Application of Copper-Chitosan Nanoparticles Stimulate Growth and Induce Resistance in Finger Millet (*Eleusine coracana* Gaertn.) Plants against Blast Disease. *J. Agric. Food Chem.* **2018**, *66*, 1784–1790. [CrossRef]
80. Stampoulis, D.; Sinha, S.K.; White, J.C. Assay-dependent phytotoxicity of nanoparticles to plants. *Environ. Sci. Technol.* **2009**, *43*, 9473–9479. [CrossRef]
81. Shende, S.; Rathod, D.; Gade, A.; Rai, M. Biogenic copper nanoparticles promote the growth of pigeon pea (*Cajanus cajan* L.). *IET Nanobiotechnol.* **2017**, *11*, 773. [CrossRef]
82. Swati; Joshi, A. Cu-Chitosan Nanoparticle Induced Plant Growth and Disease Resistance Efficiency of Soybean [*Glycine max* (L.)]. *Legum. Res.* **2022**, *1*, 6. [CrossRef]
83. Kumar, N.V.; Basavegowda, V.R.; Murthy, A.N.; Lokesh, S. Synthesis and characterization of copper-chitosan based nanofungicide and its induced defense responses in Fusarium wilt of banana. *Inorg. Nano-Met. Chem.* **2022**, 1–9. [CrossRef]

polymers

MDPI

Review

# Recent Advances in Degradation of Polymer Plastics by Insects Inhabiting Microorganisms

**Rongrong An [1], Chengguo Liu [2], Jun Wang [1,*] and Puyou Jia [2,*]**

[1] School of Geographic and Biologic Information, Nanjing University of Posts and Telecommunications, Nanjing 210023, China
[2] Institute of Chemical Industry of Forest Products, Chinese Academy of Forestry, 16 Suojin North Road, Nanjing 210042, China
* Correspondence: wangj@njupt.edu.cn (J.W.); jiapuyou@icifp.cn (P.J.)

**Abstract:** Plastic pollution endangers all natural ecosystems and living creatures on earth. Excessive reliance on plastic products and excessive production of plastic packaging are extremely dangerous for humans because plastic waste has polluted almost the entire world, whether it is in the sea or on the land. This review introduces the examination of pollution brought by non-degradable plastics, the classification and application of degradable materials, and the current situation and strategy to address plastic pollution and plastic degradation by insects, which mainly include *Galleria mellonella*, *Zophobas atratus*, *Tenebrio molitor*, and other insects. The efficiency of plastic degradation by insects, biodegradation mechanism of plastic waste, and the structure and composition of degradable products are reviewed. The development direction of degradable plastics in the future and plastic degradation by insects are prospected. This review provides effective ways to solve plastic pollution.

**Keywords:** plastic; biodegradation; polypropylene; polyethylene; polyvinyl chloride; insect

**Citation:** An, R.; Liu, C.; Wang, J.; Jia, P. Recent Advances in Degradation of Polymer Plastics by Insects Inhabiting Microorganisms. *Polymers* **2023**, *15*, 1307. https://doi.org/10.3390/polym15051307

Academic Editors: Vsevolod Aleksandrovich Zhuikov, Rosane Michele Duarte Soares and Cristiano Varrone

Received: 18 November 2022
Revised: 21 January 2023
Accepted: 2 March 2023
Published: 5 March 2023



## 1. Introduction

### 1.1. Application and Pollution of Non-Degradable Plastics

Plastic products have been widely used around the world because of low cost and easy production. Polystyrene (PS), polyvinyl chloride (PVC), polyethylene (PE), acrylonitrile butadiene styrene (ABS), and polyurethane (PU), which are usually designed as short-term and disposable products, are commonly used plastics [1–3]. Plastic products have brought great convenience to people's lives. However, a large number of plastic products have become the focus of global attention because of the environmental problems caused by their improper disposal after being used and discarded [4,5]. The global plastic output and consumption exceed 300 million tons annually, which has grown exponentially in the past 50 years [6,7]. In 27 EU countries, including Norway and Switzerland, 38% of plastic is discarded in landfills, whereas the rest is used for recycling (26%) and energy recovery through combustion (36%) [8]. The world's disposable plastic products reach 120 million tons every year, only 10% of which are recycled, 12% are burned, and more than 70% are discarded into the soil, air, and sea [9,10]. More than one trillion plastic bags are consumed every year [11]. In recent years, microplastics, as a new kind of pollutant, have attracted more attention because of their widespread distribution in the oceans and coastal waters around the world, which have polluted the marine ecological environment and marine organisms [12]. Considering that many waste plastics have not been treated scientifically and correctly, they form microplastics through a series of physical and chemical processes and enter the ocean. Some animals, such as birds, fish, and sea turtles, have been affected by plastic pollution [13]. Many plastic particles were discovered in the intestines of dead birds, fish, and turtles, indicating that plastic has caused serious damage to living creatures and bodies [14]. Animals cannot distinguish food and plastic in the environment, resulting in ingestion of plastic particles. Plastics cannot be digested and often accumulate in the body. When plastics are decomposed in the marine environment, the micro plastics, microfibers, toxic chemicals, metals, and organic micro pollutants will be transferred to the waters and sediments and finally enter the marine food chain. These substances affect the reproductive success rate and viability of marine organisms and damage the ability of "ecological engineer" corals and worms in the aquatic ecosystem to build coral reefs and change sediments through biological disturbance.

### 1.2. Classification and Application of Degradable Plastics

The scientific disposal of discarded plastic products has become a challenge. Therefore, researchers exert their energy into the study of degradation plastics. Degradable plastics meet the use requirements during the storage period and can be decomposed into harmless substances under natural environmental conditions after use. It is considered an effective way to deal with plastic pollution and has attracted attention in recent years [15]. Many kinds of degradable plastic products are available on the market, and the product performance and capacity scale are also quite different [16]. Degradable plastics include biological and petrochemical plastics based on the source of raw materials. Bio-based degradable plastics include polylactic acid (PLA), polyhydroxy fatty acid ester polymers (PHAs), starch, and cellulose; and petrochemical-based degradable plastics include carbon dioxide copolymer (PPC), polycaprolactone (PCL), and polyglycolic acid (PGA) [17]. Among them, degradable plastics based on starch are gradually eliminated because they cannot be degraded completely. PBAT, PBS, and PLA have formed industrial-scale production in the market and occupied a large market share. From the specific use of various degradable plastics, PBAT/PBS products are mainly used in packaging, tableware, cosmetic bottles, disposable medical supplies, agricultural films, pesticides, and fertilizer slow-release materials [18]. PCL is used in the production of toys, bone nails, long-acting drugs, and in other medical fields. PLA is used in general plastics, such as films, lunch boxes, and textiles. PPC is used in film bag, surgical suture, bone nail and other medical fields, membrane, and packaging materials. The global degradable plastic industry is at the initial stage of industrialization, and China's production capacity has increased significantly. In the past five years, the average growth rate of the consumption of biodegradable plastics in China was at 20% [16]. In 2019, the consumption of biodegradable plastics in China was approximately 260,000 tons. The estimated demand for biodegradable plastics in China will exceed 650,000 tons by 2024 [19].

### 1.3. Current Situation and Strategy to Address Plastic Pollution

Hundreds of years are required for plastics to be degraded naturally, including 600 years for PE and PP plastics, which will occupy land for a long time, affect crops and livestock, and cause serious white pollution [20]. A large amount of plastic waste flows into the sea after being generated. According to statistics, 3% (approximately 11 million tons in 2019) of plastic waste enters the sea every year. Approximately 14 million tons of micro plastics, which seriously affect the safety and health of marine organisms, can be found in the whole sea floor. The main disposal methods of waste plastics include landfill, incineration, chemical treatment, and recycling. Landfill method destroys soil, affects groundwater, and cannot effectively degrade waste plastics. Incineration produces a large number of toxic gases, which volatilize into the air and affect human health. The cost of chemical treatment and recycling is high. Hence, it is not suitable for wide use. Plastic waste residues do not decompose in soil for a long time, thereby damaging the physical and chemical properties of the soil, affecting the growth of plant roots and leading to crop production reduction. Waste plastics cause water pollution. According to the World Greenpeace Organization in 2016, more than 200 kg of plastics are dumped into the sea every second in the world [20,21]. The concentration of plastic fragments observed in the oceans around the world is as high as 58,000 per square kilometer. Approximately millions of marine animals die every year due to suffocation or indigestion caused by consumption of plastic by mistake. The United States is a major plastic producer in the world, and it had carried out research on the recycling of waste plastics in 1960s. The recycling rate of waste plastics exceeded 35% at the beginning of the 21st century, the average recycling rates of plastics in Europe and Germany exceeded 45% and were as high as 60% [22]. In Japan, 52% of waste plastics were recycled, of which 2% were used as chemical raw materials, 3% as remelted solid fuel, 20% as power generation fuel, and 43% as heat energy in incinerators.

Plastic pollution has endangered the natural ecosystems and living creatures on the earth, as well as human health. Recent reports on the plastic degradation by insects have drawn widespread attention (Table 1). These insects mainly include *Galleria mellonella*, *Zophobas atratus*, *Tenebrio molitor*, and other insects. In the review, the efficiency of plastic degradation by insects, biodegradation mechanism of plastic waste, and the structure and composition of degradable products are reviewed. The development direction of degradable plastics in the future and plastic degradation by insects are prospected. These degradation strategies and results provide effective ways to solve the issue of plastic pollution.

**Table 1.** The latest works on degradation of polymer plastics by insects.

| Polymer Plastic | References | Insect |
|---|---|---|
| LDPE foams | [23,24] | *Tenebrio molitor* |
| PS | [25–27] | |
| PE | [28] | |
| PVC | [29] | |
| Bio-based cross-linked polymer | [30] | |
| PU | [31] | |
| PS | [32–34] | *Zophobas atratus* |
| PU | [34,35] | |
| LDPE foams | [36,37] | |
| PP | [38] | |
| PS | [39] | *Galleria mellonella* |
| HDPE | [40] | |
| PE | [41] | |
| HDPE | [42] | |
| PE | [43] | Wax worm |
| PS | [44] | Dark mealworms |
| PVC | [45] | Black soldier fly larvae |

## 2. Plastic Degradation by Insects

### 2.1. Plastic Degradation by T. molitor

*T. molitor* is a completely metamorphosed insect [46,47]. The larvae of *T. molitor* like to flock, live, and feed at 13 °C. They can still grow at temperatures above 35 °C but cannot survive at temperatures above 50 °C. *T. molitor* can be used as an excellent feed for medicinal animals, such as scorpions, centipedes, and fish. *T. Molitor* can also be used as food raw materials, health products, and medicine for human beings [48].

Recently, the study on plastic degradation by *T. molitor* has been widely districuted [23–26,28]. Intestinal microbes of *T. molitor* larvae play a decisive role in the biodegradation of PS. The study reported that the larvae cannot degrade plastics after adding antibiotics to the food of *T. molitor* larvae to inhibit intestinal microorganisms. The researchers further successfully isolated the PS degrading bacterium exiguobacterium sp. (YT2), which can grow by using PS as the only carbon source. This strain of bacteria can grow on the surface of PS film on carbon-free agar solid medium to form a stable biofilm that can significantly erode the surface structure of the film. After the strain was cultured in liquid for 60 days, the PS chips in the liquid were decomposed, and the weight loss was up to 7.4%. The molecular weight of the residue decreased significantly, and a large number of water-soluble low-molecular intermediates were generated. Microorganisms can degrade PS. This study provides a new way to develop biodegradable polymer materials and waste plastic treatment technology [23,24]. To study the ability of *T. molitor* larvae to degrade PE and the depolymerization mode, Shan Shan Yang et al. fed two commercial low-density PS (LDPE) foam to *T. molitor* larvae at ambient temperature within 36 days. The residual PE in the feces of *T. molitor* larvae decreased by 43.3% and 31.7 ± 0.5%. The structure shows that low molecular weight PE (<5.0 kDa) is rapidly digested, whereas the long chain part (>10.0 kDa) is decomposed or cracked, indicating a widespread depolymerization. Mass balance analysis showed that 40% of ingested LDPE was digested into $CO_2$ [25]. Craig S. Criddle et al. discovered eight unique intestinal microorganisms related to PS biodegradation, including *Citrobacter freundii*, *Serratia marcescens*, and *Klebsiella aerogenes*. The intestinal microbes of *T. molitor* were helpful in accelerating the plastic biodegradation. This work provides a potential strategy for future research and large-scale cultivation of plastic-degrading microorganisms [26].

To determine whether plastics are sensitive to biodegradation in *T. molitor*, Craig S. Criddle et al. fed *T. molitor* with PE and PS, and the molecular weight (Mn) of polymer residues decreased by 40.1% and 12.8%, respectively [28]. Emmanouil Tsochatzis et al. investigated the intestinal microbes and the formation of degradation compounds of *T. molitor* larvae under different feeding strategies. The results showed that water can significantly improve the biodegradation of PS monomer and oligomer residues. Diet leads to differences in intestinal microbiota, and three potential bacterial strains were identified as the candidate strains involved in PS biodegradation [27]. Yalei Zhang et al. investigated the biodegradation of PVC by feeding *T. molitor* larvae with PVC micro plastic powder. After 16 days, the Mw, Mn, and Mz of PVC decreased by 33.4%, 32.8%, and 36.4%, respectively. The degradation products contain O-C and O-C functional groups. The survival rate of *T. molitor* larvae with PVC, as the sole food in 5 weeks, was as high as 80%, and the survival rate in three months was as low as 39%. PVC and wheat bran were fed together to *T. molitor*, and they completed the growth and pupation process within 91 days. *T. molitor* larvae can degrade PVC, but the mineralization of PVC is limited [29]. Yonghong Zhou et al. reported that *T. molitor* could degrade bio-based cross-linked polymers. Figure 1a–i shows representative images of *T. Molitor*, bio-based cross-linked polymer film used in biodegradation experiment, and biodegradation test results. A total of 8% (0.2 g) of polymer was biodegraded in the digestive system of *T. molitor*. The degradation products contain products with low molecular weight, which is the result of chain break caused by microbial attack. Biodegradation showed that the prepared biopolymer network has good degradation performance and better impact on the environment [30]. Piotr Bulak et al. studied the ability of *T. molitor* to biodegrade PE, PU, and PS foams. The results showed that the quality of PS, PU1, PU2, and PE decreased by 46.5%, 41.0%, 53.2%, and 69.7% after 58 days, which indicated that the larvae and adults of *T. molitor* could degrade plastic [31].

### 2.2. Plastic Degradation by *Z. atratus*

*Z. atratus* are mainly distributed in Central and South America, West Indies and other places, which are also known as super bread worms and super wheat barkworms [49,50]. The maximum body length of adult larvae is approximately 71 cm, which is 3–4 times larger than that of the general yellow mealworm, and the yield is 5 times higher than that of yellow mealworm. *Z. atratus* larvae are rich in protein and fat, of which the protein content can reach more than 51%, and the fat content is approximately 29%. At the same time, they also contain many nutrients, including sugars, amino acids, vitamins, and minerals. *Z. atratus* can degrade plastic similarly to *T. molitor*.

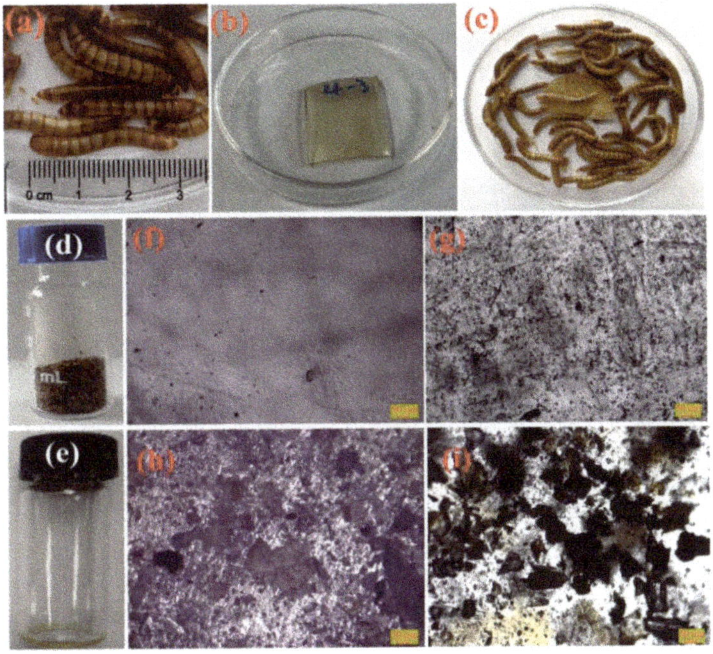


**Figure 1.** (**a**) Representative images of *Tenebrio molitor*; (**b**) bio-based cross-linked polymer film used in biodegradation experiment; (**c**) Tenebrio monitor fed with bio-based cross-linked polymer film for 30 days. (**d**) Feces of *Tenebrio molitor* collected in biodegradation experiment. (**e**) Extraction from feces of *Tenebrio molitor* using tetrahydrofuran (THF). (**f**) Microstructures of bio-based cross-linked polymer film in a reflection mode. (**g**) Microstructures of bio-based cross-linked polymer film in a transmissive mode. (**h**) Microstructure of bio-based cross-linked polymer films after feeding *Tenebrio molitor* for 30 days in a reflection mode. (**i**) Microstructures of bio-based cross-linked polymer 3 film after feeding *Tenebrio molitor* for 30 days in a transmissive mode, reprinted with permission from [30].

Yu Yang et al. determined that *Z. atratus* could ingest PS as the only food or could be fed with bran for 28 days. Figure 2a–d shows the PS foam-eating activities of *Z. atratus*, the increasing hollows in the Styrofoam block, consumption of PS foam by a group of *Z. atratus*, and survival rate of PS foam-eating and normal diet (bran)-eating *Z. atratus*. The results showed that the average consumption rate of PS foam plastic of each superworm is estimated to be 0.58 mg/d, which is four times than that of *T. molitor*. The depolymerization of long-chain PS molecules and the formation of low molecular weight products occurred in the gut of larvae. During the 16-day test period, up to 36.7% of the intake of foam plastic carbon was converted into $CO_2$. Antibiotics of intestinal microbiota inhibited the degradation of PS by *Z. atratus*, indicating that intestinal microbiota contributed to PS degradation. This new discovery extends PS degrading insects beyond Tenebrio species and indicates that the intestinal microbiota of *Z. atratus* will be a new biological source for plastic-degrading enzymes [32].

Xin Zhao et al. fed *Z. atratus* and *T. molitor* with PS or PU foam plastic for 35 days and bran as the control. Figure 3a–c shows the PS, PU, *Z. atratus*, and *T. molitor* used in the study and hollows and pits on PS. The survival rate of *Z. atratus* was 100%, but weight loss was observed after 20 days of only using plastic. In contrast, the survival rates of *T. molitor* in the PS or PU groups were 84.67% and 62.67%, respectively, and the weight of the two groups increased. The accumulative consumption of plastic by the *Z. atratus* is 49.24 mg PS/larva and 26.23 mg PU/larva, which are 18 times and 11 times higher than those of *T. molitor*, respectively [35].

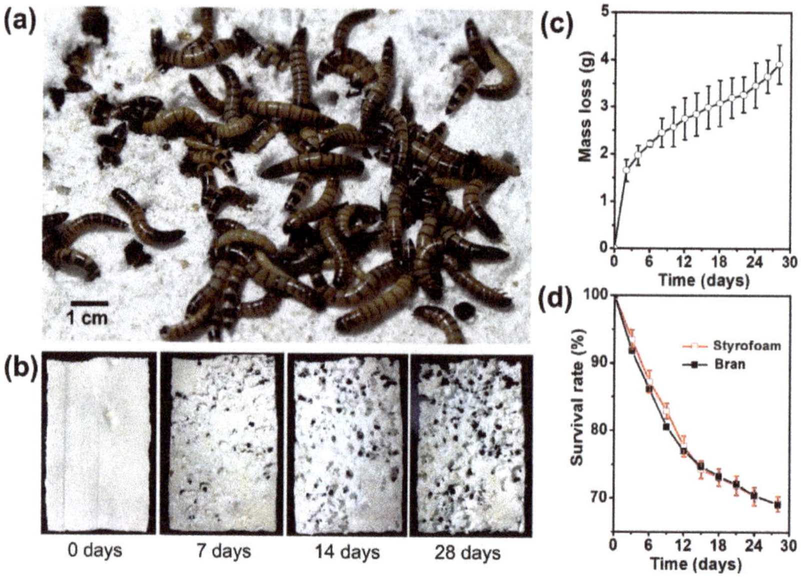


**Figure 2.** PS foam-eating activities of *Zophobas atratus*. (**a**) *Zophobas atratus* like to eat and penetrate PS foam. (**b**) Increasing hollows in the Styrofoam block; (**c**) consumption of PS foam by a group of *Zophobas atratus*; (**d**) survival rate of PS foam-eating and normal diet (bran)-eating *Zophobas atratus*, reprinted with permission from [32].

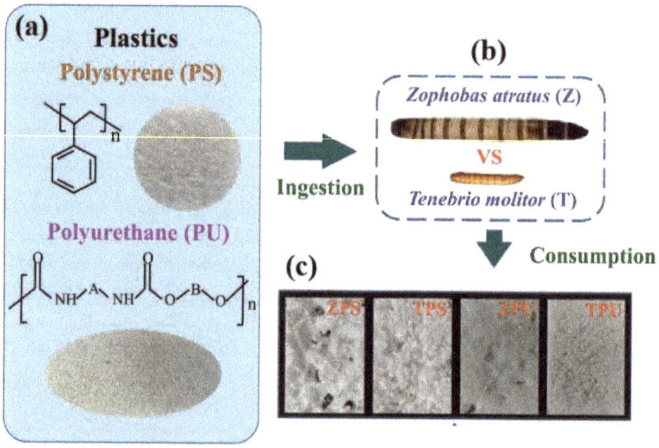


**Figure 3.** (**a**) PS and PU used in the study. (**b**) Hollows and pits on PS (**c**) *Zophobas atratus* and *Tenebrio molitor* used in the stud, adapted from [35].

Yalei Zhang et al. studied the fragmentation, larval physiology, intestinal microbiota, and microbial functional enzymes of ingested polymers through 28 days of experiments. Figure 4 shows the conceptual schematic for the biodegradation of PS and LDPE foams in *Z. atratus* larva. Larvae maintained a high survival rate, but when fed PS or LDPE, their body fat content decreased, and their consumption rates were 43.3 ± 1.5 and 52.9 ± 3.1 mg plastic/100 larvae/day, respectively. Ingested PS and LDPE were broken to an average size of 174 µM and 185 µM microplastics (by volume), size 6.3 µM and 5.9 µM particles reached the maximum number, respectively, and no nanoplastic was produced. This work provides new insights into insect-mediated biodegradation of persistent plastics for future research [36].

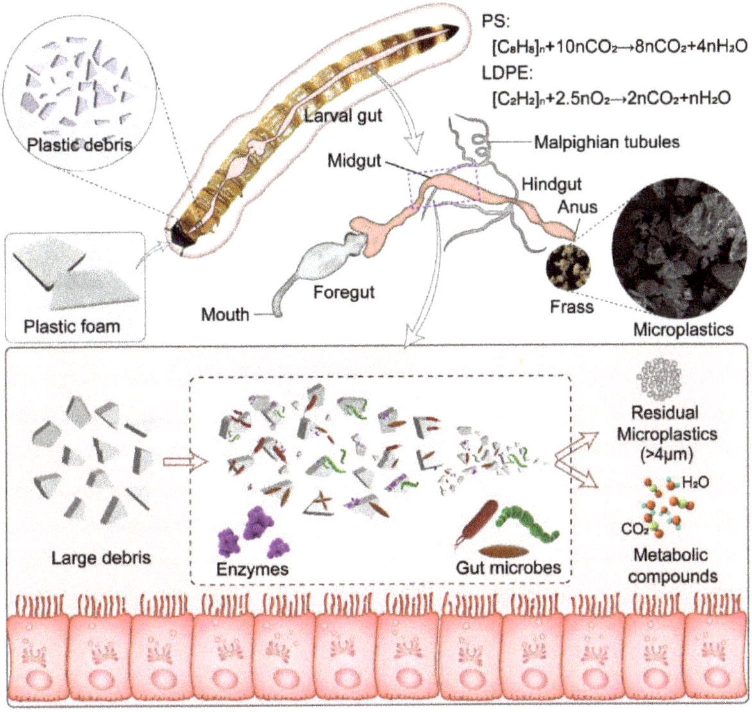


**Figure 4.** Conceptual schematic diagram for the biodegradation of PS and LDPE foams in *Zophobas atratus* larvae, reprinted with permission from [36].

Dae-Hwan Kim et al. reported that *Pseudomonas aeruginosa* strain DSM 50071, which was isolated from the gut of *Z. atratus*, could degrade PS. Pseudomonas sp. DSM50071 could effectively biodegrade PS similarly to other plastic-degrading bacteria. The conversion of PS surface from hydrophobicity to hydrophilicity through biofilm formation is crucial for PS degradation [33].

Bryan J. Cassone reported that *Z. atratus* showed more microbial abundance in the early stage (24–72 h) of LDPE feeding than caterpillars fed with starvation or natural honeycomb diet. By using PS as sole carbon source to isolate and grow intestinal bacteria for more than one year, the microorganisms in Acinetobacter participated in the biodegradation process [37].

Shan-Shan Yang et al. investigated the biodegradation of PP by feeding larvae of *Z. atratus* and *T. molitor* with PP foam. Figure 5 shows the PP foam-eating *T. molitor* larvae and *Z. atratus* larvae, PP-fed versus PP + WB-fed gut microbiome of *Z. atratus* larvae. In the study, PP foam was used as sole diet as a comparative study. When larvae of *Z. atratus* and *T. molitor* were fed with the PP foam plus wheat bran, the consumption rates were enhanced by 68.11% and 39.70%, respectively. Mw of frass decreased by 20.4% and 9.0%, respectively, which indicated that PP can be biodegraded by the larvae of *Z. atratus* and *T. molitor* via gut microbe-dependent depolymerization [38].

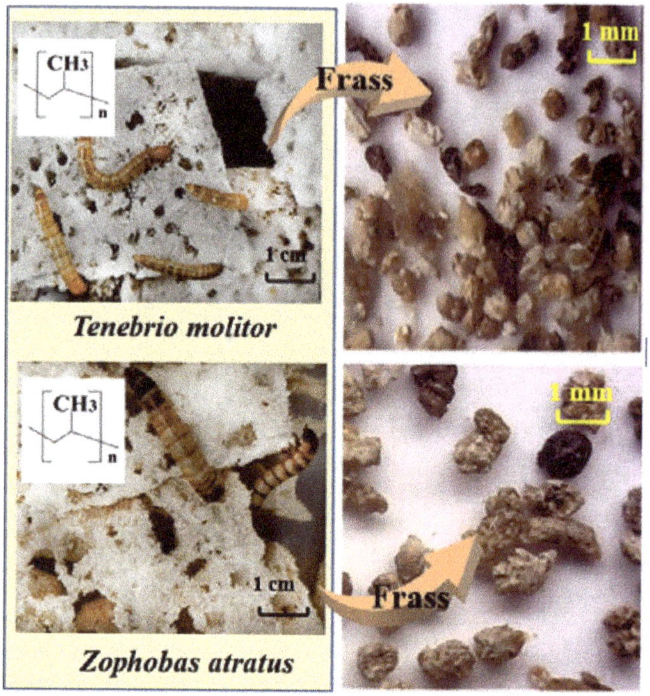


**Figure 5.** PP foam-eating *T. molitor* larvae and *Z. atratus* larvae, reprinted with permission from [38].

Liping Luo et al. used plastics, including PS, PE and PU foam, as sole feedstock to feed *T. molitor*. PS- or PU-fed larvae showed 100% survival rates, and the PE-fed and starvation larvae had decreased survival rates of 81.67% and 65%, respectively. Plastic-fed and starvation groups showed decreased larvae weight. The consumption rates of PS, PE, and PU were 1.41, 0.30, and 0.74 mg/d/larva, respectively. The results showed that *T. molitor* can partially degrade plastics [34].

### 2.3. Plastic Degradation by G. mellonella

*G. mellonella* is a completely metamorphosed insect that undergoes four stages, including egg, larva, pupa, and adult. The insects are widely distributed all over the world, especially in the tropical and subtropical regions of Southeast Asia [51–53]. *G. mellonella* is rich in protein, short in growth cycle, and easy to reproduce, eat, and feed [54]. It is one of the main insect models which is used to study innate immunity and host pathogen interactions. The giant *G. mellonella* is often used to feed freshwater fish, birds, reptiles, and amphibians, which have three generations a year, and one generation lasts approximately 60–80 days [55]. When the average temperature exceeds 13 °C, the larvae begin to awake. The activity of the giant *G. mellonella* is closely related to external temperature. Excessively low or high temperature will slow the growth and even kill giant *G. mellonella*. *G. mellonella* likes to eat beeswax, and the chemical structure of the most common hydrocarbon bond in beeswax is highly similar to that of PE. The chewed PE is digested by the wax borer and converted into small molecules of ethylene glycol, which can be degraded in the natural environment within a few weeks.

Defeng Xing et al. investigated the feasibility of enhancing larval survival and the effect of supplementing the co-diet on plastic degradation by feeding the larvae of *G. mellonella*. Figure 6 shows that plastic and supplementary nutrients are ingested by the greater wax moth larvae fed with a PS diet. Significant mass loss of plastic was observed within 21 days (i.e., 150 larvae fed only PS or PE consumed 0.88 g and 1.95 g, respectively). O-containing functional groups and long-chain fatty acids are detected in metabolic intermediates, thereby showing depolymerization and biodegradation [39]

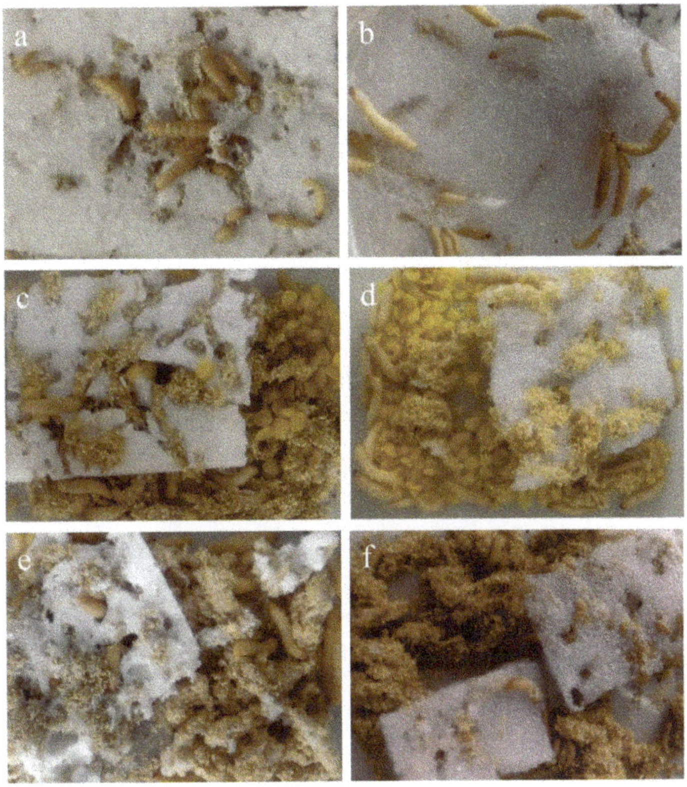


**Figure 6.** Plastic and supplementary nutrients were ingested by the greater wax moth larvae fed a PS diet (**a**), a PE diet (**b**), a PS + beeswax diet (**c**), a PE + beeswax diet (**d**), a PS + bran diet (**e**), and a PE + beeswax diet (**f**), reprinted with permission from [39].

Harsha Kundungal et al. investigated the degradation of high-density polyethylene (HDPE) by feeding the larvae of *G. mellonella*. Nutrition on PE degradation were investigated by providing wax comb as co-feed. Figure 7a–d shows *G. mellonella* larvae feeding on PE film, degraded PE films with holes after exposure to the lesser waxworm for 12 h, comparison of post-degradation weight loss percentage of waxcomb and PE after lesser waxworm consumption, and PE consumption over time. The study after degradation showed that 100 wax insects reduced the weight of PE by 43.3 $\pm$ 1.6%. In 8 days, each wax insect ingested 1.83 mg of PE every day, and the consumption of PE increased [40].

Yucheng Zhao et al. isolated a PE degrading fungus called PEDX3 from the intestine of *G. mellonella*. Figure 8a–c shows the process of PE films being degraded by PEDX3 from the intestine of *G. mellonella*, visual analysis, HT-GPC analysis, and Fourier infrared spectroscopy (FT-IR) analysis. The results showed that strain PEDX3 degraded HDPE and MPP into low molecular weight MPP after 28 days of culture. The degradation products contain carbonyl and ether groups, which verifies the degradation of PE [41].

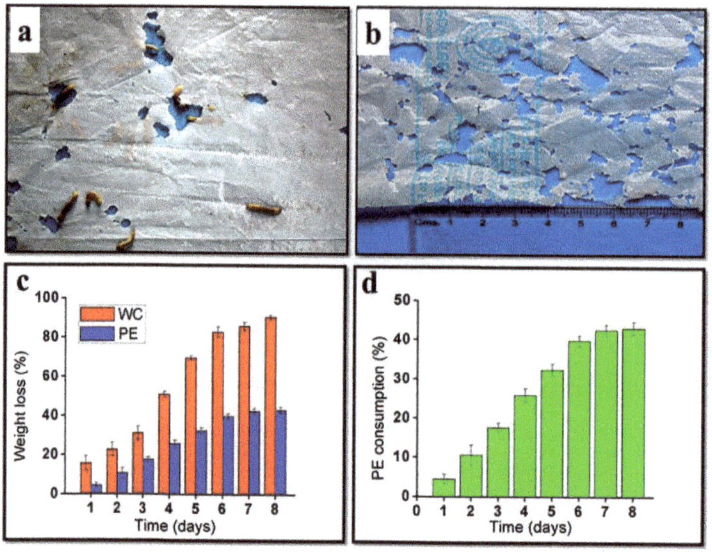


**Figure 7.** (**a**) *Galleria mellonella* larvae feeding on PE film. (**b**) Degraded PE film with holes after exposure to the lesser waxworm for 12 h. (**c**) Comparison of post-degradation weight loss percentage of waxcomb and PE after lesser waxworm consumption. (**d**) PE consumption over time, reprinted with permission from [40].

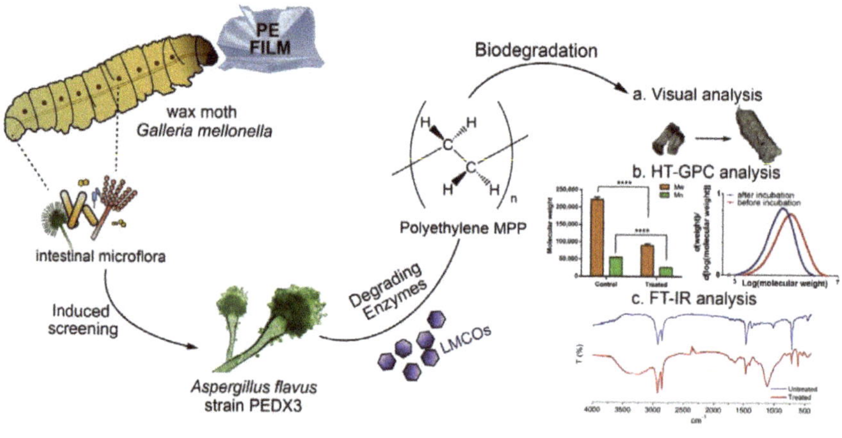


**Figure 8.** The process of PE films degraded by PEDX3(a PE degrading fungus) from the intestine of *Galleria mellonella*. (**a**) Visual analysis; (**b**) HT−GPC analysis; (**c**) FT−IR analysis, reprinted with permission from [41].

Wei-Min Wu et al. investigated the biodegradability of LDPE and HDPE by yellow and dark *G. mellonella*. The sequence of biodegradation extent showed LDPE > HDPE. The low molecular weight, high branching, and low crystallinity of PE are positive for bio-degradation. Molecular weight is the key factor that affects biodegradability [42].

### 2.4. Plastic Degradation by Other Insects

Wax insects, which are mainly distributed in more than 10 provinces such as Shandong, Hebei, Henan, Sichuan, Yunnan, Guizhou, Guangxi, and Guangdong are special resource insects in China. They are also distributed in Japan, India, Russia, and other countries. The female only undergoes three stages, namely egg, nymph, and adult, which belongs to incomplete metamorphosis type. The male undergoes four stages, namely egg, larva, pupa, and adult, hence belonging to the completely changed S type. Guocai Zhang et al. studied the mechanism of PE degradation by wax insects. They fed the wax insects with PE, separated and purified two strains with high PE degradation efficiency, and evaluated the impact of single and microbial combination on PE degradation. The results showed that PE could be degraded by *Meyerospira gilsonii* and *Serratia marcescens*. However, the degradation efficiency of microbial community is higher, and the weight loss rate of PE is 15.87%. The chemical structures of a series of PE degradation products were obtained. This study can be used to develop an effective microbial community for PE degradation and provide a basis for the reuse of PE waste [56].

Jun Yang et al. isolated two bacterial strains, namely *Escherichia coli* AST1 and Bacillus sp.YP1, which could degrade PE from the intestinal tract of wax worms. Figure 9a–c shows the PE film-eating waxworms and morphotypes of the cells in the mature biofilm on the PE sheet. After 28 days of culture on PE film, the two strains formed a living biofilm, the hydrophobicity of the PE film was reduced, and the obvious damage included pits and cavities. During the 60-day incubation period, YP1 could degrade by approximately 10.7 ± 0.2% of PE film. The molecular weight of the residual PE film is low, hence providing promising evidence for the biodegradation of PE in the environment [43].

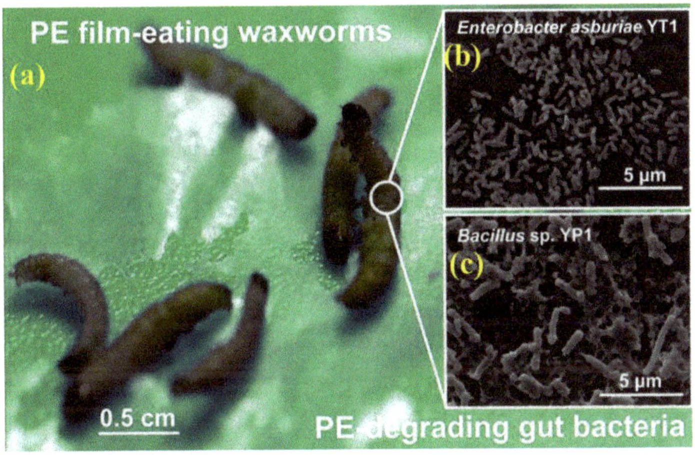


**Figure 9.** (**a**) PE film-eating waxworms. (**b**,**c**) Morphotypes of the cells in the mature biofilm on the PE sheet, reprinted with permission from [43].

Yalei Zhang showed that the degradation rate of PS in the intestine of dark mealworms was faster than that of *T. molitor*. Figure 10A–D presents the *T. molitor* and *Tenebrio obscurus* around the world and PS foam-eating behaviors. With expanded PS foam as the only diet, after 31 days, Mn of residual PS in the feces of dark mealworms decreased by 26.03%, which was significantly higher than that of whitefly (11.67%). According to the proportion of PS residues, dark mealworms can degrade PS effectively [44].

Zhang Yong Wang et al. studied the biodegradation of PS in the intestinal microbiota of *T. molitor*, Parasita magna, and Atlas Z. *atratus* larvae. The results showed that the superworm had the strongest PS consumption ability and the highest survival rate during the 30-day experiment period, able to degrade PS to different degrees. *T. molitor* strongly depolymerized PS by destroying benzene ring [57].

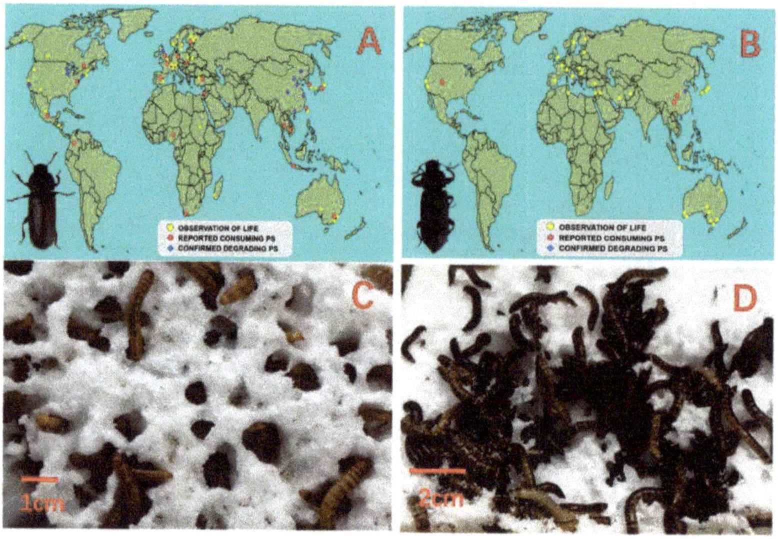


**Figure 10.** *Tenebrio molitor* and *Tenebrio obscurus* around the world and PS foam-eating behaviors. (**A**) Discovered life and original sources of *T. molitor*. (**B**) Discovered life and original sources of *T. obscurus*. (**C**) PS foam-eating *T. molitor* larvae and (**D**) *T. obscurus* larvae from Shandong Province, China, reprinted with permission from [44].

Feng Ju et al. isolated a PVC-degrading bacteria from the intestinal tract of insect larvae and studied the pathway of PVC degradation. Their research reported that the larvae of the pest *Spodoptera frugiperda* can survive by eating PVC film, which is related to the enrichment of enterococcus, Klebsiella, and other bacteria in the larval intestinal microbiota. Bacterial strains isolated from larval intestine can depolymerize PVC [58].

Mik Van Der Borgt et al. investigated the effects of PVC plastics on the growth, survival, and biotransformation of black soldier fly larvae. The growth, survival, and biotransformation parameters of larvae were measured by feeding black soldier fly larvae with artificial food waste mixed with micro, medium, and large plastics. The insects are not affected by PVC plastics in the matrix in terms of growth performance, survival rate, and biotransformation rate [45].

Defu He et al. reported the biodegradation capability of expanded PS foam in a globally distributed soil invertebrate, Achatina fulica. Figure 11a–d shows that the PS foam was uptaken by *A. fulica* and fragmented into microplastics. After a 4-week exposure, 18.5 $\pm$ 2.9 mg PS was ingested per snail in one month and microplastics in feces were egested with significant mass loss of 30.7%. A significant increase in Mw of feces-residual PS illustrated limited extent of depolymerization. Significant shifts in the gut microbiome were observed after the ingestion of PS, with an increase in families of Enterobacteriaceae, Sphingobacteriaceae, and Aeromonadaceae, which showed that gut microorganisms were associated with PS biodegradation [59].

Seongwook Woo et al. reported the PS biodegradation by the larvae of the darkling beetle *Plesiophthalmus davidis. P. davidis* ingested 34.27 mg of PS foam per larva and survived by feeding only on Styrofoam in two weeks. The ingested PS foam was oxidized. The decrease in the Mw of the residual PS in the frass compared with the feed PS foam, and C–O bonding was detected in the degradation products of PS film, which illustrated that PS foam was degraded [60].

**Figure 11.** PS foam were uptaken by *A. fulica* and fragmented into microplastics. (**a**) The exposure scene of Styrofoam and *A. fulica*. (**b**) Mark of gnawing on Styrofoam (the top right showing enlarged image in the blue box). (**c**) Microplastics in feces. (**d**) The distribution of microplastic sizes, reprinted with permission from [59].

## 3. Conclusions and Prospects

Plastic products have been widely used around the world because of its easy processing, low price, and portability. However, so far, unreasonable disposal of plastic products has led to serious pollution, and no perfect strategy can address waste plastics. It is difficult for plastics to degrade by themselves. Landfill has been proved to be infeasible for waste plastic disposal. Toxic gases are generated by incineration or chemical treatment, which cause various threats to the atmospheric environment and human health. Achieving large-scale disposal of waste plastics through recycling is also difficult. Insect-degradable plastics have not been employed in practical application. The current research reported that *T. molitor*, *Z. atratus*, wax borer, silkworm, and termite can feed on degradable plastics. Using insects to degrade plastics has the advantages of low cost and no secondary pollution. Moreover, the aforementioned insects can be used as animal feed sources, with good application and economic value. However, due to the differences in the composition and structure of plastics, the degradation efficiency of different insects varies greatly, and some plastics also affect the growth and development of insects. The research on plastic degradation by insects and isolated functional microorganisms has only been carried out in recent years, and the results are far from practical application. The researchers can conduct experiments in discovering more insects that can feed on plastic, confirming the preference of different insects for the types of plastic and further improving the efficiency of insects' feeding and degradation of plastic. The use of waste plastics as food for insects can realize the recycling of waste plastics. However, the realization of this process requires researchers to conduct toxicological analysis on plastic-eating insects to avoid the toxicological risks or cumulative effects of toxic substances in organisms. High-throughput sequencing technology is used to analyze the intestinal microbial diversity of plastic-degrading insects, detect functional microorganisms related to plastic metabolism, reveal the realization mechanism of plastic microbial degradation and new metabolic path, and isolate and purify plastic-degrading functional microorganisms.

**Author Contributions:** Conceptualization, R.A. and P.J.; investigation, R.A. and P.J.; resources, C.L.; data curation, J.W.; writing—original draft preparation, R.A. and P.J.; writing—review and editing, C.L.; visualization, C.L.; supervision, J.W.; All authors have read and agreed to the published version of the manuscript.

**Funding:** This research received no external funding.

**Institutional Review Board Statement:** Not applicable.

**Informed Consent Statement:** Not applicable.

**Data Availability Statement:** Not applicable.

**Conflicts of Interest:** The authors declare no conflict of interest. The funders had no role in the design of the study; in the collection, analyses, or interpretation of data; in the writing of the manuscript; or in the decision to publish the results.

## Abbreviations

| | |
|---|---|
| PS | Polystyrene |
| PVC | Polyvinyl chloride |
| PE | Polyethylene |
| ABS | Acrylonitrile butadiene styrene |
| PU | Polyurethane |
| PLA | Polylactic acid |
| PHAs | Polyhydroxy fatty acid ester polymers |
| PPC | Carbon dioxide copolymer |
| PCL | Polycaprolactone |
| PGA | Polyglycolic acid |
| YT2 | Exiguobacterium sp. |
| LDPE | Low-density PS |
| Mn | Molecular weight |
| THF | Tetrahydrofuran |
| FT-IR | Fourier infrared spectroscopy |
| ATR-FTIR | Fourier transform attenuated total reflection infrared spectroscopy |
| HDPE | High-density PS |

## References

1. Kim, E.; Choi, W.Z. Real-time identification of plastics by types using laser-induced Breakdown Spectroscopy. *J. Mater. Cycles Waste Manag.* **2019**, *21*, 176–180. [CrossRef]
2. Wilkes, R.A.; Aristilde, L. Degradation and metabolism of synthetic plastics and associated products by *Pseudomonas* sp.: Capabilities and challenges. *J. Appl. Microbiol.* **2017**, *123*, 582–593. [CrossRef] [PubMed]
3. Ainali, N.M.; Bikiaris, D.N.; Lambropoulou, D.A. Aging effects on low- and high-density polyethylene, polypropylene and polystyrene under UV irradiation: An insight into decomposition mechanism by Py-GC/MS for microplastic analysis. *J. Anal. Appl. Pyrolysis* **2021**, *158*, 105207. [CrossRef]
4. Shen, M.C.; Song, B.; Zeng, G.M.; Zhang, Y.X.; Huang, W.; Wen, X.F.; Tang, W.W. Are biodegradable plastics a promising solution to solve the global plastic pollution? *Environ. Pollut.* **2020**, *263*, 114469. [CrossRef]
5. Singh, P.; Sharma, V.P. Integrated Plastic Waste Management: Environmental and Improved Health Approaches. *Procedia Environ. Sci.* **2016**, *35*, 692–700. [CrossRef]
6. Halden, R.U. Plastics and Health Risks. *Annu. Rev. Public Health* **2010**, *31*, 179–194. [CrossRef]
7. Chen, G.-Q.; Patel, M.K. Plastics Derived from Biological Sources: Present and Future: A Technical and Environmental Review. *Chem. Rev.* **2012**, *112*, 2082–2099. [CrossRef] [PubMed]
8. Bombelli, P.; Howe, C.J.; Bertocchini, F. Polyethylene bio-degradation by caterpillars of the wax moth *Galleria mellonella*. *Curr. Biol.* **2017**, *27*, R292–R293. [CrossRef] [PubMed]
9. Kosior, E.; Mitchell, J. Chapter 6-Current industry position on plastic production and recycling. *Plast. Waste Recycl.* **2020**, 133–162. [CrossRef]
10. Shen, M.; Huang, W.; Chen, M.; Song, B.; Zeng, G.; Zhang, Y. (Micro)plastic crisis: Un-ignorable contribution to global greenhouse gas emissions and climate change. *J. Clean. Prod.* **2020**, *254*, 120138. [CrossRef]
11. Sharp, A.; Høj, S.; Wheeler, M. Proscription and its impact on anti-consumption behaviour and attitudes: The case of plastic bags. *J. Consum. Behav.* **2010**, *9*, 470–484. [CrossRef]
12. Auta, H.S.; Emenike, C.U.; Fauziah, S.H. Distribution and importance of microplastics in the marine environment: A review of the sources, fate, effects, and potential solutions. *Environ. Int.* **2017**, *102*, 165–176. [CrossRef] [PubMed]

13. Wilcox, C.; Mallos, N.J.; Leonard, G.H.; Rodriguez, A.; Hardesty, B.D. Using expert elicitation to estimate the impacts of plastic pollution on marine wildlife. *Mar. Policy* **2016**, *65*, 107–114. [CrossRef]
14. Alimba, C.G.; Faggio, C. Microplastics in the marine environment: Current trends in environmental pollution and mechanisms of toxicological profile. *Environ. Toxicol. Pharmacol.* **2019**, *68*, 61–74. [CrossRef]
15. Moshood, T.D.; Nawanir, G.; Mahmud, F.; Mohamad, F.; Ahmad, M.H.; AbdulGhani, A. Sustainability of biodegradable plastics: New problem or solution to solve the global plastic pollution? *Curr. Res. Green Sustain. Chem.* **2022**, *5*, 100273. [CrossRef]
16. Filiciotto, L.; Rothenberg, G. Biodegradable Plastics: Standards, Policies, and Impacts. *ChemSusChem* **2021**, *14*, 56–72. [CrossRef]
17. Chausali, N.; Saxena, J.; Prasad, R. Recent trends in nanotechnology applications of bio-based packaging. *J. Agric. Food Res.* **2022**, *7*, 100257. [CrossRef]
18. Havstad, M.R. Chapter 5-Biodegradable plastics. In *Plastic Waste and Recycling*; Letcher, T.M., Ed.; Academic Press: Cambridge, MA, USA, 2020; pp. 97–129. [CrossRef]
19. Sathyanarayanan, S.S.; Joseph, K.; Yan, B.; Karthik, O.; Palanivelu, K.; Ramachandran, A. Solid waste management practices in India and China–Sustainability issues and Opportunities. In *Waste Management Policies and Practices in BRICS Nations*; CRC Press: Boca Raton, FL, USA, 2021; p. 42. [CrossRef]
20. Alqattaf, A. (Ed.) Plastic Waste Management: Global Facts, Challenges and Solutions. In Proceedings of the 2020 Second International Sustainability and Resilience Conference: Technology and Innovation in Building Designs (51154), Sakheer, Bahrain, 11–12 November 2020. [CrossRef]
21. Tan, W. Correlation between microplastics estimation and human development index. In Proceedings of the International Conference on Statistics, Applied Mathematics, and Computing Science (CSAMCS 2021), Nanjing, China, 22 April 2022; Volume 12163, p. 1216311. [CrossRef]
22. Geyer, R. Chapter 2-Production, use, and fate of synthetic polymers. In *Plastic Waste and Recycling*; Letcher, T.M., Ed.; Academic Press: Cambridge, MA, USA, 2020; pp. 13–32. [CrossRef]
23. Yang, Y.; Yang, J.; Wu, W.-M.; Zhao, J.; Song, Y.; Gao, L.; Yang, R.; Jiang, L. Biodegradation and Mineralization of Polystyrene by Plastic-Eating Mealworms: Part 1. Chemical and Physical Characterization and Isotopic Tests. *Environ. Sci. Technol.* **2015**, *49*, 12080–12086. [CrossRef]
24. Yang, S.-S.; Ding, M.-Q.; Zhang, Z.-R.; Ding, J.; Bai, S.-W.; Cao, G.-L.; Zhao, L.; Pang, J.-W.; Xing, D.-F.; Ren, N.-Q.; et al. Confirmation of biodegradation of low-density polyethylene in dark- versus yellow- mealworms (larvae of Tenebrio obscurus versus *Tenebrio molitor*) via. gut microbe-independent depolymerization. *Sci. Total Environ.* **2021**, *789*, 147915. [CrossRef] [PubMed]
25. Brandon, A.M.; Garcia, A.M.; Khlystov, N.A.; Wu, W.-M.; Criddle, C.S. Enhanced Bioavailability and Microbial Biodegradation of Polystyrene in an Enrichment Derived from the Gut Microbiome of *Tenebrio molitor* (Mealworm Larvae). *Environ. Sci. Technol.* **2021**, *55*, 2027–2036. [CrossRef] [PubMed]
26. Brandon, A.M.; Gao, S.-H.; Tian, R.; Ning, D.; Yang, S.-S.; Zhou, J.; Wu, W.-M.; Criddle, C.S. Biodegradation of Polyethylene and Plastic Mixtures in Mealworms (Larvae of *Tenebrio molitor*) and Effects on the Gut Microbiome. *Environ. Sci. Technol.* **2018**, *52*, 6526–6533. [CrossRef]
27. Tsochatzis, E.; Berggreen, I.E.; Tedeschi, F.; Ntrallou, K.; Gika, H.; Corredig, M. Gut Microbiome and Degradation Product Formation during Biodegradation of Expanded Polystyrene by Mealworm Larvae under Different Feeding Strategies. *Molecules* **2021**, *26*, 7568. [CrossRef]
28. Yang, L.; Gao, J.; Liu, Y.; Zhuang, G.; Peng, X.; Wu, W.M.; Zhuang, X. Biodegradation of expanded polystyrene and low-density polyethylene foams in larvae of *Tenebrio molitor* Linnaeus (Coleoptera: Tenebrionidae): Broad versus limited extent depolymerization and microbe-dependence versus independence. *Chemosphere* **2021**, *262*, 127818. [CrossRef] [PubMed]
29. Peng, B.-Y.; Chen, Z.; Chen, J.; Yu, H.; Zhou, X.; Criddle, C.S.; Wu, W.-M.; Zhang, Y. Biodegradation of Polyvinyl Chloride (PVC) in *Tenebrio molitor* (Coleoptera: Tenebrionidae) larvae. *Environ. Int.* **2020**, *145*, 106106. [CrossRef] [PubMed]
30. Jia, P.; Lamm, M.E.; Sha, Y.; Ma, Y.; Buzoglu Kurnaz, L.; Zhou, Y. Thiol-ene eugenol polymer networks with chemical Degradation, thermal degradation and biodegradability. *Chem. Eng. J.* **2023**, *454*, 140051. [CrossRef]
31. Bulak, P.; Proc, K.; Pytlak, A.; Puszka, A.; Gawdzik, B.; Bieganowski, A. Biodegradation of Different Types of Plastics by *Tenebrio molitor* Insect. *Polymers* **2021**, *13*, 3508. [CrossRef] [PubMed]
32. Yang, Y.; Wang, J.; Xia, M. Biodegradation and mineralization of polystyrene by plastic-eating superworms *Zophobas atratus*. *Sci. Total Environ.* **2020**, *708*, 135233. [CrossRef]
33. Kim, H.R.; Lee, H.M.; Yu, H.C.; Jeon, E.; Lee, S.; Li, J.; Kim, D.-H. Biodegradation of Polystyrene by *Pseudomonas* sp. Isolated from the Gut of Superworms (Larvae of *Zophobas atratus*). *Environ. Sci. Technol.* **2020**, *54*, 6987–6996. [CrossRef]
34. Luo, L.; Wang, Y.; Guo, H.; Yang, Y.; Qi, N.; Zhao, X.; Gao, S.; Zhou, A. Biodegradation of foam plastics by *Zophobas atratus* larvae (Coleoptera: Tenebrionidae) associated with changes of gut digestive enzymes activities and microbiome. *Chemosphere* **2021**, *282*, 131006. [CrossRef]
35. Wang, Y.; Luo, L.; Li, X.; Wang, J.; Wang, H.; Chen, C.; Guo, H.; Han, T.; Zhou, A.; Zhao, X. Different plastics ingestion preferences and efficiencies of superworm (*Zophobas atratus* Fab.) and yellow mealworm (*Tenebrio molitor* Linn.) associated with distinct gut microbiome changes. *Sci. Total Environ.* **2022**, *837*, 155719. [CrossRef]

36. Peng, B.-Y.; Sun, Y.; Wu, Z.; Chen, J.; Shen, Z.; Zhou, X.; Wu, W.-M.; Zhang, Y. Biodegradation of polystyrene and low-density polyethylene by *Zophobas atratus* larvae: Fragmentation into microplastics, gut microbiota shift, and microbial functional enzymes. *J. Clean. Prod.* **2022**, *367*, 132987. [CrossRef]
37. Cassone, B.J.; Grove, H.C.; Elebute, O.; Villanueva, S.M.P.; LeMoine, C.M.R. Role of the intestinal microbiome in low-density polyethylene degradation by caterpillar larvae of the greater wax moth, *Galleria mellonella*. *Proc. Biol. Sci.* **2020**, *287*, 20200112. [CrossRef]
38. Yang, S.-S.; Ding, M.-Q.; He, L.; Zhang, C.-H.; Li, Q.-X.; Xing, D.-F.; Cao, G.-L.; Zhao, L.; Ding, J.; Ren, N.-Q.; et al. Biodegradation of polypropylene by yellow mealworms (*Tenebrio molitor*) and superworms (*Zophobas atratus*) via gut-microbe-dependent depolymerization. *Sci. Total Environ.* **2021**, *756*, 144087. [CrossRef]
39. Lou, Y.; Ekaterina, P.; Yang, S.-S.; Lu, B.; Liu, B.; Ren, N.; Corvini, P.F.X.; Xing, D. Biodegradation of Polyethylene and Polystyrene by Greater Wax Moth Larvae (*Galleria mellonella* L.) and the Effect of Co-diet Supplementation on the Core Gut Microbiome. *Environ. Sci. Technol.* **2020**, *54*, 2821–2831. [CrossRef] [PubMed]
40. Kundungal, H.; Gangarapu, M.; Sarangapani, S.; Patchaiyappan, A.; Devipriya, S.P. Efficient biodegradation of polyethylene (HDPE) waste by the plastic-eating lesser waxworm (Achroia grisella). *Environ. Sci. Pollut. Res. Int.* **2019**, *26*, 18509–18519. [CrossRef] [PubMed]
41. Zhang, J.; Gao, D.; Li, Q.; Zhao, Y.; Li, L.; Lin, H.; Bi, Q.; Zhao, Y. Biodegradation of polyethylene microplastic particles by the fungus Aspergillus flavus from the guts of wax moth *Galleria mellonella*. *Sci. Total Environ.* **2020**, *704*, 135931. [CrossRef]
42. Yang, S.-S.; Ding, M.-Q.; Ren, X.-R.; Zhang, Z.-R.; Li, M.-X.; Zhang, L.-L.; Pang, J.-W.; Chen, C.-X.; Zhao, L.; Xing, D.-F.; et al. Impacts of physical-chemical property of polyethylene on depolymerization and biodegradation in yellow and dark mealworms with high purity microplastics. *Sci. Total Environ.* **2022**, *828*, 154458. [CrossRef]
43. Yang, J.; Yang, Y.; Wu, W.-M.; Zhao, J.; Jiang, L. Evidence of Polyethylene Biodegradation by Bacterial Strains from the Guts of Plastic-Eating Waxworms. *Environ. Sci. Technol.* **2014**, *48*, 13776–13784. [CrossRef] [PubMed]
44. Peng, B.; Su, Y.; Chen, Z.; Chen, J.; Zhou, X.; Benbow, M.; Criddle, C.; Wu, W.; Zhang, Y. Biodegradation of Polystyrene by Dark (*Tenebrio obscurus*) and Yellow (*Tenebrio molitor*) Mealworms (Coleoptera: Tenebrionidae). *Environ. Sci. Technol.* **2019**, *53*, 5256–5265. [CrossRef]
45. Lievens, S.; Poma, G.; Frooninckx, L.; Van der Donck, T.; Seo, J.W.; De Smet, J.; Covaci, A.; Van Der Borght, M. Mutual Influence between Polyvinyl Chloride (Micro) Plastics and Black Soldier Fly Larvae (*Hermetia illucens* L.). *Sustainability* **2022**, *14*, 12109. [CrossRef]
46. Hong, J.; Han, T.; Kim, Y.Y. Mealworm (*Tenebrio molitor* Larvae) as an Alternative Protein Source for Monogastric Animal: A Review. *Animals* **2020**, *10*, 2068. [CrossRef]
47. Zepeda-Bastida, A.; Ocampo-López, J.; Alarcón-Sánchez, B.R.; Idelfonso-García, O.G.; Rosas-Madrigal, S.; Aparicio-Bautista, D.I.; Pérez-Carreón, J.I.; Villa-Treviño, S.; Arellanes-Robledo, J. Aqueous extracts from *Tenebrio molitor* larval and pupal stages inhibit early hepatocarcinogenesis in vivo. *J. Zhejiang Univ.-Sci. B* **2021**, *22*, 1045–1052. [CrossRef] [PubMed]
48. Wu, Q.; Tao, H.; Wong, M.H. Feeding and metabolism effects of three common microplastics on *Tenebrio molitor* L. *Environ. Geochem. Health* **2019**, *41*, 17–26. [CrossRef]
49. Johnston, A.S.A.; Sibly, R.M.; Hodson, M.E.; Alvarez, T.; Thorbek, P. Effects of agricultural management practices on earthworm populations and crop yield: Validation and application of a mechanistic modelling approach. *J. Appl. Ecol.* **2015**, *52*, 1334–1342. [CrossRef]
50. Sharma, S.; Dhaliwal, S.S. Conservation agriculture based practices enhanced micronutrients transformation in earthworm cast soil under rice-wheat cropping system. *Ecol. Eng.* **2021**, *163*, 106195. [CrossRef]
51. Gohl, P.; LeMoine, C.M.R.; Cassone, B.J. Diet and ontogeny drastically alter the larval microbiome of the invertebrate model *Galleria mellonella*. *Can. J. Microbiol.* **2022**, *68*, 594–604. [CrossRef]
52. Zhang, Z.; Zhu, S.; De Mandal, S.; Gao, Y.; Yu, J.; Zeng, L.; Huang, J.; Zafar, J.; Jin, F.; Xu, X. Combined transcriptomic and proteomic analysis of developmental features in the immune system of Plutella xylostella during larva-to-adult metamorphosis. *Genomics* **2022**, *114*, 110381. [CrossRef] [PubMed]
53. Mukherjee, K.; Dobrindt, U. The emerging role of epigenetic mechanisms in insect defense against pathogens. *Curr. Opin. Insect Sci.* **2022**, *49*, 8–14. [CrossRef] [PubMed]
54. Pereira, M.F.; Rossi, C.C. Overview of rearing and testing conditions and a guide for optimizing *Galleria mellonella* breeding and use in the laboratory for scientific purposes. *Apmis* **2020**, *128*, 607–620. [CrossRef]
55. Kong, H.G.; Kim, H.H.; Chung, J.-H.; Jun, J.; Lee, S.; Kim, H.-M.; Jeon, S.; Park, S.G.; Bhak, J.; Ryu, C.-M. The *Galleria mellonella* Hologenome Supports Microbiota-Independent Metabolism of Long-Chain Hydrocarbon Beeswax. *Cell Rep.* **2019**, *26*, 2451–2464.e5. [CrossRef]
56. Lou, H.; Fu, R.; Long, T.; Fan, B.; Guo, C.; Li, L.; Zhang, J.; Zhang, G. Biodegradation of polyethylene by Meyerozyma guilliermondii and Serratia marcescens isolated from the gut of waxworms (larvae of *Plodia interpunctella*). *Sci. Total Environ.* **2022**, *853*, 158604. [CrossRef] [PubMed]
57. Jiang, S.; Su, T.; Zhao, J.; Wang, Z. Biodegradation of Polystyrene by *Tenebrio molitor*, *Galleria mellonella*, and *Zophobas atratus* Larvae and Comparison of Their Degradation Effects. *Polymers* **2021**, *13*, 3539. [CrossRef]
58. Zhang, Z.; Peng, H.; Yang, D.; Zhang, G.; Zhang, J.; Ju, F. Polyvinyl chloride degradation by a bacterium isolated from the gut of insect larvae. *Nat. Commun.* **2022**, *13*, 5360. [CrossRef] [PubMed]

59. Song, Y.; Qiu, R.; Hu, J.; Li, X.; Zhang, X.; Chen, Y.; Wu, W.-M.; He, D. Biodegradation and disintegration of expanded polystyrene by land snails Achatina fulica. *Sci. Total Environ.* **2020**, *746*, 141289. [CrossRef]
60. Woo, S.; Song, I.; Cha, H.J. Fast and Facile Biodegradation of Polystyrene by the Gut Microbial Flora of Plesiophthalmus davidis Larvae. *Appl. Environ. Microbiol.* **2020**, *86*, e01361-20. [CrossRef] [PubMed]

polymers

MDPI

Article

# Effect of Varying Curing Conditions on the Strength of Biopolymer Modified Sand

Kehinde Lemboye [1,2,*] and Abdullah Almajed [1]

1 Department of Civil Engineering, College of Engineering, King Saud University, Riyadh 11421, Saudi Arabia
2 School of Civil and Mechanical Engineering, Curtin University, Bentley, WA 6102, Australia
* Correspondence: 438105781@student.ksu.edu.sa or kehinde.lemboye@postgrad.curtin.edu.au

**Abstract:** Recently, the improvement of the engineering properties of soil has been centered on using sustainable and eco-friendly materials. This study investigates the efficacy of three biopolymers: Acacia, sodium alginate, and pectin, on the unconfined compressive strength (UCS) of dune sand. The UCS test measured the effects of the biopolymer type and concentration, curing intervals and temperature, and moisture loss. The changes in the morphology caused by the biopolymer addition were examined via scanning electron microscopy (SEM). Results indicate that the UCS of the biopolymer-modified sand increased with biopolymer concentration and curing intervals. Varying the curing temperature from 25–110 °C, slightly affected the strength of the acacia-modified sand specimen, increased that of the sodium alginate-modified sand specimen up to a temperature of 85 °C, and continued to decrease that of the pectin-modified sand specimen as the temperature was increased from 25 to 110 °C. The SEM images indicated that the biopolymer's presence within the sand pores significantly contributed to the strength. Bond decomposition occurs at temperatures greater than 110 °C for sodium alginate and pectin-modified sands, whereas bonds remain stable at higher temperatures for the acacia-modified sand. In conclusion, all three biopolymers show potential as robust and economic dune stabilisers.

**Keywords:** acacia; biopolymer; properties; gelling; pectin; sodium alginate; soil stabilization unconfined compressive strength

**Citation:** Lemboye, K.; Almajed, A. Effect of Varying Curing Conditions on the Strength of Biopolymer Modified Sand. *Polymers* **2023**, *15*, 1678. https://doi.org/10.3390/polym15071678

Academic Editors: Vsevolod Aleksandrovich Zhuikov and Rosane Michele Duarte Soares

Received: 6 March 2023
Revised: 20 March 2023
Accepted: 21 March 2023
Published: 28 March 2023



## 1. Introduction

Soil properties such as strength, volume stability, durability, compressibility, and permeability are essential to geotechnical applications, from dune stabilisation to foundation design [1]. Traditional methods for improving the geotechnical properties of soil include mechanical and chemical techniques. Conventional methods for chemical stabilisation, which employ materials such as cement, lime, fly ash, cement kiln dust, and blast-furnace slag, are the oldest and have gained popularity among researchers [2–21]. These materials either aid compaction, repel water, or bind to the soil particles when incorporated into the soil matrix. Soil treated with chemicals, except sodium silicate, is often prone to environmental hazards [22]. Cement and lime can change the pH of the soil, pollute the groundwater, and pose a threat to human health. The cement production industry is one of the most significant sources of carbon dioxide ($CO_2$) emissions, accounting for approximately 8% of global emissions [23]. About 60% of these emissions result from chemical reactions, whereas the remaining 40% are due to the combustion of fossil fuels during cement production [24]. Attempts to reduce the massive carbon footprint of cement production have given rise to adopting nontraditional stabilisation agents, such as waste materials, lignosulfonate, epoxy resins, polymers, geopolymers, and biopolymers [25–33]. These additives are cost-and-time-effective alternatives to conventional materials.

Biopolymers have many applications in various fields, including water treatment, medicine, paper production, energy, oil exploration, and textile and food processing.

They are produced from renewable resources and are biodegradable and non-toxic, making them suitable for use in geotechnical engineering. In this field, biopolymers have been reported to improve soil properties [22,34–46]. Taytak et al. (2012) [46] reported an increase in the maximum dry density and a reduction in the permeability of the bentonite–kaolin–sand mixture with an increase in the biopolymer concentration. Latifi et al. (2016) [43] investigated the effect of curing time and biopolymer concentration on the shear strength of xanthan-gum-modified organic peat. The curing periods were 3, 7, 28, and 90 days, with biopolymer concentrations of 0.5%, 1.0%, 1.5%, and 2.0%. The cohesion, strength, and friction angle increased significantly with increases in curing time during the first 28 days at an optimum concentration of 2%. Qureshi et al. (2017) [45] investigated the UCS and unconsolidated undrained shear strength of sand modified with xanthan gum (concentrations of 0%, 1%, 2%, 3%, and 5%) at confining pressures of 50, 100, and 150 kPa. This study reported an optimal concentration of 2% to achieve maximum increases in the UCS strength. However, strength decreased at higher concentrations. A study by Fatehi et al. (2019) [38] used sodium alginate in concentrations of up to 5% with different curing times and temperatures. The results indicated a correlation between UCS increases and biopolymer concentration. The strength of the sand modified with a biopolymer concentration of 2% increased rapidly for the first 14 days of curing, with a slight increase observed after 28 days. The strength increased as the curing temperature rose to 45 °C and decreased at temperatures higher than 45 °C after 14 days.

Most geotechnical engineering studies have focused on using xanthan and guar gums to improve soil mechanical properties. This research investigates the effects of acacia, sodium alginate, and pectin on soil properties. Laboratory tests, including unconfined UCS and SEM techniques, used different concentrations, curing periods, and temperatures to provide insights into the mechanical properties of the modified sand.

## 2. Materials and Methods

### 2.1. Soil

The soil used in this study was obtained locally from Thadiq, located approximately 120 km north of Riyadh, Saudi Arabia. Several soil specimens were collected at depths ranging from 0–250 mm. Before use, the soil was air-dried in the laboratory and passed through a sieve with an opening diameter of 1.18 mm to remove the crop shaft, leaves, and sticks. Table 1 presents the basic properties of the sand; it is classified as poorly graded according to the Unified Soil Classification System. Figure 1 shows the particle-size distribution, with 99.2% sand and 0.8% fines.

**Table 1.** Basic properties of the Thadiq sand.

| Specific Gravity, Gs | $e_{min}$ | $e_{max}$ | D50 (mm) | D10 (mm) | Coefficient of Uniformity, Cu | Coefficient of Curvature, Cc | pH |
|---|---|---|---|---|---|---|---|
| 2.66 | 0.51 | 0.76 | 0.15 | 0.09 | 1.99 | 0.88 | 8.33 |

### 2.2. Biopolymers

Experiments used acacia powder, pectin, and sodium alginate mixed with natural Thadiq sand to investigate the effects on the UCS measurements. The biopolymer was selected based on film formation, thickening, gelling capability, ease of use, availability, and sustainability.

Acacia gum, known as gum Arabic, is derived from exudates of mature Acacia senegal or Acacia seyal trees. The major components of this exudate are polysaccharides and glycoproteins [47]. Acacia gum is highly soluble in water and forms solutions with various concentrations, less viscous than solutions formed by other gums. It consists of D-galactose, D-glucuronic acid, arabinose, and rhamnose. The acacia used in this study was obtained from Qualikems Chem Pvt. It has a purity of 79.25%.

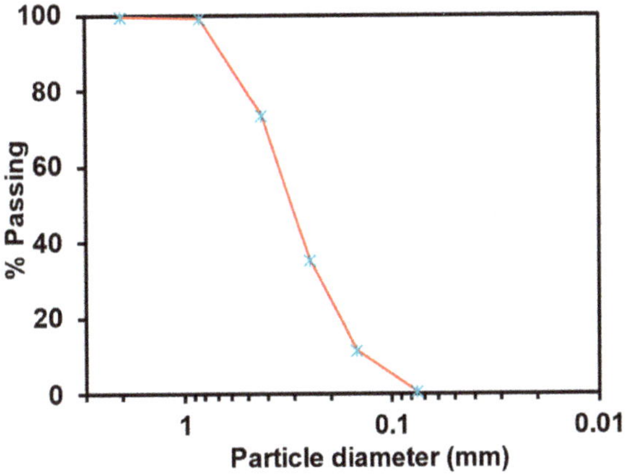


**Figure 1.** Particle-size distribution of the Thadiq sand.

Sodium alginate is a hydrophilic and water-soluble anionic polysaccharide that can be extracted from the cell walls of various species of brown marine algae and the capsular component of bacteria such as Azotobacter vinelandii and Pseudomonas species [48,49]. It is comprised of (1,4)-β-D-mannuronate (M) and α-L-guluronate (G) residues arranged as M blocks, G blocks, and alternating M and G blocks [50]. The sodium alginate used in this study was obtained from Techno Pharmchem. It has a molecular weight of 216.121 g/mol and a purity of 57%.

Pectin is a significant component of the middle lamella of the primary cell wall and is mainly isolated commercially from citrus fruit peels and apple pomace cell walls [51]. Other sources of pectin include apricots, cherries, carrots, grapes, mangoes, pawpaw, and strawberries. It is a polysaccharide that contains significant amounts of galacturonic acid and smaller amounts of L-arabinose, D-galactose, and L-rhamnose. α-(1–4) linkages connect the galacturonic acid residues and form a linear chain. The pectin used in this study was obtained from Acros Organics.

### 2.3. Specimen Preparation

The biopolymers were mixed directly with the required amount of deionised water to form the treatment solutions with concentrations of 0%, 0.5%, 1%, 2%, 3%, and 5%. The slow addition of the biopolymer powder to the deionised water at the target concentration prevented clump formation. A high-speed mechanical stirrer was used to mix the solution for 10–15 min until the mixture became homogeneous. The ratio of the dry biopolymer weight to the total weight of the resulting treatment solution defined the target concentration.

Fabrication of UCS test specimens with a diameter of 49.80 mm and a height of 102.34 mm used a mould. The mixture had a moisture content of 10% and was compacted lightly with 25 blows and then compressed between two end plates using a hydraulic jack to achieve a wet density of 1.64 g/cm$^3$. Specimens underwent different curing conditions before testing.

The specimens used to determine the variation in the moisture of the biopolymer specimen with curing days were prepared in a container that had a diameter of 20 mm and a height of 30 mm. The specimens were compacted to achieve a wet density similar to the UCS. The resulting weight of the specimen was monitored as the curing proceeded over seven days. To ensure the repeatability of the test, three UCS samples were prepared for each test condition under consideration.

Unscathed chunks of the sand specimens collected from the UCS columns were examined via SEM to study the influence of the biopolymer. The piece was placed on a stopper and kept under vacuum conditions for 10 min, followed by the deposition of a platinum coating for 60 s to a thickness of 25 nm to prevent charging under the electron beam during the imaging process.

### 2.4. Unconfined Compressive Strength Tests (UCS)

UCS tests complied with ASTM D2166-06 (2007) [52], which specifies the standard test method for cohesive soil. The specimens were cured between 7 and 56 days at room temperature (25 °C) to examine the effect of biopolymer concentration on the UCS of the modified sand. Tests also investigated the effect of curing time on the UCS using stabilised sand specimens tested after curing periods of 0, 1, 3, 7, 14, 21, 28, and 56 days at room temperature (25 °C). Additionally, to evaluate the effect of the curing temperature (25, 40, 60, 85, and 110 °C) on the UCS, specimens were cured for 7, 14, 28, and 56 days. The specimens were allowed to cool for 15 min before testing. The test was conducted at an applied strain rate of 0.3 mm/min.

### 2.5. Scanning Electron Microscope Imaging

The surface morphology of the modified sand was characterised via field-emission SEM (JEOL; JSM7600F) together with energy-dispersive X-ray spectroscopy. For the SEM, electron beams were utilised to examine the surface morphology in detail. The main signals detected were from backscattered and secondary electrons, which generated a high-resolution image of the specimen.

## 3. Results

### 3.1. Effect of Biopolymer Concentration

Figure 2 presents the strength of Thadiq sand stabilised with the biopolymer at varying concentrations of 0%, 0.5%, 1%, 2%, 3%, and 5% after seven days of curing. The UCS increased significantly with an increase in the biopolymer concentration. There were insufficient bonds between the sand grains without adding the biopolymer (0% to represent the control specimen), resulting in a low compressive strength of 12.43 kPa after seven days. The obtained strength is due to compaction during specimen preparation. Incorporating 0.5% biopolymer concentration into the soil matrix improved the strength to 28.79 kPa, 54.66 kPa, and 92.38 kPa for acacia, sodium alginate, and pectin, respectively. However, the biopolymer concentration of 5.0% increases the compressive strength to 692.38 kPa, 916.68 kPa, and 1771.97 kPa for acacia, sodium alginate, and pectin, respectively. This increase in strength corresponds to a rise of approximately 54-, 71-, and 138-folds, respectively, compared to the control specimen. The results of this research complement the study by Fatehi et al. (2018) [39] and Fatehi et al. (2019) [38] utilising casein and sodium caseinate and sodium biopolymers, respectively. Ayeldeen et al. (2016) [34] made a similar observation with xanthan gum, guar gum, and modified starch on silty soil.

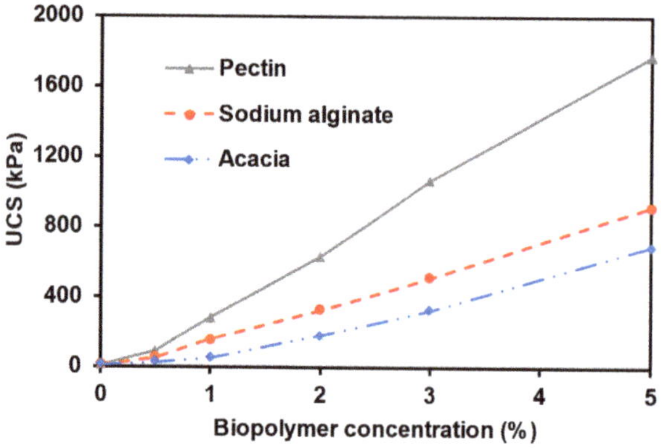


**Figure 2.** Effect of the biopolymer concentration on the UCS after a curing period of seven days.

Remarkable improvement in strength with biopolymer concentration can be attributed to the increases in the viscosity and the binding capability of the biopolymers. Pectin biopolymer yielded the most significant improvement in the compressive strength of the Thadiq sand, followed by sodium alginate and acacia. The low strength of acacia-modified specimens can be attributed to the low viscosity in water compared to the other biopolymers. The low viscosity of acacia biopolymer results from the highly branched galactose/arabinose/rhamnose/glucuronic acid side chains attached to the galactan main chain [53]. The gelling property responsible for the strength of sodium alginate is a function of the ratio of β-D-mannuronate to α-L-guluronate units (M/G) within the structure. A lower M/G ratio corresponds to a stronger and more brittle gel [54]. α-L-guluronate units play a significant role in the gelation process, in contrast to β-D-mannuronate, which comprises alternating guluronate and mannuronate units. The linear sequence of (1,4)-α-D-galacturonic acid residues within the pectin structure forms aggregates held together by secondary valence bonds primarily responsible for the gelation property [55], which is responsible for the pectin strength.

Figure 3 presents the elastic modulus for the biopolymer-modified sand specimens to the concentration of the biopolymer after seven days of curing. The stiffness of the modified sand was determined in the inelastic region of the stress-strain curve. The secant modulus of elasticity, E50, was calculated from the slope of the line connecting 50% of the maximum UCS to the origin. It is defined as minor stress-dependent [56]. Results showed that increased biopolymer concentration increased the modulus of elasticity. Among the biopolymers, acacia powder yielded the least improvement in the compressive strength for all the concentrations tested; however, the stiffness at a concentration of 5% was higher than sodium alginate. Adding 5% pectin resulted in the most significant improvement in compressive strength and stiffness at the end of the 7-day curing period. These results agree with previous research by Ayeldeen et al. (2016) [34], which concluded that the elastic modulus increased as the biopolymer (xanthan gum, guar gum, or modified starch) concentration increased. Figure 4 shows the shear failure results for the control specimen and sand modified with acacia, sodium alginate, and pectin at a concentration of 5%. The failure planes of the control and acacia specimens were vertical, while that of the sodium alginate and pectin specimens were inclined at approximately 60°.

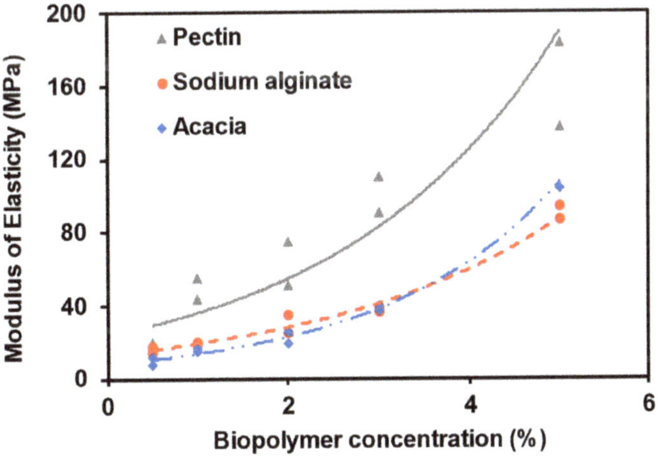


**Figure 3.** Effect of the biopolymer concentration on the stiffness of the modified sand after a curing period of seven days.

**Figure 4.** Failure planes for (**a**) control, (**b**) 5% Acacia, (**c**) 5% Sodium alginate, and (**d**) 5% Pectin, after a curing period of seven days.

### *3.2. Effect of Curing Intervals*

The impact of varying curing time over 0, 1, 3, 7, 14, 21, 28, and 56 days was examined on the UCS of the biopolymer-modified sand at a room temperature of 25 °C. Since the UCS of the modified sand increases continually with the biopolymer concentration, a concentration of 2% was selected to study the effect of the curing time of the UCS. Figure 5 presents the influence of curing time on the UCS of sand modified with acacia, sodium alginate, and pectin biopolymer. When the modified sand was not allowed to cure and was tested immediately after preparation, a low UCS of 9.18 kPa, 12.11 kPa, and 9.66 kPa were obtained for acacia, sodium alginate, and pectin, respectively. The strength of the biopolymer-modified sand increased significantly after one day of curing to 142.31 kPa, 102.32 kPa, and 149.33 kPa for acacia, sodium alginate, and pectin, respectively. The results indicate that the curing time is an essential parameter for the strength development of biopolymer-modified sand.

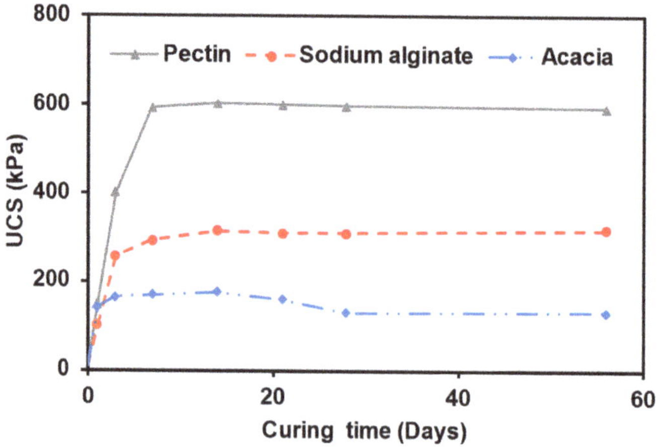


**Figure 5.** Effect of the curing time on the UCS of the modified sand.

The highest rate of strength increases in the biopolymer-modified sands occurs within the first seven days of curing. The significant effect of curing time on the strength development of the modified soil within the first seven days of curing agrees with previous studies by Latifi et al. (2017, 2016) [42,43]. This significant improvement in the strength was attributed to reduced moisture content within the specimens. After one day of curing, the modified specimen was observed to have achieved approximately 80, 32, and 25% of the maximum UCS achievable for acacia, sodium alginate, and pectin, respectively. All the biopolymers utilised reached maximum UCS gain after a curing period of 14 days. However, the UCS of specimens modified with acacia declined by 9% after 21 days and 8% after a curing period of 28 days. After that, the UCS maintained its value between 28 to 56 days of curing. However, the UCS of sodium alginate- and pectin-modified specimens remained relatively constant over the 14–56 days of curing.

As shown in Figure 6, the loss of moisture from the biopolymer-modified specimen played a crucial role in the rate of strength development. The rate of moisture loss was observed to be proportional to the increase in the UCS. The sodium alginate- and pectin-modified sand retained more moisture longer than the acacia-modified sand after one day of curing. The amount of water lost after one day of curing was 75, 47, and 41% for acacia, sodium alginate, and pectin, respectively. Furthermore, after three days of curing, the modified specimen was observed to have lost 96, 83, and 84%, respectively. The modified specimens were observed to lose all the moisture after seven days of curing completely. Furthermore, for both sodium alginate and pectin specimens, it was observed that the UCS decreased as the curing day was increased from 7 to 56 days, irrespective of the curing temperature.

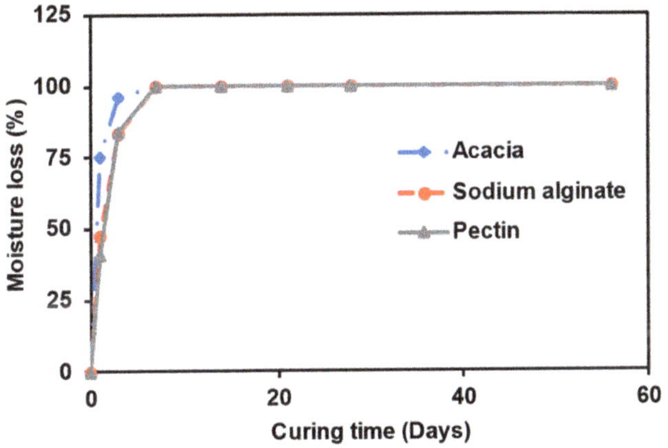


**Figure 6.** Variation of moisture loss with curing days for 2% biopolymer-modified sand.

### 3.3. *Effect of Curing Temperature on UCS*

Figure 7 presents the results of the impact of curing temperature on the UCS of specimens modified with 2% biopolymer after curing periods of 7, 14, 28, and 56 days. Curing temperatures varied as 25, 40, 50, 85, and 110 °C. The strength development pattern of the modified specimen varied with the curing temperature. The UCS of the acacia-modified sand exhibited an erratic pattern as the curing temperature was increased from 25 to 110 °C. However, it is fair to conclude that increasing the curing temperature from 25 to 110 °C did not have a significant impact on the UCS compared to the other biopolymers. The UCS of the sodium alginate-modified sand increased as the curing temperature rose to 85 °C. A further increase in the temperature to 110 °C resulted in a significant reduction in the UCS. This strength reduction can be attributed to the thermal decomposition of the sodium alginate—biopolymer bond between adjacent soil grains. The thermal decomposition resulted in a breakdown of the linear polymer α-L-guluronic (G) acid residue responsible for gelation within the structure. The thermal decomposition of sodium alginate is due to dehydration at 103 °C, the destruction of glycosidic bonds at 212 °C, and the conversion of alginate to carbonaceous residue [57]. This finding contradicts the finding reported by Fatehi et al. (2019) [38], where UCS was reported to increase with curing temperatures up to 45 °C with further increase in the temperature resulting in a reduction in UCS. This discrepancy in results may result from the type of sodium alginate used. In contrast, the UCS of the pectin-modified sand decreased significantly as the curing temperature was increased from 25 to 110 °C. The reduction in the UCS can mainly be ascribed to the breakdown of primary chemical valence linkages between the galacturonic acid polymeric units of the pectin biopolymer network within the soil matrix.

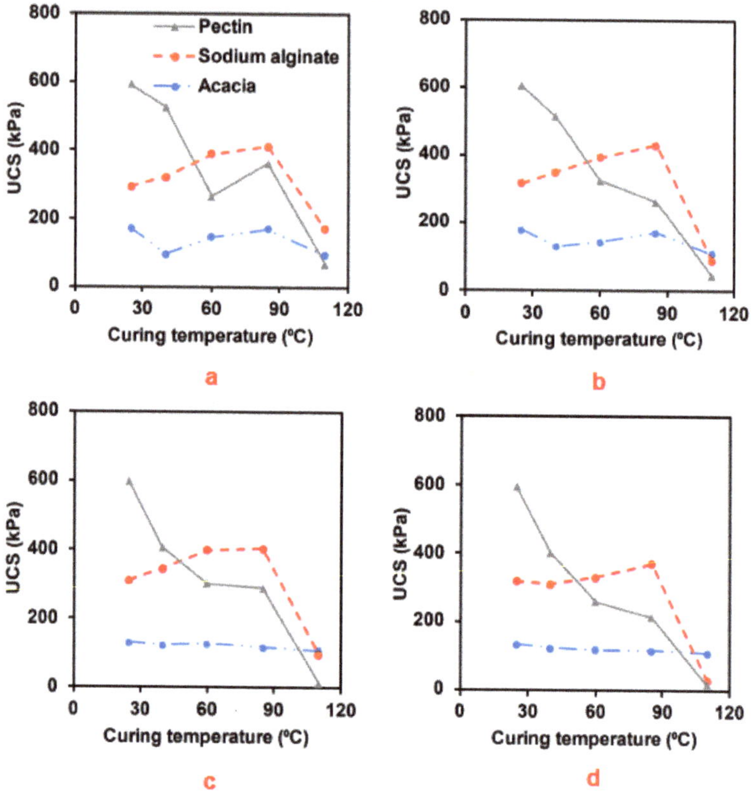


**Figure 7.** Effect of the curing temperature on the UCS of 2% biopolymer-modified sand at (**a**) 7 days, (**b**) 14 days, (**c**) 28 days, and (**d**) 56 days.

For instance, at 56 days, the UCS of the acacia-modified specimen decreased by 8%, 11%, 13%, and 17%; the pectin-modified specimen decreased by 33%, 56%, 64%, and 97%, and the sodium-alginate modified specimen increased by −3, 4, 16, and −91% as the curing temperature was varied from 25 °C to 40 °C, 60 °C, 85 °C, and 110 °C, respectively.

### 3.4. Effect of Moisture Loss on UCS

Moisture loss significantly resulted in the strength development of the biopolymer-modified sand. Therefore, additional analyses were conducted to further examine the impact of preventing moisture loss with curing days on the strength. Figure 8 presents the effect of no moisture loss on the UCS of sand treated with 2% biopolymer concentration and cured for 0, 7, 4, 28, and 56 days. After seven days of curing, the UCS was 8.16 kPa, 12.17 kPa, and 9.02 kPa for the sand modified with acacia, sodium alginate, and pectin, respectively. However, when the specimens were allowed to lose moisture, the UCS rose to 171.46 kPa, 293.57 kPa, and 592.62 kPa, respectively. As the curing time increased from 0 to 56 days, the strength of the acacia-modified specimens increased significantly to 74 kPa, while the other two biopolymer-modified specimens both maintained a low strength of approximately 10 kPa, respectively. The acacia-modified specimen lost much moisture over the 56-day curing period. The moisture loss was faster than the other biopolymer-modified specimens owing to their low viscosity. Thus, moisture loss significantly affects the UCS of biopolymer-modified sand.

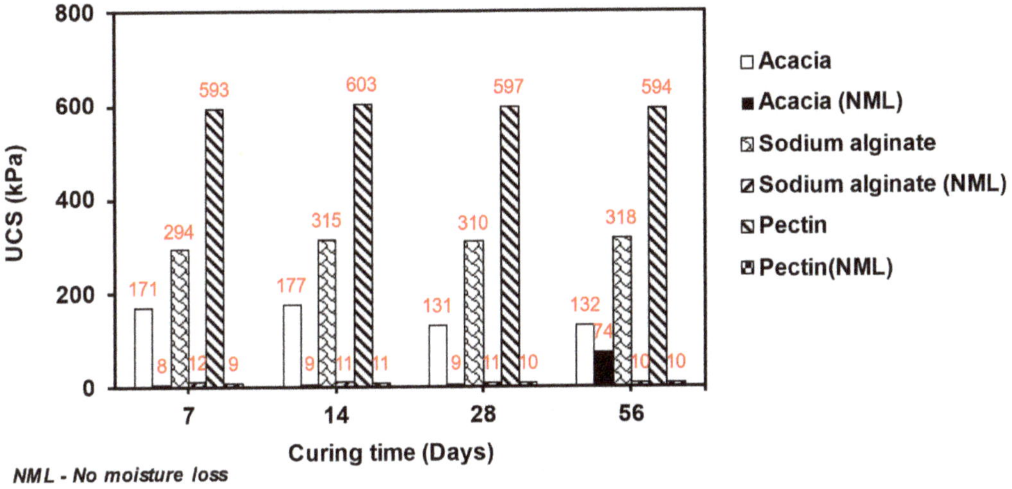


**Figure 8.** Effect of moisture loss on the UCS of the 2% biopolymer-modified sand.

Furthermore, an in-depth investigation was conducted to monitor the moisture loss from the biopolymer-modified specimens cured under laboratory conditions. The result is illustrated in Figure 9. The time required by the biopolymer-modified specimen to reach complete dehydration depend on the type and concentration of the biopolymer used. The acacia-modified specimens were observed to achieve complete dehydration after three days of curing across all biopolymer concentrations. However, sodium alginate-modified specimens with a concentration of 0.5% and 1% required three days, 2% took four curing days, and those with 3% and 5% took five curing days to complete dehydration. The pectin-modified specimens required 3, 5, and 6 curing days at a biopolymer concentration of 0.5%, 1–3%, and 5%, respectively.

### 3.5. Microstructure Scaling Analysis

Figure 10a–d present the micrographs of Thadiq sand in its natural state and a biopolymer-modified specimen after UCS testing. Images of the natural sand display dispersed grains of sand with noticeable pockets of voids and pores, as shown in Figure 10a. The soil particles had no significant link, resulting in a low UCS. Figure 10b–d shows the micrographs of the specimens modified with a biopolymer concentration of 2.0%. Accumulation of the biopolymer between the sand particles was observed, which was responsible for the binding effect within the matrix of the modified sand. Thus, the UCS of the modified sand is dependent on the strength of the biopolymer linkages or the binding force of the sand grains. Generally, the biopolymer stabilization's reinforcement mechanism depends on moisture loss from the biopolymer-sand admixture. The mixture of biopolymer with water results in the formation of a hydrogel absorbed onto the surface of the sand particles via hydrogen and intermolecular bond. The hydrogel becomes a thin and hardened membrane network as the moisture evaporates, thus increasing strength and elasticity modulus. As a result, with more biopolymer concentration, the polymer membrane gets harder, which results in more stiffness.

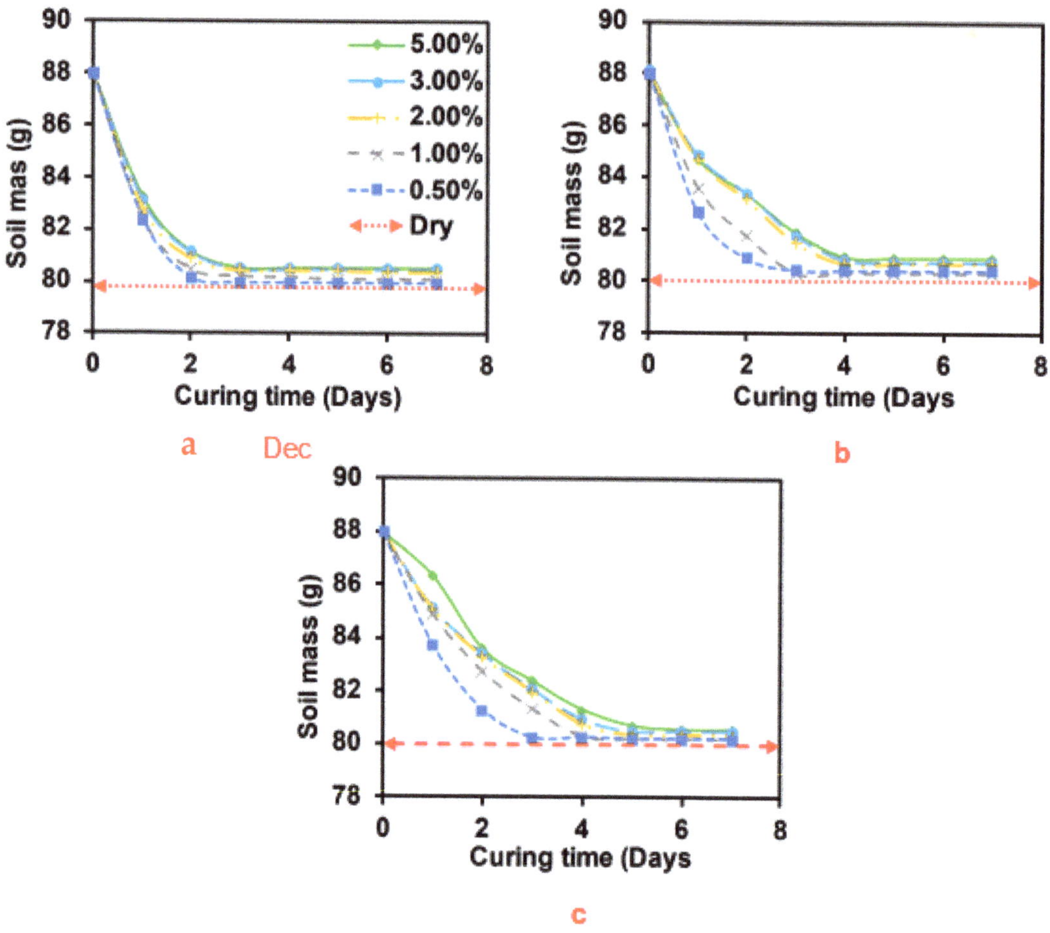


**Figure 9.** Variation of moisture content of soil specimen modified with 2% concentration of (**a**) Acacia, (**b**) Sodium alginate, and (**c**) Pectin.

Figure 11a–c presents micrographs of the biopolymer-modified sand subjected to an elevated temperature of 110 °C for 56 days after UCS testing. As shown in Figure 11a, the acacia biopolymer links between the sand particles were not significantly affected by the temperature. This signifies why there were no significant changes in the UCS values obtained after curing at 7, 14, 28 and 56 days. However, Figure 11b,c show the scanty sight of sodium alginate and pectin biopolymer films observed on the surface and within the pore spaces of the sand grains. The links between the sand grains were observed to be decomposed after exposure to a temperature of 110 °C, leaving thin biopolymer film residues around the sand grains, which resulted in a reduction in the UCS compared to the specimens cured at a room temperature of 25 °C.

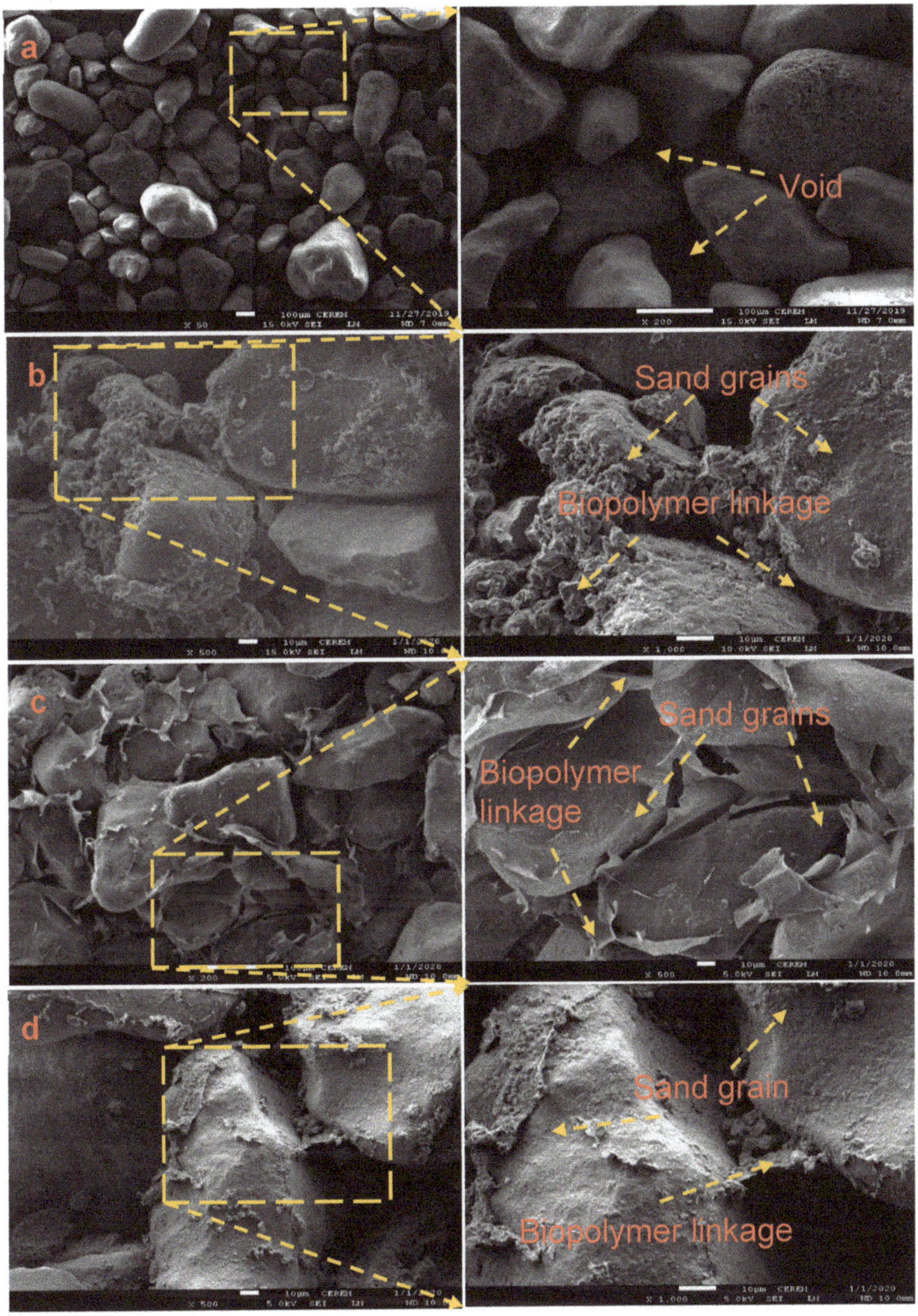


**Figure 10.** SEM images of (**a**) natural sand; (**b**) 2% Acacia-modified; (**c**) 2% Sodium alginate-modified; (**d**) 2% Pectin-modified.

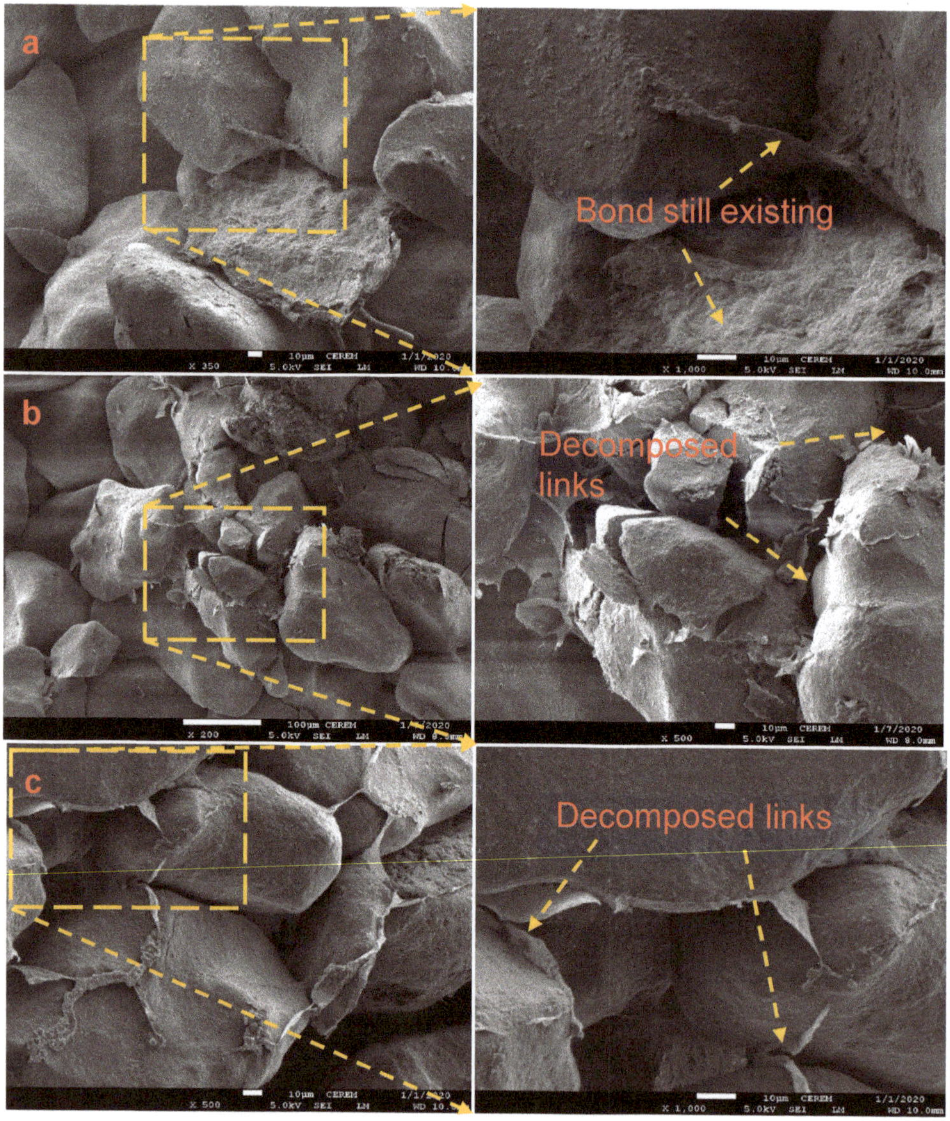


**Figure 11.** SEM images of modified specimens after curing at 110 °C for 2% (**a**) Acacia, (**b**) Sodium alginate, and (**c**) Pectin.

## 4. Conclusions

This research compared the effects of acacia, sodium alginate, and pectin on the soil properties of Thadiq dune sand. A control specimen and three biopolymer-modified sand specimens were tested at different conditions, including five concentrations, curing periods, and temperature, to examine these effects on the UCS and shear wave velocity. Salient conclusions are below:

- Incorporating the biopolymers within the sand matrix improved the strength and stiffness of the sand. An increase in the biopolymer concentration increased the strength and stiffness.

- Pectin yielded more significant improvements in the strength and stiffness of the sand compared to the acacia and sodium alginate biopolymers.
- The increased strength of the modified sand significantly improved within the first seven days of curing compared with the more extended curing period from 7 to 56 days, in which the increase in strength was minor.
- Moisture loss over time significantly affected the strength of the biopolymer-modified sand. UCS results for specimens protected against moisture loss indicated that the strength remained relatively constant as the curing time increased from 0 to 56 days, except for the acacia-modified sand, whose strength increased significantly after 56 days.
- The temperature significantly affected the strength of the specimens. The acacia-modified sand exhibited an irregular strength development pattern, with optimum results obtained at room temperature. The optimum curing temperature for the sodium alginate-modified sand was 85 °C; higher temperatures had detrimental effects on strength.
- SEM images indicated that incorporating the biopolymer within the sand matrix led to the formation of a bond between the soil particles, which enhanced the strength and stiffness of the Thadiq sand. The SEM images indicated that cementation occurred even when the acacia-modified sand was subjected to a high temperature of 110 °C. In contrast, the films on the surface and within the grains were mostly decomposed for the pectin-modified sand.
- The durability of biopolymers is a major concern. Hence, further studies should be conducted on repeated wet-dry cycles and other environmental conditions.

**Author Contributions:** Conceptualisation, K.L. and A.A.; methodology, K.L.; formal analysis, K.L.; investigation, K.L.; resources, K.L. and A.A.; data curation, K.L.; writing—original draft preparation, K.L.; writing—review and editing, K.L. and A.A.; supervision, A.A.; project administration, A.A.; funding acquisition, A.A. All authors have read and agreed to the published version of the manuscript.

**Funding:** Researchers Supporting Project number RSP-2023/279, King Saud University, Riyadh, Saudi Arabia.

**Institutional Review Board Statement:** Not applicable.

**Informed Consent Statement:** Not applicable.

**Data Availability Statement:** All data related to this manuscript is available upon request.

**Acknowledgments:** The authors gratefully acknowledge the Researchers Supporting Project number RSP-2023/279, King Saud University, Riyadh, Saudi Arabia for their financial support for the research work reported in this article.

**Conflicts of Interest:** The authors declare no conflict of interest. The funders had no role in the design of the study; in the collection, analyses, or interpretation of data; in the writing of the manuscript; or in the decision to publish the results.

## References

1. Harkes, M.P.; Van Paassen, L.A.; Booster, J.L.; Whiffin, V.S.; van Loosdrecht, M.C. Fixation and distribution of bacterial activity in sand to induce carbonate precipitation for ground reinforcement. *Ecol. Eng.* **2010**, *36*, 112–117. [CrossRef]
2. Al-Aghbari, M.Y.; Mohamedzein, Y.-A.; Taha, R. Stabilisation of desert sands using cement and cement dust. *Proc. Inst. Civ. Eng.-Ground Improv.* **2009**, *162*, 145–151. [CrossRef]
3. Al-Homidy, A.A.; Al-Amoudi, O.; Maslehuddin, M.; Saleh, T.A. Stabilisation of dune sand using electric arc furnace dust. *Int. J. Pavement Eng.* **2017**, *18*, 513–520. [CrossRef]
4. Al-Khafaji, R.; Dulaimi, A.; Jafer, H.; Mashaan, N.S.; Qaidi, S.; Obaid, Z.S.; Jwaida, Z. Stabilization of Soft Soil by a Sustainable Binder Comprises Ground Granulated Blast Slag (GGBS) and Cement Kiln Dust (CKD). *Recycling* **2023**, *8*, 10. [CrossRef]
5. Bahar, R.; Benazzoug, M.; Kenai, S. Performance of compacted cement-stabilised soil. *Cem. Concr. Compos.* **2004**, *26*, 811–820. [CrossRef]
6. Chenari, R.J.; Fatahi, B.; Ghorbani, A.; Alamoti, M.N. Evaluation of strength properties of cement stabilized sand mixed with EPS beads and fly ash. *Geomech. Eng.* **2018**, *14*, 533–544.

7. Cristelo, N.; Cunha, V.M.; Gomes, A.T.; Araújo, N.; Miranda, T.; de Lurdes Lopes, M. Influence of fibre reinforcement on the post-cracking behaviour of a cement-stabilised sandy-clay subjected to indirect tensile stress. *Constr. Build. Mater.* **2017**, *138*, 163–173. [CrossRef]
8. Cristelo, N.; Glendinning, S.; Miranda, T.; Oliveira, D.; Silva, R. Soil stabilisation using alkaline activation of fly ash for self compacting rammed earth construction. *Constr. Build. Mater.* **2012**, *36*, 727–735. [CrossRef]
9. Hossain, K.M.A.; Mol, L. Some engineering properties of stabilized clayey soils incorporating natural pozzolans and industrial wastes. *Constr. Build. Mater.* **2011**, *25*, 3495–3501. [CrossRef]
10. Jan, O.Q.; Mir, B.A. Strength behaviour of cement stabilised dredged soil. *Int. J. Geosynth. Ground Eng.* **2018**, *4*, 1–14. [CrossRef]
11. Kaniraj, S.R.; Havanagi, V.G. Behavior of cement-stabilized fiber-reinforced fly ash-soil mixtures. *J. Geotech. Geoenviron. Eng.* **2001**, *127*, 574–584. [CrossRef]
12. Kogbara, R.B.; Al-Tabbaa, A. Mechanical and leaching behaviour of slag-cement and lime-activated slag stabilised/solidified contaminated soil. *Sci. Total Environ.* **2011**, *409*, 2325–2335. [CrossRef] [PubMed]
13. Mashizi, M.N.; Bagheripour, M.H.; Jafari, M.M.; Yaghoubi, E. Mechanical and Microstructural Properties of a Stabilized Sand Using Geopolymer Made of Wastes and a Natural Pozzolan. *Sustainability* **2023**, *15*, 2966. [CrossRef]
14. Obuzor, G.N.; Kinuthia, J.M.; Robinson, R.B. Soil stabilisation with lime-activated-GGBS—A mitigation to flooding effects on road structural layers/embankments constructed on floodplains. *Eng. Geol.* **2012**, *151*, 112–119. [CrossRef]
15. Onishi, K.; Tsukamoto, Y.; Saito, R.; Chiyoda, T. Strength and small-strain modulus of lightweight geomaterials: Cement-stabilised sand mixed with compressible expanded polystyrene beads. *Geosynth. Int.* **2010**, *17*, 380–388. [CrossRef]
16. Pathak, A.K.; Pandey, V.; Murari, K.; Singh, J.P. Soil stabilisation using ground granulated blast furnace slag. *Int. J. Eng. Res. Appl.* **2014**, *4*, 164–171.
17. Shalabi, F.I.; Mazher, J.; Khan, K.; Alsuliman, M.; Almustafa, I.; Mahmoud, W.; Alomran, N. Cement-stabilized waste sand as sustainable construction materials for foundations and highway roads. *Materials* **2019**, *12*, 600. [CrossRef]
18. Shooshpasha, I.; Shirvani, R.A. Effect of cement stabilization on geotechnical properties of sandy soils. *Geomech. Eng.* **2015**, *8*, 17–31. [CrossRef]
19. Shuja, D.; Rollakanti, C.R.; Poloju, K.K.; Joe, A. An experimental investigation on -stabilization of sabkha soils with cement and Cement Kiln Dust (CKD) in Sultanate of Oman. *Mater. Today Proc.* **2022**, *65*, 1033–1039. [CrossRef]
20. White, D.J.; Harrington, D.; Ceylan, H.; Rupnow, T. *Fly Ash Soil Stabilization for Non-Uniform Subgrade Soils, Volume I: Engineering Properties and Construction Guidelines*; Iowa Department of Transportation, Highway Division: Ames, IA, USA, 2005.
21. Yadu, L.; Tripathi, R.K. Effects of granulated blast furnace slag in the engineering behaviour of stabilized soft soil. *Procedia Eng.* **2013**, *51*, 125–131. [CrossRef]
22. Khatami, H.R.; O'Kelly, B.C. Improving mechanical properties of sand using biopolymers. *J. Geotech. Geoenviron. Eng.* **2013**, *139*, 1402–1406. [CrossRef]
23. Andrew, R.M. Global $CO_2$ emissions from cement production, 1928–2018. *Earth Syst. Sci. Data* **2019**, *11*, 1675–1710. [CrossRef]
24. Ishak, S.A.; Hashim, H.J. Low carbon measures for cement plant–a review. *Clean. Prod.* **2015**, *103*, 260–274. [CrossRef]
25. Ameta, N.K.; Wayal, A.S.; Hiranandani, P. Stabilization of dune sand with ceramic tile waste as admixture. *Am. J. Eng. Res.* **2013**, *2*, 133–139.
26. Consoli, C.N.; Daniel, W.; Batista, L.H. Durability, Strength, and Stiffness of Green Stabilized Sand. *Am. Soc. Civ. Eng.* **2018**, *144*, 04018057. [CrossRef]
27. Devaraj, V.; Mangottiri, V.; Balu, S. Sustainable utilization of industrial wastes in controlled low-strength materials: A review. *Environ. Sci. Pollut. Res.* **2023**, *30*, 14008–14028. [CrossRef]
28. Georgiannou, V.N.; Pavlopoulou, E.-M.; Bikos, Z. Mechanical behaviour of sand stabilised with colloidal silica. *Geotech. Res.* **2017**, *4*, 1–11. [CrossRef]
29. Liu, J.; Bai, Y.; Song, Z.; Lu, Y.; Qian, W.; Kanungo, D.P. Mechanical behaviour of sand stabilised with colloidal silica. *Polymers* **2018**, *10*, 287. [CrossRef]
30. Naeini, S.A.; Ghorbanali, M.J. Effect of wet and dry conditions on strength of silty sand soils stabilized with epoxy resin polymer. *Appl. Sci.* **2010**, *10*, 2839–2846. [CrossRef]
31. Rahgozar, M.A.; Saberian, M. Geotechnical properties of peat soil stabilised with shredded waste tyre chips. *Mires Peat* **2016**, *18*, 1–12. [CrossRef]
32. Santoni, R.L.; Tingle, J.S.; Nieves, M. Accelerated strength improvement of silty sand with nontraditional additives. *Transp. Res. Rec.* **2005**, *1936*, 34–42. [CrossRef]
33. Seda, J.H.; Lee, J.C.; Carraro, J.A.H. Beneficial Use of Waste Tire Rubber for Swelling Potential Mitigation in Expansive Soils. In *Soil Improvement*; American Society of Civil Engineers: Reston, VA, USA, 2012; pp. 1–9.
34. Ayeldeen, M.K.; Negm, A.M.; El Sawwaf, M.A. Evaluating the physical characteristics of biopolymer/soil mixtures. *Arab. J. Geosci.* **2016**, *9*, 371. [CrossRef]
35. Chang, I.; Cho, G.-C. Geotechnical behavior of a beta-1,3/1,6-glucan biopolymer-treated residual soil. *Geomech. Eng.* **2014**, *7*, 633–647. [CrossRef]
36. Chang, I.; Cho, G.-C. Strengthening of Korean residual soil with β-1,3/1,6-glucan biopolymer. *Constr. Build. Mater.* **2012**, *30*, 30–35. [CrossRef]

37. Chen, R.; Ramey, D.; Weiland, E.; Lee, I.; Zhang, L.J. Experimental Investigation on Biopolymer Strengthening of Mine Tailings. *Geotech. Geoenviron. Eng.* **2016**, *142*, 06016017. [CrossRef]
38. Fatehi, H.; Bahmani, M.; Noorzad, A. Strengthening of Dune Sand with Sodium Alginate Biopolymer. In *Geo-Congress 2019: Soil Improvement*; American Society of Civil Engineers: Reston, VA, USA, 2019; pp. 157–166.
39. Fatehi, H.; Abtahi, S.M.; Hashemolhosseini, H.; Hejazi, S.M. A novel study on using protein based biopolymers in soil strengthening. *Constr. Build. Mater.* **2018**, *167*, 813–821. [CrossRef]
40. Hataf, N.; Ghadir, P.; Ranjbar, N.J. Investigation of soil stabilization using chitosan biopolymer. *Clean. Prod.* **2018**, *170*, 1493–1500. [CrossRef]
41. Khaleghi, M.; Heidarvand, M.J. A novel study on hydro-mechanical characteristics of biopolymer-stabilized dune sand. *Clean. Prod.* **2023**, *398*, 136518. [CrossRef]
42. Latifi, N.; Horpibulsuk, S.; Meehan, C.L.; Abd Majid, M.Z.; Tahir, M.M.; Mohamad, E.T. Improvement of Problematic Soils with Biopolymer—An Environmentally Friendly Soil Stabilizer. *J. Mater. Civ. Eng.* **2017**, *29*, 04016204. [CrossRef]
43. Latifi, N.; Meehan, C.; Abd Majid, M.Z.; Rashid, A.S.A. Xanthan gum biopolymer: An eco-friendly additive for stabilization of tropical organic peat. *Environ. Earth Sci.* **2016**, *75*, 1–10. [CrossRef]
44. Muguda, S.; Booth, S.J.; Hughes, P.N.; Augarde, C.E.; Perlot, C.; Bruno, A.W.; Gallipoli, D. Mechanical properties of biopolymer-stabilised soil-based construction materials. *Géotech. Lett.* **2017**, *7*, 309–314. [CrossRef]
45. Qureshi, M.; Chang, I.; AlSadarani, K. Strength and durability characteristics of biopolymer-treated desert sand. *Geomech. Eng.* **2017**, *12*, 785–801. [CrossRef]
46. Taytak, B.; Pulat, H.F.; Yukselen-Aksoy, Y. Improvement of engineering properties of soils by biopolymer additives. Near East University, Nicosia, North Cyprus, June 2012.
47. Goodrum, L.J.; Patel, A.; Leykam, J.F.; Kieliszewski, M.J. Gum arabic glycoprotein contains glycomodules of both extensin and arabinogalactan-glycoproteins. *Phytochemistry* **2000**, *54*, 99–106. [CrossRef] [PubMed]
48. Remminghorst, U.; Rehm, B.H.A. Bacterial alginates: From biosynthesis to applications. *Biotechnol. Lett.* **2006**, *28*, 1701–1712. [CrossRef]
49. Zhang, H.; Cheng, J.; Ao, Q. Preparation of alginate-based biomaterials and their applications in biomedicine. *Mar. Drugs* **2021**, *19*, 264. [CrossRef]
50. Aarstad, O.A.; Tøndervik, A.; Sletta, H.; Skjåk-Bræk, G. Alginate Sequencing: An Analysis of Block Distribution in Alginates Using Specific Alginate Degrading Enzymes. *Biomacromolecules* **2012**, *13*, 106–116. [CrossRef]
51. Nesic, A.R.; Seslija, S.I. The influence of nanofillers on physical–chemical properties of polysaccharide-based film intended for food packaging. In *Food Packaging*; Elsevier: Amsterdam, The Netherlands, 2017; pp. 637–697.
52. *ASTM D2166-06*; Standard Test Method for Unconfined Compressive Strength of Cohesive Soil. ASTM International: Conshohocken, PA, USA, 2007.
53. Williams, P.A.; Phillips, G.O. *Gums and Stabilisers for the Food Industry 9*; Elsevier: Amsterdam, The Netherlands, 1998.
54. Nishinari, K.; Doi, E. *Food Hydrocolloids: Structures, Properties, and Functions*; Springer Science & Business Media: Berlin, Germany, 2012.
55. Telis, V.R.N. *Biopolymer Engineering in Food Processing*; CRC Press: Boca Raton, FL, USA, 2012.
56. Obrzud, R.F.; Truty, A. *The Hardening Soil Model—A Practical Guidebook*; Zace Services Ltd.: Preverenges, Switzerland, 2018.
57. Flores-Hernández, C.G.; de los Cornejo-Villegas, M.A.; Moreno-Martell, A.; Del Real, A. Synthesis of a biodegradable polymer of poly (sodium alginate/ethyl acrylate). *Polymers* **2021**, *13*, 504. [CrossRef]

*polymers*

*Article*

# Study of the Plasticization Effect of 1-Ethyl-3-methylimidazolium Acetate in TPS/PVA Biodegradable Blends Produced by Melt-Mixing

**Jennifer M. Castro [1,2,*], Mercedes G. Montalbán [3], Daniel Domene-López [2], Ignacio Martín-Gullón [1,2] and Juan C. García-Quesada [1,2,*]**

1 Chemical Engineering Department, University of Alicante, Apartado 99, 03080 Alicante, Spain
2 Institute of Chemical Process Engineering, University of Alicante, Apartado 99, 03080 Alicante, Spain
3 Chemical Engineering Department, Faculty of Chemistry, Regional Campus of International Excellence "Campus Mare Nostrum", University of Murcia, 30071 Murcia, Spain
* Correspondence: jennifer.martinez@ua.es (J.M.C.); jc.garcia@ua.es (J.C.G.-Q.)

**Abstract:** The first step towards the production and marketing of bioplastics based on renewable and sustainable materials is to know their behavior at a semi-industrial scale. For this reason, in this work, the properties of thermoplastic starch (TPS)/polyvinyl alcohol (PVA) films plasticized by a green solvent, as the 1-ethyl-3-methylimidazolium acetate ($[Emim^+][Ac^-]$) ionic liquid, produced by melt-mixing were studied. These blends were prepared with a different content of $[Emim^+][Ac^-]$ (27.5–42.5 %wt.) as a unique plasticizer. According to the results, this ionic liquid is an excellent plasticizer due to the transformation of the crystalline structure of the starch to an amorphous state, the increase in flexibility, and the drop in $T_g$, as the $[Emim^+][Ac^-]$ amount increases. These findings show that the properties of these biomaterials could be modified in the function of $[Emim^+][Ac^-]$ content in the formulations of TPS, depending on their final use, thus becoming a functional alternative to conventional polymers.

**Keywords:** thermoplastic starch/polyvinyl alcohol blend; ionic liquid; 1-ethyl-3-methylimidazolium acetate; melt-mixing; plasticizer

**Citation:** Castro, J.M.; Montalbán, M.G.; Domene-López, D.; Martín-Gullón, I.; García-Quesada, J.C. Study of the Plasticization Effect of 1-Ethyl-3-methylimidazolium Acetate in TPS/PVA Biodegradable Blends Produced by Melt-Mixing. *Polymers* **2023**, *15*, 1788. https://doi.org/10.3390/polym15071788

Academic Editors: Vsevolod Aleksandrovich Zhuikov and Rosane Michele Duarte Soares

Received: 3 February 2023
Revised: 27 March 2023
Accepted: 2 April 2023
Published: 4 April 2023



## 1. Introduction

Ionic liquids (ILs) are gaining interest in the field of plastic production because they have the potential to act as plasticizers in polymer matrices [1] due to properties such as high thermal and chemical stability, non-flammability, low vapor pressure, tunable solubility, and good hydraulic properties [2–7]. In addition, they can be found as liquids in a wide range of temperatures because they are organic salts with melting points below 100 °C, formed by a bulky inorganic cation (with a low degree of symmetry) and as an anion resulting in a little compact structure with large and non-uniform ions [8,9]. Some authors have previously reported the use of ILs as plasticizers of different conventional polymers. Scott et al. [10] studied the properties of poly(methyl methacrylate) (PMMA) plasticized by the IL 1-butyl-3-methylimidazolium hexafluorophosphate and concluded that the material was comparable with those that contain conventional plasticizers, noting that the low volatile IL created a more flexible polymer over a long lifetime which prevented the material from brittleness. Similar conclusions were obtained by Rahman et al. [9], who reported that the addition of a variety of ILs to different PVC formulations improved mechanical properties (increasing the flexibility) at the same time that films had a longer life with a decrease of plasticizer migration.

In industry, ILs are commonly used as catalysts for the production and storage [11] of energy, or as an extracting agent for organic compounds [12]. ILs can be considered "green" solvents because they are environmentally friendly, with low explosion risk, and with easily recovered and recycled procedures [13,14]. Therefore, ILs can play an interesting role in the new environmental context, where research oriented toward the development of new materials from renewable and sustainable sources is experiencing significant growth. In this sense, the plastic industry is one of the most affected sectors and exchanging conventional fossil-based polymers with renewable polymers is a solution for some up-to-date problems due to the enormous consumption of "limited" resources, the problems related to plastic waste mismanagement [15–18], and their harmful effects on the environment [19–21].

In the past, some researchers have investigated the effect of ILs as a solvent in the research of biodegradable and renewable materials. However, recently, ILs have been used to study their plasticizer effect in biodegradable materials [22–31], with starch being one of the most promising materials due to its abundance, low cost, and biodegradability [15]. Starch is a biopolymer composed of two polysaccharides: amylose and amylopectin [32,33]; and the presence of plasticizers is needed in order for it to be transformed into thermoplastic starch (TPS) [34] by a gelatinization process [35]. There are a number of studies on the use of ILs as plasticizers in biodegradable starch blends. Wilpiszewska et al. [8] concluded that the presence of ILs on TPS could decrease the glass transition temperature more than when using glycerol, suggesting the better plasticization efficiency of ILs in comparison to polyols which are conventional plasticizers for these biomaterials [36–39]. These conclusions were supported by Xie et al. [40], Domene-López et al. [41], and Abera et al. [42], who studied the behavior of films prepared with maize, potato, and anchote starch, respectively, using 1-ethyl-3-methylimidazolium acetate as a plasticizer by a solution casting method in the presence of water.

The use of 1-ethyl-3-methylimidazolium acetate ($[Emim^+][Ac^-]$), among other ILs, as a TPS plasticizer is interesting because it has a more biodegradable character and is less toxic than other ILs due to its carboxylic anion [43,44]. Furthermore, it is capable of forming chemical bonds with $CO_2$ [45]. Chen et al. [46] studied the process of atmospheric water sorption and $CO_2$ capture in a system of $[Emim^+][Ac^-]$ and biopolymers (cellulose, chitin, and chitosan). Their findings indicate that these biomaterials could be employed for food packaging and storage applications. In our research group, we have studied the prospective use of $[Emim^+][Ac^-]$ with different botanical sources of starch, and have concluded that the starch characteristics are crucial in order to obtain good blend properties and the potato blend values were especially very promising [47]. Moreover, the incorporation of multi-walled carbon nanotubes (MWCNT) to the starch blends plasticized by $[Emim^+][Ac^-]$ showed great electrical conductivity, proving that these types of biomaterials could be used in a wide range of applications, such as lithium batteries, fuel cells, or dye-sensitized solar cells [26,41].

The use of the casting method is common in laboratory-scale TPS research because this allows for the study of the starch films plasticized under ideal gelatinization conditions, and promoted by an excess of plasticizer. Nevertheless, for the purposes of industrial research, these materials should be processed in semi-industrial scale equipment, considering real conditions, and without an excess of plasticizer. For example, Sankri et al. [25] prepared TPS from maize starch plasticized by water, glycerol, and 1-butyl-3-methylimidazolium chloride using melt processing and they confirmed that the IL had strong plasticized effects through the mechanical property values achieved. Decaen et al. [48] studied the rheology properties of TPS from maize starch plasticized by three ILs in the presence of water by melt rheology at 120 °C. They concluded that the $[Emim^+][Ac^-]$ is a promising plasticizer due to the lowest shear viscosity and limited macromolecular degradation.

TPS is usually formulated with water since it is considered the best starch plasticizer, but its use has several limitations such as the evaporation of the water during the process (temperatures above 100 °C are needed in the industrial process), and the interaction between the water and the ILs acetate anions can limit the plasticization effect of the IL studied [49,50]. TPS pure films are brittle due to the strong chain interactions of the starch polymer via hydrogen bonding which increases the cohesion forces of the matrix [51]. Therefore, the use of other adjuvant polymers such as polyvinyl alcohol (PVA) [52] is a good option to confer resistance to starch-based materials. PVA is currently employed in the food, medical, and hygiene sectors, among others [15]. PVA is a synthetic polymer with a high degree of biocompatibility, solubility in water, chemical resistance, biodegradability, and other notable physical properties by its hydroxyl groups, which promotes the formation of hydrogen bonds [53] and, in addition, this thermoplastic has remarkably superior features as an oxygen barrier, compared with the other biopolymers [54].

Thus, the aim of this work is the study of the plasticization effect of $[Emim^+][Ac^-]$ in thermoplastic starch/PVA blends by melt-mixing using semi-industrial scale equipment. Samples have been prepared without water to prevent interaction with the ILs. Mechanical and hydration properties, migration degree of plasticizer, dynamic mechanical analysis, morphology, and XRD and ATR-FTIR spectra of the TPS films are discussed to determine the potential capability of these materials to replace synthetic and non-biodegradable polymers. To the best of our knowledge, no studies can be found in the literature about starch/PVA mixtures plasticized by a single plasticizer consisting in IL, without an additional water supply, and produced by melt-mixing.

## 2. Materials and Methods

### *2.1. Materials*

The potato starch was provided by Across Organics (Geel, Belgium) and characterized in our previous works [47,55]. PVA hydrolyzed (Mw: 89,000–98,000 Da) was purchased from Sigma-Aldrich (Madrid, Spain). The ionic liquid 1-ethyl-3-methylimidazolium acetate ($[Emim^+][Ac^-]$) (>95% purity) was supplied by loLiTec-Ionic Liquids Technologies GmbH (Heilbronn, Germany). Zinc stearate, used as a lubricant, was also provided by Sigma-Aldrich (Madrid, Spain). All chemicals were used without further purification.

### *2.2. Preparation of the Starch/PVA Films Using the Ionic Liquid as Plasticized*

Starch films were prepared by the method of starch melt-compounding previously described by Domene-Lopéz et al. [15], with some modifications. The processing was carried out with the HAAKE TM PolyLab TM QC Modular Torque Rheometer (ThermoFisher Scientific, Waltham, MA, USA). The samples were processed at 130 °C for 14 min; the first 8 min at 50 rpm at the rest of the time at 100 rpm. The blend obtained was then hot pressed at 160 °C for 15 min at a pressure of 10 tons obtaining a 1 mm thick sheet. Samples were stored in a controlled atmosphere of relative humidity of 50% for 48 h prior to their further characterization, making sure to avoid the TPS properties being changed by the humidity effect [56].

Starch, PVA, and ionic liquid were manually premixed at room temperature for 3 min and then introduced in the HAAKE rheometer. The weight ratio between the potato starch and PVA in the sample was 1:1, adding different amounts of ionic liquid as plasticizer: 27.5, 30, 32.5, 37.5, 40, and 42.5 %wt. of ionic liquid; the studied formulations were labeled as IL_27.5, IL_30, IL_32.5, IL_37.5, IL_40, and IL_42.5, respectively, and 0.5 %wt. of zinc stearate was also added to all blends.

### *2.3. Characterization of the Films*

#### 2.3.1. Scanning Electron Microscopy (SEM) Analysis

SEM images were obtained with a Hitachi Scanning Electron Microscope (Hitachi S3000N, Tokyo, Japan) using an accelerating voltage of 10 kV. To observe the microstructure and morphology, the samples were cryo-fractured by immersion in liquid nitrogen and cooled with gold.

#### 2.3.2. Mechanical Properties

Mechanical properties of the films were determined with an Instron 3344 Universal Test Instrument (Norwood, MA, USA) equipped with a 2000 N load cell and operated at 25 mm/min following ASTM D882-12(2012) [57] standard recommendations. Each film was cut into dumbbell-shaped specimens. The tensile properties studied were Young's Modulus (MPa), tensile strength at break (MPa), and elongation at break (%), which were calculated using the average thickness of the specimen and at least eight specimens were tested.

#### 2.3.3. Hydration Properties

The hydration properties were determined following the method previously described in the literature [15,41,47,55,58]. The films were cut into 1 × 1 $cm^2$ specimens. On the one hand, the water content was determined by measuring the lost weight after drying the samples in an oven at 110 °C for 5 h. The moisture content was calculated using Equation (1) and was expressed as a percentage (grams of water in 100 g of sample):

$$H(\%) = \left(\frac{m_0 - m_1}{m_0}\right) \times 100 \quad (1)$$

where $m_0$ is the initial mass and $m_1$ is the mass after drying.

On the other hand, the solubility in the samples of water was determined by placing the films individually in 10 mL tubes filled with 9 mL of distilled water, which were capped and stored at 25 °C for 24 h. After that, the samples were taken out and dried again at 110 °C for 5 h. Solubility in water was calculated using Equation (2):

$$Solubility\ (\%) = \left(\frac{m_0 - m_f}{m_0}\right) \times 100 \quad (2)$$

where $m_f$ is the final dry mass.

The water content and solubility values were taken as the average of at least five repetitions.

#### 2.3.4. Migration Degree

The degree of migration of plasticizer to the surface of films is determined as the weight loss of the sample after the migration test, which was carried out following a modified procedure based on methods described by Marcilla et al. [59] and Rahman et al. [9]. The films were cut into small circles with dimensions of 7 mm in diameter and 1 mm thickness, which were placed between two Petri dishes with absorbent papers and kept under a pressure of 16.5 kPa in an oven at 60 °C for 1 week. The values of plasticizer migration were obtained by taking out the samples at different times (1, 2, 5, and 7 days), and then calculating the weight loss using Equation (3):

$$Migration\ Degree\ (\%) = \left(\frac{m_0 - m_f}{m_0}\right) \times 100 \quad (3)$$

where $m_0$ and $m_f$ are the initial and final weights of the sample.

The results of the migration of plasticized were obtained from an average of at least three repetitions.

#### 2.3.5. Attenuated Total Reflection-Fourier Transform Infrared (ATR-FTIR) Spectroscopy

A Bruker Spectrometer (IFS 66/S model, Ettlingen, Germany) with an ATR accessory was used to obtain the ATR-FTIR spectra of films between 400 and 4000 $cm^{-1}$. ATR-FTIR spectra were processed using the FITYK software (1.3.1 version) and the ratio of band intensities at 995 and 1022 $cm^{-1}$ was determined to study the film's molecular rearrangement.

#### 2.3.6. X-ray Diffraction (XRD) Studies

A Bruker diffractometer (D8-Advance model, Ettlingen, Germany) was used to obtain the diffractograms of the films. This equipment has a KRISTALLOFLEX K 76080F X-Ray generator (power = 3000 W, voltage = 20–60 kV, intensity = 5–80 mA), that has an X-Ray tube with a copper anode. The dimensions of specimens were 1 × 1 $cm^2$ and the equipment was operated at 40 kV and 40 mA with 2θ varying from 4 to 50° with a step size of 0.05°.

### 2.3.7. Dynamic Mechanical Analysis (DMA)

The DMA analysis was carried out in a DMA 1 Instrument (Mettler-Toledo, Barcelona, Spain). The dimensions of the samples used for this analysis were 8.5 × 25 × 1 mm and the temperature range was −100 °C to 60 °C at a rate of 3 °C/min and a constant frequency of (1 Hz).

## 3. Results and Discussion

The visual appearance of the films is shown in Figure 1. As can be seen, all are homogeneous and transparent. However, the higher the [Emim$^+$][Ac$^-$] percentage, the higher the transparency of the samples. In addition, they show a yellowish color attributed to the color of the [Emim$^+$][Ac$^-$].

**Figure 1.** The visual aspect of starch/PVA blends plasticized by [Emim$^+$][Ac$^-$].

### 3.1. *Scanning Electron Microscopy (SEM) Analysis*

SEM images of the surface of the samples are shown in Figure 2. These provide useful information about the homogeneity, morphology, etc. of the films [55], features that are closely related to the final mechanical properties of the material [34].

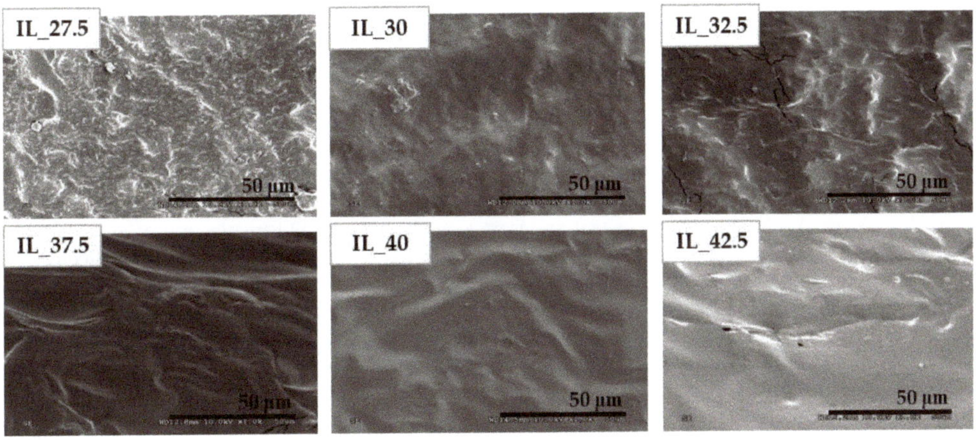


**Figure 2.** SEM images of the cryogenic fracture surface of sheets (magnification 1000×).

SEM micrographs showed a heterogeneous matrix with some cracks and some readily visible starch granules that had not been dissolved during mixing and hot pressing at IL concentrations of up to 32.5%. At higher concentrations, starch is clearly destructured, dissolved, and incorporated into a continuous phase that exhibits a smooth surface in SEM images, especially in samples with 40 and 42.5% of IL.

### 3.2. Mechanical Properties

The mechanical properties of starch/PVA with IL as a plasticizer are shown in Figure 3. From these properties, it is possible to know the potential use of the films in future applications [47,60].

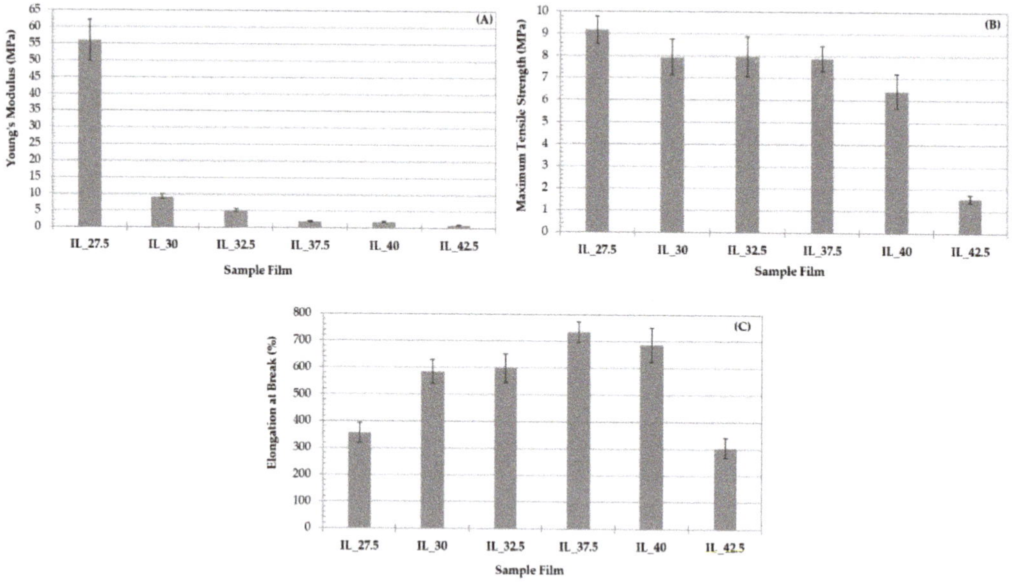


**Figure 3.** Mechanical properties of the films studied. (**A**) Young's modulus; (**B**) Maximum Tensile Strength; and (**C**) Elongation at break.

According to Figure 3, the values are heavily dependent on the content of IL; the results of Young's modulus vary from 56.1 MPa to 0.911 MPa, and the maximum tensile strength from 9.17 MPa to 1.56 MPa. In both cases, the maximum and minimum values correspond to the IL_27.5 and IL_42.5 samples, respectively. Elongation at break ranged from 735% to 304% at IL concentrations of 37.5 and 42.5%, respectively.

Results obtained reveal a sharp drop in Young's modulus when IL concentration is around 30%. At higher concentrations, Young's modulus progressively decreases coinciding with a raise in elongation at break which indicates a clear plasticizing effect of the IL. Nevertheless, concentrations up to approximately 40% involve a clear excess of plasticizer, since above this value tensile strength and elongation at break markedly decrease. In fact, at these concentrations, the material starts to become a gel, extremely soft and sticky.

In order to understand this behavior, it is important to bear in mind the effect of IL in starch. As reported in the literature, ILs are good starch solvents, but at the same time they could provoke a certain degree of depolymerization, and, more specifically, those including the ion chloride. The use of acetate does not prevent this depolymerization process since it seems to occur, even to a low extent, at concentrations of around 50% of [$Emim^+$][$Ac^-$] [61].

Thus, the results shown in Figure 3 could be the result of the contribution of several effects. On the one hand, as the IL content increases, starch is progressively dissolved and the blend becomes more ductile [40] as expected by the presence of the plasticizer:

tensile strength drops while elongation at break increases. Above 37.5% of IL content, the complete gelatinization of the material for having an excess of plasticizer contributes to a weak interaction between the starch chains which is responsible for the drop in mechanical properties.

### *3.3. Hydration Properties*

The hydration properties are the water content and the solubility in water, which are shown in Table 1. These properties are limiting and important factors of the film to determine its use in the future [62].

**Table 1.** Hydration properties of the sheets studied.

| Sample | Water Content (%) | Solubility in Water (%) |
|---|---|---|
| IL_27.5 | 9.41 ± 0.25 | 35.2 ± 0.2 |
| IL_30 | 9.30 ± 0.26 | 37.0 ± 0.4 |
| IL_32.5 | 9.93 ± 0.29 | 40.8 ± 0.2 |
| IL_37.5 | 9.97 ± 0.16 | 49.3 ± 1.2 |
| IL_40 | 11.4 ± 0.2 | 50.8 ± 1.0 |
| IL_42.5 | 14.9 ± 0.9 | 57.6 ± 1.5 |

On the one hand, as can be observed in the table, the values obtained of moisture of films with $[Emim^+][Ac^-]$ as a plasticizer show a plateau with a value close to 10% of moisture content, but drastically increases at a 40% content of IL reaching a value of moisture around 15%. On the other hand, solubility values progressively increase with IL concentration, observing solubility changes not only attributable to the IL extraction and dissolution. When IL concentration is 27.5%, around 35% of starch formulation dissolves, but when IL concentration raises 15 units up to 42.5% (i.e., 15% more), the solubility increases to 58%. In general, this behavior could be attributed in part to the hygroscopic nature of the ILs and, in particular, $[Emim^+][Ac^-]$ has a high hygroscopicity and water solubility [63]. In addition, the IL could contribute to higher and stronger water–ion interactions [41,64] and, at the same time, to a certain starch depolymerization [61] which forms monosaccharides and disaccharides at high IL contents [65]. This phenomenon is due to the alkyl chain length of cations, meaning the hydrophobicity of the ILs increases as the increase in the alkyl chain length of the cations [2,5].

It is worth mentioning that water solubility is related to the biodegradability of the material; the material's biodegradability should increase with the solubility [58]. However, there are applications where the material has to be insoluble to preserve the integrity of the final product [66].

### *3.4. Migration Degree*

The migration degree of plasticizers is an interesting parameter to bear in mind because it can determine the life of the material since this phenomenon is an undesirable effect that causes difficulties in its commercialization and uses in long-term applications [59]. The migration is due to starch retrogradation or recrystallization [67], which involves changes from the amorphous state of starch to new ordered structures [68]. In addition, the migration process depends on nature, molecular weight, and amount of plasticizer, and this in turn results in the loss of material properties [59] because when it is produced for plasticizer loss in TPS, this causes a decrease of elongation at break and an increase in the elastic modulus [69,70].

The values of IL migration are shown in Figure 4. These results follow a clear trend since the migration of IL increases as the plasticizer amount increases. Hence, a small content of $[Emim^+][Ac^-]$ as a plasticizer in the starch/PVA blends helps to reduce this effect (IL_42.5 and IL_27.5 have 11.8% and 7.46% of migration degree of plasticizer, respectively). It must be taken into account that the superficial moisture of the samples has probably been eliminated during this test. However, this fact does not change the results or the trend obtained from the migration test. Analyzing these, the plasticizer migration mainly occurs in the early 24 h of the test.

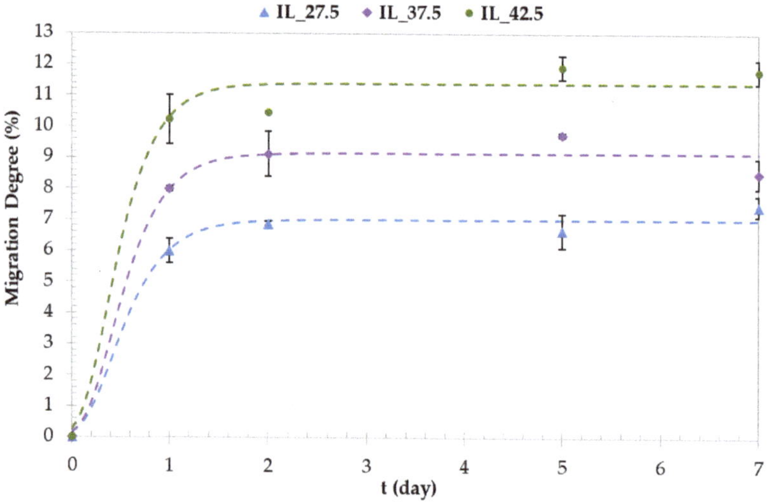


**Figure 4.** Migration degree as a function of time and [Emim$^+$][Ac$^-$] content of starch/PVA blends.

As far as we know, there are no scientific findings in the literature on the migration of starch/PVA films plasticized by any IL. Nevertheless, Rahman et al. [9] studied the migration in samples of PVC plasticized by different ILs and concluded that the PVC with traditional plasticizers had a greater weight loss than those that use ILs as a plasticizer, which is due to the bulky structure of some ILs that hinders their migration to the surface of the material. They also carried out the study of plasticizer loss due to leaching. These can be considered similar methods and their obtained results are comparable because the aim in both cases is to quantify the plasticizer loss; the water is used as an extractant in the leaching experiment, while migration samples are subjected to a determinate pressure. The same trend was observed in the percentage of plasticizer loss to that shown in Figure 4 and they confirmed that using IL as a plasticizer reduced the leaching.

Therefore, we could conclude that the problem of plasticizer migration to the surface of the films is reduced when using a small percentage of certain ILs as plasticizers in starch/PVA blends.

### 3.5. Attenuated Total Reflection-Fourier Transform Infrared (ATR-FTIR) Spectroscopy

Figure 5 shows the ATR-FTIR spectra of the samples. This technique reflects the changes in molecular interactions in the starch/PVA films with [Emim$^+$][Ac$^-$] as a plasticizer. As can be seen, the spectra are similar and, as expected, show the corresponding peaks to IL and starch.

First of all, it is worth mentioning a wide band between 3000–3600 cm$^{-1}$ where two different bands overlap: on the one hand, the stretching vibration of the –OH groups of the glucose units and water [60] and the vibration C–H of the imidazolium ring in [Emim$^+$][Ac$^-$] [71].

Other bands shared by both components (starch and [Emim$^+$][Ac$^-$]) are localized between 2800–3000 cm$^{-1}$, and this is attributed to asymmetric and symmetric stretching of –$CH_2$ and –$CH_3$, and aliphatic asymmetric C–H stretching vibrations [47].

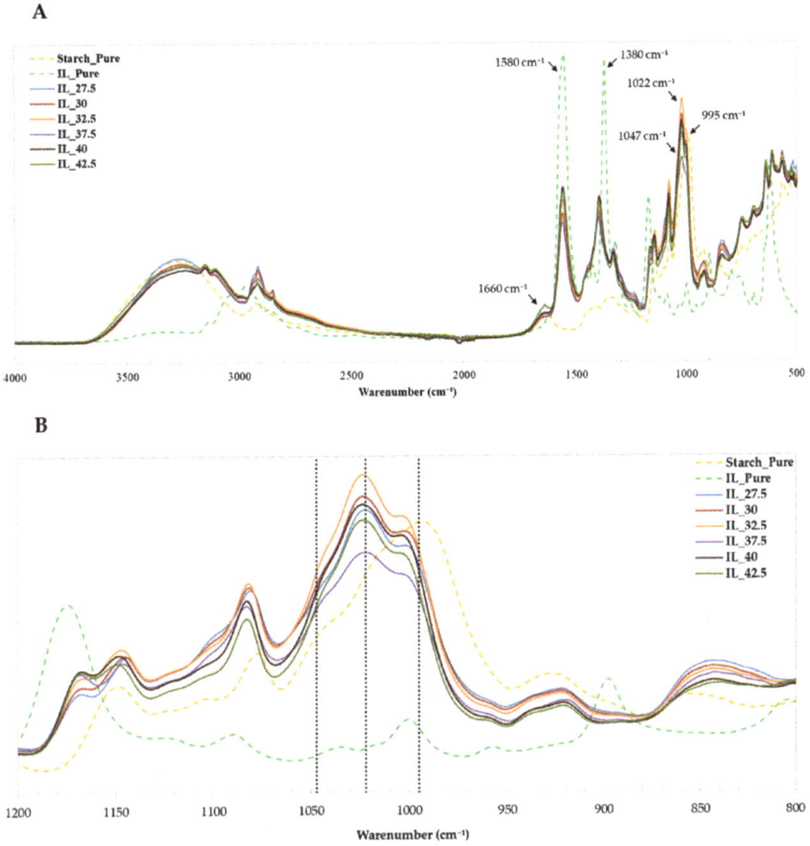


**Figure 5.** ATR-FTIR spectra of starch/PVA blends plasticized by [Emim$^+$][Ac$^-$]. (**A**) 500–4000 cm$^{-1}$ range; (**B**) 800–1200 cm$^{-1}$ range.

The infrared spectrum of the pure IL reveals two peaks at 1380 and 1580 cm$^{-1}$ corresponding to symmetric and asymmetric O–C–O stretches of [Ac$^-$] anion of the IL, respectively [72]. However, in Figure 5, those peaks are shown at 1400 and 1560 cm$^{-1}$, slightly shifted to the position in the pure IL, probably due to the result of the hydrogen bonding between the IL and water molecules (which presents in the starch as moisture) [73]. In addition, the small peak around 1660 cm$^{-1}$ is assigned to water [47] or the –C–H stretching of the vinyl group [74].

The band at 1148–925 cm$^{-1}$ shows the stretching of –C–O– situated in –C–O–H and –C–O–C– bonds of the glucose of starch [47]. More concretely, the peaks at around 995 cm$^{-1}$, 1022 cm$^{-1}$, and 1047 cm$^{-1}$ are related to the degree of the rearrangements of the starch after the process of plasticization. While the band at 995 cm$^{-1}$ corresponds to the hydrogen bonds and the regularity given by molecular rearrangements, the peak at 1022 cm$^{-1}$ is associated with the amorphous phase of the films, and a decrease in crystallinity [26,33,50,74]; the degree of order in starch can be quantified in consequence by the ratio of the areas of the bands at 995 and 1022 cm$^{-1}$ [47].

The results are shown in Table 2. According to the ratio values calculated, it can be seen that $R_{995/1022}$ decreases when increasing the IL content (IL_27.5 and IL_42.5 have values of 0.824 and 0.536, respectively), indicating the ability of this IL to hinder molecular rearrangements after plasticization. These results are in good agreement with the XRD and DMA, as shown below.

**Table 2.** ATR-FTIR band intensity ratio ($R_{995/1022}$) of the starch/PVA blends obtained.

| | Films Sample | | | | | | |
|---|---|---|---|---|---|---|---|
| | **Starch_Pure** | **IL_27.5** | **IL_30** | **IL_32.5** | **IL_37.5** | **IL_40** | **IL_42.5** |
| $R_{995/1022}$ | 1.39 | 0.824 | 0.673 | 0.659 | 0.637 | 0.613 | 0.536 |

### *3.6. X-ray Diffraction (XRD) Studies*

The XRD spectra of the starch/PVA blends plasticized by [Emim$^+$][Ac$^-$] are shown in Figure 6. This analysis allows the study of the crystalline structure of films. All samples showed a similar pattern, where it is possible to observe peaks at around 20, 23, and 40° with different intensities; the peak at 2θ = 20° is less prominent as [Emim$^+$][Ac$^-$] content increases, so a loss of crystallinity is observed. This fact is due to the plasticization process since when the IL content increases the disruption of hydrogen bonds between the starch molecules [75,76], this increases the mobility of the starch chains and contributes to the creation of a more amorphous structure, according to the conclusions of Domene-López et al. [47,55] and Luchese et al. [32], who also studied the native starch and the starch films using [Emim$^+$][Ac$^-$] as a plasticizer, obtaining similar results. Xie et al. [40] concluded that using IL as a plasticizer contributed to a decrease in crystallinity, in comparison with glycerol, because the [Emim$^+$][Ac$^-$] decreases the B-type crystallinity and the V-type crystallinity, hindering the formation of the single-helical structure due to interaction between the hydroxyl groups of starch and acetate anion in [Emim$^+$][Ac$^-$].

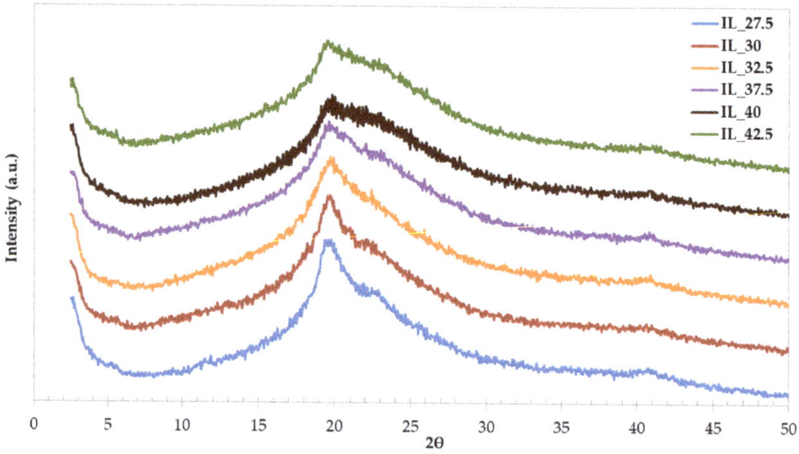


**Figure 6.** XRD spectra of the starch/PVA blends plasticized by [Emim$^+$][Ac$^-$].

It is worth mentioning that various authors such as Zhang et al. [26], Bhagabati et al. [77], and Domene-López et al. [47] state that the transparency of the final material is related to its crystallinity; more amorphous polymeric films had higher transparency due to the disruption of starch grains, and this effect was observed in the obtained films (see Figure 1); when increasing the IL content, the final films lose their crystalline structure, but they are more transparent.

### *3.7. Dynamic Mechanical Analysis (DMA)*

The DMA data of the starch/PVA[Emim$^+$][Ac$^-$] are shown in Figure 7. The curves are similar, showing a peak corresponding to the glass transition of the system [15,78]. As can be seen, the glass transition of samples is shifted towards lower temperatures when IL content increases; for example, the $T_g$ of IL_27.5 was −4 °C while the $T_g$ of IL_42.5 was −30 °C. Jaramillo et al. [58] concluded that this was associated with the plasticization process. In addition, the peak is more prominent as the IL content increases, in good agreement with Xie et al. [40]. Thus, a higher content of IL resulted in a more amorphous structure, hence with greater mobility. This loss of crystallinity is also reflected in ATR-FTIR and XRD spectra, as commented above. These data are consistent with the mechanical properties since a decrease in the glass transition temperature with IL content produces less stiff samples with lower Young's modulus (Figure 3A).

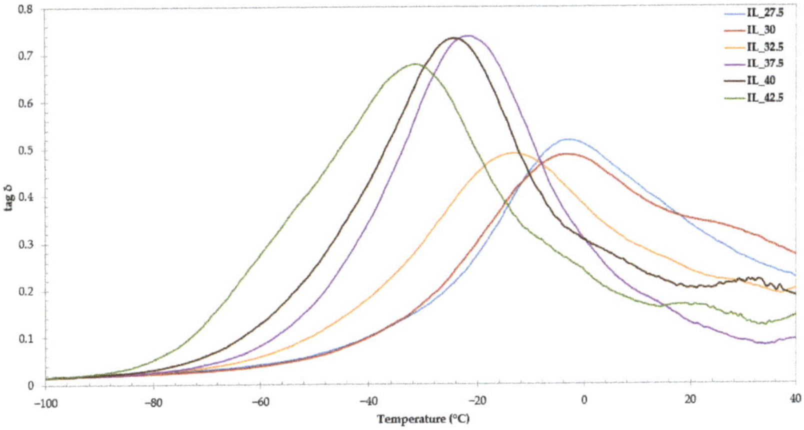


**Figure 7.** DMA spectra of the studied sheets. Dependence of loss tangent (tan δ) on the temperature at a constant frequency of 1 Hz.

It should be kept in mind that moisture content (favored by the presence of IL) could also affect glass transition, as Madrigal et al. [78] reported, showing a relation between the moisture contents and the value of tan δ; they concluded that the curves showed a more prominent peak as the water content increased. As is possible to observe in Table 1, the highest results of moisture correspond to the samples with more prominent peaks (IL_37.5, IL_40, and IL_42.5).

It is worth mentioning that there is not much literature on the DMA of starch films plasticized by ILs. However, the DMA analysis of the TPS plasticized by glycerol among other plasticizers has been extensively studied in the literature [79–82] and all the curves of DMA showed two peaks (one attributed to a phase rich in plasticizer and the other in starch) while that in the [$Emim^+$][$Ac^-$] plasticized-starch films appear in only one transition. As mentioned above, water and conventional starch plasticizers do not bring about a full starch dissolution and gelation; it could be the reason for the presence of two different glass transitions readily observable in glycerol-starch compounds. In the case of the IL-starch, they seem to be fully dissolved and it could be the reason for the presence of one single glass transition.

## 4. Conclusions

In the present work and, for the first time, starch/PVA films have been plasticized by a single plasticizer being an ionic liquid by melt-mixing.

According to the results obtained, it is possible to conclude that the 1-ethyl-3-methylimidazolium acetate ([$Emim^+$][$Ac^-$]) is an efficient plasticizer to obtain these biomaterials. After the analysis of the results, it has been possible to prove that this IL provokes the disruption of hydrogen bonds between the starch molecules, changing its crystalline structure to an amorphous state, which is possible to observe in the XRD spectra. This was also confirmed by the SEM images, which show that the presence of IL involves the dissolution of the starch granules, even when the [$Emim^+$][$Ac^-$] content in the films was low. Furthermore, a clear trend has been found in accordance with the IL amount added to the blends that the mechanical properties were significantly changed; when there are more crystalline domains than amorphous, the samples were stiffer (increasing Young's modulus and maximum tensile strength). By contrast, the samples showed more flexibility when the amorphous phase was predominant, due to the increase in chain movement. [$Emim^+$][$Ac^-$] is a good starch plasticizer as revealed by the drop in $T_g$ when the IL content increases; the IL_27.5 sample has a glass transition temperature at −3.17 °C, while the IL_42.5 sample has −31.4 °C, finding a difference of 28.2 °C between both. The same occurs in the diffusion and hydration properties, which are higher as the IL content increases; the values of IL_27.5 and IL_42.5 for the migration degree of plasticizer and solubility in water are 7.46–11.8% and 35.2–57.6%, respectively.

For all these reasons, the [$Emim^+$][$Ac^-$] could be considered an interesting plasticizer to obtain biodegradable polymeric blends from starch, constituting an alternative to other conventional plasticizers such as water or polyols. The results achieved through a technique prior to industrial development, such as melt-mixing, show the high potential of these materials to be a real replacement to conventional polymer in a wide range of applications such as packaging and provide the possibility to be used in applications which previously had no place in this type of biomaterial, such as in dye-sensitized solar cells, fuel cells, among others.

**Author Contributions:** J.M.C.: methodology, procedures, formal analysis, writing—original draft; M.G.M.: conceptualization, formal analysis, supervision, writing; D.D.-L.: conceptualization, formal analysis, supervision, writing; I.M.-G.: project administration, funding acquisition, supervision; J.C.G.-Q.: supervision, formal analysis, writing, project administration. All authors have read and agreed to the published version of the manuscript.

**Funding:** This work was partially supported by the Spanish Ministry of Science and Ministry of Economy, Industry, and competitiveness (PID2019-108632RB-I00) and Generalitat Valenciana (PROMETEO CIPROM/2021/027). This work was partially supported by the TED2021-130389B-C21 research project, funded by MCIN/AEI/10.13039/501100011033 and by the European Union NextGenerationEU/PRTR.

**Institutional Review Board Statement:** Not applicable.

**Informed Consent Statement:** Not applicable.

**Data Availability Statement:** Not applicable.

**Conflicts of Interest:** The authors declare no conflict of interest.

## References

1. Wypych, G. (Ed.) Plasticizers Use and Selection for Specific Polymers. In *Handbook of Plasticizers*; ChemTec Publishing: Scarborough, ON, Canada, 2012; pp. 307–419.
2. Shi, W.; Damodaran, K.; Nulwala, H.B.; Luebke, D.R. Theoretical and Experimental Studies of Water Interaction in Acetate Based Ionic Liquids. *Phys. Chem. Chem. Phys.* **2012**, *14*, 15897–15908. [CrossRef] [PubMed]
3. Ghandi, K. A Review of Ionic Liquids, Their Limits and Applications. *Green Sustain. Chem.* **2014**, *4*, 44–53. [CrossRef]
4. Crowhurst, L.; Mawdsley, P.R.; Perez-Arlandis, J.M.; Salter, P.A.; Welton, T. Solvent-Solute Interactions in Ionic Liquids. *Phys. Chem. Chem. Phys.* **2003**, *5*, 2790–2794. [CrossRef]
5. Ren, F.; Wang, J.; Xie, F.; Zan, K.; Wang, S.; Wang, S. Applications of Ionic Liquids in Starch Chemistry: A Review. *Green Chem.* **2020**, *22*, 2162–2183. [CrossRef]
6. Montalbán, M.G.; Hidalgo, J.M.; Collado-González, M.; Díaz Baños, F.G.; Víllora, G. Assessing Chemical Toxicity of Ionic Liquids on Vibrio Fischeri: Correlation with Structure and Composition. *Chemosphere* **2016**, *155*, 405–414. [CrossRef]
7. Díaz Alvarez, J.C.; Martínez Rey, R.; Barrero Acosta, R. Ionic Liquids: Physicochemical Properties and Potential Application in Upgrading of Heavy Crude Oils. *Rev. Ion* **2012**, *25*, 61–87.
8. Wilpiszewska, K.; Spychaj, T. Ionic Liquids: Media for Starch Dissolution, Plasticization and Modification. *Carbohydr. Polym.* **2011**, *86*, 424–428. [CrossRef]
9. Rahman, M.; Brazel, C.S. Ionic Liquids: New Generation Stable Plasticizers for Poly(Vinyl Chloride). *Polym. Degrad. Stab.* **2006**, *91*, 3371–3382. [CrossRef]
10. Scott, M.P.; Brazel, C.S.; Benton, M.G.; Mays, J.W.; Holbrey, J.D.; Rogers, R.D. Application of Ionic Liquids as Plasticizers for Poly(Methyl Methacrylate). *Chem. Commun.* **2002**, *2*, 1370–1371. [CrossRef]

11. Wong, S.I.; Lin, H.; Sunarso, J.; Wong, B.T.; Jia, B. Optimization of Ionic-Liquid Based Electrolyte Concentration for High-Energy Density Graphene Supercapacitors. *Appl. Mater. Today* **2020**, *18*, 100522. [CrossRef]
12. Montalbán, M.G.; Collado-González, M.; Lozano-Pérez, A.A.; Díaz Baños, F.G.; Víllora, G. Extraction of Organic Compounds Involved in the Kinetic Resolution of Rac-2-Pentanol from n-Hexane by Imidazolium-Based Ionic Liquids: Liquid-Liquid Equilibrium. *J. Mol. Liq.* **2018**, *252*, 445–453. [CrossRef]
13. Rojas, O.G.; Hall, S.R. On the Synergistic Interaction of an Ionic Liquid and Biopolymers in the Synthesis of Strontium Niobate. *Mater. Chem. Phys.* **2017**, *202*, 220–224. [CrossRef]
14. Lucas, M.; MacDonald, B.A.; Wagner, G.L.; Joyce, S.A.; Rector, K.D. Ionic Liquid Pretreatment of Poplar Wood at Room Temperature: Swelling and Incorporation of Nanoparticles. *ACS Appl. Mater. Interfaces* **2010**, *2*, 2198–2205. [CrossRef] [PubMed]
15. Domene-López, D.; Guillén, M.M.; Martin-Gullon, I.; García-Quesada, J.C.; Montalbán, M.G. Study of the Behavior of Biodegradable Starch/Polyvinyl Alcohol/Rosin Blends. *Carbohydr. Polym.* **2018**, *202*, 299–305. [CrossRef] [PubMed]
16. Geyer, R.; Jambeck, J.R.; Law, K.L. Production, Use, and Fate of All Plastics Ever Made. *Sci. Adv.* **2017**, *3*, 25–29. [CrossRef]
17. Kaza, S.; Yao, L.; Bhada-Tata, P.; Van Woerden, F. *What a Waste 2.0: A Global Snapshot of Solid Waste Management to 2050*; World Bank Publications: Washington, DC, USA, 2018.
18. Scaffaro, R.; Citarrella, M.C.; Gulino, E.F.; Morreale, M. Hedysarum Coronarium-Based Green Composites Prepared by Compression Molding and Fused Deposition Modeling. *Materials* **2022**, *15*, 465. [CrossRef]
19. Kumar, M.; Xiong, X.; He, M.; Tsang, D.C.W.; Gupta, J.; Khan, E.; Harrad, S.; Hou, D.; Ok, Y.S.; Bolan, N.S. Microplastics as Pollutants in Agricultural Soils. *Environ. Pollut.* **2020**, *265*, 114980. [CrossRef]
20. Li, W.; Wufuer, R.; Duo, J.; Wang, S.; Luo, Y.; Zhang, D.; Pan, X. Microplastics in Agricultural Soils: Extraction and Characterization after Different Periods of Polythene Film Mulching in an Arid Region. *Sci. Total Environ.* **2020**, *749*, 141420. [CrossRef]
21. Lusher, A.L.; Bråte, I.L.N.; Munno, K.; Hurley, R.R.; Welden, N.A. Is It or Isn't It: The Importance of Visual Classification in Microplastic Characterization. *Appl. Spectrosc.* **2020**, *74*, 1139–1153. [CrossRef] [PubMed]
22. Park, S.; Oh, Y.; Yun, J.; Yoo, E.; Jung, D.; Park, K.S.; Oh, K.K.; Lee, S.H. Characterization of Blended Cellulose/Biopolymer Films Prepared Using Ionic Liquid. *Cellulose* **2020**, *27*, 5101–5119. [CrossRef]
23. Zan, K.; Wang, J.; Ren, F.; Yu, J.; Wang, S.; Xie, F.; Wang, S. Structural Disorganization of Cereal, Tuber and Bean Starches in Aqueous Ionic Liquid at Room Temperature: Role of Starch Granule Surface Structure. *Carbohydr. Polym.* **2021**, *258*, 117677. [CrossRef] [PubMed]
24. Wang, J.; Ren, F.; Yu, J.; Copeland, L.; Wang, S.; Wang, S. Toward a Better Understanding of Different Dissolution Behavior of Starches in Aqueous Ionic Liquids at Room Temperature. *ACS Omega* **2019**, *4*, 11312–11319. [CrossRef] [PubMed]
25. Sankri, A.; Arhaliass, A.; Dez, I.; Gaumont, A.C.; Grohens, Y.; Lourdin, D.; Pillin, I.; Rolland-Sabaté, A.; Leroy, E. Thermoplastic Starch Plasticized by an Ionic Liquid. *Carbohydr. Polym.* **2010**, *82*, 256–263. [CrossRef]
26. Zhang, B.; Xie, F.; Shamshina, J.L.; Rogers, R.D.; McNally, T.; Wang, D.K.; Halley, P.J.; Truss, R.W.; Zhao, S.; Chen, L. Facile Preparation of Starch-Based Electroconductive Films with Ionic Liquid. *ACS Sustain. Chem. Eng.* **2017**, *5*, 5457–5467. [CrossRef]
27. Bendaoud, A.; Chalamet, Y. Plasticizing Effect of Ionic Liquid on Cellulose Acetate Obtained by Melt Processing. *Carbohydr. Polym.* **2014**, *108*, 75–82. [CrossRef]
28. Mahmood, H.; Moniruzzaman, M.; Yusup, S.; Welton, T. Ionic Liquids Assisted Processing of Renewable Resources for the Fabrication of Biodegradable Composite Materials. *Green Chem.* **2017**, *19*, 2051–2075. [CrossRef]
29. Mateyawa, S.; Xie, D.F.; Truss, R.W.; Halley, P.J.; Nicholson, T.M.; Shamshina, J.L.; Rogers, R.D.; Boehm, M.W.; McNally, T. Effect of the Ionic Liquid 1-Ethyl-3-Methylimidazolium Acetate on the Phase Transition of Starch: Dissolution or Gelatinization? *Carbohydr. Polym.* **2013**, *94*, 520–530. [CrossRef]
30. Zdanowicz, M.; Spychaj, T. Ionic Liquids as Starch Plasticizers or Solvents. *Polimery* **2011**, *56*, 861–864. [CrossRef]
31. Yang, X.; Qiao, C.; Li, Y.; Li, T. Dissolution and Resourcfulization of Biopolymers in Ionic Liquids. *React. Funct. Polym.* **2016**, *100*, 181–190. [CrossRef]
32. Luchese, C.L.; Benelli, P.; Spada, J.C.; Tessaro, I.C. Impact of the Starch Source on the Physicochemical Properties and Biodegradability of Different Starch-Based Films. *J. Appl. Polym. Sci.* **2018**, *135*, 46564. [CrossRef]
33. Lopez-Rubio, A.; Flanagan, B.M.; Shrestha, A.K.; Gidley, M.J.; Gilbert, E.P. Molecular Rearrangement of Starch during in vitro Digestion: Toward a Better Understanding of Enzyme Resistant Starch Formation in Processed Starches. *Biomacromolecules* **2008**, *9*, 1951–1958. [CrossRef] [PubMed]
34. Lekube, B.M.; Fahrngruber, B.; Kozich, M.; Wastyn, M.; Burgstaller, C. Influence of Processing on the Mechanical Properties and Morphology of Starch-Based Blends for Film Applications. *J. Appl. Polym. Sci.* **2019**, *136*, 47990. [CrossRef]
35. Valero-Valdivieso, M.F.; Ortegón, Y.; Uscategui, Y. Biopolímeros: Avances y Perspectivas. *Dyna* **2013**, *80*, 171–180.
36. Biliaderis, C.G.; Lazaridou, A.; Arvanitoyannis, I. Glass Transition and Physical Properties of Polyol-Plasticized Pullulan-Starch Blends at Low Moisture. *Carbohydr. Polym.* **1999**, *40*, 29–47. [CrossRef]
37. Talja, R.A.; Helén, H.; Roos, Y.H.; Jouppila, K. Effect of Various Polyols and Polyol Contents on Physical and Mechanical Properties of Potato Starch-Based Films. *Carbohydr. Polym.* **2007**, *67*, 288–295. [CrossRef]
38. Sreekumar, P.A.; Al-Harthi, M.A.; De, S.K. Effect of Glycerol on Thermal and Mechanical Properties of Polyvinyl Alcohol/Starch Blends. *J. Appl. Polym. Sci.* **2012**, *123*, 135–142. [CrossRef]
39. Da Róz, A.L.; Carvalho, A.J.F.; Gandini, A.; Curvelo, A.A.S. The Effect of Plasticizers on Thermoplastic Starch Compositions Obtained by Melt Processing. *Carbohydr. Polym.* **2006**, *63*, 417–424. [CrossRef]

40. Xie, F.; Flanagan, B.M.; Li, M.; Sangwan, P.; Truss, R.W.; Halley, P.J.; Strounina, E.V.; Whittaker, A.K.; Gidley, M.J.; Dean, K.M.; et al. Characteristics of Starch-Based Films Plasticised by Glycerol and by the Ionic Liquid 1-Ethyl-3-Methylimidazolium Acetate: A Comparative Study. *Carbohydr. Polym.* **2014**, *111*, 841–848. [CrossRef]
41. Domene-López, D.; Delgado-Marín, J.J.; García-Quesada, J.C.; Martín-Gullón, I.; Montalbán, M.G. Electroconductive Starch/Multi-Walled Carbon Nanotube Films Plasticized by 1-Ethyl-3-Methylimidazolium Acetate. *Carbohydr. Polym.* **2020**, *229*, 115545. [CrossRef]
42. Abera, G.; Woldeyes, B.; Demash, H.D.; Miyake, G. The Effect of Plasticizers on Thermoplastic Starch Films Developed from the Indigenous Ethiopian Tuber Crop Anchote (*Coccinia abyssinica*) Starch. *Int. J. Biol. Macromol.* **2020**, *155*, 581–587. [CrossRef]
43. Gomez-Coma, L.; Garea, A.; Irabien, A. Carbon Dioxide Capture by [Emim][Ac] Ionic Liquid in a Polysulfone Hollow Fiber Membrane Contactor. *Int. J. Greenh. Gas Control* **2016**, *52*, 401–409. [CrossRef]
44. Ostadjoo, S.; Berton, P.; Shamshina, J.L.; Rogers, R.D. Scaling-up Ionic Liquid-Based Technologies: How Much Do We Care about Their Toxicity? Prima Facie Information on 1-Ethyl-3-Methylimidazolium Acetate. *Toxicol. Sci.* **2018**, *161*, 249–265. [CrossRef] [PubMed]
45. Gómez-Coma, L.; Garea, A.; Irabien, A. Non-Dispersive Absorption of $CO_2$ in [Emim][EtSO4] and [Emim][Ac]: Temperature Influence. *Sep. Purif. Technol.* **2014**, *132*, 120–125. [CrossRef]
46. Chen, Y.; Sun, X.; Yan, C.; Cao, Y.; Mu, T. The Dynamic Process of Atmospheric Water Sorption in [EMIM][Ac] and Mixtures of [EMIM][Ac] with Biopolymers and $CO_2$ Capture in These Systems. *J. Phys. Chem. B* **2014**, *118*, 11523–11536. [CrossRef] [PubMed]
47. Domene-López, D.; Delgado-Marín, J.J.; Martin-Gullon, I.; García-Quesada, J.C.; Montalbán, M.G. Comparative Study on Properties of Starch Films Obtained from Potato, Corn and Wheat Using 1-Ethyl-3-Methylimidazolium Acetate as Plasticizer. *Int. J. Biol. Macromol.* **2019**, *135*, 845–854. [CrossRef]
48. Decaen, P.; Rolland-Sabaté, A.; Colomines, G.; Guilois, S.; Lourdin, D.; Della Valle, G.; Leroy, E. Influence of Ionic Plasticizers on the Processing and Viscosity of Starch Melts. *Carbohydr. Polym.* **2020**, *230*, 115591. [CrossRef]
49. Tan, X.; Li, X.; Chen, L.; Xie, F. Solubility of Starch and Microcrystalline Cellulose in 1-Ethyl-3-Methylimidazolium Acetate Ionic Liquid and Solution Rheological Properties. *Phys. Chem. Chem. Phys.* **2016**, *18*, 27584–27593. [CrossRef]
50. Liu, K.; Tan, X.; Li, X.; Chen, L.; Xie, F. Characterization of Regenerated Starch from 1-Ethyl-3-Methylimidazolium Acetate Ionic Liquid with Different Anti-Solvents. *J. Polym. Sci. Part B Polym. Phys.* **2018**, *56*, 1231–1238. [CrossRef]
51. Gunawardene, O.H.P.; Amaraweera, S.M.; Wannikayaka, W.M.D.B.; Fernando, N.M.L.; Gunathilake, C.A.; Manamperi, W.A.; Kulatunga, A.K.; Manipura, A. Role of Compatibilization of Phthalic Acid in Cassava Starch/Poly Vinyl Alcohol Thin Films. In Proceedings of the 12th International Conference on Structural Engineering and Construction Management: ICSECM 2021, Kandy, Sri Lanka, 2–5 December 2021; Springer Nature: Singapore, 2022; pp. 665–689.
52. Narancic, T.; Cerrone, F.; Beagan, N.; O'Connor, K.E. Recent Advances in Bioplastics: Application and Biodegradation. *Polymers* **2020**, *12*, 920. [CrossRef]
53. Gulino, E.F.; Citarrella, M.C.; Maio, A.; Scaffaro, R. An Innovative Route to Prepare in Situ Graded Crosslinked PVA Graphene Electrospun Mats for Drug Release. *Compos. Part A* **2022**, *155*, 106827. [CrossRef]
54. Gaaz, T.S.; Sulong, A.B.; Akhtar, M.N.; Kadhum, A.A.H.; Mohamad, A.B.; Al-amiery, A.A. Properties and Applications of Polyvinyl Alcohol, Halloysite Nanotubes and Their Nanocomposites. *Molecules* **2015**, *20*, 22833–22847. [CrossRef]
55. Domene-López, D.; García-Quesada, J.C.; Martin-Gullon, I.; Montalbán, M.G. Influence of Starch Composition and Molecular Weight on Physicochemical Properties of Biodegradable Films. *Polymers* **2019**, *11*, 1084. [CrossRef] [PubMed]
56. Eaton, M.D.; Domene-López, D.; Wang, Q.; Montalbán, M.G.; Martin-Gullon, I.; Shull, K.R.; Martin-Gullon, I.; Shull, K.R. Exploring the Effect of Humidity on Thermoplastic Starch Films Using the Quartz Crystal Microbalance. *Carbohydr. Polym.* **2021**, *261*, 117727. [CrossRef] [PubMed]
57. *ASTM D882-12*; Standard Test Method for Tensile Properties of Thin Plastic Sheeting. ASTM International: West Conshohocken, PA, USA, 2012. Available online: https://www.astm.org/DATABASE.CART/HISTORICAL/D882-02.htm (accessed on 28 March 2018).
58. Jaramillo, C.M.; González Seligra, P.; Goyanes, S.; Bernal, C.; Famá, L. Biofilms Based on Cassava Starch Containing Extract of Yerba Mate as Antioxidant and Plasticizer. *Starch/Staerke* **2015**, *67*, 780–789. [CrossRef]
59. Marcilla, A.; García, S.; García-Quesada, J.C. Study of the Migration of PVC Plasticizers. *J. Anal. Appl. Pyrolysis* **2004**, *71*, 457–463. [CrossRef]
60. Dai, L.; Zhang, J.; Cheng, F. Effects of Starches from Different Botanical Sources and Modification Methods on Physicochemical Properties of Starch-Based Edible Films. *Int. J. Biol. Macromol.* **2019**, *132*, 897–905. [CrossRef]
61. Sciarini, L.S.; Rolland-Sabaté, A.; Guilois, S.; Decaen, P.; Leroya, E.; Le Bail, P. Understanding the Destructuration of Starch in Water–Ionic Liquid Mixtures. *Green Chem.* **2015**, *17*, 291–299. [CrossRef]
62. Zhong, Y.; Li, Y.; Liang, W.; Liu, L.; Li, S.; Xue, J.; Guo, D. Comparison of Gelatinization Method, Starch Concentration, and Plasticizer on Physical Properties of High-Amylose Starch Films. *J. Food Process Eng.* **2018**, *41*, 1–8. [CrossRef]
63. Otero-Mato, J.M.; Lesch, V.; Montes-Campos, H.; Smiatek, J.; Diddens, D.; Cabeza, O.; Gallego, L.J.; Varela, L.M. Solvation in Ionic Liquid-Water Mixtures: A Computational Study. *J. Mol. Liq.* **2019**, *292*, 111273. [CrossRef]
64. Ismail, S.; Mansor, N.; Majeed, Z.; Man, Z. Effect of Water and [Emim][OAc] as Plasticizer on Gelatinization of Starch. *Procedia Eng.* **2016**, *148*, 524–529. [CrossRef]

65. Li, X.; Gao, B.; Zhang, S. Adjusting Hydrogen Bond by Lever Principle to Achieve High Performance Starch-Based Biodegradable Films with Low Migration Quantity. *Carbohydr. Polym.* **2022**, *298*, 120107. [CrossRef] [PubMed]
66. Edhirej, A.; Sapuan, S.M.; Jawaid, M.; Zahari, N.I. Cassava/Sugar Palm Fiber Reinforced Cassava Starch Hybrid Composites: Physical, Thermal and Structural Properties. *Int. J. Biol. Macromol.* **2017**, *101*, 75–83. [CrossRef] [PubMed]
67. Liu, H.; Xie, F.; Yu, L.; Chen, L.; Li, L. Thermal Processing of Starch-Based Polymers. *Prog. Polym. Sci.* **2009**, *34*, 1348–1368. [CrossRef]
68. Li, C.; Hu, Y.; Li, E. Effects of Amylose and Amylopectin Chain-Length Distribution on the Kinetics of Long-Term Rice Starch Retrogradation. *Food Hydrocoll.* **2021**, *111*, 106239. [CrossRef]
69. Schmitt, H.; Guidez, A.; Prashantha, K.; Soulestin, J.; Lacrampe, M.F.; Krawczak, P. Studies on the Effect of Storage Time and Plasticizers on the Structural Variations in Thermoplastic Starch. *Carbohydr. Polym.* **2015**, *115*, 364–372. [CrossRef]
70. Nazrin, A.; Sapuan, S.M.; Ilyas, R.A. Water Barrier and Mechanical Properties of Sugar Palm Crystalline Nanocellulose Reinforced Thermoplastic Sugar Palm Starch (TPS)/Poly(Lactic Acid) (PLA) Blend Bionanocomposites. *Nanotechnol. Rev.* **2021**, *10*, 431–442. [CrossRef]
71. Zhang, B.; Chen, L.; Xie, F.; Li, X.; Truss, R.W.; Halley, P.J.; Shamshina, J.L.; Rogers, R.D.; McNally, T. Understanding the Structural Disorganization of Starch in Water-Ionic Liquid Solutions. *Phys. Chem. Chem. Phys.* **2015**, *17*, 13860–13871. [CrossRef]
72. Delgado, J.M.; Rodes, A.; Orts, J.M. B3LYP and in Situ ATR-SEDIRAS Study of the Infrared Behavior and Bonding Mode of Adsorbed Acetate Anions on Silver Thin-Film Electrodes. *J. Phys. Chem. C* **2007**, *111*, 14476–14483. [CrossRef]
73. Zhang, B.; Xie, F.; Zhang, T.; Chen, L.; Li, X.; Truss, R.W.; Halley, P.J.; Shamshina, J.L.; McNally, T.; Rogers, R.D. Different Characteristic Effects of Ageing on Starch-Based Films Plasticised by 1-Ethyl-3-Methylimidazolium Acetate and by Glycerol. *Carbohydr. Polym.* **2016**, *146*, 67–79. [CrossRef]
74. Das, K.; Ray, D.; Bandyopadhyay, N.R.; Gupta, A.; Sengupta, S.; Sahoo, S.; Mohanty, A.; Misra, M. Preparation and Characterization of Cross-Linked Starch/Poly(Vinyl Alcohol) Green Films with Low Moisture Absorption. *Ind. Eng. Chem. Res.* **2010**, *49*, 2176–2185. [CrossRef]
75. Phetwarotai, W.; Potiyaraj, P.; Aht-Ong, D. Characteristics of Biodegradable Polylactide/Gelatinized Starch Films: Effects of Starch, Plasticizer, and Compatibilizer. *J. Appl. Polym. Sci.* **2010**, *116*, 2658–2667. [CrossRef]
76. Bendaoud, A.; Chalamet, Y. Effects of Relative Humidity and Ionic Liquids on the Water Content and Glass Transition of Plasticized Starch. *Carbohydr. Polym.* **2013**, *97*, 665–675. [CrossRef] [PubMed]
77. Bhagabati, P.; Hazarika, D.; Katiyar, V. Tailor-Made Ultra-Crystalline, High Molecular Weight Poly(ε-Caprolactone) Films with Improved Oxygen Gas Barrier and Optical Properties: A Facile and Scalable Approach. *Int. J. Biol. Macromol.* **2019**, *124*, 1040–1052. [CrossRef]
78. Madrigal, L.; Sandoval, A.J.; Müller, A.J. Effects of Corn Oil on Glass Transition Temperatures of Cassava Starch. *Carbohydr. Polym.* **2011**, *85*, 875–884. [CrossRef]
79. Sreekumar, P.A.; Al-Harthi, M.A.; De, S.K. Studies on Compatibility of Biodegradable Starch/Polyvinyl Alcohol Blends. *Polym. Eng. Sci.* **2012**, *52*, 2167–2172. [CrossRef]
80. López, O.V.; Lecot, C.J.; Zaritzky, N.E.; García, M.A. Biodegradable Packages Development from Starch Based Heat Sealable Films. *J. Food Eng.* **2011**, *105*, 254–263. [CrossRef]
81. Zhang, S.; He, Y.; Yin, Y.; Jiang, G. Fabrication of Innovative Thermoplastic Starch Bio-Elastomer to Achieve High Toughness Poly(Butylene Succinate) Composites. *Carbohydr. Polym.* **2019**, *206*, 827–836. [CrossRef]
82. Jha, P.; Dharmalingam, K.; Nishizu, T.; Katsuno, N.; Anandalakshmi, R. Effect of Amylose–Amylopectin Ratios on Physical, Mechanical, and Thermal Properties of Starch-Based Bionanocomposite Films Incorporated with CMC and Nanoclay. *Starch/Staerke* **2020**, *72*, 1900121. [CrossRef]

polymers

MDPI

Article

# Mechanical, Thermal, and Fire Retardant Properties of Rice Husk Biochar Reinforced Recycled High-Density Polyethylene Composite Material

**Atta ur Rehman Shah [1], Anas Imdad [2], Atiya Sadiq [2], Rizwan Ahmed Malik [3], Hussein Alrobei [4,*] and Irfan Anjum Badruddin [5]**

1 Department of Mechanical Engineering, COMSATS University Islamabad, Wah Campus, Wah Cantt 47040, Pakistan
2 Department of Mechanical Engineering, HITEC University, Taxila 47050, Pakistan
3 Department of Metallurgy & Materials Engineering, Faculty of Mechanical and Aeronautical Engineering, University of Engineering and Technology, Taxila 47050, Pakistan
4 Department of Mechanical Engineering, College of Engineering, Prince Sattam bin Abdullaziz University, AlKharj 11942, Saudi Arabia
5 Department of Mechanical Engineering, College of Engineering, King Khalid University, Abha 61421, Saudi Arabia
* Correspondence: h.alrobei@psau.edu.sa

**Citation:** Shah, A.u.R.; Imdad, A.; Sadiq, A.; Malik, R.A.; Alrobei, H.; Badruddin, I.A. Mechanical, Thermal, and Fire Retardant Properties of Rice Husk Biochar Reinforced Recycled High-Density Polyethylene Composite Material. *Polymers* **2023**, *15*, 1827. https://doi.org/10.3390/polym15081827

Academic Editors: Rosane Michele Duarte Soares and Vsevolod Aleksandrovich Zhuikov

Received: 18 February 2023
Revised: 30 March 2023
Accepted: 31 March 2023
Published: 9 April 2023



**Abstract:** This study concentrated on the influence of rice husk biochar on the structural, thermal, flammable, and mechanical properties of recycled high-density polyethylene (HDPE). The percentage of rice husk biochar with recycled HDPE was varied between 10% and 40%, and the optimum percentages were found for the various properties. Mechanical characteristics were evaluated in terms of the tensile, flexural, and impact properties. Similarly, the flame retardancy of the composites was observed by means of horizontal and vertical burning tests (UL-94 tests), limited oxygen index, and cone calorimetry. The thermal properties were characterized using thermogravimetric analysis (TGA). For detailed characterization, Fourier transform infrared spectroscopy (FTIR) and scanning electron microscopy (SEM) tests were performed, to elaborate on the variation in properties. The composite with 30% rice husk biochar demonstrated the maximum increase in tensile and flexural strength, i.e., 24% and 19%, respectively, compared to the recycled HDPE, whereas the 40% composite showed a 22.5% decrease in impact strength. Thermogravimetric analysis revealed that the 40% rice husk biochar reinforced composite exhibited the best thermal stability, due to having the highest amount of biochar. In addition, the 40% composite also displayed the lowest burning rate in the horizontal burning test and the lowest V-1 rating in the vertical burning test. The 40% composite material also showed the highest limited oxygen index (LOI), whereas it had the lowest peak heat release rate (PHRR) value (52.40% reduced) and total heat release rate (THR) value (52.88% reduced) for cone calorimetry, when compared with the recycled HDPE. These tests proved that rice husk biochar is a significant additive for enhancing the mechanical, thermal, and fire-retardant properties of recycled HDPE.

**Keywords:** polymer composites; recycled high density polyethylene; rice husk biochar; mechanical properties; thermal properties

## 1. Introduction

Global demand for fossil fuels is constantly increasing, but a huge challenge that the world has to resolve is their rapid depletion with the passage of time, along with the excessive pollution caused [1,2]. This is why people are shifting towards the use of renewable materials, particularly due to environmental concerns and the future scarcity of petroleum-based products [3]. On the other hand, the demand for food for the world's population is increasing day by day, which has caused the accumulation of agricultural waste, and the most common way to get rid of this is to dispose of or incinerate it [4]. Disposing of this waste poses serious threats to environmental safety and hygiene. However, the biomass can be converted into products and energy [5]. Employing these various agricultural wastes in different composite materials could help alleviate this problem and represents a great solution for recycling and resource conservation [6]. Several types of organic waste, such as waste paper sludge and wool, have already been used to produce biocomposites [7–10]. However, researchers are now inclining towards the development of biochar from agricultural waste and utilizing it in composite materials, to improve the desirable mechanical, thermal, and fire-retardant properties. When any type of biomass is heated with an absence/limited supply of oxygen (pyrolysis), it will leave behind a porous carbonaceous material, with the volatile gases absent, thus resulting in biochar formation [11]. Biochar is composed of highly ordered turbostatically crystalline regions, accompanied by some random amorphous regions [12], owing to the presence of cellulose. This isotropic structure results in spaces that form a porous structure, just like a honeycomb structure [13]. The reviews carried out by Väisänen et al. and Mohanty et al. gave a fair idea of how effective biochar can be in a composite material [14,15].

Ayrilmis et al. analyzed the mechanical properties of a wood–plastic composite (Polypropylene with Maleic Anhydride-grafted Polypropylene and different mixtures of wood flour and charcoal flour) [16]. An investigation was performed to find the optimum biomass for biochar-based polypropylene composites with rice husk, course wool, coffee husk, and landfilled wood as biomasses and biochar made from landfill wood [17]. The wood-based composite exhibited the best mechanical properties, with rice husk coming second. Zhang et al. examined the incorporation of rice husk biochar into high-density polyethylene and then compared this with wood–plastic composites [18]. It was found that the biochar-reinforced plastic composites had much superior mechanical properties to the wood–plastic composites. Sundarakannan et al. reinforced sugarcane biochar with polyester resin [19]. The study also focused on the development of a biocomposite by reinforcing a cashew nut shell biochar with unsaturated polyester resin at different loadings [20]. Huber et al. investigated the effect of the particle size of miscanthus biochar on Polyamide polymer and found that particle size varied the mechanical properties of the composite to an appreciable extent [21]. Khan et al. studied the influence of biochar addition to epoxy resin and compared this with the addition of carbon nanotubes in an epoxy matrix [22]. The biochar-reinforced epoxy composites outperformed the carbon nanotube-reinforced epoxy composites, in terms of the mechanical properties. Bartoli et al. tested the mechanical characteristics of an epoxy-based composite with five different types of commercial biochar (rice husk, mixed softwood, miscanthus, oil seed rape, and wheat straw) [23]. Pandey et al. focused on the study of a biochar hybrid composite with sisal fiber, softwood biochar, and epoxy as resin [24]. Similarly, Ketabchi et al. carried out an analysis of varied amounts of biochar (made of oil palm empty fruit bunch fiber) in polypropylene/ethylene-vinyl acetate, to form a hybrid composite [25]. They found that 30% biochar was preferable for enhancing the mechanical properties without degrading the thermal properties. Zhang et al. evaluated the influence of rice husk biochar on the dynamic mechanical properties of a composite [26]. The study revealed that biochar had a positive effect on the creep resistance, dynamic viscoelasticity, and stress relaxation properties of the composite. However, as the testing temperature increased, the stress relaxation and creep resistance started to reduce. Bajwa et al. treated high-density polyethylene–wood fiber composites with maleic anhydride polyethylene (MAPE) and found an improvement in the mechanical properties of the biocomposite [27]. However, concern about the undesirable increase in the overall cost of production was also reported, owing to the expensive compatibilizers used.

Das et al. reported that the addition of biochar to a composite could increase the thermal stability and flame retardance to an excellent extent compared to the neat composite [1,2]. Time to ignition (TTI) and PHRR were significantly reduced with increased biochar contents, whereas the THR did not vary significantly. Das et al. also introduced wood and biochar into polypropylene, with two flame retardants (FRs): magnesium hydroxide ($Mg(OH)_2$) and ammonium polyphosphate (APP) [28]. They concluded that less wood and a higher proportion of biochar was optimum for reducing flammability. Another research work by Das et al. considered the addition of wood-waste-derived activated biochar to polypropylene and revealed that the biochar was extremely effective in increasing the thermal stability and reducing the flammability of the PP. Dahal et al. converted wood pellets into biochar and then added this to epoxy. The study concluded that an increased biochar loading had a better tendency to retard fire [29]. Li et al. studied the addition of nano charcoal in polypropylene composites and found that an increase in the degradation temperature led to an increase in the overall thermal properties of the composites [30]. Zhang et al. also observed an enhancement in the thermal degradation temperature with rice husk biochar filled composites [31]. Bartoli and Arrigo et al. performed TGA on polylactic acid composites (PLA) [32]. They concluded that the addition of biochar increased the thermal properties of the composites, but at higher filler loadings, they started to decrease, due to agglomeration.

The quality of a biochar can adversely affect the mechanical properties of biocomposites. Das et al. stated that the most crucial factor in the performance of biochar-based composites, and what makes them so promising, is the carbon content and the surface area of the biochar [1,2]. Similar results were derived by Ho et al., in which they stated that a high surface area of biochar, which results from high temperature pyrolysis, usually assists in achieving the optimum dispersion of biochar particles in the polymer matrix [3]. Tomczyk et al. stated that increasing the pyrolysis temperature increased the carbon content and specific surface area of biochar [33]. Das et al. studied the quality of biochars made in different pyrolysis reactors [34]. They revealed that different types of pyrolysis reactors yielded different qualities of biochar, and the utilization of these reactors to produce biochar should be aligned with the intended application. However, among the operating conditions, temperature is most important, as it primarily controls the properties and quality, and adversely affects the yield of biochar produced during the process [35–38]. The effect of pyrolysis temperature on the quality of biochar was studied by Elnour et al., in terms of the morphological properties, physical properties, and biochar structure [39]. Zhang et al. also conducted research regarding the addition of rice husk biochar to high-density polyethylene, varying the pyrolysis temperature of the biochar [40]. Ayadi et al. investigated the effect of pyro-gasification temperature on the mechanical characteristics of biochar–polymer biocomposites [41].

The aim of this study was to achieve desirable properties of recycled HDPE close to those of virgin HDPE. For this purpose, particles of rice husk were converted into biochar. Three types of property were studied, i.e., mechanical, thermal, and fire retardant. Higher amounts of cellulose and proper dispersion of the rice husk biochar caused an enhancement in the mechanical properties of the recycled HDPE, whereas the formation of inorganic substances, and char layers increased the thermal and flame retardancy of the recycled HDPE. HDPE was found to fill the voids on the rough surfaces of biochar particles which has caused improvement in the mechanical properties, along with the improvement in the thermal and fire-retardant properties. Moreover, very little research has been done on recycled polymers. This works represents an attempt to move the recycled polymer industry into the limelight, so that more research will be carried out on this topic in the future.

## 2. Materials and Methods

Recycled HDPE pellets having a density of 0.96 $g/cm^3$ and melt flow index of 0.55 g/10 min were obtained from NEWTECH pipes, Islamabad, Pakistan. The HDPE was recycled by the company from faulty pieces of pipe. It is a post-industrial material obtained from the crushing of HDPE pipes and re-extrusion. Rice husk with a bulk density of 0.12 $g/cm^3$ was purchased from Allied Industry, Lahore, Pakistan.

The raw rice husk was crushed in a masala grinder (Geepas Masala grinder—RPM 4500) multiple times, to convert it into rice husk powder. The rice husk powder obtained from this masala grinder was then passed through a sieve of mesh size 50, to obtain a rice husk powder particle size of less than 300 μm. This rice husk powder was then put through the process of pyrolysis in a muffle furnace (Thermolyne Benchtop muffle furnace) at a temperature of 550 °C, in order to produce rice husk biochar. The pyrolysis temperature was increased from 25 °C (ambient temperature) to 550 °C, at a heating rate of 15 °C/min and maintained for 3 h under a 20 mL/min nitrogen environment.

The rice husk biochar and recycled HDPE were then dried under sunlight for 24 h. The recycled HDPE was mixed with rice husk biochar in a high-speed mixer for 30 min, to obtained the blends. To mix the composite blends, a micro twin-screw extruder (SHJ-Omega-20, Shanghai, China) with screw outer diameter of 19.8 mm and length/diameter ratio of 38:1 was used. Both zones of the extruder, i.e., the extruding zone and blending zone, were operated at 180 °C.

The resulting pellets were then put in a manual injection molding machine, with the purpose of making the various samples for the mechanical and other types of test. The temperature of the manual injection molding was set at 185 °C. The percentage of rice husk biochar that was mixed with recycled HDPE was varied between 10% and 40%. These percentages with their nomenclature are shown in Table 1.

**Table 1.** Nomenclature of the samples.

| Sr. No.# | Recycled Pellets wt. % | Rice Husk Biochar wt. % | Nomenclature |
|---|---|---|---|
| 1 | 100 | 0 | Rec-HDPE |
| 2 | 90 | 10 | C10 |
| 3 | 80 | 20 | C20 |
| 4 | 70 | 30 | C30 |
| 5 | 60 | 40 | C40 |

FTIR spectroscopy was carried out, in order to investigate and analyze the possible interactions between the matrix (Recycled HDPE) and filler (Rice husk biochar). This was performed with FTIR equipment (Scientific Nicolet 6700 model, Waltham, MA, USA), using a KBr disc method, with spectra ranging from 500 to 4000 $cm^{-1}$ (32 scans at 4 $cm^{-1}$ resolution).

Mechanical characterization of all the samples was performed using tensile, bending, and impact tests. The tensile test was performed using a Testometric Inc., Manchester, UK (Load cell: 100 kN), with reference to the D638-14 ASTM standard and a cross-head speed of 1.5 mm/s. A gauge length of 25 mm was used. The bending test was performed using a Testometric Inc., UK (Load cell: 100 kN), with reference to the D790-10 ASTM standard and a cross-head speed of 1.5 mm/s. A span length of 51 mm was used. The test was a three-point bending test. An Izod impact test was performed using a TM235, Bangkok, Thailand (Load arm: 16 kg Model) with reference to the D256-10 ASTM standard. The V-notch in the testing specimens was made using a milling machine (Hermle UWF, Gosheim, Germany). Mechanical tests were carried out on 5 samples for each composition.

Scanning electron microscopy (SEM) was conducted with a JEOL JSM5910 (Tokyo, Japan) (maximum magnification: 300,000×; maximum resolving power: 2.3 nm) on specimens that had undergone tensile testing. The dispersion of the filler particles, possible interactions between the matrix and filler particles, as well as the reasons behind the variation in tensile properties were observed with a microscope.

The fire retardancy of all composite samples was checked using a horizontal and vertical burning tests, cone calorimetric tests, and limited oxygen index test (LOI test). Horizontal and vertical burning tests were carried out, in order to investigate the reaction-to-fire properties according to the UL-94 standards with reference to the D635-03 and D3801-19 ASTM standards. The dimensions of the samples for both tests were 125 × 13 × 3.6 mm.

The composite burning time and rates were determined in a horizontal burning test, whereas the flame time, afterglow time, extent of after flame or afterglow up to the holding clamp, and dripping of cotton due to flaming drops for all types of composite specimen were analyzed in a vertical burning test. After the horizontal burning tests, the composites were awarded an HB rating based on the burning rate, and they were awarded V-0, V-1, or V-2 on the basis of the abovementioned parameters observed in the vertical burning test. Cone calorimeter (CC) tests were also performed, in order to evaluate the reaction-to-fire characteristics of the specimens using the E1354-17 ASTM standard. Samples of 100 × 100 × 3.2 mm were conditioned at 50% relative humidity and 23 °C. An external heat flux of 50 $kW/m^2$ was used to determine the various flammable properties. The type of cone calorimeter employed for this purpose was from FTT Limited, East Grinstead, UK. The limited oxygen index (LOI) of all samples was determined using an oxygen index tester. The standard ASTM D 2863-17 was followed, with a sample size of 100 mm × 6.5 mm × 3.2 mm. This was performed in a Sataton Limited oxygen index tester. The flow rate of oxygen and oxygen–nitrogen mixture were 3 L/min and 20 L/min, respectively.

Thermal stability was investigated in a Perkin Elmer, Waltham, MA, USA (Pyris Diamond Series TG/DTA model) using TG and DTG curves. This was carried out by heating the sample at a constant rate in an inert atmosphere. Heating was performed at a rate of 20 °C/min at a flow rate of 35 mL/min in an inert atmosphere of nitrogen. Weight loss vs. temperature plots were recorded as a result of the TGA. Similarly, in the case of DTG, the rate of material loss (weight loss) with heating was plot against the temperature and used to simplify the readings of the TG curves, which were quite close together.

## 3. Results

The FTIR spectra of the composite samples, as well as the recycled HDPE and rice husk biochar, are shown in Figure 1. Different peaks were observed at different wavenumbers, which are shown in Table 2. However, the intensity of these peaks varied from sample to sample. C-H stretching vibration, which is attributed to aliphatic structures, was responsible for the peak at 2924 $cm^{-1}$ [42]. Similarly, the peak at 1492 $cm^{-1}$ reflected C-H bending, which is a sign of alkanes [31]. However, if we focus on the wavenumber 955 $cm^{-1}$, peaks can be observed for the composite samples, but no peak is observed for the recycled HDPE sample. These peaks in the biochar-derived composite sample were actually caused by C-O stretching vibration in the composite samples only, which indirectly indicates a lack of C-O stretching vibration in the recycled HDPE [43]. This reveals that the peak at this wavenumber was caused by a functional group that is only present in rice husk biochar, similarly to the presence of carbohydrates only found in rice husk. Similarly, the minor peak at 3611 $cm^{-1}$ is related to $H_2O$, which is caused by O-H bond stretching vibration, and was weakly detected and only contributed by the biochar [44–47]. The peak at 1614 $cm^{-1}$ demonstrates C=C stretching [48], whereas that at 778 $cm^{-1}$ represents the silica functional group [49]. Both of these groups were also contributed by the rice husk biochar. The above study also showed that these functional groups were either contributed by the recycled HDPE or rice husk biochar, without any new functional groups being present. This revealed the potential of the physical combination of HDPE and rice husk biochar.

Figure 2 compares the tensile properties of the rice husk biochar derived composite samples with the recycled HDPE. The tensile strength of recycled HDPE was found to be 14.99 MPa. The tensile strengths of C10, C20, C30, and C40 were 15.71, 16.43, 18.54, and 15.97 MPa, respectively. The maximum tensile strength was exhibited by C30. At C30, a 24% increase in tensile strength was observed, compared to the 14.99 MPa of recycled HDPE. Regarding the elastic modulus, the elastic modulus of the recycled HDPE was found to be 1.01 GPa, and the elastic moduli of C10, C20, C30, and C40 were 1.11, 1.18, 1.35, and 1.1 GPa, respectively. The maximum elastic modulus was displayed by C30, which showed an 34% enhancement in tensile modulus.

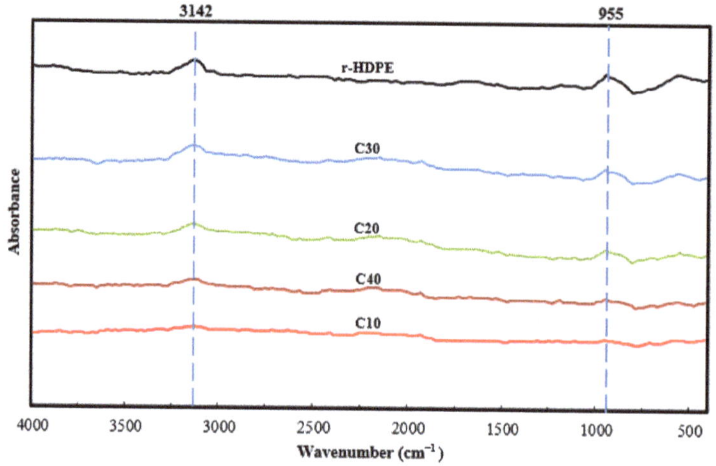


**Figure 1.** FTIR of recycled HDPE and all composites.

**Table 2.** Functional groups in the composites and filler particles, corresponding to their wavelength.

| Material | Wavelength ($cm^{-1}$) | Bonds |
|---|---|---|
| Recycled HDPE | 2924<br>1492 | C-H Stretching of Hydrocarbons<br>C-H Bending of Alkanes |
| Rice husk biochar | 3611<br>1614<br>955<br>778 | O-H Stretching of Phenolic hydroxyl and Alcohol hydroxyl<br>C=C Stretching of Hydroxyl Functional groups<br>C-O Stretching of Carbohydrates<br>Silicon Hydrogen Single Bond (Si-H) |
| C10,C20,C30,C40 | 3611<br>2924<br>1614<br>1492<br>955<br>778 | O-H Stretching of Phenolic hydroxyl and Alcohol hydroxyl<br>C-H Stretching of Hydrocarbons<br>C=C Stretching of Hydroxyl Functional groups<br>C-H Bending of Alkanes<br>C-O Stretching of Carbohydrates<br>Silicon Hydrogen Single Bond (Si-H) |

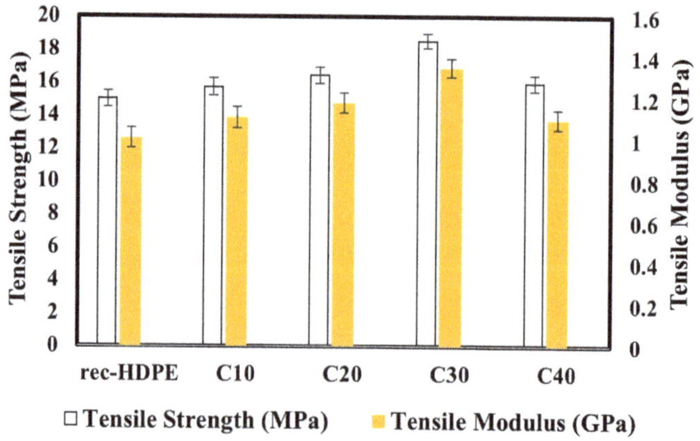


**Figure 2.** Tensile properties of the recycled HDPE and composites.

The mechanical properties of composites are dependent on the homogeneous dispersion of filler particles in the matrix and the extent of agglomeration of the particles with each other. Another factor that reflects the tensile strength of a composite is the stress transfer mechanism of interfacial bonding [50]. If the recycled HDPE uniformly and effectively filled the pores of the biochar and transferred the stress productively, then the chance of an increase in mechanical properties was extremely high, as this would provide rigidity to the composite.

With reference to our case, the high tensile strength of the 30% rice husk biochar derived composite could be attributed to the uniform and homogenous dispersion of biochar in the HDPE matrix, less agglomeration of particles, and a good stress transfer mechanism, which was caused by good interfacial bonding, due to proper filling of the biochar pores by the recycled HDPE, which is also shown in the SEM image of C30. With the addition of 40% biochar, the tensile strength decreased because of the non-uniform dispersion of biochar particles, greater agglomeration of particles, and weak interfacial bonding [51]. Regarding the elastic modulus, biochar particles are subject to less deformation and are extremely rigid. On account of this property, the mobility of the recycled HDPE macromolecules was severely restricted, and this led to an increase in elastic modulus with the addition of biochar particles to the recycled HDPE, from 0 to 30% [52]. The elastic modulus of the C40 might have been low due to the agglomeration of particles and poor stress transfer between the filler particles and matrix. The SEM images further illustrate the causes of the variation in the tensile properties.

SEM images of the tensile cross-sectional fracture surface of the recycled HDPE, as well as the rice husk biochar composites, are shown in Figure 3. The smooth structure of recycled HDPE is shown in Figure 3a. For the 30% rice husk reinforced composite, i.e., in the case of C30, the porous structure of biochar was evident, as can be seen in Figure 3b. Most of the pores of the biochar were uniformly filled by the recycled HDPE. The better distribution of the filler particles can be viewed in this figure. This led to a good physical–mechanical interlocking structure [44–46]. This superior dispersion also caused an efficient transfer of stress between the rice husk biochar particles and recycled HDPE. This phenomenon revealed that the highest tensile strength among all samples was for C30. In the SEM image of C40 in Figure 3c, there are some signs of agglomeration and clusters of rice husk biochar, which lowered the tensile strength.

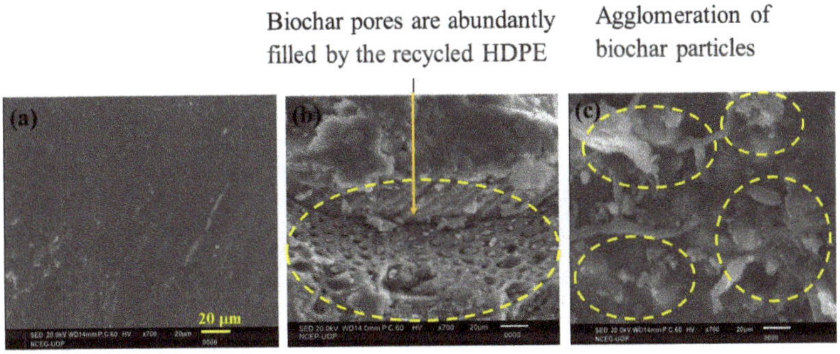


**Figure 3.** SEM images of (**a**) unmodified recycled HDPE, (**b**) 30% rice husk reinforced recycled HDPE, and (**c**) 40% rice husk reinforced recycled HDPE.

Furthermore, C30 also displayed the highest tensile modulus, due to the having the most interlocking of biochar pores with the recycled HDPE, which gave it rigidity and stiffness. C10 displayed the lowest tensile modulus, owing to the lower rice husk biochar loading, whereas the agglomeration in C40 also caused a decreased tensile modulus.

Figure 4 compares the flexural properties of the recycled HDPE with the biochar-derived composite samples. The flexural property of the recycled HDPE was discovered to be 20.12 MPa. The flexural properties of C10, C20, C30, and C40 were 20.56, 21.94, 23.91, and 22.12 MPa, respectively. C30 demonstrated an enhancement of 19% in flexural strength compared with the recycled HDPE and showed the maximum flexural strength. Moreover, C30 also recorded an increase of 80% for the flexural modulus compared with the recycled HDPE, which displayed a 0.77 GPa flexural modulus. C10, C20, C30, and C40 displayed 0.99, 1.14, 1.39, and 1.09 GPa flexural moduli. The trend in the flexural properties was found to be similar to that of the tensile properties.

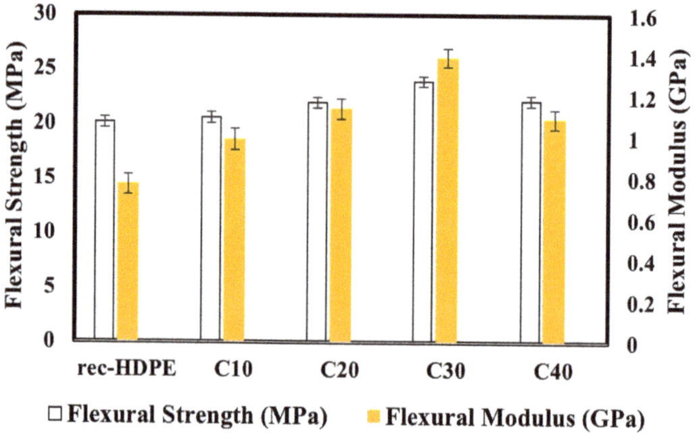


**Figure 4.** Flexural properties of the recycled HDPE and composites.

This enhancement in flexural properties was due to the porous structure of the biochar and the physical interlocking between the biochar and recycled HDPE. When screw extrusion was performed, the HDPE (which was in a fluid state) could fill the pores of the biochar, so that a rigid structure with a good interaction between them was achieved; and after cooling, a strong physically interlocking structure was secured [11].This was the main reason behind the enhancement in the flexural properties with the addition of biochar to the recycled HDPE, since it allowed transferring the stress efficiently. Another factor responsible for the variation in the flexural properties, and reported in many documents, was the good particle dispersion in the matrix [53,54]. This could also have been the reason behind the decrease in the flexural strength and flexural modulus in the C40 composite. The biochar particles may not have been uniformly dispersed in the HDPE matrix or these particles might have become agglomerated, as a result of which there was appreciable reduction in the flexural characteristics.

The impact strengths of the recycled HDPE and composite materials are shown in Figure 5. The impact strength of the recycled HDPE, C10, C20, C30, and C40 were 3.43, 3.77, 5.76, 5.11, and 4.51 KJ/m$^2$, respectively. The impact strengths all displayed a different trend compared to the tensile strength and flexural strength. Its decrease was consistent with the fiber–polymer composites [18,55]. The decrease of the impact strength from the rec-HDPE to the C40 sample was around 22.5%.

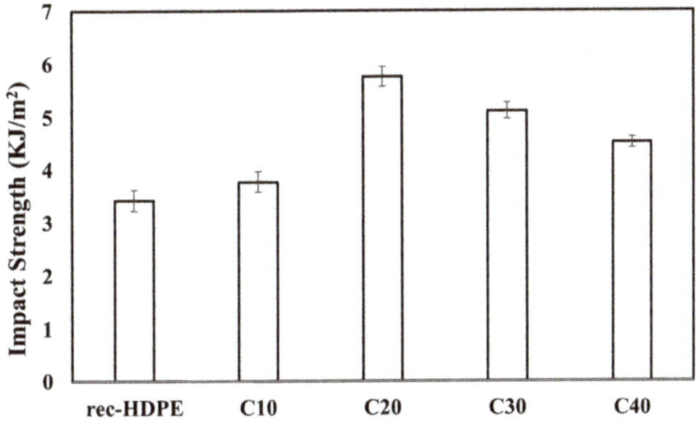


**Figure 5.** Impact properties of the recycled HDPE and composites.

The impact strength of a composite material is entirely dependent on the rigidity and toughness [31]. The lesser the rigidity of the material, the greater its toughness, and the greater its impact strength, and vice versa. The rigidity of the materials was dependent on the mobility of the recycled HDPE molecules, which in turn was dependent on the physical–mechanical interlocking structure of the biochar with the HDPE molecules. When the content of the rice husk biochar in the recycled HDPE matrix was 10%, there was less chance of the recycled HDPE flowing into the pores of biochar, and thus the mobility of the recycled HDPE was very high. Its ability to absorb energy during fracture propagation was excellent, and thus its toughness was high. There were less regions of stress concentration, which required less energy to induce cracks in the composite, and thus the impact strength was very high. However, as the content of rice husk biochar was increased from 20% to 40%, the recycled HDPE started filling the pores of the biochar and the ability of the recycled HDPE to move began to be limited. This caused the increase in the rigidity of the composite and decrease in the impact strength of the samples [31]. As such, less energy was required to resist a sudden impact as the content was increased to 40%. Hence, the impact energy started to decrease. Another reason might have been that there were abundant regions of poor stress concentrations, and this led to easier crack propagation in the composite samples [31]. Table 3 compares the mechanical properties of the samples.

**Table 3.** Mechanical properties of the recycled HDPE and composites.

| Samples | Tensile Strength (MPa) | Tensile Modulus (GPa) | Flexural Strength (MPa) | Flexural Modulus (GPa) | Impact Strength (kJ/m$^2$) |
|---|---|---|---|---|---|
| Rec-HDPE | 14.99 | 1.01 | 20.12 | 0.77 | 3.43 |
| C10 | 15.71 | 1.11 | 20.56 | 0.99 | 3.77 |
| C20 | 16.43 | 1.18 | 21.94 | 1.14 | 5.76 |
| C30 | 18.54 | 1.35 | 23.91 | 1.39 | 5.11 |
| C40 | 15.97 | 1.10 | 22.12 | 1.09 | 4.51 |

Horizontal and vertical burning tests were conducted on all specimens, i.e., with and without the inclusion of the rice husk biochar in the recycled HDPE. The results were then compared with the UL-94 standard for horizontal and vertical burning tests. For both the tests, five specimens for each sample were tested, and the average of the results was taken. In the case of the horizontal burning test, the time needed to reach from 25 mm to 100 mm (burning time) was increased as the content of rice husk biochar in the recycled HDPE was increased. This caused a decrease in the burning rate from the recycled HDPE to C40. All the specimens received an HB rating based on the burning rate, except the recycled HDPE sample, whose burning rate was greater than 40 mm/min and, thus, not awarded an HB rating. The decrease in the average burning rate from the recycled HDPE (45.10 mm/min) to C40 (25.99 mm/min) was 42.37%.

In the case of the vertical burning test, the flame and afterglow times after the 1st and 2nd flame applications progressively decreased with the increase in the biochar loading, from C20 to C40. This was why the recycled HDPE did not receive a rating, due to its poor flame retardancy characteristics. C10 also did not obtain a rating, due to the small amount of biochar. The after-flame time for C10 was greater than for the recycled HDPE, because the content of biochar was less. The rice husk biochar partially slowed down the flame speed but could not stop it, and the flame eventually reached the holding clamp. C20 obtained the rating of V-2, owing to the relatively high content of biochar, which resisted the flame and did not allow it to reach the holding clamp. C30 and C40 obtained a V-1 rating, owing to the shorter after flame and afterglow times, during which the rice husk biochar acted against the flame, slowed down the process of combustion, and stopped the flame before it reached the holding clamp. Cotton dripping was also not recorded with the C20, C30, and C40 samples. Finally, C40 had the lowest chance of burning. It is expected that, following this trend, recycled HDPE having 50% biochar would achieve a rating of V-0. Table 4 compares the fire-retardant properties of all samples.

**Table 4.** Fire retardant properties of the recycled HDPE and composites.

| | Cone Calorimetric Test | | | | Horizontal Burning Test | | | | | Vertical Burning Test | | | | LOI Test |
|---|---|---|---|---|---|---|---|---|---|---|---|---|---|---|
| **Material Types** | **TTI (s)** | **TPHRR (s)** | **PHRR (kW/m$^2$)** | **THR (MJ/m$^2$)** | **Avg. Burning Time (min)** | **Avg. Burning Rate (mm/min)** | **Rating** | **Max. After Flame Time (s)** | **Total After Flame Time (s)** | **Max After Flame + After Glow Time(s)** | **Flame up to the Holding Clamp** | **Cotton Ignited by Flaming Drops** | **Rating** | **LOI (%)** |
| Rec-HDPE | 59 | 210 | 633 | 144 | 1.66 | 45.10 | Nil | 61 | 281 | - | Yes | Yes | Nil | 16.97 |
| C10 | 32 | 214 | 434 | 120 | 2.01 | 37.19 | HB | 63 | 304 | - | Yes | Yes | Nil | 18.82 |
| C20 | 28 | 222 | 399 | 103 | 2.30 | 32.66 | HB | 26 | 235 | 34 | No | Yes | V-2 | 20.73 |
| C30 | 30 | 229 | 344 | 83 | 2.64 | 28.38 | HB | 16 | 152 | 22 | No | No | V-1 | 23.01 |
| C40 | 31 | 233 | 301 | 68 | 2.89 | 25.99 | HB | 7 | 59 | 8 | No | No | V-1 | 25.28 |

Cone calorimeter tests were performed for the purpose of elaborating the principal fire retardant characteristics of the recycled HDPE, as well as the rice husk biochar derived composite samples. Figures 6 and 7 show the HRR and THR curves of all samples with the response time, whereas Table 4 displays the time to ignition (TTI), time to peak heat release rate (TPHRR), heat release rate (HRR), and total heat release rate (THR) of all samples.

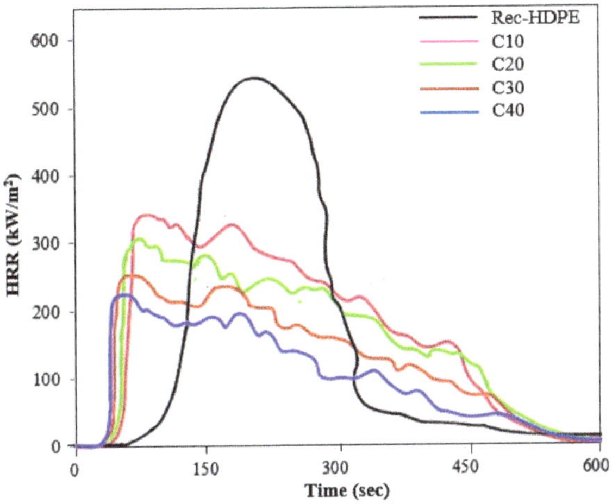


**Figure 6.** Heat release rate of the recycled HDPE and composites.

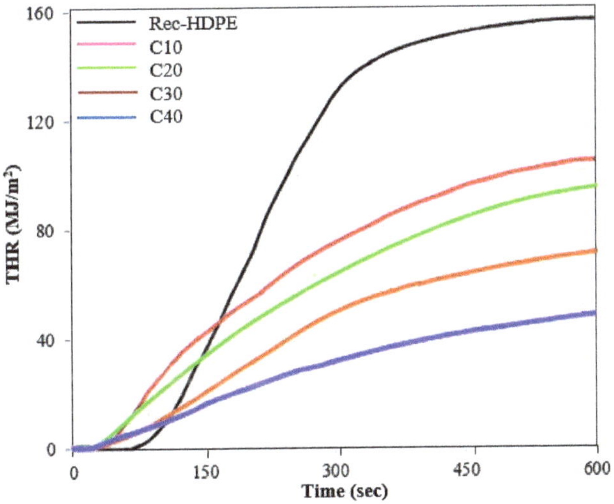


**Figure 7.** Total heat release rate of the recycled HDPE and composites.

Time to ignition (TTI) is one of the crucial factors that dictates the flammable characteristics of composites. The recycled HDPE took some time to ignite and catch fire (59 s), whereas the TTI of the rice husk biochar derived samples first decreased for C10 and C20 and then increased for C30 and C40 as the content of the biochar increased. This means that the rice husk biochar samples were ignited in less time compared to the recycled HDPE. This decrease in TTI was immediate when the biochar was added to the composites, i.e., from the recycled HDPE to C10. This decrease in TTI with the addition of biochar was consistent with the report of Das et al. [11]. Similarly, the peak heat release rate (PHRR) and total heat release rate (THR) are also critical factors in judging the fire-retardant characteristics of composites and describe the overall combustion behavior of the material [56]. On the other hand, TPHRR tells us about the time it takes to produce the maximum amount of heat. The peak heat release rate (PHRR) and total heat release rate (THR) of the recycled HDPE were highest at 633 $kW/m^2$ and 144 $MJ/m^2$, due to its poor flammable properties and the intense combustion, whereas both of these properties showed a decline with the addition of rice husk biochar to the composite samples, with the 40% rice husk biochar derived composite sample displaying the lowest PHRR (301 kW/m$^2$) and lowest THR (68 MJ/m$^2$).

This meant that C40 exhibited a large decrease of 52.40% in PHRR value and 52.88% in THR value. C40 also had the longest TPHRR (233 s) among all samples, which is also a positive attribute. All these desirable shifts in properties occurred with the increase in the content of biochar in the composite samples. This means that C40 displayed the best flammable properties, and the trend in best fire-retardant properties was as follows:

$$C40 > C30 > C20 > C10 > \text{Recycled HDPE}$$

The presence of biochar, which also possessed excellent thermal properties, acted as a hindering agent and became a barrier to the transfer of heat between the recycled HDPE and the source of heat [11,17]. Moreover, the formation of char layers from the rice husk biochar played a pivotal role in reducing the HRR and PHRR values [11].

Limited oxygen index (LOI) tests were performed, for the purpose of elaborating and further characterizing the flame retardancy behavior of the recycled HDPE as well as the biochar-derived composite samples. The greater the LOI of a specimen, the better its flame retardant properties. The limited oxygen index basically refers to the volume ratio of oxygen in a mixture of nitrogen and oxygen. It is the minimum concentration of oxygen that is required to sustain a flame or to support the combustion of a material.

With reference to the LOI of the recycled HDPE and the biochar-derived samples, as shown in Figure 8, the LOI of the recycled HDPE was the lowest, whereas it increased as the content of rice husk biochar in the composite samples was increased.

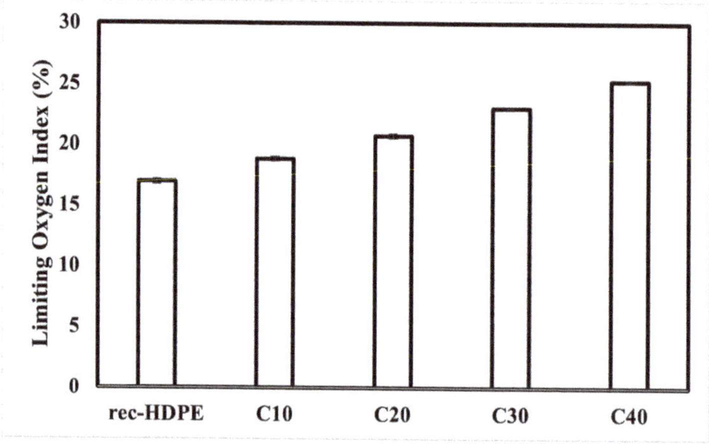


**Figure 8.** Limited oxygen index of the recycled HDPE and composites.

The recycled HDPE showed the LOI of 16.97%, whereas the LOI of C40 was 25.28%, which was highest among all samples. This trend of the enhancement in LOI is consistent with previous reports [44–46]. This means that an increase of 48.97% was recorded in the LOI from the recycled HDPE to C40. The prime reason behind this was the presence of biochar, which enhanced the flame-retardant properties of the composites. Biochar has a very high thermal stability, so inclusion of biochar repelled the heat transport between the source of heat and the recycled HDPE [11]. It was also reported in the literature that, during the process of pyrolysis or carbonization, many metal oxides and inorganic substances are formed, which have the ability to slow down the process of combustion, as they are nonflammable [57–59].

Thermogravimetric and derivative thermogravimetric curves are shown in Figures 9 and 10. The TG curves depict the percentage weight loss of the samples, whereas the DTG displays the derivative of those curves. These curves show that, as the heating temperature was increased, the samples initially responded well, with no degradation, but as the temperature reached 402.34 °C, the recycled HDPE sample started to thermally degrade, with degradation occurring at 522.16 °C. As the heating temperature was further increased, all samples started to degrade one by one, with C40 degraded last. We can also notice from the TGA that, as the biochar content in the composite samples was increased, the delay in thermal degradation increased, and this delay increased as the biochar content in the recycled HDPE was increased. This means that the increase in the biochar loading was extremely important in repelling the heat from decomposing the composite samples and in increasing the time of decomposition. In addition, the onset degradation temperature ($T_0$) also rose with the increase in rice husk biochar loading. This increase was recorded as 4.05% from the recycled HDPE (425.09 °C) to C40 (442.36 °C). Table 5 shows the onset degradation temperature ($T_0$), start degradation temperature ($T_1$), finish degradation temperature ($T_2$), point of inflection temperature ($T_{peak}$), and residue % at 600 °C.

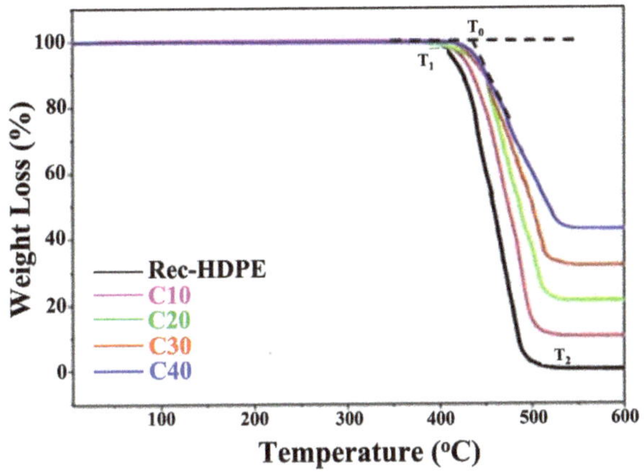


**Figure 9.** TG curves of the recycled HDPE and composites.

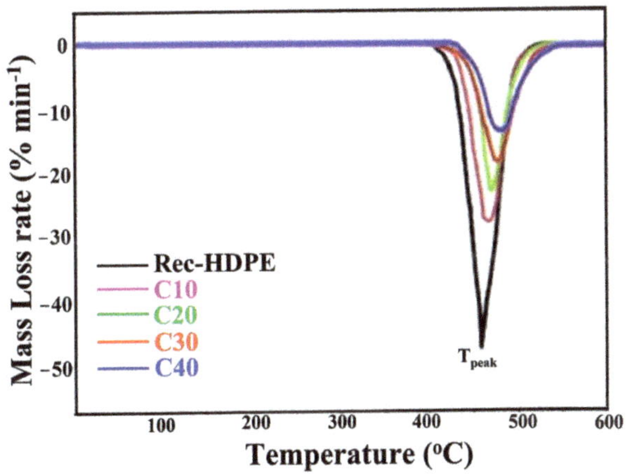


**Figure 10.** DTG curves of the recycled HDPE and composites.

**Table 5.** Thermal properties of the recycled HDPE and composites.

| Type of Sample | $T_0$ (Onset Degradation Temperature) (°C) | $T_1$ (Start Degradation Temperature) (°C) | $T_2$ (Finish Degradation Temperature) (°C) | $T_{peak}$ (Point of Inflection Temperature) (°C) | Residue % at 600 °C |
|---|---|---|---|---|---|
| Rec-HDPE | 425.09 | 402.34 | 522.12 | 459.63 | 0.39% |
| C10 | 435.86 | 406.12 | 528.05 | 466.41 | 10.95% |
| C20 | 440.93 | 410.87 | 535.67 | 471.18 | 21.47% |
| C30 | 438.14 | 416.03 | 541.31 | 477.35 | 32.09% |
| C40 | 442.36 | 421.21 | 548.88 | 481.22 | 43.51% |

Moreover, if we look at the TG curves, we come to the conclusion that the amount of residue left after heating was directly proportional to the biochar loading. Das reported that this increase in residue generation is due to the increase in stable biochar loading [11]. De Bhowmick et al. also reported that the main reason behind the increase in the residue is the enrichment of stable $SiO_2$ in stable biochar loading [57]. This is why the recycled HDPE was almost completely decomposed after heating and why this decomposition decreased as we moved towards higher loadings of biochar. C40 retained the maximum residue after heating, owing to a greater percentage of biochar. This increase in residues at 600 °C also proved the fact that the thermal stability was improved with the addition of biochar. The following trend of residue retention was followed by the samples:

$$C40 > C30 > C20 > C10 > \text{Recycled HDPE}$$

Hence, we come to the conclusion that biochar is an extremely important additive for increasing the thermal stability of composites.

## 4. Conclusions

The influence of variation of the rice husk biochar content in recycled HDPE was evaluated, in order to maximize the use of agricultural waste. All functional groups were supplied either by the recycled HDPE or rice husk biochar, which was revealed using FTIR.

The 30% biochar loading had the highest tensile strength (18.54 MPa), tensile modulus (1.35 GPa), flexural strength (23.91 MPa), and flexural modulus (1.39 GPa). The tensile and flexural strength were increased by 24% and 19%, respectively, compared with the recycled HDPE, whereas the tensile and flexural moduli were improved by 34% and 80%, respectively. However, the impact strength decreased (22.5%) when adding rice husk biochar to the recycled HDPE, due to the increase in rigidity of the composites compared with the recycled HDPE (4.63 $KJ/m^2$). This variation in properties was due to the extent of agglomeration, level of homogeneous dispersion of the filler particles in the matrix, and the extent of the recycled HDPE filling the pores of the rice husk biochar, which was also displayed in the SEM images. Overall, this revealed that the addition of rice husk biochar had a positive effect on the mechanical properties of the composites.

With the increase in the content of rice husk biochar, the delay in thermal decomposition was enhanced, the ability to repel the heat from decomposing the composite samples was increased, and the time of decomposition was also increased. The C40 composite with 40% rice husk biochar displayed the best thermal stability and longest thermal decay time among all samples. Moreover, C40 showed an increase of 4.05% in the temperature of the onset of degradation compared to the recycled HDPE.

The composite with 40% rice husk biochar demonstrated the best flame retardancy characteristics. The horizontal burning test revealed a decrease in the burning rate of the recycled HDPE from 45.10 mm/min to 25.99 mm/min for C40. The vertical burning test also showed a change in the flammability rating, from the no rating of recycled HDPE to the V-1 of C40. The limited oxygen tests also showed that C40 had the highest LOI (25.28%). Moreover, the cone calorimetry also showed a decrease of 52.40% for the PHRR value and 52.88% for the THR value compared with the recycled HDPE. This was due to the inherent capacity of biochar to halt the transport of heat from the heat source to the matrix, as well as due to the formation of inflammable metal oxides and inorganic substances during pyrolysis.

**Author Contributions:** Conceptualization, A.u.R.S., A.I. and A.S.; methodology, A.u.R.S., A.I. and A.S.; software, A.u.R.S., A.I. and A.S.; formal analysis, A.u.R.S., A.I. and A.S.; investigation, A.u.R.S., A.I. and A.S.; writing—original draft preparation, A.u.R.S., A.I., A.S., R.A.M., H.A. and I.A.B.; writing—review and editing, R.A.M., H.A. and I.A.B.; supervision, A.u.R.S. and H.A. All authors have read and agreed to the published version of the manuscript.

**Funding:** This research was funded by Deputyship for Research & Innovation, Ministry of Education in Saudi Arabia through project number IFPSAU-2021/01/18816.

**Institutional Review Board Statement:** Not applicable.

**Informed Consent Statement:** Not applicable.

**Data Availability Statement:** All data are included in the manuscript.

**Acknowledgments:** The authors extend their appreciation to the Deputyship for Research & Innovation, Ministry of Education in Saudi Arabia for funding this research work through project number IFPSAU-2021/01/18816.

**Conflicts of Interest:** The authors declare no conflict of interest.

## References

1. Das, O.; Sarmah, A.K.; Bhattacharyya, D. A novel approach in organic waste utilization through biochar addition in wood/polypropylene composites. *Waste Manag.* **2015**, *38*, 132–140. [CrossRef] [PubMed]
2. Das, O.; Sarmah, A.K.; Bhattacharyya, D. Structure–mechanics property relationship of waste derived biochars. *Sci. Total Environ.* **2015**, *538*, 611–620. [CrossRef] [PubMed]
3. Ho, M.P.; Lau, K.T.; Wang, H.; Hui, D. Improvement on the properties of polylactic acid (PLA) using bamboo charcoal particles. *Compos. Part B Eng.* **2015**, *81*, 14–25. [CrossRef]
4. Nan, N.; DeVallance, D.B.; Xie, X.; Wang, J. The effect of bio-carbon addition on the electrical, mechanical, and thermal properties of polyvinyl alcohol/biochar composites. *J. Compos. Mater.* **2016**, *50*, 1161–1168. [CrossRef]
5. Behazin, E.; Misra, M.; Mohanty, A.K. Sustainable biocarbon from pyrolyzed perennial grasses and their effects on impact modified polypropylene biocomposites. *Compos. Part B Eng.* **2017**, *118*, 116–124. [CrossRef]
6. DeVallance, D.B.; Oporto, G.S.; Quigley, P. Investigation of hardwood biochar as a replacement for wood flour in wood–polypropylene composites. *J. Elastomers Plast.* **2016**, *48*, 510–522. [CrossRef]
7. Ahmetli, G.; Kocaman, S.; Ozaytekin, I.; Bozkurt, P. Epoxy composites based on inexpensive char filler obtained from plastic waste and natural resources. *Polym. Compos.* **2013**, *34*, 500–509. [CrossRef]
8. Bilal, A.; Lin, R.; Jayaraman, K. Analysis of the mechanical properties of rice husk reinforced polyethylene composites using experiments with mixtures. In *Advanced Materials Research*; Trans Tech Publications Ltd.: Stafa-Zurich, Switzerland, 2013; Volume 747, pp. 395–398.
9. Ashori, A.; Nourbakhsh, A. Bio-based composites from waste agricultural residues. *Waste Manag.* **2010**, *30*, 680–684. [CrossRef]
10. Kim, N.K.; Dutta, S.; Bhattacharyya, D. A review of flammability of natural fibre reinforced polymeric composites. *Compos. Sci. Technol.* **2018**, *162*, 64–78. [CrossRef]
11. Das, O.; Sarmah, A.K.; Bhattacharyya, D. Biocomposites from waste derived biochars: Mechanical, thermal, chemical, and morphological properties. *Waste Manag.* **2016**, *49*, 560–570. [CrossRef]
12. Xie, X.; Goodell, B.; Qian, Y.; Peterson, M.; Jellison, J. Significance of the heating rate on the physical properties of carbonized maple wood. *Holzforschung* **2008**, *62*, 591–596. [CrossRef]
13. Marsh, H. CHAPTER 6-activation processes (chemical). In *Activated Carbon*; Marsh, H., Rodríguez-Reinoso, F., Eds.; Elsevier: Amsterdam, The Netherlands, 2006; pp. 322–365. [CrossRef]
14. Väisänen, T.; Haapala, A.; Lappalainen, R.; Tomppo, L. Utilization of agricultural and forest industry waste and residues in natural fiber-polymer composites: A review. *Waste Manag.* **2016**, *54*, 62–73. [CrossRef] [PubMed]
15. Mohanty, A.K.; Vivekanandhan, S.; Pin, J.M.; Misra, M. Composites from renewable and sustainable resources: Challenges and innovations. *Science* **2018**, *362*, 536–542. [CrossRef] [PubMed]
16. Ayrilmis, N.; Kwon, J.H.; Han, T.H.; Durmus, A. Effect of wood-derived charcoal content on properties of wood plastic composites. *Mater. Res.* **2015**, *18*, 654–659. [CrossRef]
17. Das, O.; Kim, N.K.; Hedenqvist, M.S.; Lin, R.J.; Sarmah, A.K.; Bhattacharyya, D. An attempt to find a suitable biomass for biochar-based polypropylene biocomposites. *Environ. Manag.* **2018**, *62*, 403–413. [CrossRef] [PubMed]

18. Zhang, Q.; Yi, W.; Li, Z.; Wang, L.; Cai, H. Mechanical properties of rice husk biochar reinforced high density polyethylene composites. *Polymers* **2018**, *10*, 286. [CrossRef]
19. Sundara, K.R.; Arumugaprabu, V.; Poomari Muthukumar, G.; Vigneshwaran, S.; Deepan Saravana Kumar, S.R.; Muneesh Raj, R. Biochar from Sugarcane Waste in Polymer Matrix Composite. *Int. J. Innov. Technol. Explor. Eng.* **2019**, *9*, 350–352.
20. Sundarakannan, R.; Arumugaprabu, V.; Manikandan, V.; Vigneshwaran, S. Mechanical property analysis of biochar derived from cashew nut shell waste reinforced polymer matrix. *Mater. Res. Express* **2020**, *6*, 125349. [CrossRef]
21. Huber, T.; Misra, M.; Mohanty, A.K. The effect of particle size on the rheological properties of polyamide 6/biochar composites. In *AIP Conference Proceedings 2015, Proceedings of the 30th International Conference of the Polymer Processing Society (PPS-30), Cleveland, OH, USA, 6–12 June 2014*; AIP Publishing: New York, NY, USA, 2015; Volume 1664, p. 150004. [CrossRef]
22. Khan, A.; Savi, P.; Quaranta, S.; Rovere, M.; Giorcelli, M.; Tagliaferro, A.; Jia, C.Q. Low-cost carbon fillers to improve mechanical properties and conductivity of epoxy composites. *Polymers* **2017**, *9*, 642. [CrossRef]
23. Bartoli, M.; Giorcelli, M.; Rosso, C.; Rovere, M.; Jagdale, P.; Tagliaferro, A. Influence of commercial biochar fillers on brittleness/ductility of epoxy resin composites. *Appl. Sci.* **2019**, *9*, 3109. [CrossRef]
24. Pandey, A.; Telang, A.; Rana, R. Mechanical characterization of bio-char made hybrid composite. *Int. J. Eng. Res. Appl.* **2016**, *6*, 26–31.
25. Ketabchi, M.R.; Khalid, M.; Walvekar, R. Effect of oil palm EFB-biochar on properties of PP/EVA composites. *J. Eng. Sci. Technol.* **2017**, *12*, 797–808.
26. Zhang, Q.; Cai, H.; Ren, X.; Kong, L.; Liu, J.; Jiang, X. The dynamic mechanical analysis of highly filled rice husk biochar/high-density polyethylene composites. *Polymers* **2017**, *9*, 628. [CrossRef] [PubMed]
27. Bajwa, D.S.; Adhikari, S.; Shojaeiarani, J.; Bajwa, S.G.; Pandey, P.; Shanmugam, S.R. Characterization of bio-carbon and ligno-cellulosic fiber reinforced bio-composites with compatibilizer. *Constr. Build. Mater.* **2019**, *204*, 193–202. [CrossRef]
28. Das, O.; Kim, N.K.; Kalamkarov, A.L.; Sarmah, A.K.; Bhattacharyya, D. Biochar to the rescue: Balancing the fire performance and mechanical properties of polypropylene composites. *Polym. Degrad. Stab.* **2017**, *144*, 485–496. [CrossRef]
29. Dahal, R.K.; Acharya, B.; Saha, G.; Bissessur, R.; Dutta, A.; Farooque, A. Biochar as a filler in glassfiber reinforced composites: Experimental study of thermal and mechanical properties. *Compos. Part B Eng.* **2019**, *175*, 107169. [CrossRef]
30. Li, S.; Xu, Y.; Jing, X.; Yilmaz, G.; Li, D.; Turng, L.S. Effect of carbonization temperature on mechanical properties and biocompatibility of biochar/ultra-high molecular weight polyethylene composites. *Compos. Part B Eng.* **2020**, *196*, 108120. [CrossRef]
31. Zhang, Q.; Khan, M.U.; Lin, X.; Cai, H.; Lei, H. Temperature varied biochar as a reinforcing filler for high- density polyethylene composites. *Compos. Part B Eng.* **2019**, *175*, 107151. [CrossRef]
32. Arrigo, R.; Bartoli, M.; Malucelli, G. Poly (lactic acid)–biochar biocomposites: Effect of processing and filler content on rheological, thermal, and mechanical properties. *Polymers* **2020**, *12*, 892. [CrossRef] [PubMed]
33. Tomczyk, A.; Sokołowska, Z.; Boguta, P. Biochar physicochemical properties: Pyrolysis temperature and feedstock kind effects. *Rev. Environ. Sci. Bio/Technol.* **2020**, *19*, 191–215. [CrossRef]
34. Das, O.; Hedenqvist, M.S.; Johansson, E.; Olsson, R.T.; Loho, T.A.; Capezza, A.J.; Holder, S. An all- gluten biocomposite: Comparisons with carbon black and pine char composites. *Compos. Part A Appl. Sci. Manuf.* **2019**, *120*, 42–48. [CrossRef]
35. Bridgwater, A.V. Review of fast pyrolysis of biomass and product upgrading. *Biomass Bioenergy* **2012**, *38*, 68–94. [CrossRef]
36. Lua, A.C.; Yang, T.; Guo, J. Effects of pyrolysis conditions on the properties of activated carbons prepared from pistachio-nut shells. *J. Anal. Appl. Pyrolysis* **2004**, *72*, 279–287. [CrossRef]
37. Özçimen, D.; Ersoy-Meriçboyu, A. A study on the carbonization of grapeseed and chestnut shell. *Fuel Process. Technol.* **2008**, *89*, 1041–1046. [CrossRef]
38. Hasan, M.M.; Bachmann, R.T.; Loh, S.K.; Manroshan, S.; Ong, S.K. Effect of pyrolysis temperature and time on properties of palm kernel shell-based biochar. *IOP Conf. Ser. Mater. Sci. Eng.* **2019**, *548*, 012020. [CrossRef]
39. Elnour, A.Y.; Alghyamah, A.A.; Shaikh, H.M.; Poulose, A.M.; Al-Zahrani, S.M.; Anis, A.; Al-Wabel, M.I. Effect of pyrolysis temperature on biochar microstructural evolution, physicochemical characteristics, and its influence on biochar/polypropylene composites. *Appl. Sci.* **2019**, *9*, 1149. [CrossRef]
40. Zhang, Q.; Zhang, D.; Lu, W.; Khan, M.U.; Xu, H.; Yi, W.; Zou, R. Production of high-density polyethylene biocomposites from rice husk biochar: Effects of varying pyrolysis temperature. *Sci. Total Environ.* **2020**, *738*, 139910. [CrossRef] [PubMed]
41. Ayadi, R.; Koubaa, A.; Braghiroli, F.; Migneault, S.; Wang, H.; Bradai, C. Effect of the pyro-gasification temperature of wood on the physical and mechanical properties of biochar-polymer biocomposites. *Materials* **2020**, *13*, 1327. [CrossRef] [PubMed]
42. Hahn, A.; Gerdts, G.; Völker, C.; Niebühr, V. Using FTIRS as pre-screening method for detection of microplastic in bulk sediment samples. *Sci. Total Environ.* **2019**, *689*, 341–346. [CrossRef]
43. Song, B.; Chen, M.; Zhao, L.; Qiu, H.; Cao, X. Physicochemical property and colloidal stability of micron- and nano-particle biochar derived from a variety of feedstock sources. *Sci. Total Environ.* **2019**, *661*, 685–695. [CrossRef]
44. Zhang, Q.; Zhang, D.; Xu, H.; Lu, W.; Ren, X.; Cai, H.; Mateo, W. Biochar filled high-density polyethylene composites with excellent properties: Towards maximizing the utilization of agricultural wastes. *Ind. Crops Prod.* **2020**, *146*, 112185. [CrossRef]
45. Zhang, Q.; Lei, H.; Cai, H.; Han, X.; Lin, X.; Qian, M.; Zhao, Y.; Huo, E.; Villota, E.M.; Mateo, W. Improvement on the properties of microcrystalline cellulose/polylactic acid composites by using activated biochar. *J. Clean. Prod.* **2020**, *252*, 119898. [CrossRef]
46. Zhang, Q.; Xu, H.; Lu, W.; Zhang, D.; Ren, X.; Yu, W.; Lei, H. Properties evaluation of biochar/high- density polyethylene composites: Emphasizing the porous structure of biochar by activation. *Sci. Total Environ.* **2020**, *737*, 139770. [CrossRef] [PubMed]

47. Kai, X.; Li, R.; Yang, T.; Shen, S.; Ji, Q.; Zhang, T. Study on the co-pyrolysis of rice straw and high density polyethylene blends using TG-FTIR-MS. *Energy Convers. Manag.* **2017**, *146*, 20–33. [CrossRef]
48. Jindo, K.; Mizumoto, H.; Sawada, Y.; Sanchez-Monedero, M.A.; Sonoki, T. Physical and chemical characterizations of biochars derived from different agricultural residues. *Biogeosci. Discuss.* **2014**, *11*, 6613–6621. [CrossRef]
49. Wei, L.; Huang, Y.; Li, Y.; Huang, L.; Mar, N.N.; Huang, Q.; Liu, Z. Biochar characteristics produced from rice husks and their sorption properties for the acetanilide herbicide metolachlor. *Environ. Sci. Pollut. Res.* **2017**, *24*, 4552–4561. [CrossRef] [PubMed]
50. Fu, S.Y.; Feng, X.Q.; Lauke, B.; Mai, Y.W. Effects of particle size, particle/matrix interface adhesion and particle loading on mechanical properties of particulate–polymer composites. *Compos. Part B Eng.* **2008**, *39*, 933–961. [CrossRef]
51. Nasser, J.; Lin, J.; Steinke, K.; Sodano, H.A. Enhanced interfacial strength of aramid fiber reinforced composites through adsorbed aramid nanofiber coatings. *Compos. Sci. Technol.* **2019**, *174*, 125–133. [CrossRef]
52. Siebert, H.M.; Wilker, J.J. Deriving commercial level adhesive performance from a bio-based mussel mimetic polymer. *ACS Sustain. Chem. Eng.* **2019**, *7*, 13315–13323. [CrossRef]
53. Goud, V.; Alagirusamy, R.; Das, A.; Kalyanasundaram, D. Influence of various forms of polypropylene matrix (fiber, powder and film states) on the flexural strength of carbon-polypropylene composites. *Compos. Part B Eng.* **2019**, *166*, 56–64. [CrossRef]
54. Oliveira, L.Á.; Santos, J.C.; Panzera, T.H.; Freire, R.T.; Vieira, L.M.; Scarpa, F. Evaluation of hybrid-short-coir-fibre-reinforced composites via full factorial design. *Compos. Struct.* **2018**, *202*, 313–323. [CrossRef]
55. Kaymakci, A.; Ayrilmis, N. Investigation of correlation between Brinell hardness and tensile strength of wood plastic composites. *Compos. Part B Eng.* **2014**, *58*, 582–585. [CrossRef]
56. Manzello, S.L. (Ed.) *Encyclopedia of Wildfires and Wildland-Urban Interface (WUI) Fires*; Springer International Publishing: Cham, Switzerland, 2020. [CrossRef]
57. De Bhowmick, G.; Sarmah, A.K.; Sen, R. Production and characterization of a value added biochar mix using seaweed, rice husk and pine sawdust: A parametric study. *J. Clean. Prod.* **2018**, *200*, 641–656. [CrossRef]
58. Chen, C.; Yan, X.; Xu, Y.; Yoza, B.A.; Wang, X.; Kou, Y.; Ye, H.; Wang, Q.; Li, Q.X. Activated petroleum waste sludge biochar for efficient catalytic ozonation of refinery wastewater. *Sci. Total Environ.* **2019**, *651*, 2631–2640. [CrossRef] [PubMed]
59. Chen, J.; Wang, J.; Ni, A.; Chen, H.; Shen, P. Synthesis of a novel phosphorous-nitrogen based charring agent and its application in flame-retardant HDPE/IFR composites. *Polymers* **2019**, *11*, 1062. [CrossRef] [PubMed]

polymers

MDPI

Review

# Polyhydroxybutyrate Metabolism in *Azospirillum brasilense* and Its Applications, a Review

**María de los Ángeles Martínez Martínez, Lucía Soto Urzúa, Yovani Aguilar Carrillo, Mirian Becerril Ramírez and Luis Javier Martínez Morales ***

Centro de Investigaciones en Ciencias Microbiológicas, Instituto de Ciencias, Benemérita Universidad Autónoma de Puebla, Av. San Claudio y Av. 24 Sur, Col. San Manuel Ciudad Universitaria, Puebla 72570, Mexico; angeles.martinezm@correo.buap.mx (M.d.l.Á.M.M.); lucia.soto@correo.buap.mx (L.S.U.); yovani.aguilarc@alumno.buap.mx (Y.A.C.); mirian.becerrilr@alumno.buap.mx (M.B.R.)
* Correspondence: luis.martinez@correo.buap.mx; Tel.: +52-222-229-5500

**Abstract:** Gram-negative *Azospirillum brasilense* accumulates approximately 80% of polyhydroxybutyrate (PHB) as dry cell weight. For this reason, this bacterium has been characterized as one of the main microorganisms that produce PHB. PHB is synthesized inside bacteria by the polymerization of 3-hydroxybutyrate monomers. In this review, we are focusing on the analysis of the PHB production by *A. brasilense* in order to understand the metabolism during PHB accumulation. First, the carbon and nitrogen sources used to improve PHB accumulation are discussed. *A. brasilense* accumulates more PHB when it is grown on a minimal medium containing a high C/N ratio, mainly from malate and ammonia chloride, respectively. The metabolic pathways to accumulate and mobilize PHB in *A. brasilense* are mentioned and compared with those of other microorganisms. Next, we summarize the available information to understand the role of the genes involved in the regulation of PHB metabolism as well as the role of PHB in the physiology of *Azospirillum*. Finally, we made a comparison between the properties of PHB and polypropylene, and we discussed some applications of PHB in biomedical and commercial areas.

**Keywords:** polyhydroxybutyrate; *Azospirillum brasilense*; PHB genes; PHB regulation; PHB metabolism

**Citation:** Martínez, M.d.l.Á.M.; Urzúa, L.S.; Carrillo, Y.A.; Ramírez, M.B.; Morales, L.J.M. Polyhydroxybutyrate Metabolism in *Azospirillum brasilense* and Its Applications, a Review. *Polymers* **2023**, *15*, 3027. https://doi.org/10.3390/polym15143027

Academic Editors: Vsevolod Aleksandrovich Zhuikov and Rosane Michele Duarte Soares

Received: 23 March 2023
Revised: 28 June 2023
Accepted: 6 July 2023
Published: 13 July 2023



## 1. Introduction

Gram-negative *Azospirillum brasilense* belongs to the α-proteobacteria class. It is a motile, vibrio-shaped bacterium of 2.0–4.0 μm length [1]. *Azospirillum* promotes plant growth. Also, it produces high quantities of bioplastic called poly-β-hydroxybutyrate (PHB) [2]. PHB is part of a cluster of bioplastics called polyhydroxyalkanoates (PHA). There are more than 150 different PHAs discovered. The two PHAs most studied are polyhydroxyvalerate (PHV) and PHB [3].

PHB is a biodegradable and biocompatible plastic characterized to have a methyl radical in the β-position of the carbon skeleton of PHA [4]. It has been shown that *A. brasilense* produces only 3-hydroxybutyrate monomers [5,6]. Fourier-transform infrared spectroscopy (FTIR) analyses have shown an ester band ν(C=O) at 1727 $cm^{-1}$, which is compatible with PHB [7]. This review aims to summarize the most important factors to consider for understanding PHB metabolism in *A. brasilense*. Throughout the text, we discuss the best carbon and nitrogen sources for improving PHB production. The regulation of PHB metabolism and the functions of PHB are analyzed. Finally, a comparison between the characteristics of PHB and polypropylene (PP) is reviewed, and some examples of uses of PHB in the medical industry are described.

## 2. The Role of the Carbon Source in PHB Production by *A. brasilense*

Previously, it was demonstrated that *A. brasilense* accumulates large quantities of PHB when it grows on a medium supplemented with high concentrations of carbon with minimal quantities of nitrogen (high C/N ratio) [2,8,9]. *Azospirillum* can use a wide range of carbon and nitrogen sources. In terms of carbon, it grows well in fructose, malate, succinate, oxaloacetate, pyruvate, glycerol, lactate, and β-hydroxybutyric acid, among others [1,10,11]. Amino acids are poorly used as carbon, and glucose cannot support the growth of *A. brasilense* [12,13]. $N_2$, amino acids, $NH_3$, $NH_4$, and $NO_3^-$ have been reported as good nitrogen sources for this bacterium [11,14,15]. *Azospirillum* can use a wide spectrum of carbon and nitrogen sources because in this bacterium occur tricarboxylic acid (TCA), glyoxylate, and Entner–Duodoroff cycles, but it lacks Embden–Meyerhof–Parnas and hexose monophosphate pathways [12,13,16].

To improve PHB synthesis, several carbon and nitrogen sources have been evaluated. The highest quantities of PHB were produced when malic acid and ammonia chloride were used as carbon and nitrogen sources, respectively [5,9,17]. When *A. brasilense* grows on malate and ammonia chloride, it accumulates up to 88% of dry cell weight (DCW) as PHB. The carbon source, malate, enters the TCA cycle to produce both primary and secondary metabolites.

On fructose or lactate, Azospirillum reaches 40 and 50% of PHB as dry cell weight, respectively [5]. Another nitrogen source that allowed high PHB accumulation was sodium nitrate [18]. *A. brasilense* fixes nitrogen when the nitrogen source is depleted. Under nitrogen-fixing conditions, it accumulates from 30 to 75% of PHB as dry cell weight [8,18,19]. Oxygen is also important for PHB synthesis. Data showed that the use of malate and ammonium chloride in addition to high oxygen levels inhibits PHB accumulation by *A. brasilense* [8]. However, low oxygen availability leads *Azospirillum* to accumulate more than 70% of dry cell weight as PHB [5,9]. Previous studies found the highest accumulation of PHB when a 70–140 C/N ratio was used [2,8].

Another *Rhodospirillaceae*, *Rhodospirillum rubrum*, uses acetate for PHB synthesis and prefers anaerobic conditions to improve it [20]. Pseudomonads turn acetate, ethanol, fructose, glucose, gluconate, and glycerol into acetyl-CoA for PHA synthesis. In this bacterium, PHA metabolites are obtained through β-oxidation and de novo fatty acid synthesis pathways [21]. Other carbon sources used by bacteria to produce PHB are methane for *Methylobacterium* strains [22], mannitol for *Bradirhizobium diazoefficiens* [23], and glucose for *R. eutropha*. The latter accumulates up to 90% of PHB (Table 1) [24]. *R. eutropha*, *Azotobacter* spp., *Bacillus*, *Pseudomonas*, and *Azospirillum* spp. are the most studied microorganisms in terms of PHB production [2,24].

**Table 1.** PHB accumulation by common strains.

| Strain | Carbon Source | %PHB/DCW | Reference |
|---|---|---|---|
| *A. brasilense* Sp7 | Malate, fructose, pyruvate | 70–88% | [2,8,9,18] |
| *R. eutropha* | Glucose | 80–90% | [25] |
| *R. rubrum* | Acetate | | [20] |
| *Pseudomonas extremaustralis* | Octanoate, fructose, glucose, glycerol | 70–80% | [21,26,27] |
| *Methylocystis hirsuta* | Methanol:ethanol, methane | 73–85% | [28] |
| *Bradyrhizobium diazoefficiens* | Mannitol, glucose, and glycerol | 68% | [29] |
| *A. vinelandii* | Sucrose | 85% | [30] |
| *Bacillus subtilis* | Various sources | 60% | [31] |
| *Rhizobium nepotum* | Pyruvate | 62% | [28] |

## 3. PHB Synthesis and Degradation by *A. brasilense*

Biopolymer synthesis by *A. brasilense* involves three enzymatic reactions. The first is catalyzed by β-ketothiolase (coded by the *phb*A gene), which condenses two acetyl-CoA molecules and synthesizes acetoacetyl-CoA. Afterward, it is reduced to β-hydroxybutyryl-CoA by an NAD(P)-dependent acetoacetyl-CoA reductase (coded by the *phb*B gene). Finally, the β-hydroxybutyryl-CoA is polymerized into PHB by the PHB polymerase coded by the *phb*C gene (Figure 1) [32,33]. This pathway occurs similarly in *A. beijerinckii*, *R. eutropha*, and *Sinorhizobium meliloti*, among others [34–36]. Other microorganisms such as *P. putida* can synthesize PHB and PHV and copolymers, for example, PHB-*co*-PHV. In PHA synthesis by *P. putida*, the roles of PhaJ, epimerase, and FabG have been described. The PhaJ oxidizes acyl-CoAs into enoyl-CoA. The latter is converted into 3-hydroxyacyl-CoA by an epimerase. Then, 3-hydroxyacyl-CoA is reduced by FabG to form (R)-3-ketoacyl-CoA. Finally, a PhaC polymerizes (R)-3-ketoacyl-CoA into PHA. Another pathway to synthesize PHA in *P. putida* begins with the transacylation of malonyl-CoA and acetyl-CoA with the acyl carrier protein (ACP). The resulting malonyl-ACP and acyl-ACP are condensed into ketoacyl-ACP. Afterward, it is reduced to (R)-3-hydroxyacyl-ACP. Next, PhaG elongates (R)-3-hydroxyacyl-ACP with two carbon units into PHA monomers. To finish, a PhaC polymerizes monomers into PHA [21].

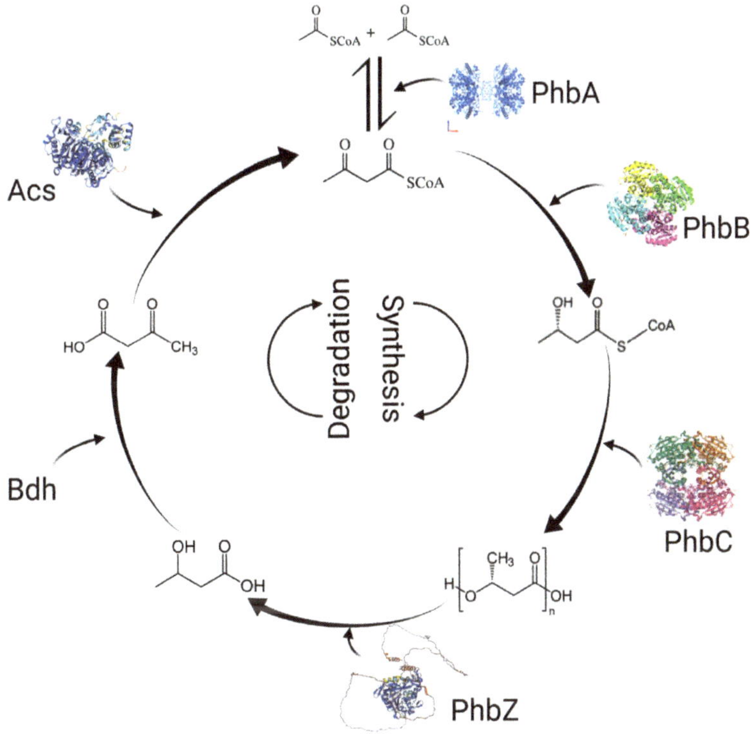


**Figure 1.** The metabolic pathway for PHB synthesis and degradation. The enzymes involved in PHB synthesis are PhbA (β-ketothiolase), PhbB (Acetoacetyl-CoA reductase), and PhbC (PHB synthase). The enzymes involved in PHB degradation are PhbZ (PHB depolymerase), Bdh (β-hydroxybutyrate dehydrogenase), Acs (Acetyl-CoA synthetase), and PhbA (β-ketothiolase) (Created with data previously reported [32,37,38]).

PHB (and PHA in general) is a hydrophobic material that needs to be stabilized in the cytoplasm. The proteins involved in stabilizing it are known as granule-associated proteins (GAPs). A single PHB granule (carbonosome) contains 98% polymer and 2% GAPs [39–42]. GAPs include PHB synthases, PHB depolymerases, regulators, and phasins (Figure 2) [43]. PHB synthase and PHB depolymerase initiate PHB synthesis or degradation, respectively [32,38]. Phasins coat and stabilize PHB chains inside bacteria and control the size of the PHB granules [2,43–45]. Finally, regulator proteins regulate the expression level of phasins when PHB is synthesized or degraded [46].

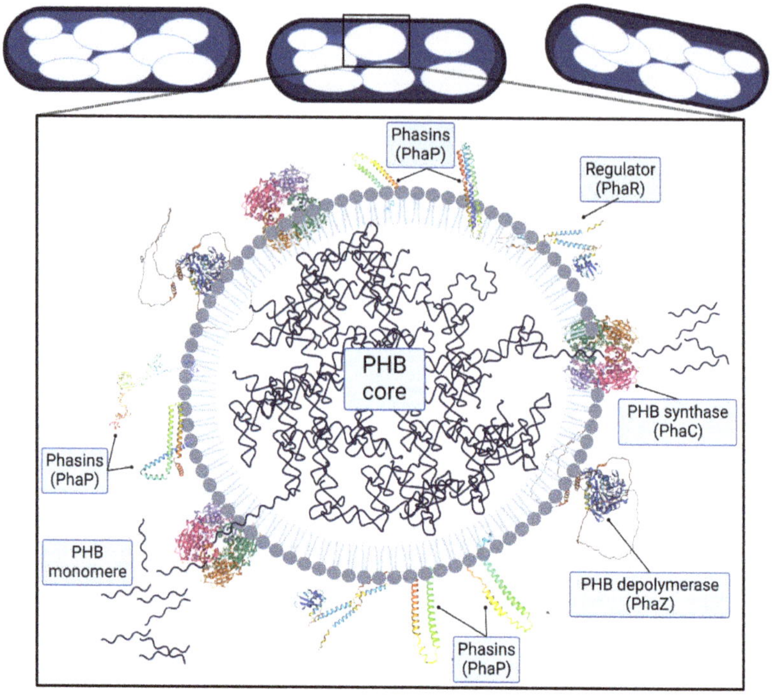


**Figure 2.** PHB granules in bacteria. Granules contain growing PHB chains at the core and are surrounded by GAPs (Created with data previously reported [39,43,47]).

PHB degradation occurs when bacteria enter a state of starvation, and the exogenous carbon source is depleted. The resulting products can support bacterial growth, serving as a carbon and energy source [5,38]. PHB mobilization in *A. brasilense* involves a PHB depolymerase (PhaZ) that cuts PHB into β-hydroxybutyrate monomers. Then, an NAD(P)-dependent β-hydroxybutyrate dehydrogenase oxidizes β-hydroxybutyrate monomers to acetoacetate. The subsequent step is to convert acetoacetate into acetoacetyl-CoA by an acetoacetyl-CoA synthetase, and, finally, the acetoacetyl-CoA is hydrolyzed by a β-ketothiolase that releases two acetyl-CoA molecules (Figure 1) [18,37,48]. Acetyl-CoA can enter the TCA cycle, β-oxidation, or glyoxylate pathways, among others, and be used to produce metabolic intermediates and energy to sustain the growth of the bacterium [2,8,29]. *S. meliloti*, *R. eutropha*, and other microorganisms share the same mobilization pathway as *Azospirillum* [36]. In *R. eutropha*, PHB is poorly degraded in the absence of nitrogen [49]. In contrast, nitrogen-fixing bacteria such as *A. brasilense* can mobilize PHB when the nitrogen source is depleted, due to the nitrogenase complex. Previous studies suggest PHB mobilization provides enough energy to sustain nitrogen fixation and two binary fissions [50]. Most PHA-producer microorganisms code for several depolymerase isoenzymes [49]. Since *A. brasilense* can use β-hydroxybutyric acid for growth, this bacterium may have extracellular and intracellular depolymerases. However, more studies are needed to provide us with more information [38].

PHB is accumulated at the middle and the end of the logarithmic growth phases. At the stationary phase, when the carbon source is depleted, PHB begins to be degraded, and it is used to sustain bacterial growth [5,8]. Martínez-Martínez et al. [2] observed that PHB was mainly accumulated after 72 h of growth when a high C/N ratio and microaerophilic conditions were used.

## 4. Studies on PHB Synthesis and Degradation Genes

In *A. brasilense* Sp7, some genes involved in PHB synthesis and degradation have been analyzed. It was shown that a *phb*C mutant strain was unable to accumulate PHB granules after 48 h of growth [32]. On the contrary, a *phb*Z mutant strain accumulated the highest quantities of biopolymer and was incapable of using it [37]. Until now, there are no studies evaluating the effect of deleting the *phb*A gene on PHB synthesis; it may probably be because PhbA has other functions than polymer synthesis. However, a *phb*B mutant strain was reported to be unable to produce PHB. PhbB is the only enzyme implicated in PHB synthesis, and it uses NADH and NAD(P)H as coenzymes [51]. The enzyme β-hydroxybutyrate dehydrogenase has been reported to function as an NAD-dependent tetramer formed by four similar subunits [17]. It was found that NADH, NADPH, pyruvate, and acetyl-CoA inhibit β-hydroxybutyrate dehydrogenase activity [17,37].

Kadouri et al. [32] were the first to report the genetic sequence of genes involved in PHB synthesis by *A. brasilense*. The *pha*A and *pha*B genes were co-transcribed, whereas the *phb*C gene was located in the complementary strand. Recently, the whole available genomic sequence of *A. baldaniorum* Sp245 (formerly named *A. brasilense* Sp245) has shown several copies of PHB genes. In the A. baldaniorum chromosome, a copy of the *phb*C gene was located. The *phb*CAB operon was in plasmid 4. A phbA homolog was found in plasmid 1, whereas copies of phbB were in plasmids 1 and 2 (Figure 3) [52]. In most microorganisms, genes encoding PhbA, PhbB, and PhbC are commonly clustered in an operon [36]. *R. eutropha* and *A. brasilense* contain the *phb*CAB operon, whereas *Azotobacter vinelandii* contains the *phb*BAC operon [36,52,53]. In *P. putida*, the PHA cluster is organized into two operons, *pha*C1ZC2D and *pha*IF [21]. However, there are bacteria with several copies of homologous genes randomly distributed throughout bacterial chromosomes and plasmids [52,54].

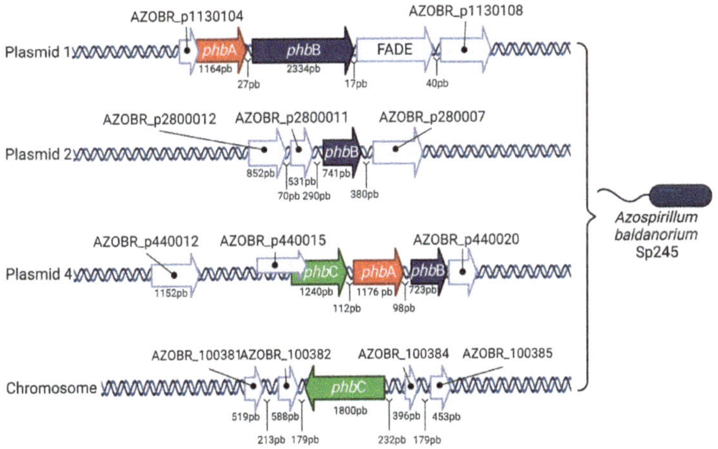


**Figure 3.** Genes involved in PHB synthesis in *A. baldaniorum* Sp245 (Created with data modified from reference [52]).

Martínez-Martínez et al. [2] analyzed phasin content in the *A. brasilense* Sp7 genome by looking for a phasin_2 domain (PF09361). It was found that this bacterium contains six genes that encode for phasins. The genes were named *pha*P1 (AMK58_RS17065), *pha*P2 (AMK58_RS04265), *pha*P3 (AMK58_RS04270), *pha*P4 (AMK58_RS07520), *pha*P5 (AMK58_RS13850), and *pha*P6 (AMK58_RS20955). Deletion of the *pha*P1 gene showed a phenotype compatible with phasins in other microorganisms. The *pha*P1 mutant strain accumulated fewer PHB granules of a higher size in comparison with the wild-type strain, which accumulated more PHB granules of a lower size.

## 5. PHB Metabolism Regulation in *A. brasilense* Sp7

The PHB pathway is closely related to the TCA cycle because acetyl-CoA molecules are metabolic intermediates. Depending on bacterial needs, acetyl-CoA may be used for one or another [55]. When bacteria are grown on a medium with a high C/N ratio, acetyl-CoA molecules are turned towards the TCA cycle to synthesize the metabolic intermediates required to produce macromolecules such as carbohydrates, proteins, nucleic acids, and lipids. As the TCA cycle is active, high quantities of citrate are synthesized and inhibit the enzymatic activity of citrate synthase. Then, acetyl-CoA molecules are condensed into acetoacetyl-CoA by β-ketothiolase, beginning PHB synthesis [9,54]. The increment of CoA-SH released from acetyl-CoA condensation inhibits β-ketothiolase activity [9]. Excessive acetoacetyl-CoA also inhibits PhbA activity, the key enzyme for PHB biosynthesis [9,56].

Another factor involved in switching on/off PHB synthesis is the redox state of bacteria [39]. Large quantities of NAD(P)H are produced mainly during the TCA cycle and cell respiration. The increase of NAD(P)H is eliminated by acetoacetyl-CoA reductase, which requires NAD(P)H to reduce acetoacetyl-CoA into β-hydroxybutyryl-CoA monomers [39,57]. Then, PHB synthesis leads to an increase in NAD(P) levels. The latter favors the activity of PhaZ that starts the PHB degradation [9,17,21,34,37,39,55,58,59]. *Azotobacter beijerinckii* controls PHB accumulation and mobilization in a similar way as *Azospirillum* [34].

The enzymes β-ketothiolase, acetoacetyl-CoA reductase, PHB synthase, β-hydroxybutyrate dehydrogenase, and acetoacetyl-CoA synthetase are constitutively expressed in *A. brasilense* [9,37]. It seems to be that phasins and regulator proteins are also constitutive [2,60]. For PHB accumulation, *A. brasilense* Sp7 prefers growth in a medium with low oxygen availability. Microaerophilic conditions protect the nitrogenase complex [60]. A high C/N ratio environment allows bacterial growth even if nitrogen is depleted from the medium, due to the nitrogenase complex. Studies by Tal et al. [9] demonstrated that enzymatic activities of β-ketothiolase, acetoacetyl-CoA reductase, and β-hydroxybutyrate dehydrogenase were higher in *Azospirilla* grown in a medium with oxygen limitation.

Although *A. brasilense* Sp7 grows well in ammonia chloride, it has been demonstrated that high ammonia chloride decreases the citrate synthase, isocitrate dehydrogenase, and succinate dehydrogenase activities of *Azospirillum lipoferum*. The effect increases with lower dissolved oxygen (DO). Then, carbon metabolism is restricted to PHB synthesis [61].

## 6. Nitrogen's Role in PHB Metabolism

*Azospirillum* grows in a medium with a low ammonium concentration. Ammonium assimilation by *Azospirilla* occurs through glutamate dehydrogenase activity that converts glutamate into α-ketoglutarate and ammonia [61]. After ammonium depletion, this bacterium continues growing because it fixes atmospheric nitrogen [62,63]. This process expends a lot of energy, supported by PHB catabolism [50]. The presence of exogenous nitrogen, such as ammonium, nitrate, and nitrite, represses nitrogen fixation [64].

The regulation of nitrogen metabolism in *A. brasilense* Sp7 includes the well-known proteins: GlnD, GlnB, GlnZ, NtrB, and NtrC and other genes involved in the nitrogen fixation process. In the PII-Pz sensing nitrogen system, *gln*B encodes for the PII protein, whereas the Pz protein is coded by *gln*Z. GlnD transfers uridyl groups to GlnB and GlnZ in a medium with nitrogen deficiency [65–67]. Under non-nitrogen-limiting conditions, GlnD removes uridyl groups of GlnB and GlnZ. When nitrogen is absent in the growth medium, uridylated GlnB cannot interact with NtrB. Then, NtrB phosphorylates NtrC. NtrC, phosphorylated, induces transcription of genes involved in sensing nitrogen-alternative sources [66,67].

Sun et al. [65] evaluated PHB accumulation by an *A. brasilense* Sp7 *gln*D mutant strain. This mutant was unable to sense nitrogen cell status. Under no nitrogen-limiting conditions, the PII-Pz system in the *gln*D mutant strain was not uridylated/deuridylated. Furthermore, the *gln*D mutant strain accumulated higher amounts of PHB in comparison with the wild-type strain. The *A. brasilense* Sp7 *gln*B–*gln*Z double mutant strain also synthesized higher amounts of PHB than the wild-type strain [65]. In both the *gln*D single mutant strain and the glnB–glnZ double mutant strain, PHB was accumulated during the logarithmic phase of growth.

Sun et al. [58] analyzed the PHB production of *A. brasilense* Sp7 *ntr*B and *ntr*C single mutant strains when ammonium chloride was used as a nitrogen source. The *ntr*B and *ntr*C mutant strains were unable to sense the nitrogen levels when a medium with a low C/N ratio was used for growth, in comparison with the *A. brasilense* Sp7 wild type. As a result, the mutant strains accumulated up to 45% of PHB as dry cell weight, whereas the wild-type strain produced only 10% of PHB under the same conditions. The *ntr*C mutant of *Herbaspirillum seropedicae* accumulated more PHB than the wild-type strain. Also, it was more resistant to oxidative stress [68].

Kukolj et al. [66] analyzed the gene expression profile of *A. brasilense* grown in a medium with low and high nitrogen availability. Bacteria grown under nitrogen-limiting conditions increased the expression of proteins involved in energy production and conversion, signal transduction, and amino acid metabolism. However, when there was no nitrogen limiting, the bacterium increased the expression of proteins related to signal transduction, cell wall biogenesis, coenzyme metabolism, and energy metabolism.

## 7. Flocculation and Cyst Involvement in PHB Production, the Role of Oxygen

Apart from inducing PHB accumulation, media with a high C/N ratio can also stimulate the flocculation of bacteria [11,69–72]. However, the flocculation phenomenon is present mainly under high oxygen concentrations [11,69,70,73]. For generating energy and fixing nitrogen, the appropriate oxygen concentration for *A. brasilense* is 3–5 μM [62,74]. Under elevated DO *A. brasilense* Sp7 tends to clump. In clumping, motile cells interact between them in response to an elevated aeration rate. Clumped bacteria keep a microaerophilic environment to protect them from excessive oxygen [75]. When high aeration is maintained, clumped cells form irreversible macromolecular aggregates known as floccules, but when high aeration decreases, clumped cells return to a vegetative state [11,55,60,69].

Floccules are macroscopic bacterial aggregates that are clumped by a high accumulation of exopolysaccharides (EPS). EPS form a fibrilar matrix surrounding bacteria. Floccules protect *A. brasilense* Sp7 against the stresses caused by high temperatures (up to 40 °C), desiccation, oxygen availability, and pH [8,11,72,75–79]. *A. brasilense* floccules contain PHB granules with high biopolymer levels [80–83].

The EPS of *A. brasilense* Sp7 contain glucose during the exponential growth phase. However, in the stationary phase growth, they contain mainly arabinose [70,71,84]. EPS, whether glucose-containing or arabinose-rich, served as a carbon source when exogenous carbon was depleted. Interestingly, *A. brasilense* Sp7 cannot grow in media with arabinose or glucose as unique exogenous carbon sources, but it uses its arabinose or glucose as a carbon source [81].

Low nitrogen conditions and a high aeration rate led to *fcl*A overexpression. The *fcl*A gene controls the morphological transition from vegetative to cystic in response to environmental changes. Also, *fcl*A may control nitrogen assimilation by downregulating glutamine synthetase, which synthesizes glutamine from glutamate and ammonia. GlnA (citrate synthase) was also implicated in EPS production. It seems that *fcl*A promotes sugar assimilation by synthesizing carbohydrates for EPS. Also, it was noted that acetyl-CoA was mainly addressed to PHB synthesis rather than the TCA cycle in the *A. brasilense fcl*A mutant strain due to the *phb*A gene being overexpressed in an *fcl*A mutant strain [55].

Bible et al. [60] analyzed protein expression during growth under clumping conditions in *A. brasilense che*Y1 and the *che*B1–*che*R1 mutants. The *A. brasilense che*Y1 mutant strain clumped more than the wild-type strain. On the contrary, the *A. brasilense che*B1–*che*R1 double mutant strain was unable to clump. Both *che*Y1 and the *che*B1–*che*R1 mutants showed an increase in the expression of genes involved in PHB metabolism, such as the acetoacetyl-CoA reductase gene and the *pha*P1, *pha*P2, and *pha*P6 genes.

Malinich and Bauer [85] analyzed the transcription profile of encysted *A. brasilense.* Cysts repress genes involved in amino acid biosynthesis, ribosomal biogenesis, and translation.

In *A. brasilense* cysts, the *pha*R (AMK58_RS26785) and 3-hydroxyacyl-CoA dehydrogenase (AMK58_19430) genes were also repressed. Genes required for nitrogen metabolism, such as NasT, a response regulator required for growth under nitrate, nitrite/nitrite transporter (AMK58_21400, AMK58_21405, and AMK58_21410), and nitrate reductase (AMK58_21395) were upregulated as well as genes involved in nitrogen fixation [85].

## 8. PHB and Biofilm in *A. brasilense*

Depending on the growth conditions, *A. brasilense* produces a biofilm [86]. Vieruega et al. [87] have reported that the *Azospirillum* biofilm contains proteins such as polar flagellin and OmaA (outer membrane protein), extracellular DNA, and EPS. According to Tugarova et al. [88] and Shelud'ko et al. [89], the *Azospirillum* biofilm also contains high quantities of PHB. PHB would allow bacterial stabilization during biofilm formation.

## 9. Functions of PHB in *Azospirillum brasilense*

PHB accumulation and utilization by bacteria allow its establishment and survival in competitive environments. In such conditions, PHB functions as a carbon and energy source to sustain bacterial growth [18,32,38,84,90]. Also, it was demonstrated that bacteria with high PHB content colonize the rhizosphere to exert beneficial effects on plant growth and crop yield [38,72,78]. The use of inoculants based on *Azospirillum* with a high PHB content can prolong their useful life [38,72,83,90–92]. Previous studies have reported that the inability to synthesize or degrade PHB affects the resistance of *A. brasilense* to osmotic and oxidant stresses, as well as its resistance to UV radiation and high temperatures [23]. In *Pseudomonas extremaustralis*, better UV resistance was observed in agreement with a higher PHB accumulation [93]. PHB provides enough energy to sustain growth under nitrogen-fixing conditions. Previous studies have suggested PHB degradation can supply enough energy to sustain nitrogen fixation and spore development [94]. Similarly, PHB synthesis controls the redox state of bacteria, serving as an electron sink [95].

## 10. PHB Properties and Applications

A comparison between the characteristics of PHB and polypropylene (PP) is listed in Table 2. PHB shares similar properties with PP, in areas such as tensile modulus, tensile strength, and melting temperature. However, PHB is biodegradable and biocompatible, and its degradation does not release toxic products, which does not occur with PP [47]. The above characteristics would make PHB suitable to replace the use of PP in the industry. However, its high production costs, in addition to low thermal stability, a high degree of crystallinity, hydrophobicity, and brittleness, make PHB less suitable for being used in commercial applications [96,97]. To improve the quality of PHB, it has been combined with other materials, which has enhanced its mechanical and physical properties. Some examples of PHB blending are polylactic acid (PLA), hyaluronic acid (HA), polycaprolactone (PCL), polyethylene glycol (PEG), chitosan, and cellulose pectin, among others [97]. Functionalization of PHB by adding epoxy, hydroxyl, carbonyl, phenyl groups, and halogen atoms improves PHB properties [97,98]. It makes it possible to expand the uses of PHB, for example, in drug delivery.

**Table 2.** Mechanical properties of PHB and PP (modified from [47,97–100]).

| Parameter | PP | PHB |
|---|---|---|
| Tensile modulus (GPa) | 1.95 | 3–3.5 |
| Tensile strength (Mpa) | 31–45 | 20–40 |
| Elongation at break (%) | 50–400 | 5–10 |
| Crystallinity (%) | 42.6–70 | 50–60 |
| Melting temperature (°C) | 160–176 | 165–180 |
| Glass transition (°C) | −20–−5 | 2–9 |
| Density (g/cm$^2$) | 0.905 | 1.25 |
| UV resistance | Poor | Good |
| Biodegradability | No | Yes |
| Biocompatibility | No | Yes |

Drug delivery has been one of the most important uses of PHB. Previous studies have shown that PHB can be implanted in the human body, where it causes a mild inflammatory reaction that does not lead to fibrosis or necrosis [101]. Macrophage-mediated inflammation results in an exposition of PHB to extracellular liquids and cells, which results in a slow biopolymer degradation into 3-hydroxybutytyrate monomers and oligomers [102]. These properties suggest PHB is a good candidate for drug delivery [103]. Pandian et al. [104] loaded ursolic acid (an inhibitory agent against the proliferation of tumors) into PHB nanoparticles and evaluated the delivery, availability, and activity of the ursolic acid released from PHB nanoparticles against HeLa cells. In agreement with the number of dead tumor cells, it was concluded that the PHB nanoparticles had released ursolic acid, and they were more effective at 96 h. Another example of drug delivery was reported by Parsian et al. [105], who designed PHB-coated magnetic nanoparticles loaded with gemcitabine (GEM-PHB-MNPs) in order to release gemcitabine for treating breast cancer. Results showed that gemcitabine was released in an acidic microenvironment, like tumors. It was also shown that PHB-MNPs were not cytotoxic to cells. PHB/chitosan blends have been impregnated with ketoprofen [106]. PHB nanospheres and PHB microspheres loaded with extended-spectrum antibiotics have been used to prevent post-surgery infections. Sulbactam ampicillin/cefoperazone, and gentamicin have been loaded into PHB-co-PHV for drug delivery [107]. Other studies on drug release from PHB can be reviewed [97,108–110].

PHB can be used for tissue engineering. Deng et al. [111] developed an extracellular matrix of rabbit chondrocytes grown on PHB-co-PHH (polyhybroxybutyrate-co-hydroxyhexanoate) scaffolds. The results showed better seeding on PHB-co-PHH scaffolds than on single PHB scaffolds. It was also shown that more collagen was produced on PHB-co-PHH than on PHB. Temporary stents, bone plates, patches, and screws have been fabricated [111,112]. PHB-based composites were suitable for wound dressing and ocular implants [113]. Also, several artificial tissues, such as retinal, bone, tendon, cartilage, and muscle, have been developed [114,115]. Biopolymers can support cell growth [111,116]. PHB blends can be used as scaffolds and bone implants [97,117].

To be used for packaging, PHB must be stable, flexible, and highly resistant. A good biopolymer must provide a barrier against water vapor, oxygen, and carbon dioxide [118]. PHB blends are promising materials for packaging because they exhibit good barrier properties, and their degradation products are non-toxic for the environment [119]. PHB blends are used in bottles, jars, films, et cetera [120]. Composites of PHB prepared with coconut fibers showed good thermal stability and better tensile properties, making them suitable to be used as plastic bags for recovered seeds and planted for agriculture [121].

## 11. PHB Biodegradability

PHB biodegradability occurs in soils, water, and aerobic and anaerobic environments. It occurs in microorganisms that have extracellular depolymerases. PHB is degraded upon exposure to soil, compost, or marine sediment. It is supposed that there are approximately 10% of PHB-degrading microbes. PHB can be degraded in aerobic and anaerobic environments. In aerobic environments, PHB degradation results in $CO_2$ and $H_2O$, whereas methane is released in anaerobic environments. PHB degradation depends on microbial activity, moisture, temperature, pH, molecular weight of PHB, et cetera. PHB degradation units can be processed throughout the β-oxidation and TCA cycles [122].

## 12. Future Outlooks

The metabolic abilities of *A. brasilense* are extensive. It can grow in minimal to rich media and in a wide range of carbon and nitrogen sources. *A. brasilense* accumulates up to 80% of its dry cell weight as PHB, a biopolymer characterized as biodegradable and biocompatible. In bacteria, PHB serves as carbon and energy reserves. Also, it provides resistance to several stressful conditions. Understanding PHB metabolism and regulation in *A. brasilense* may help exploit PHB properties to improve quality of life. In this review were summarized the requirements to improve PHB accumulation by *A. brasilense*. Studies have shown this bacterium accumulates more PHB when it is grown at a high C/N ratio, ranging from 70 to 90. Also, the best carbon and nitrogen sources were malate and ammonia chloride, respectively. A microaerophilic environment is important too.

Although there are some studies on the genes involved in PHB metabolism, these genes have not been fully understood. Further studies are needed to identify other genes implicated in PHB synthesis and degradation. Now that the *A. brasilense* genome is available, it will be possible to analyze other genes that may be involved in PHB metabolism. The National Center for Biotechnology Information (NCBI), the Kyoto Encyclopedia of Genes and Genomes (KEGG), and the Clusters of Orthologous Genes (COG) databases showed that *A. brasilense* contains several copies of genes involved in PHB synthesis and degradation. It would be interesting to know when these genes are expressed and the interactions that occur between each gene, as well as the transcription factors involved in regulating genetic expression. Bioinformatics and in vitro analyses will help us improve our understanding.

PHB is a biodegradable and biocompatible plastic with potential utility for medical and environmental purposes. However, recovering PHB from bacterial cultures is highly expensive. Given that *A. brasilense* is one of the most important microorganisms that produce PHB, it is necessary to develop environmentally friendly techniques that allow us to increase the recovery and purity of the polymer. Also, it is important to make more efforts toward functionalizing PHB in order to expand its uses and to decrease the use of polypropylene.

**Author Contributions:** Conceptualization, L.J.M.M., L.S.U. and M.d.l.Á.M.M.; methodology, M.d.l.Á.M.M. and Y.A.C.; software, Y.A.C., M.d.l.Á.M.M. and M.B.R.; validation, L.S.U. and M.d.l.Á.M.M.; formal analysis, L.J.M.M. and L.S.U.; investigation, M.d.l.Á.M.M., Y.A.C., M.B.R. and L.S.U.; resources, L.J.M.M. and L.S.U.; data curation, L.J.M.M. and L.S.U.; writing—original draft preparation, L.J.M.M. and L.S.U.; writing—review and editing, M.d.l.Á.M.M. and L.S.U.; visualization, M.d.l.Á.M.M., Y.A.C. and M.B.R.; supervision, L.S.U.; project administration, L.J.M.M.; funding acquisition, L.J.M.M. and L.S.U. All authors have read and agreed to the published version of the manuscript.

**Funding:** This research was funded by: Vicerrectoría de Investigación y Estudios de Posgrado, Benemérita Universidad Autónoma de Puebla (VIEP-BUAP), grant number: 00243. The APC was funded by VIEP-BUAP.

**Institutional Review Board Statement:** Not applicable.

**Data Availability Statement:** Not applicable.

**Acknowledgments:** The authors gratefully acknowledge financial support from the VIEP-BUAP projects: MAML-NAT17-1, SOUL-NAT17-1. The authors are very grateful to Cristina Guadalupe Aguilera Chapital for helping with the English language.

**Conflicts of Interest:** The authors declare no conflict of interest. The funders had no role in the design of the study; in the collection, analyses, or interpretation of data; in the writing of the manuscript; or in the decision to publish the results.

## Abbreviations

PHB: poly-β-hydroxybutyrate; PHA: polyhydroxyalkanoates; PHV: polyhydroxyvalerate; FTIR: Fourier-transform infrared spectroscopy; PP: polypropylene; C/N: carbon to nitrogen ratio; N2: nitrogen gas; NH3: ammonia; NH4: ammonium; $NO_3^-$: ammonium nitrate; TCA: tricarboxylic acid; DCW: dry cell weight; NAD(P: nicotinamide adenine dinucleotide (phosphate); ACP: acyl carrier protein; GAP: granule-associated protein; h:hours; DO: dissolved oxygen; PHB-co-PHH: Polyhydroxybutyrate-co-hydroxyhexanoate; NCBI: National Center for Biotechnology Information; KEGG: Kyoto Encyclopedia of Genes and Genomes; COG: Clusters of Orthologous Genes.

## References


1. Pereg, L. *Azospirillum* Cell Aggregation, Attachment, and Plant Interaction. In *Handbook for Azospirillum Technical Issues and Protocols*, 1st ed.; Cassán, F.D., Okon, Y., Creus, C.M., Eds.; Springer International Publishing: Cham, Switzerland, 2015; Volume 1, pp. 181–197. [CrossRef]
2. Martínez-Martínez, M.D.L.A.; González-Pedrajo, B.; Dreyfus, G.; Soto-Urzúa, L.; Martínez-Morales, L.J. Phasin PhaP1 is involved in polyhydroxybutyrate granules morphology and in controlling early biopolymer accumulation in *Azospirillum brasilense* Sp7. *AMB Express* **2019**, *9*, 1–15. [CrossRef] [PubMed]
3. Volova, T.G.; Boyandin, A.N.; Vasiliev, A.D.; Karpov, V.A.; Prudnikova, S.V.; Mishukova, O.V.; Boyarskikh, U.A.; Filipenko, M.L.; Rudnev, V.P.; Xuân, B.B.; et al. Biodegradation of polyhydroxyalkanoates (PHAs) in tropical coastal waters and identification of PHA-degrading bacteria. *Polym. Degrad. Stab.* **2010**, *95*, 2350–2359. [CrossRef]
4. Halevas, E.G.; Pantazaki, A.A. Polyhydroxyalkanoates: Chemical structure. In *Polyhydroxyalkanoates: Biosynthesis, Chemical Structure and Applications*, 1st ed.; Williams, H., Kelly, P., Eds.; Nova Science Publishers Inc.: New York, NY, USA, 2018; Volume 1, pp. 133–166.
5. Itzigsohn, R.; Yarden, O.; Okon, Y. Polyhydroxyalkanoate analysis in *Azospirillum brasilense*. *Can. J. Micrcobiol.* **1995**, *41*, 73–76. [CrossRef]
6. Patiño, I.M.E.; Soto, U.L.; Orea, F.M.L.; López, V.D.; Martínez-Morales, L.J. Extraction and NMR determinantion of PHB from *Azospirillum brasilense* Sp7. *JCBPS Spec. Issue* **2014**, *4*, 26–32.
7. Kamnev, A.A.; Antonyuk, L.P.; Tugarova, A.V.; Tarantilis, P.A.; Polissiou, M.G.; Gardiner, P.H.E. Fourier transform infrared spectroscopic characterisation of heavy metal-induced metabolic changes in the plant-associated soil bacterium *Azospirillum brasilense* Sp7. *J. Mol. Struct.* **2002**, *610*, 127–131. [CrossRef]
8. Tal, S.; Okon, Y. Production of the reserve material poly-β-hydroxybutyrate and its function in *Azospirillum brasilense* Cd. *Can. J. Microbiol.* **1985**, *31*, 608–613. [CrossRef]
9. Tal, S.; Smirnoff, P.; Okon, Y. The regulation of poly-β-hydroxybutyrate metabolism in *Azospirillum brasilense* during balanced growth and starvation. *Microbiology* **1990**, *136*, 1191–1196. [CrossRef]
10. Westby, C.A.; Cutshall, D.S.; Vigil, G.V. Metabolism of various carbon sources by *Azospirillum brasilense*. *J. Bacteriol.* **1983**, *156*, 1369–1372. [CrossRef]
11. Sadasivan, L.; Neyra, C.A. Flocculation in *Azospirillum brasilense* and *Azospirillum lipoferum*: Exopolysaccharides and cyst formation. *J. Bacteriol.* **1985**, *163*, 716–723. [CrossRef]
12. Loh, W.H.; Randles, C.I.; Sharp, W.R.; Miller, R.H. Intermediary carbon metabolism of *Azospirillum brasilense*. *J. Bacteriol.* **1984**, *158*, 264–268. [CrossRef]
13. Alexandre, G.; Greer, S.E.; Zhulin, I.B. Energy taxis is the dominant behavior in *Azospirillum brasilense*. *J. Bacteriol.* **2000**, *182*, 6042–6048. [CrossRef]
14. Okon, Y.; Albrecht, S.L.; Burris, R.H. Factors affecting growth and nitrogen fixation of *Spirillum lipoferum*. *J. Bacteriol.* **1976**, *127*, 1248–1254. [CrossRef] [PubMed]
15. Neyra, C.A.; Döbereiner, J.; Lalande, R.; Knowles, R. Denitrification by $N_2$-fixing *Spirillum lipoferum*. *Can. J. Microbiol.* **1977**, *23*, 300–305. [CrossRef] [PubMed]
16. Martinez-Drets, G.; Del Gallo, M.; Burpee, C.; Burris, R.H. Catabolism of carbohydrates and organic acids and expression of nitrogenase by *Azospirilla*. *J. Bacteriol.* **1984**, *159*, 80–85. [CrossRef] [PubMed]
17. Tal, S.; Smirnoff, P.; Okon, Y. Purification and characterization of d-β-hydroxybutyrate dehydrogenase from *Azospirillum brasilense* Cd. *Microbiology* **1990**, *136*, 645–649. [CrossRef]
18. Okon, Y.; Itzigsohn, R. Poly-β-hydroxybutyrate metabolism in *Azospirillum brasilense* and the ecological role of PHB in the rhizosphere. *FEMS Microbiol.* **1992**, *9*, 131–139. [CrossRef]
19. Manna, A.; Pal, S.; Paul, A.K. Occurrence of poly-3-hydroxybutyrate in *Azospirillum* sp. *Folia Microbiol.* **1997**, *42*, 629–634. [CrossRef]
20. Narancic, T.; Scollica, E.; Kenny, S.T.; Gibbons, H.; Carr, E.; Brennan, L.; Cafney, G.; Wynne, K.; Murphy, C.; Raberg, M.; et al. Understanding the physiological roles of polyhydroxybutyrate (PHB) in Rhodospirillum rubrum S1 under aerobic chemoheterotrophic conditions. *Appl. Microbiol. Biotechnol.* **2016**, *100*, 8901–8912. [CrossRef]
21. Prieto, A.; Escapa, I.F.; Martínez, V.; Dinjaski, N.; Herencias, C.; de la Peña, F.; Tarazona, N.; Revelles, O. A holistic view of polyhydroxyalkanoate metabolism in *Pseudomonas putida*. *Environ. Microbiol.* **2016**, *18*, 341–357. [CrossRef]
22. Ochsner, A.M.; Sonntag, F.; Buchhaupt, M.; Schrader, J.; Vorholt, J.J. *Methylobacterium extorquens*: Methylotrophy and biotechnological applications. *Appl. Microbiol. Biotechnol.* **2015**, *99*, 517–534. [CrossRef]
23. Quelas, J.I.; Mesa, S.; Mongiardini, E.J.; Jendrossek, D.; Lodeiro, A.R. Regulation of polyhydroxybutyrate synthesis in the soil bacterium *Bradyrhizobium diazoefficiens*. *Appl. Environ. Microbiol.* **2016**, *82*, 4299–4308. [CrossRef]
24. Sreedevi, S.; Unni, K.N.; Sajith, S.; Priji, P.; Josh, M.S.; Benjamin, S. Bioplastics: Advances in polyhydroxybutyrate research. In *Advances in Polymer Science*; Abe, A., Albertsson, A.C., Coates, G.W., Genzer, J., Kobayashi, S., Lee, K.S., Leibler, L., Long, T.E., Möller, M., Okay, O., et al., Eds.; Springer: Berlin/Heidelberg, Germany, 2016; Volume 1, pp. 1–30. [CrossRef]
25. Soto, L.R.; Byrne, E.; Van Niel, E.W.; Sayed, M.; Villanueva, C.C.; Hatti-Kaul, R. Hydrogen and polyhydroxybutyrate production from wheat straw hydrolysate using *Caldicellulosiruptor* species and *Ralstonia eutropha* in a coupled process. *Bioresour. Technol.* **2009**, *272*, 259–266. [CrossRef] [PubMed]

26. Ayub, N.D.; Pettinari, M.J.; Ruiz, J.A.; López, N.I. A polyhydroxybutyrate-producing *Pseudomonas* sp. isolated from Antarctic environments with high stress resistance. *Curr. Microbiol.* **2004**, *49*, 170–174. [CrossRef] [PubMed]
27. Ayub, N.D.; Pettinari, M.J.; Méndez, B.S.; López, N.I. Impaired polyhydroxybutyrate biosynthesis from glucose in *Pseudomonas* sp. 14-3 is due to a defective beta-ketothiolase gene. *FEMS Microbiol. Lett.* **2006**, *264*, 125–131. [CrossRef] [PubMed]
28. Ghoddosi, F.; Golzar, H.; Yazdian, F.; Khosravi-Darani, K.; Vasheghani-Farahani, E. Effect of carbon sources for PHB production in bubble column bioreactor: Emphasis on improvement of methane uptake. *J. Environ. Chem. Eng.* **2019**, *7*, 102978. [CrossRef]
29. Manju, J.; Prabakaran, P. Effect of carbon sources in the production of polyhydroxybutyrate (PHB) by *Bradyrhizobium* and *Rhizobium* sp. from *Aeschynomene indica*. *Int. J. Res. Anal. Rev.* **2019**, *6*, 823–827.
30. Torres-Pedraza, A.J.; Salgado-Lugo, H.; Segura, D.; Díaz-Barrera, A.; Peña, C. Composition control of poly (3-hydroxybutyrate-co-3-hydroxyvalerate) copolymerization by oxygen transfer rate (OTR) in *Azotobacter vinelandii* OPNA. *J. Chem. Technol. Biotechnol.* **2021**, *96*, 2782–2791. [CrossRef]
31. Hassan, M.A.; Bakhiet, E.K.; Hussein, H.R.; Ali, S.G. Statistical optimization studies for polyhydroxybutyrate (PHB) production by novel *Bacillus subtilis* using agricultural and industrial wastes. *Int. J. Environ. Sci. Technol.* **2019**, *16*, 3497–3512. [CrossRef]
32. Kadouri, D.; Burdman, S.; Jurkevitch, E.; Okon, Y. Identification and isolation of genes involved in poly (β-hydroxybutyrate) biosynthesis in *Azospirillum brasilense* and characterization of a *phb*C mutant. *Appl. Environ. Microbiol.* **2002**, *68*, 2943–2949. [CrossRef]
33. Kadouri, D.; Jurkevitch, E.; Okon, Y.; Castro-Sowinski, S. Ecological and Agricultural Significance of Bacterial Polyhydroxyalkanoates. *Crit. Rev. Microbiol.* **2005**, *31*, 55–67. [CrossRef]
34. Senior, P.J.; Dawes, E.A. The regulation of poly-β-hydroxybutyrate metabolism in *Azotobacter beijerinckii*. *Biochem. J.* **1973**, *134*, 225–238. [CrossRef]
35. Volova, T.G.; Kalacheva, G.S.; Gorbunova, O.V.; Zhila, N.O. Dynamics of Activity of the Key Enzymes of Polyhydroxyalkanoate Metabolism in *Ralstonia eutropha*. *Appl. Biochem. Microbiol.* **2004**, *40*, 170–177. [CrossRef]
36. Trainer, M.A.; Charles, T.C. The role of PHB metabolism in the symbiosis of rhizobia with legumes. *Appl. Microbiol. Biotechnol.* **2006**, *71*, 377–386. [CrossRef] [PubMed]
37. Edelshtein, Z.; Kadouri, D.; Jurkevitch, E.; Vande Broek, A.; Vanderleyden, J.; Okon, Y. Characterization of genes involved in poly-β-hydroxybutyrate metabolism in *Azospirillum brasilense*. *Symbiosis* **2003**, *34*, 157–170.
38. Kadouri, D.; Jurkevitch, E.; Okon, Y. Involvement of the reserve material poly-β-hydroxybutyrate in *Azospirillum brasilense* stress endurance and root colonization. *Appl. Environ. Microbiol.* **2003**, *69*, 3244–3250. [CrossRef]
39. Müller-Santos, M.; Maltempi de Souza, E.; de Oliveira-Pedrosa, F.; Chubatsu, L.S. Polyhydroxybutyrate in *Azospirillum brasilense*. In *Handbook for Azospirillum Technical Issues and Protocols*, 1st ed.; Cassán, F.D., Okon, Y., Creus, C.M., Eds.; Springer International Publising: Cham, Switzerland, 2015; Volume 1, pp. 241–250. [CrossRef]
40. Jendrossek, D. Polyhydroxyalkanoate granules are complex subcellular organelles (carbonosomes). *J. Bacteriol.* **2009**, *191*, 3195–3202. [CrossRef] [PubMed]
41. Bresan, S.; Sznajder, A.; Hauf, W.; Forchhammer, K.; Pfeiffer, D.; Jendrossek, D. Polyhydroxyalkanoate (PHA) granules have no phospholipids. *Sci. Rep.* **2016**, *6*, 1–13. [CrossRef]
42. Bresan, S.; Jendrossek, D. New Insights into PhaM-PhaC-Mediated Localization of Polyhydroxybutyrate Granules in *Ralstonia eutropha* H16. *Appl. Environ. Microbiol.* **2017**, *83*, e00505–17. [CrossRef]
43. Jurasek, L.; Marchessault, R.H. The role of phasins in the morphogenesis of poly (3-hydroxybutyrate) granules. *Biomacromolecules* **2002**, *3*, 256–261. [CrossRef]
44. Pötter, M.; Steinbüchel, A. Poly (3-hydroxybutyrate) granule-associated proteins: Impacts on poly (3-hydroxybutyrate) synthesis and degradation. *Biomacromolecules* **2005**, *6*, 552–560. [CrossRef]
45. Tirapelle, E.F.; Müller-Santos, M.; Tadra-Sfeir, M.Z.; Kadowaki, M.A.S.; Steffens, M.B.R.; Monteiro, R.A.; Souza, E.M.; Pedrosa, F.O.; Chubatsu, L.S. Identification of proteins associated with polyhydroxybutyrate granules from *Herbaspirillum seropedicae* SmR1-old partners, new players. *PLoS ONE* **2013**, *8*, e75066. [CrossRef]
46. Maehara, A.; Doi, Y.; Nishiyama, T.; Takagi, Y.; Ueda, S.; Nakano, H.; Yamane, T. PhaR, a protein of unknown function conserved among short-chain-length polyhydroxyalkanoic acids producing bacteria, is a DNA-binding protein and represses *Paracoccus denitrificans* phaP expression in vitro. *FEMS Microbiol. Lett.* **2001**, *200*, 9–15. [CrossRef] [PubMed]
47. Chai, J.M.; Amelia, T.S.M.; Mouriya, G.K.; Bhubalan, K.; Amirul, A.A.A.; Vigneswari, S.; Ramakrishna, S. Surface-modified highly biocompatible bacterial-poly (3-hydroxybutyrate-co-4-hydroxybutyrate): A review on the promising next-generation biomaterial. *Polymers* **2020**, *13*, 51. [CrossRef] [PubMed]
48. Aneja, P.; Charles, T.C. Poly-3-hydroxybutyrate degradation in *Rhizobium* (*Sinorhizobium*) *meliloti*: Isolation and characterization of a gene encoding 3-hydroxybutyrate dehydrogenase. *J. Bacteriol.* **1999**, *181*, 849–857. [CrossRef]
49. Handrick, R.; Reinhardt, S.; Jendrossek, D. Mobilization of poly (3-hydroxybutyrate) in *Ralstonia eutropha*. *J. Bacteriol.* **2000**, *182*, 5916–5918. [CrossRef]
50. Bashan, Y.; Holguin, G.; De-Bashan, L.E. *Azospirillum*-plant relationships: Physiological, molecular, agricultural, and environmental advances (1997–2003). *Can. J. Microbiol.* **2004**, *50*, 521–577. [CrossRef] [PubMed]
51. Vieille, M.; Elmerich, C. Characterization of an *Azospirillum brasilense* Sp7 gene homologous to *Alcaligenes eutrophus phb*B and *Rhizobium meliloti nod*G. *Mol. Gen. Genet.* **1992**, *231*, 375–384. [CrossRef]

52. Aguilar, G.G. Análisis del Efecto de la Co-Transcripción de los Genes *phb*ABC Sobre la Producción de PHB de *Azospirillum brasilense* Sp245. Master's Thesis, Benemérita Universidad Autónoma de Puebla, Puebla, México, 2016.
53. Peralta-Gil, M.; Segura, D.; Guzman, J.; Servin-Gonzalez, L.; Espin, G. Expression of the *Azotobacter vinelandii* poly-3-hydroxybutyrate biosynthetic phbBAC operon is driven by two overlapping promoters and is dependent on the transcriptional activator PhbR. *J. Bacteriol.* **2002**, *184*, 5672–5677. [CrossRef]
54. Aneja, P.; Dai, M.; Lacorre, D.A.; Pillon, B.; Charles, T.C. Heterologous complementation of the exopolysaccharide synthesis and carbon utilization phenotypes of *Sinorhizobium meliloti* Rm1021 polyhydroxyalkanoate synthesis mutants. *FEMS Microbiol. Lett.* **2004**, *239*, 277–283. [CrossRef]
55. Hou, X.; McMillan, M.; Coumans, J.V.; Poljak, A.; Raftery, M.J.; Pereg, L. Cellular responses during morphological transformation in *Azospirillum brasilense* and its *flc*A knockout mutant. *PLoS ONE* **2014**, *9*, e114435. [CrossRef]
56. Akhlaq, S.; Singh, D.; Mittal, N.; Srivastava, G.; Siddiqui, S.; Faridi, S.A.; Siddiqui, M.H. Polyhydroxybutyrate biosynthesis from different waste materials, degradation, and analytic methods: A short review. *Polym. Bull.* **2023**, *80*, 5965–5997. [CrossRef]
57. Paul, E.; Mulard, D.; Blanc, P.; Fages, J.; Goma, G.; Pareilleux, A. Effects of partial $O_2$ pressure, partial $CO_2$ pressure, and agitation on growth kinetics of *Azospirillum lipoferum* under fermentor conditions. *Appl. Environ. Microbiol.* **1990**, *56*, 3235–3239. [CrossRef] [PubMed]
58. de Eugenio, L.I.; Escapa, I.F.; Morales, V.; Dinjaski, N.; Galán, B.; García, J.L.; Prieto, M.A. The turno-ver of medium-chain-length polyhydroxyalkanoates in *Pseudomonas putida* KT2442 and the fundamental role of PhaZ depolymerase for the metabolic balance. *Environ. Microbiol.* **2010**, *12*, 207–221. [CrossRef] [PubMed]
59. Hauf, W.; Schlebusch, M.; Hüge, J.; Kopka, J.; Hagemann, M.; Forchhammer, K. Metabolic changes in *Synechocystis* PCC6803 upon nitrogen starvation: Excess NADPH sustains polyhydroxybutyrate accumulation. *Metabolites* **2013**, *3*, 101–118. [CrossRef]
60. Bible, A.N.; Khalsa-Moyers, G.K.; Mukherjee, T.; Green, C.S.; Mishra, P.; Purcell, A.; Aksenova, A.; Hurst, G.B.; Alexandre, G. Metabolic adaptations of *Azospirillum brasilense* to oxygen stress by cell-to-cell clumping and flocculation. *Appl. Environ. Microbiol.* **2015**, *81*, 8346–8357. [CrossRef]
61. Kefalogianni, I.; Aggelis, G. Metabolic activities in *Azospirillum lipoferum* grown in the presence of $NH_4$. *Appl. Microbiol. Biotechnol.* **2003**, *62*, 574–578. [CrossRef]
62. Kefalogianni, I.; Aggelis, G. Modeling growth and biochemical activities of *Azospirillum spp. Appl. Microbiol. Biotechnol.* **2002**, *58*, 352–357. [CrossRef]
63. Kamnev, A.A.; Sadovnikova, J.N.; Tarantilis, P.A.; Polissiou, M.G.; Antonyuk, L.P. Responses of *Azospirillum brasilense* to nitrogen deficiency and to wheat lectin: A diffuse reflectance infrared Fourier transform (DRIFT) spectroscopic study. *Microb. Ecol.* **2008**, *56*, 615–624. [CrossRef]
64. Gallori, E.; Bazzicalupo, M. Effect of nitrogen compounds on nitrogenase activity in *Azospirillum brasilense. FEMS Microbiol. Lett. 1985. 28*, 35–38. [CrossRef]
65. Sun, J.; Van Dommelen, A.; Van Impe, J.; Vanderleyden, J. Involvement of *gln*B, *gln*Z, and *gln*D genes in the regulation of poly-3-hydroxybutyrate biosynthesis by ammonia in *Azospirillum brasilense* Sp7. *Appl. Environ. Microbiol.* **2002**, *68*, 985–988. [CrossRef]
66. Kukolj, C.; Pedrosa, F.O.; de Souza, G.A.; Sumner, L.W.; Lei, Z.; Sumner, B.; Lei, Z.; Summer, B.; do Amaral, F.P.; Juexin, W.; et al. Proteomic and metabolomic analysis of *Azospirillum brasilense ntr*C mutant under high and low nitrogen conditions. *J. Proteome Res.* **2020**, *19*, 92–105. [CrossRef] [PubMed]
67. Sun, J.; Peng, X.; Van Impe, J.; Vanderleyden, J. The *ntr*B and *ntr*C genes are involved in the regulation of poly-3-hydroxybutyrate biosynthesis by ammonia in *Azospirillum brasilense* Sp7. *Appl. Environ. Microbiol.* **2000**, *66*, 113–117. [CrossRef] [PubMed]
68. Sacomboio, E.N.M.; Kim, E.Y.S.; Ruchaud-Correa, H.L.; Bonato, P.; de Oliveira-Pedrosa, F.; de Souza, E.M.; Chubatsu, L.S.; Müller-Santos, M. The transcriptional regulator NtrC controls glucose-6-phosphate dehydrogenase expression and polyhydroxybutyrate synthesis through NADPH availability in *Herbaspirillum seropedicae. Sci. Rep.* **2017**, *7*, 13546. [CrossRef] [PubMed]
69. Burdman, S.; Jurkevitch, E.; Schwartsburd, B.; Hampel, M.; Okon, Y. Aggregation in *Azospirillum brasilense*: Effects of chemical and physical factors and involvement of extracellular components. *Microbiology* **1998**, *144*, 1989–1999. [CrossRef]
70. Burdman, S.; Jurketvitch, E.; Soria-Diaz, M.E.; Gil-Serrano, A.M.; Okon, Y. Extracellular polysaccharide composition of *Azospirillum brasilense* and its relation with cell aggregation. *FEMS Microbiol. Lett.* **2000**, *489*, 259–264. [CrossRef]
71. Fischer, S.E.; Marioli, J.M.; Mori, G. Effect of root exudates on the exopolysaccharide composition and the lipopolysaccharide profile of *Azospirillum brasilense* Cd under saline stress. *FEMS Microbiol. Lett.* **2003**, *219*, 53–62. [CrossRef]
72. Oliveira, A.L.; Santos, O.J.; Marcelino, P.R.; Milani, K.M.; Zuluaga, M.Y.; Zucareli, C.; Gonçalves, L.S. Maize inoculation with *Azospirillum brasilense* Ab-V5 cells enriched with exopolysaccharides and polyhydroxybutyrate results in high productivity under low N fertilizer input. *Front. Microbiol.* **2017**, *8*, 1873. [CrossRef]
73. Joe, M.; Karthikeyan, M.B.; Sekar, C.; Deiveekasundaram, M. Optimization of biofloc production in *Azospirillum brasilense* (MTCC-125) and evaluation of its adherence with the roots of certain crops. *Indian J. Microbiol.* **2010**, *50*, 21–25. [CrossRef]
74. Zhulin, I.B.; Bespalov, V.A.; Johnson, M.S.; Taylor, B.L. Oxygen taxis and proton motive force in *Azospirillum brasilense. J. Bacteriol.* **1996**, *178*, 5199–5204. [CrossRef]
75. Konnova, S.A.; Makarov, O.E.; Skvortsov, I.M.; Ignatov, V.V. Isolation, fractionation and some properties of polysaccharides produced in a bound form by *Azospirillum brasilense* and their possible involvement in *Azospirillum*–wheat root interactions. *FEMS Microbiol. Lett.* **1994**, *118*, 93–99. [CrossRef]

76. Konnova, S.A.; Brykova, O.S.; Sachkova, O.A.; Egorenkova, I.V.; Ignatov, V.V. Protective role of the polysaccharide-containing capsular components of *Azospirillum brasilense*. *Microbiology* **2001**, *70*, 436–440. [CrossRef]
77. Vendan, R.T.; Thangaraju, M. Development and standardization of cyst based liquid formulation of *Azospirillum* bioinoculant. *Acta Microbiol. Immunol. Hung.* **2007**, *54*, 167–177. [CrossRef] [PubMed]
78. Fadel-Picheth, C.M.T.; Souza, E.M.; Rigo, L.U.; Yates, M.G.; Pedrosa, F.O. Regulation of *Azospirillum brasilense nif*A gene expression by ammonium and oxygen. *FEMS Microbiol. Lett.* **1999**, *179*, 281–288. [CrossRef] [PubMed]
79. Hartmann, A.; Burris, R.H. Regulation of nitrogenase activity by oxygen in *Azospirillum brasilense* and *Azospirillum lipoferum*. *J. Bacteriol.* **1987**, *169*, 944–948. [CrossRef]
80. Bleakley, B.H.; Gaskins, M.H.; Hubbell, D.H.; Zam, S.G. Floc formation by *Azospirillum lipoferum* grown on poly-β-hydroxybutyrate. *Appl. Environ. Microbiol.* **1988**, *54*, 2986–2995. [CrossRef] [PubMed]
81. Sadasivan, L.; Neyra, C.A. Cyst production and brown pigment formation in aging cultures of *Azospirillum brasilense* ATCC 29145. *J. Bacteriol.* **1987**, *169*, 1670–1677. [CrossRef] [PubMed]
82. Li, H.; Cui, Y.; Wu, L.; Tu, R.; Chen, S. cDNA-AFLP analysis of differential gene expression related to cell chemotactic and encystment of *Azospirillum brasilense*. *Microbiol. Res.* **2011**, *166*, 595–605. [CrossRef]
83. Santos, M.S.; Hungria, M.; Nogueira, M.A. Production of polyhydroxybutyrate (PHB) and biofilm by *Azospirillum brasilense* aiming at the development of liquid inoculants with high performance. *Afr. J. Bitechnol.* **2017**, *16*, 1855–1862. [CrossRef]
84. Bahat-Samet, E.; Castro-Sowinski, S.; Okon, Y. Arabinose content of extracellular polysaccharide plays a role in cell aggregation of *Azospirillum brasilense*. *FEMS Microbiol. Lett.* **2004**, *237*, 195–203. [CrossRef]
85. Malinich, E.A.; Bauer, C.E. Transcriptome analysis of *Azospirillum brasilense* vegetative and cyst states reveals large-scale alterations in metabolic and replicative gene expression. *Microb. Genom.* **2018**, *4*. [CrossRef]
86. Ramírez-Mata, A.; López-Lara, L.I.; Xiqui-Vázquez, M.L.; Jijón-Moreno, S.; Romero-Osorio, A.; Baca, B.E. The cyclic-di-GMP diguanylate cyclase CdgA has a role in biofilm formation and exopolysaccharide production in *Azospirillum brasilense*. *Res. Microbiol.* **2016**, *167*, 190–201. [CrossRef]
87. Viruega-Góngora, V.I.; Acatitla-Jácome, I.S.; Zamorano-Sánchez, D.; Reyes-Carmona, S.R.; Xiqui-Vázquez, M.L.; Baca, B.E.; Ramírez-Mata, A. The GGDEF-EAL protein CdgB from *Azospirillum baldaniorum* Sp245, is a dual function enzyme with potential polar localization. *PLoS ONE* **2022**, *17*, e0278036. [CrossRef]
88. Tugarova, A.V.; Scheludko, A.V.; Dyatlova, Y.A.; Filip'echeva, Y.A.; Kamnev, A.A. FTIR spectroscopic study of biofilms formed by the rhizobacterium *Azospirillum brasilense* Sp245 and its mutant *Azospirillum brasilense* Sp245. 1610. *J. Mol. Struct.* **2017**, *1140*, 142–147. [CrossRef]
89. Shelud'ko, A.V.; Mokeev, D.I.; Evstigneeva, S.S.; Filip'echeva, Y.A.; Burov, A.M.; Petrova, L.P.; Ponomareva, E.G.; Katsy, E.I. Cell Ultrastructure in *Azospirillum brasilense* Biofilms. *Microbiology* **2020**, *89*, 50–63. [CrossRef]
90. Sivasakthivelan, P.; Saranraj, P.; Al-Tawaha, A.R.; Amala, K.; Al Tawaha, A.R.; Thangadurai, D.; Sangeetha, J.; Rauf, A.; Khalid, S.; Alsultan, W.; et al. Adaptation of *Azospirillum* to stress conditions: A review. *Adv. Environ. Biol.* **2021**, *15*, 1–5. [CrossRef]
91. Fibach-Paldi, S.; Burdman, S.; Okon, Y. Key physiological properties contributing to rhizosphere adaptation and plant growth promotion abilities of *Azospirillum brasilense*. *FEMS Microbiol. Lett.* **2012**, *326*, 99–108. [CrossRef] [PubMed]
92. Carrasco-Espinosa, K.; García-Cabrera, R.I.; Bedoya-López, A.; Trujillo-Roldán, M.A.; Valdez-Cruz, N.A. Positive effect of reduced aeration rate on growth and stereospecificity of DL-malic acid consumption by *Azospirillum brasilense*: Improving the shelf life of a liquid inoculant formulation. *J. Biotechnol.* **2015**, *195*, 74–81. [CrossRef] [PubMed]
93. Tribelli, P.M.; Pezzoni, M.; Brito, M.G.; Montesinos, N.V.; Costa, C.S.; López, N.I. Response to lethal UVA radiation in the Antarctic bacterium *Pseudomonas extremaustralis*: Polyhydroxybutyrate and cold adaptation as protective factors. *Extremophiles* **2020**, *24*, 265–275. [CrossRef]
94. López, J.A.; Naranjo, J.M.; Higuita, J.C.; Cubitto, M.A.; Cardona, C.A.; Villar, M.A. Biosynthesis of PHB from a new isolated *Bacillus megaterium* strain: Outlook on future developments with endospore forming bacteria. *Biotechnol. Bioprocess Eng.* **2012**, *17*, 250–258. [CrossRef]
95. Koch, M.; Forchhammer, K. Polyhydroxybutyrate: A useful product of chlorotic cyanobacteria. *Microb. Physiol.* **2021**, *31*, 67–77. [CrossRef]
96. Majerczak, K.; Wadkin-Snaith, D.; Magueijo, V.; Mulheran, P.; Liggat, J.; Johnston, K. Polyhydroxybutyrate: A review of experimental and simulation studies of the effect of fillers on crystallinity and mechanical properties. *Polym. Int.* **2022**, *71*, 1398–1408. [CrossRef]
97. Raza, Z.A.; Khalil, S.; Abid, S. Recent progress in development and chemical modification of poly (hydroxybutyrate)-based blends for potential medical applications. *Int. J. Biol. Macromol.* **2020**, *160*, 77–100. [CrossRef] [PubMed]
98. McAdam, B.; Brennan, F.M.; McDonald, P.; Mojicevic, M. Production of polyhydroxybutyrate (PHB) and factors impacting its chemical and mechanical characteristics. *Polymer* **2020**, *12*, 2908. [CrossRef] [PubMed]
99. Verlinden, R.A.; Hill, D.J.; Kenward, M.; Williams, C.D.; Radecka, I. Bacterial synthesis of biodegradable polyhydroxyalkanoates. *J. Appl. Microbiol.* **2007**, *102*, 1437–1449. [CrossRef] [PubMed]
100. Hankermeyer, C.R.; Tjeerdema, R.S. Polyhydroxybutyrate: Plastic made and degraded by microorganisms. *Rev. Environ. Contam. Toxicol.* **1999**, *159*, 1–24. [PubMed]
101. Nair, L.S.; Laurencin, C.T. Biodegradable polymers as biomaterials. *Prog. Polym. Sci.* **2007**, *32*, 762–798. [CrossRef]

102. Bonartsev, A.P.; Bonartseva, G.A.; Reshetov, I.V.; Kirpichnikov, M.P.; Shaitan, K.V. Application of polyhydroxyalkanoates in medicine and the biological activity of natural poly (3-hydroxybutyrate). *Acta Nat.* **2019**, *11*, 4–16. [CrossRef]
103. Shishatskaya, E.I.; Voinova, O.N.; Goreva, A.V.; Mogilnaya, O.A.; Volova, T.G. Biocompatibility of polyhydroxybutyrate microspheres: In vitro and in vivo evaluation. *J. Mat. Sci.* **2008**, *19*, 2493–2502. [CrossRef]
104. Pandian, S.R.K.; Kunjiappan, S.; Pavadai, P.; Sundarapandian, V.; Chandramohan, V.; Sundar, K. Delivery of Ursolic Acid by Polyhydroxybutyrate Nanoparticles for Cancer Therapy: In silico and in vitro Studies. *Drug Res.* **2022**, *72*, 72–81.
105. Parsian, M.; Mutlu, P.; Yalcin, S.; Gunduz, U. Characterization of gemcitabine loaded polyhydroxybutyrate coated magnetic nanoparticles for targeted drug delivery. *Anticancer Agents Med. Chem.* **2020**, *20*, 1233–1240. [CrossRef]
106. Lins, L.C.; Bazzo, G.C.; Barreto, P.L.; Pires, A.T. Composite PHB/chitosan microparticles obtained by spray drying: Effect of chitosan concentration and crosslinking agents on drug relesase. *J. Braz. Chem. Soc.* **2014**, *25*, 1462–1471. [CrossRef]
107. Yagmurlu, M.F.; Korkusuz, F.; Gürsel, I.; Korkusuz, P.; Örs, Ü.; Hasirci, V. Sulbactam-cefoperazone polyhydroxybutyrate-co-hydroxyvalerate (PHBV) local antibiotic delivery system: In vivo effectiveness and biocompatibility in the treatment of implant-related experimental osteomyelitis. *J. Biomed. Mater. Res.* **1999**, *46*, 494–503. [CrossRef]
108. Shrivastav, A.; Kim, H.Y.; Kim, Y.R. Advances in the applications of polyhydroxyalkanoate nanoparticles for novel drug delivery system. *Biomed. Res. Int.* **2013**, *2013*, 581684. [CrossRef]
109. Barouti, G.; Jaffredo, C.G.; Guillaume, S.M. Advances in drug delivery systems based on synthetic poly (hydroxybutyrate)(co) polymers. *Prog. Polym. Sci.* **2017**, *73*, 1–31. [CrossRef]
110. Prakash, P.; Lee, W.H.; Loo, C.Y.; Wong, H.S.J.; Parumasivam, T. Advances in polyhydroxyalkanoate nanocarriers for effective drug delivery: An overview and challenges. *Nanomaterial* **2022**, *12*, 175. [CrossRef] [PubMed]
111. Deng, Y.; Lin, X.S.; Zheng, Z.; Deng, J.G.; Chen, J.C.; Ma, H.; Chen, G.Q. Poly (hydroxybutyrate-co-hydroxyhexanoate) promoted production of extracellular matrix of articular cartilage chondrocytes in vitro. *Nanomaterial* **2003**, *24*, 4273–4281. [CrossRef] [PubMed]
112. Lootz, D.; Behrend, D.; Kramer, S.; Freier, T.; Haubold, A.; Benkiesser, G.; Schmitz, K.P.; Becher, B. Laser cutting: Influence on morphological and physicochemical properties of polyhydroxybutyrate. *Nanomaterial* **2001**, *22*, 2447–2452. [CrossRef]
113. Sharma, N. Polyhydroxybutyrate (PHB) production by bacteria and its application as biodegradable plastic in various industries. *Acad. J. Polym. Sci.* **2019**, *2*, 001–003. [CrossRef]
114. Lee, C.W.; Horiike, M.; Masutani, K.; Kimura, Y. Characteristic cell adhesion behaviors on various derivatives of poly (3-hydroxybutyrate) (PHB) and a block copolymer of poly (3-[RS]-hydroxybutyrate) and poly (oxyethylene). *Polym. Degrad. Stab.* **2015**, *111*, 194–202. [CrossRef]
115. Garcia-Garcia, D.; Ferri, J.M.; Boronat, T.; López-Martínez, J.; Balart, R. Processing and characterization of binary poly (hydroxybutyrate)(PHB) and poly (caprolactone)(PCL) blends with improved impact properties. *Polym. Bull.* **2016**, *73*, 3333–3350. [CrossRef]
116. Chu, C.F.; Lu, A.; Liszkowski, M.; Sipehia, R. Enhanced growth of animal and human endothelial cells on biodegradable polymers. *Biochim. Biophys. Acta* **1999**, *1472*, 479–485. [CrossRef] [PubMed]
117. Sanhueza, C.; Acevedo, F.; Rocha, S.; Villegas, P.; Seeger, M.; Navia, R. Polyhydroxyalkanoates as biomaterial for electrospun scaffolds. *Int. J. Biol. Macromol.* **2019**, *124*, 102–110. [CrossRef] [PubMed]
118. Rhim, J.W.; Park, H.M.; Ha, C.S. Bio-nanocomposites for food packaging applications. *Prog. Polym Sci.* **2013**, *38*, 1629–1652. [CrossRef]
119. Popa, M.S.; Frone, A.N.; Panaitescu, D.M. Polyhydroxybutyrate blends: A solution for biodegradable packaging? *Int. J. Biol. Macromol.* **2022**, *207*, 263–277. [CrossRef]
120. Atta, O.M.; Manan, S.; Shahzad, A.; Ul-Islam, M.; Ullah, M.W.; Yang, G. Biobased materials for active food packaging: A review. *Food Hydrocoll.* **2022**, *125*, 107419. [CrossRef]
121. Hosokawa, M.N.; Darros, A.B.; Moris, V.A.D.; Paiva, J.M.F.D. Polyhydroxybutyrate composites with random mats of sisal and coconut fibers. *Mater. Res.* **2016**, *20*, 279–290. [CrossRef]
122. Rajan, K.P.; Thomas, S.P.; Gopanna, A.; Chavali, M. Polyhydroxybutyrate (PHB): A standout biopolymer for environmental sustainability. In *Handbook of Ecomaterials*, 1st ed.; Torres-Martínez, L.M., Vasilievna-Kharissova, O., Ildusovich-Kharisov, B., Eds.; Springer: Cham, Switzerland, 2020; pp. 2803–2825.

MDPI
St. Alban-Anlage 66
4052 Basel
Switzerland
www.mdpi.com

*Polymers* Editorial Office
E-mail: polymers@mdpi.com
www.mdpi.com/journal/polymers

www.ingramcontent.com/pod-product-compliance
Lightning Source LLC
LaVergne TN
LVHW070222100526
838202LV00015B/2074